Conference on Geodynamics and Stress State of the Earth's Interior (GSSEI 2017)

IOP Conference Series: Earth and Environmental Science
Volume 134

Novosibirsk, Russia
2 - 6 October 2017

Editor:

Mikhail Kurlenya

ISBN: 978-1-5108-6058-2
ISSN: 1755-1307

Printed from e-media with permission by:

Curran Associates, Inc.
57 Morehouse Lane
Red Hook, NY 12571

Some format issues inherent in the e-media version may also appear in this print version.

Printed by Curran Associates, Inc. (2018)

For permission requests, please contact the Institute of Physics
at the address below.

Institute of Physics
Dirac House, Temple Back
Bristol BS1 6BE UK

Phone: 44 1 17 929 7481
Fax: 44 1 17 920 0979

techtracking@iop.org

Additional copies of this publication are available from:

Curran Associates, Inc.
57 Morehouse Lane
Red Hook, NY 12571 USA
Phone: 845-758-0400
Fax: 845-758-2633
Email: curran@proceedings.com
Web: www.proceedings.com

Table of contents

Volume 134

Geodynamics and Stress State of the Earth's Interior (GSSEI 2017)

2–6 October 2017, Novosibirsk, Russian Federation

Accepted papers received: 9 March 2018
Published online: 28 March 2018

Preface

Papers

IOP Conf. Series: Earth and Environmental Science **134** (2018) 011001 doi:10.1088/1755-1315/134/1/011001

2017 Conference Proceedings

Geodynamics and stress state of the Earth's interior

October 2–6 2017

Novosibirsk, Russia

It is a pleasure for us to have this opportunity and offer our foreign readers the selected papers presented at the annual 2017 Conference on Geodynamics and stress state of the Earth's interior.

The Conference has been hosted by the Institute of Mining of the Siberian Branch of the Russian Academy of Sciences under auspices of the Presidium of the Siberian Branch RAS and the Scientific Council on Mining of the Russian Academy of Sciences.

The Conference has become a forum of academic scientists and practitioners from Russia's top mining companies such as Norilsk Nickel, Evrazruda, ALROSA, SUEK and Siberian Association for Industrial Hydraulics and Pneumatics.

The Conference provided a platform for the discussion of the latest theoretical and experimental research findings in the field of diagnostics and forecasting of the stress state of the rock massif. The scope of the Conference encompassed such topics as:

- Geodynamic, geomechanical and technogenic processes in the development of mineral deposits
- Experimental methods, instruments, equipment, robotic systems and technologies for studying the stress state and mechanical properties of rocks
- Modeling of the physical state of the rock massif and technogenic processes in the vicinity of mine workings
- Theory and practice of controlling the physical state of the rock massif

Mikhail Kurlenya

Academician RAS

Editor-in-Chief

Peer review statement

All papers published in this volume of *IOP Conference Series: Earth and Environmental Science* have been peer reviewed through processes administered by the proceedings Editors. Reviews were conducted by expert referees to the professional and scientific standards expected of a proceedings journal published by IOP Publishing.

Stress state estimation in multilayer support of vertical shafts, considering off-design cross-sectional deformation

SV Antsiferov*, AS Sammal and PV Deev

Tula State University, Tula, Russia

E-mail: *antsser@mail.ru

Abstract. To determine the stress-strain state of multilayer support of vertical shafts, including cross-sectional deformation of the tubing rings as against the design, the authors propose an analytical method based on the provision of the mechanics of underground structures and surrounding rock mass as the elements of an integrated deformable system. The method involves a rigorous solution of the corresponding problem of elasticity, obtained using the mathematical apparatus of the theory of analytic functions of a complex variable. The design method is implemented as a software program allowing multivariate applied computation. Examples of the calculation are given.

Monitoring of shaft support sometimes reveals that the shape of tubing rings deviates from the designed circle. Deviation from the circular cross-section may result from low quality assembling of the tubing ring, use of different model tubings in the same ring, nonuniform deformation of of rocks during ice wall creation behind the installed lining, excessive local pressure of grouting mortar injection in the process of water imperviousness recovery of shaft support, tectonics, etc. Naturally, departure from the circular cross-section can, under certain conditions, reduce load-bearing capacity of shaft support and make the shaft unserviceable.

Developed at the Tula State University, the analytical method of combination (multi-layered) shaft support design based on the underground structure mechanics principle on interaction of rock mass and support as a uniform deformable system makes it possible to analyze load-bearing capacity of mine shaft support with regard to deviation of the actual geometry of tubing rings from the design crops-section.

The proposed approach extends the methodology of accounting for the influence exerted by factual cross-section of underground structures proposed by Dr Eng, Prof Fotieva and the theory of underground structure mechanics by Dr Eng, Prof Bulychev in the context of the multi-layered shaft support design: the support loads are not set 'a priory' but determined as radial and shear stresses during calculation of the whole rock mass–support system depending on geometrical and deformation characteristics of the underground structure constituents. The method rests on the earlier obtained and adapted analytical solution to the plane problem of elasticity on stress state of a noncircular multi-layered ring which simulates shaft support composed of cast iron tubes and concrete fill in a hole in an infinitely linearly deformed medium simulating rock mass [2, 3].

The mathematical model includes such factors that essentially influence stress state and load-bearing capacity of the compound support as: support design; dimension and actual shape of cross-section; deformation, rheological (within the linear hereditary creep theory) and permeability characteristics of rock mass and concrete layer; initial stress field conditioned by weight of overburden (rock pressure), concrete support and tubing, as well as by groundwater pressure (with regard to seepage through the concrete fill).

Cast iron tubing is composed of elements with flat joints with bolted connections that are tension braces and is for this reason is modeled as a two-layer tube. The introduction of the additional inner layer is conditioned by the ring stiffeners installed at regular intervals; the influence of the radial stiffeners is disregarded as they only prevent liner plates between the ring stiffeners (Figure 1).

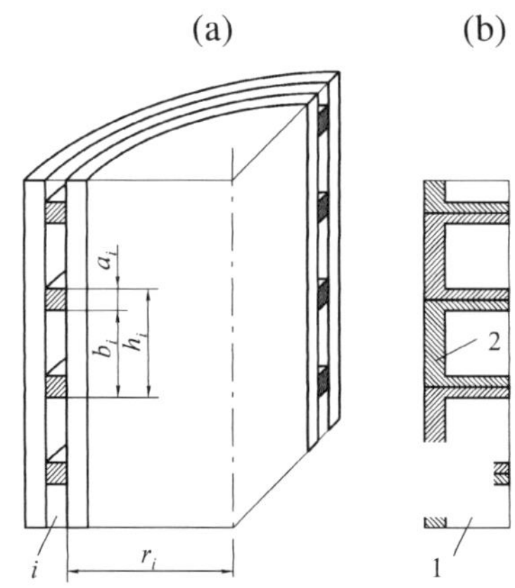

Figure 1. Schematics of multilayer shaft support with periodically nonuniform layers: (a) nonuniform layer i; (b) nonuniform layer 1 and uniform layer 2 simulating stiffeners and plates, respectively.

The analytical model (Figure 2) presents the shaft support as a three-layer ring where the outward layer with a thickness Δ models concrete fill and the other two layers model the tubing cross-section. The boundaries of the layers differ from circle in the general case (the inside design circle boundary is shown by the dashed line).

The ring support the circular hole in the infinite ponderable linearly deformable isotropic (or viscoelastic) medium S_0 modeling surrounding rock mass with the deformation characteristics represented by the deformation modulus E_0, Poisson's ratio ν_0 and linear hereditary creep parameters α_0, δ_0. The layers in S_i have different deformation characteristics—deformation moduli E_i and Poisson's ratios ν_i $(i = 1, 2, 3)$.

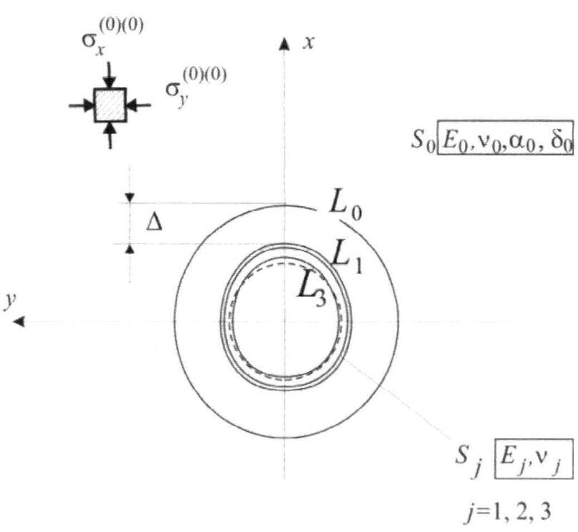

Figure 2. Analytical model.

So, the model assumes that the inside nonuniform layer formed by the ring stiffeners works as a quasi-uniform layer with the reduced deformation characteristics, and the displacements of the interface between plates and stiffeners are the same. The validity of such assumptions is proved by the research [12], including the shaft support designs.

The deformation characteristics of the quasi-uniform inside layer S_3 are governed by the reduced deformation modulus given by [2]:

$$E_3 = \mu E_{ci}, \tag{1}$$

where E_{ci}—deformation modulus of cast iron, MPa; μ—reduction factor: $\mu = \frac{4a}{h}$; a—width of stiffeners (tubing design includes 4 stiffeners), m; $h = 1.5$ m—tubing height.

Poisson's ratio of the cast iron layer S_3 is assumed $\nu_3 = 0.3$.

Gravity conditioned by the weight of overlying rocks and support, or by groundwater level is modeled by setting relevant initial compressive stresses in the medium S_0 and in the layers S_i $(i = 1, 2, 3)$.

To calculate the load due to overlying rock weight, the initial stresses in the medium S_0 are given by:

$$\sigma_x^{(0)(0)} = -\gamma H \lambda \alpha^*, \ \sigma_y^{(0)(0)} = -\xi \sigma_x^{(0)(0)}, \tag{2}$$

where γ—averaged weight of overlying rocks; H—depth of the analyzed cross-section of the shaft; λ—lateral earth pressure coefficient; ξ—factor of nonuniformity of initial stress field; α^*—correction factor to account for the process influences (lag between the support and the bottomhole advance, sequence of cast iron tubing and concrete fill installation, etc.) [2, 3].

The weight of the support is included in the model by setting initial stresses in the domains S_i $(i = 1, 2, 3)$ simulating the corresponding layer of the support structure:

$$\sigma_x^{(i)(0)} = \sigma_y^{(i)(0)} = -\gamma_i H \lambda_i, \tag{3}$$

where γ_i—specific weigh of material of the support layer S_i; $\lambda_i = \frac{\nu_i}{1-\nu_i}$—later earth pressure coefficient in the given medium.

In the layer S_3 formed by the ring stiffeners of the tubing, the specific weight is calculated as $\gamma_3 = \gamma_{ci} \cdot \mu$, where γ_{ci}—specific weight of cast iron.

In the calculation of influence of the groundwater pressure, it is assumed that as a consequence of water seepage through the outward concrete layer, total water head recovers on the water-impermeable tubing column. Therefore, the initial stresses acting in the domains S_i $(i = 0, 1)$ are given by:

$$\sigma_x^{(i)(0)} = \sigma_y^{(i)(0)} = -\gamma_w H_e \ (i = 0, 1), \tag{4}$$

where $\gamma_w = 9.8$ kN/m^3—specific weight of groundwater; H_e—groundwater head, m.

Total stresses in the analyzed domains are presented as the sums of initial and secondary stresses conditioned by the formation of the supported hole. The initial stresses in rock mass before the shaft sinking are neglected.

The tubing layers S_i $(i = 1, 2, 3)$ and the medium S_0 deform jointly, i.e. the conditions of continuous vectors of displacements and total stresses are satisfied in in the lines L_i $(i = 0, 1, 2)$.

The boundary conditions in the formulated problem are:

—at the boundaries L_i $(i = 0, 1, 2)$

$$\tilde{\sigma}_n^{(i,i+1)} = \tilde{\sigma}_n^{(i,i)},\ \tilde{\tau}_{n,\varphi}^{(i,i+1)} = \tilde{\tau}_{n,\varphi}^{(i,i)};\ u_n^{(i,i+1)} = u_n^{(i,i)},\ u_\varphi^{(i,i+1)} = u_\varphi^{(i,i)}; \tag{5}$$

—at the inner boundary L_3

$$\tilde{\sigma}_n^{(3,3)} = 0,\ \tilde{\tau}_{n,\varphi}^{(3,3)} = 0, \tag{6}$$

where $\tilde{\sigma}_n^{(i,j)}$, $\tilde{\tau}_{n,\varphi}^{(i,j)}$ $(i = 0, 1, 2;\ j = i, i+1)$—components of the total normal and shear stresses at the points in the domain S_j at the boundary L_i; $u_n^{(i,j)}$, $u_\varphi^{(i,j)}$—components of displacement vectors at the points of the contacting domains S_j along the normal n and tangent φ drawn to the boundary L_i.

The formulations are linear, which allows using principle of superposition. Each problem is solved in terms of the fractions of the initial stress $\sigma_x^{(0)(0)}$. After the calculated stress is multiplied by the value of $\sigma_x^{(0)(0)}$, the results are summed up with regard to the most unfavorable combinations of loads.

The calculation of the shaft support cross-section in watered rock mass includes the ascending force of groundwater [5].

When stresses at the internal and external boundaries of the quasi-uniform inside layer S_3 are found, the stresses on the stiffeners are calculated. The this effect, the stresses in S_3 are multiplied by the factor $k = \dfrac{1}{\mu}$.

The influence of viscoelastic deformation of rocks susceptible to creep can be taken into account based on the theory of linear hereditary creep using the method of variable moduli, according to which the the deformation characteristics of the medium S_0, included in the elasticity problems solution, are presented as functions of time [1].

After introduction of complex potentials $\varphi_j(\zeta)$, $\psi_j(\zeta)$, characterizing stress state in the domains S_j $(j = 0, 1, 2, 3)$ connected with the additional stresses and displacements by the Kolosov–Muskhelishvili formulas [4], the set problem reduces to the boundary value problem of the theory of functions of a complex variable and is solved using the apparatus of the conformal mapping and complex series.

The described solution is implemented in a computer program for multi-variant calculations in practical shaft support design.

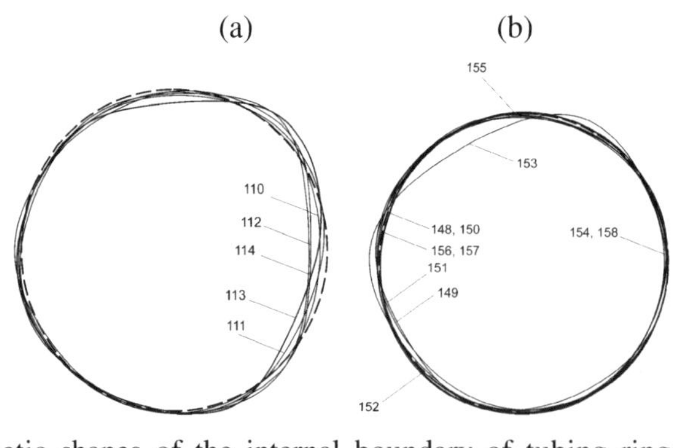

Figure 3. Schematic shapes of the internal boundary of tubing rings: (a) first section (rings 110–115) and (b) second section (rings 148–158).

The resultant maximum deviation from the design circle is depicted in Figure 3 in terms of two sections of rubbing rings (magnification by 10 times): section 1 (Figure 3a) lies in chalky clay, tubing 7.0–50, concrete fill thickness 960 mm; section 2 (Figure 3b) lies in rock salt, tubing 7.0–60, concrete fill layer 1160 mm thick.

The calculated normal and normal tangential stresses in the shaft support in case of departure from the design circular cross-section are only compressive; shearing stresses in the normal (radial) cross-sections of the support are insignificant.

The criterion of the real ring shape was assumed the value $\varepsilon = \dfrac{\delta}{R} \times 100\%$, where δ—decrease in the design radius R at the point of the local displacement.

The resultant curves of increase in the normal tangential stresses $\Delta\sigma$ (%) in the tubing and concrete fill in sections 1 and 2 (as in Figure 1) versus ε are plotted in Figure 4 as the solid and dashed lines, respectively. The lines marked by 1 are for the cast iron tubing, the lines marked by 2 are for the concrete fill.

The calculations show that with the thinner layer of the concrete fill, the departure of the tubing ring from the design circle has higher influence on the stress state of the support than in the case of the thicker concrete layer. It follows from the analysis of the plots in Figure 4 that the higher displacement of the support inward the shaft results in the increase $\Delta\sigma$ in the tubing ring and concrete fill almost by the linear law; the maximum $\Delta\sigma$ due to the deviation of the design shape of tubing ring is not higher than 20% (ring 153) and is within 7% in all other cases.

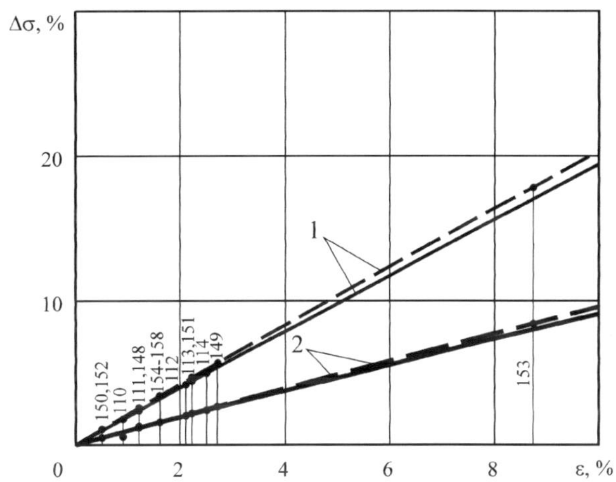

Figure 4. Relative increase in the stress $\Delta\sigma$, %, in the support in sections 1 and 2 of the shaft as against the decrease in the radius ε.

The highest deviation of the tubing ring shape from the design in ring 153 conforms with the maximum increase in stresses.

When the support loads are set in accord with [5], the stresses in the support do not exceed limiting values and the required load-bearing capacity of the support is ensured.

The scientific provisions included in the described calculation method have many times been approved in practical support designs for cage and skip hosting shafts in potassium and potash-magnesium salt mines in the area Perm, Bashkiria, Volgograd and Kaliningrad.

Conclusions

For the purpose of calculating stress state in the multilayer support of mine shafts with regard to the deviation of the tubing ring cross-section from the design geometry, the analytical method has been developed. The method is implemented as a computer program for the multi-variant practical calculations.

References

[1] Amusin BZ and Linkov AM 1974 On the use of the method of variable modules for solving a class of linear hereditary creep problems Izv. AN SSSR. Mekh. Tverd. Tela No 6 pp 162–166

[2] Bulychev NS 1994 *Mechanics of Underground Structures: University Textbook* Moscow: Nedra (in Russian)

[3] Bulychev NS, Fotieva NN and Streltsov EV 1986 *Designing and Computation of Support for Permanent Roadways* Moscow: Nedra (in Russian)

[4] Muskhelishvili NI 1966 *Some Basic Problems of Mathematical Elastic Theory* Moscow: Nauka (in Russian)

[5] *Mine and Support Design Manual* Moscow: Stroyizdat 1983 (in Russian)

Seismicity of rock rupture filled with fluid

AV Azarov* and SV Serdyukov

Chinakal Institute of Mining, Siberian Branch, Russian Academy of Sciences, Novosibirsk, Russia

E-mail: *antonazv@mail.ru

Abstract. The task of generation of seismic radiation by ruptures of rocks filled with a fluid is considered in the work. Possible causes of the occurrence of seismic waves due to the interaction of rock and fluid are considered. Examples of phenomena are given where the observed seismicity is associated with fluid motion. The models describing these phenomena are analyzed. Conclusions are drawn about the possibility of generation of seismic waves by rock ruptures with fluid and about radiation characteristics.

1. Introduction

Ruptures of rock, filled with fluid, can be a source of seismic radiation. In this case the observed radiation arises due to interaction between the fluid and the rock. Numerous studies show that the interaction generates signals with one or more dominating fundamental harmonics, i.e. the signal is harmonic. Under certain conditions periodic oscillations can be observed, for example, when the fluid flows through a rupture. In other cases, a decaying harmonic signal can be observed, for example, when a crack filled with fluid gets in resonance by the action of a impulse source of unknown nature.

There are many examples in which the recorded seismic signals are associated with the result of the interaction of a fluid and rock. For instance, fluid-induced oscillations of channel walls or excitation and resonance of fluid-filled cracks are possible causes of volcanic tremor and long-period (LP) events [1]. These phenomena are observed during seismic monitoring of volcanoes. Tremor and LP events have similar spectral characteristics of the signal, but differ in their duration. LP event is a single low-frequency event, the duration of which can reach up to 3 minutes. The duration of volcanic tremor can range from several minutes to several days or months. For these phenomena a steady periodic low-frequency signal with a spectral peak of 0.1-7 Hz is observed. There have been many works studying the dynamics of cracks filled with fluid and the dynamics of ruptures with a fluid flow to explain the volcanic tremor and LP events [2, 3, 4, 5, 6].

Another example is the signals observed during seismic monitoring of subglacial flows, which are similar to volcanic tremor by their characteristics. A number of studies have established a correlation between the existence of a subglacial water flow and the presence of a harmonic signal [7, 8,9]. Thus, for example, work [9] deals with low-frequency radiation (3-11 Hz) observed in the Greenland glaciers area. The work presents observations of harmonic oscillations under the moving ice of MacAyeal (MacAyeal Ice Stream) in the western Antarctic. There were two recorded events of a tremor with a steady frequency of 3 Hz energy arrival lasting for 10 minutes. The harmonic nature of the tremor is interpreted as the result of the resonance of the subglacial cracks and channels filled with water. Duration, monochromatic character and tremor dynamics indicate that the source mechanism is most likely associated with the water flow in the subglacial water system, which flows from a small subglacial lake.

Harmonic signals are also observed during seismic monitoring of hydraulic fracturing. Usually, such signals are low-frequency and, in certain cases, they are associated with oscillations of the hydraulic rupture.

Work [10] reports on the recorded harmonic low-frequency signals during hydraulic fracturing at Fenton Hill in New Mexico. The authors use a three-dimensional crack model filled with fluid [3] to explain the spectral characteristics of the signal. In the model the occurring seismic waves are a consequence of crack resonance, which is maintained by very slow dispersion waves on the crack surface ("crack waves"). A comparison of the seismograms of the fracturing and volcanic tremor leads to the conclusion that the two signals are similar. The only difference is in scale. So the spectral peaks registered during hydraulic fracturing at Fenton Hill are approximately at frequencies equal to 112 Hz, 185 Hz, 268 Hz and higher, while the spectral peaks of volcanic tremor can be at frequencies ranging from a few tenths Hz.

Work [11] also includes the study of hydraulic fracturing experiments at Felton Hill. As well as in work [10] the causes of long-period events are associated with the action of the source, and not with effects arising from the seismic waves' transition through the medium. There is a source proposed to explain the spectral peaks of the signal. It consists of two cracks of different sizes with different fluid pressures and the channel connecting them, which suddenly opens. Each peak in the spectrum is associated with one of the cracks, and one or more peaks may correspond to each crack.

Work [12] studies two hydraulic fracturing processes during seismic monitoring of which steady-state periodic signals are recorded. It is said that a reason for the observed signal is the resonance of cracks filled with fluid. But in the case of the first hydraulic fracturing, the most probable cause of oscillations in the range 8-11 Hz is the pumping equipment and / or vibration within the vertical part of the injection well. For the second hydraulic rupture the situation is more complicated. Spectrograms show peaks at frequencies of 17, 35 and 51 Hz. The authors explain the results by the stationary, but not laminar flow inside the open fracture, which can be performed near the injection well. Such a flow leads to resonance of the crack. Less likely causes are related to fluid properties and regular crack growth.

It is important to note that the registered harmonic low-frequency signals can not always be associated with the source mechanism (oscillation of the rupture). Such signals can be formed after elastic waves pass through a complex medium, or they can arise due to the oscillations' excitation on surface or in a wellbore with installed receivers.

In recent years a number of works describing another group of low-frequency events registered during seismic monitoring of hydraulic fracturing have been written [13, 14]. They are called long period, long-duration (LPLD) seismic events and can be observed both during and after the fracturing. The LPLD event is characterized by a low-amplitude tremor-like signal lasting from 10 seconds to a minute with dominant frequencies between 10 and 80 Hz. In this case seismicity is similar to the tectonic tremor occurring during subduction but different from volcanic tremor. The main difference is in the signal spectrum. Spectrum of volcanic tremor has dominant harmonics. The spectrum of the LPLD tremor is flat at low frequencies, and it drops like the spectra of regular earthquakes. For this reason, it is impossible to create connection between these events and oscillations in the rupture filled with fluid. Works [13, 14] make hypotheses that the possible causes of LPLD events are slow shifts along already existing ruptures. The above studies are important for the development of microseismic monitoring technologies [15, 16].

2. Mathematical models of fluid-filled ruptures' dynamics

Work [2] is one of the first studies on the dynamics of a crack filled with fluid. This paper considered a dry or fluid-filled two-dimensional crack of finite size located in an unbounded elastic space. The excitation of a crack is initiated by its sharp opening at the apex. The study of the crack behavior was carried out using numerical methods, where the elasticity equations were solved by finite-difference methods. A seismic field in a long distance from the crack was calculated using a numerical solution of the crack surface motion and analytical formulas. Also, the author calculated the spectral radiation peaks in the far zone, depending on the size of the crack and properties of the fluid (the bulk modulus) and the elastic medium.

The study of the dynamics of a crack filled with fluid continued in [3, 17]. Work [17] develops Aki's studies [2] by improving the boundary conditions and studying a viscous fluid supporting the propagation of acoustic waves. The study of the crack dynamics is carried out by simultaneous solution of the equations of the fluid motion inside a crack and the elasticity equations outside a crack using numerical methods, similar to that used in [2]. In contrast to [17] work [3] investigates the dynamics of a three-dimensional fluid-filled crack excited to resonance by effect on the surface element. The paper studies the crack response to the impulse action for various ratios of width to crack length and for different values of crack stiffness ($C=(b/mu)/(L/d)$), where b – the bulk fluid compression modulus, mu – the medium rigidity, L and d are the length and width of the crack, respectively.

A similar problem is solved in [18], where a numerical method is proposed for modeling the dynamic response of a crack filled with fluid. In case of this method, the motion of the fluid inside the crack and the motion of the solid elastic body outside the crack are registered in the form of boundary integrals in the frequency domain. It is noted that this method is more effective than the finite-difference method, which was used in [2, 3, 17]. Advantages of the proposed method are the ability to conduct research for a wide range of parameters and the possibility of immediate calculation of the damping factor for the crack resonance.

The observed waves mentioned above are surface waves, and they are called Stoneley waves. They can make a crack the source of low-frequency radiation. The waves run along the crack's surface, reflecting from the peaks to maintain its resonance. Due to the exponential attenuation of the Stoneley wave, it is difficult or impossible to record them at a great distance. But due to their presence volumetric waves can be generated, which makes it possible to register the crack oscillations. As the results of numerical simulation [19] show, volumetric waves can form at the crack ape, when a surface wave is reflected.

The source of seismic oscillations can be the fluid flowing through a crack in the rock. In this case, it is possible to lose stability of the system with the formation of self-oscillations of the crack walls. The resulting oscillations are maintained by the energy of a uniform constant fluid flow.

Julian [5] proposed a model of the harmonic oscillations' occurrence in the course of a uniform fluid flow through a crack in the rock. Julian suggested that instability may be caused by the interaction of the moving flow and the walls of the crack. Instability is achieved at certain threshold values of the system parameters, which leads to an self-oscillating process. The paper considers a model with lumped parameters, where a layer of fluid lies between two walls of finite length. The walls simulate a crack and have such parameters as elasticity, mass and damping. At the edges of the crack pressure difference is maintained, so that the fluid that fills the crack is set to flow. The width of the crack opening does not depend on the spatial coordinates. Due to the simultaneous solution of the equations of fluid motion and the elasticity equations, Julian obtained a system of nonlinear ordinary differential equations describing the dynamics of the crack wall. The solution of the system presents stable harmonic oscillations of the walls when a certain threshold value of the flow velocity is reached. The resulting system, in addition to modeling linear resonances, shows additional nonlinear effects, such as the doubling period.

The problem of the interaction of a moving fluid and an elastic body is solved in [6]. In contrast to [5], we study not a crack, but a cylindrical cavity in an elastic three-dimensional space. As in [5] the motion of the wall does not depend on the spatial coordinates. The statement of the problem makes it possible to obtain a seismic field from a system of connected cylindrical cavities with different diameters. Nonlinear differential equations describing the motion of the cavity wall and the fluid in it are derived. The obtained mathematical model shows that the interaction of a moving fluid and an elastic body generates oscillations.

The possible way of generating oscillations in the interaction of a moving fluid and an elastic body is also studied in [4, 20]. In contrast to [5], in works [4, 20] they adhere to a more rigorous mathematical approach. Another difference is that the width of the crack opening depends not only on time, but also on the spatial coordinate. The paper considers a two-dimensional model in which a relatively thin layer of a viscous fluid is located between two elastic half-spaces. The authors determine the conditions for the occurrence of instability of the flow. Also, they study the oscillations of the crack walls caused by the instability. A linear analysis of the stability of the system shows that the oscillations induced by the flow can occur at arbitrarily low Reynolds numbers. Moreover, the stability of the system depends on the size of the crack, the velocity of the fluid, the velocity of the elastic waves in a continuous medium, and does not depend explicitly on the fluid viscosity.

The obtained results confirm that the mechanisms of harmonic oscillations proposed by Julian are plausible. In addition, a cylindrical cavity with fluid passing through it is considered in [4]. The stability of the obtained system is studied in a similar way, and conclusions are made about the possibility of oscillations' occurrence.

Another approach to modeling of the oscillations associated with the flow of fluid through a channel is proposed in [21]. The paper considers a model of a one-dimensional flow of fluid moving through a crack with a partition. The partition opens if the fluid pressure becomes higher than a certain threshold value and closes in all other cases. The proposed model neglects the deformation of the crack walls and the thickness of the partition. The constant flow velocity at one end of the crack and the constant pressure at the other end are boundary conditions imitating the fluid flowing from the reservoir into the free space. Numerical calculations show fluctuations in the fluid pressure, and the frequency of the oscillations depends on the flow velocity.

3. Conclusion

Taking into consideration results of the work, it can be concluded that rock ruptures filled with fluid can be sources of seismic radiation. This is confirmed by the numerous examples given in the article. The paper shows that the observed seismic signal is harmonic and its characteristics depend on the parameters of the rupture, the parameters of the rock and the fluid. The source mechanisms of seismicity mechanisms may vary. The article presents many models explaining the mechanisms generating oscillations of the ruptures and the characteristics of the signal generated by them.

The results of the presented work can be used to develop technologies for seismic monitoring of rock ruptures with fluid and fluid filtration zones.

References

[1] Konstantinou KI, Schlindwein V 2003 Nature, wavefield properties and source mechanism of volcanic tremor: a review *Journal of Volcanology and Geothermal Research* Vol 119 No 1 pp 161-187

[2] Aki K, Fehler M, Das S 1977 Source mechanism of volcanic tremor: Fluid-driven crack models and their application to the 1963 Kilauea eruption *Journal of volcanology and geothermal research* Vol 2 No 3 pp 259-287

[3] Chouet B 1986 Dynamics of a fluid-driven crack in three dimensions by the finite difference method *Journal of Geophysical Research: Solid Earth* Vol 91 No B14 pp 13967-13992

[4] Balmforth N. J., Craster R. V., Rust A. C. 2005 Instability in flow through elastic conduits and volcanic tremor *Journal of Fluid Mechanics* T. 527 pp 353-377

[5] Julian BR 1994 Volcanic tremor: nonlinear excitation by fluid flow *Journal of Geophysical Research: Solid Earth* Vol 99 No B6 pp 11859-11877

[6] Corona-Romero P, Arciniega-Ceballos A, Sánchez-Sesma FJ 2012 Simulation of LP seismic signals modeling the fluid–rock dynamic interaction *Journal of Volcanology and Geothermal Research* Vol 211 pp 92-111

[7] Winberry JP, Anandakrishnan S., Alley R. B. 2009 Seismic observations of transient subglacial water-flow beneath MacAyeal Ice Stream, West Antarctica *Geophysical Research Letters*

Vol 36 No 11

[8] Lawrence WST, Qamar A. 1979 Hydraulic transients: A seismic source in volcanoes and glaciers *Science* Vol 203 No 4381 pp 654-656

[9] Röösli C et al. 2014 Sustained seismic tremors and icequakes detected in the ablation zone of the Greenland ice sheet *Journal of Glaciology* Vol 60 No 221 pp 563-575

[10] Ferrazzini V et al. 1990 Quantitative analysis of long-period events recorded during hydrorupture experiments at Fenton Hill, New Mexico *Journal of Geophysical Research: Solid Earth* Vol 95 No B13 pp 21871-21884

[11] Bame D, Fehler M 1986 Observations of long period earthquakes accompanying hydraulic fracturing *Geophysical Research Letters* Vol 13 No 2 pp 149-152.

[12] Tary JB, Baan M, Eaton DW 2014 Interpretation of resonance frequencies recorded during hydraulic fracturing treatments *Journal of Geophysical Research: Solid Earth* Vol 119 No 2 pp 1295-1315

[13] Das I, Zoback MD 2013 Long-period, long-duration seismic events during hydraulic stimulation of shale and tight-gas reservoirs—Part 1: Waveform characteristics *Geophysics* Vol 78 No 6 pp KS97-KS108

[14] Das I, Zoback MD 2011 Long-period, long-duration seismic events during hydraulic rupture stimulation of a shale gas reservoir *The Leading Edge* Vol 30 No 7 pp 778-786.

[15] Kurlenya MV, Serdyukov AS, Azarov AV, Nikitin AA 2015 Numerical modeling of wavefields of microseismic events in underground mining *Journal of Mining Science* Vol 54 No 4 pp 689-695

[16] Serdyukov SV, Azarov AV, Dergach PA, Duchkov AA 2015 Equipment for microseismic monitoring of geodynamic processes in underground hard mineral mining *Journal of Mining Science* Vol 51 No 3 pp 634-640

[17] Chouet B, Julian BR 1985 Dynamics of an expanding fluid-filled crack *Journal of Geophysical Research: Solid Earth* Vol 90 No B13 pp 11187-11198.

[18] Yamamoto M, Kawakatsu H 2008 An efficient method to compute the dynamic response of a fluid-filled crack *Geophysical Journal International* Vol 174 No 3 pp 1174-1186.

[19] Frehner M., Schmalholz S. M. 2010 Finite-element simulations of Stoneley guided-wave reflection and scattering at the tips of fluid-filled ruptures *Geophysics* Vol 75 No 2 pp T23-T36

[20] Rust A. 2004 Flow-induced oscillations: A source mechanism for volcanic tremor? *2003 Program of Study Non-Newtonian Geophysical Fluid Dynamics* p 113

[21] Honda S., Yomogida K. 1993 Periodic magma movement in the conduit with a barrier: a model for the volcanic tremor *Geophysical research letters* 1993 Vol 20 No 3 pp 229-232

IOP Conf. Series: Earth and Environmental Science **134** (2018) 012003 doi:10.1088/1755-1315/134/1/012003

Excitation of seismic waves at the motion of fluid in the rock massive crack

AV Azarov* and AS Serdyukov

Chinakal Institute of Mining, Siberian Branch, Russian Academy of Sciences, Novosibirsk, Russia

E-mail: *antonazv@mail.ru

Abstract. The paper deals with the problem of generation of seismic radiation during the motion of a fluid in the rock massive crack. As the reason for the occurrence of seismic vibrations, a process is considered in which the interaction of a fluid and a rock causes instability of the crack walls, which leads to self-oscillations. The mathematical model describing the given process is investigated. Conclusions are made about the conditions for the occurrence of oscillations in the flow of fluid through the crack and the characteristics of the generated seismic signal.

1. Introduction

There are examples, when the recorded seismic signals are associated with the motion of the fluid through the ruptures in the rock. For example, fluid-induced oscillations of the channel walls or the excitation and resonance of fluid-filled cracks are possible causes of volcanic tremor and long-period (LP) events [1]. These phenomena are observed during seismic monitoring of volcanoes, during which steady periodic low-frequency signals with a spectral peak in the range 0.1–7 Hz are recorded. Another example is the signals observed during seismic monitoring of subglacial flows, which in its characteristics are similar to volcanic tremor. A number of studies have established a correlation between the existence of a water flow under ice and the presence of a harmonic signal [2–4]. For example, work [4] informs us about low-frequency radiation (3–11 Hz) observed in the Greenland glaciers area. The work presents observations of harmonic oscillations under the flow of moving ice in the western Antarctic. The harmonic character of the tremor is interpreted as a result of the resonance of subglacial cracks and channels filled with water. Duration, monochromatic character and tremor dynamics indicate that the source mechanism is most likely associated with the water flow in the subglacial water system, which flows from a small subglacial lake.

Also, there are examples in which harmonic low-frequency signals are observed during seismic monitoring of fracturing. In some cases, the causes of the recorded signals are associated with fluid flow within the fracture cracks [5, 6].

The mentioned above examples show that the fluid flowing through the crack in the rock can cause seismic oscillations. In this case low-frequency harmonic signals are observed. The source mechanisms of these signals may vary. In this article a process of a fluid and rock interaction causes instability of the crack walls, this leads to self-oscillations. These studies can be used to model microseismic processes [7] and the development of microseismic monitoring technologies [8].

2. Mathematical model

Let us consider the problem of the fluid flow through a crack located in an elastic space (Figure 1). In [9] a mathematical model describing the behavior of such system was proposed to explain the volcanic tremor. Similar problems were addressed in the works [10–12].

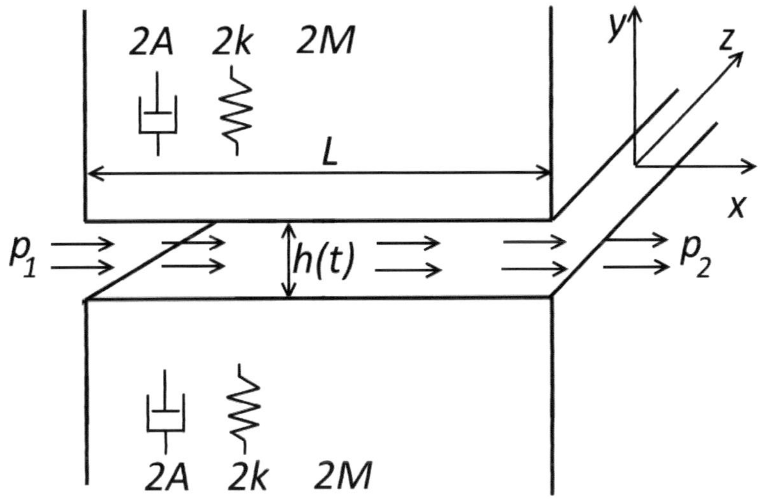

Figure 1. Fluid flowing through a crack in the elastic space

Work [9] proposes a mathematical model describing the behavior of the considered system (Figure 1) as follows. The fluid flows through a crack the walls of which are elastic half-spaces. At the crack edges pressures p_1 and p_2 are maintained. The half-space is characterized by elasticity (coefficient *k*) and viscosity (coefficient *A*). The gap has a length *L* and a width *h (t)*. *h (t)* does not depend on *x*, i.e. the whole crack wall either contracts or moves apart. In the *z* direction, the crack's length is bigger than the width, so the problem is considered to be two-dimensional. Then, the equations describing the system are as follows:

$$\left[M+\frac{\rho L^3}{12h}\right]\ddot{h}+\left[A+\frac{L^3}{12h}\left(\frac{12\eta}{h^2}-\frac{\rho}{2}\frac{\dot{h}}{h}\right)\right]\dot{h}+k\left(h-h_0\right)=L\left[\frac{p_1+p_2}{2}-\frac{\rho v^2}{2}\right] \tag{1}$$

$$\rho\dot{v}+\frac{12\eta}{h^2}v=\frac{p_1-p_2}{L}$$

where h_0 – the crack opening without influence of external forces, ρ – the fluid density, η – the fluid viscosity, A – the attenuation coefficient, and M – the attached mass.

It is easy to introduce the rock pressure force into the system of equations by putting in an additional term to the right-hand side of the first equation of the system (1). Also, we specify the crack opening without influence of external forces, after which the system of equations will have the following form:

$$\left[\mathrm{M}+\frac{\rho L^3}{12h}\right]\ddot{h}+\left[A+\frac{L^3}{12h}\left(\frac{12\eta}{h^2}-\frac{\rho}{2}\frac{\dot{h}}{h}\right)\right]\dot{h}+kh=L\left[\frac{p_1+p_2}{2}-\frac{\rho v^2}{2}-F_r\right] \tag{2}$$

$$\rho\dot{v}+\frac{12\eta}{h^2}v=\frac{p_1-p_2}{L}$$

When this method of introducing rock pressure is employed in the case $\frac{p_1 + p_2}{2} > F_r$, the model shows the crack opening and the oscillations of the walls. If $\frac{p_1 + p_2}{2} \le F_r$, the crack goes from the initial data state to the closed state.

3. Numerical studies of the mathematical model

Let us consider the problem of the fluid flow through a crack located in an elastic space (Figure In this paper we consider the problem in which there is a continuous uniform fluid flow through a crack in the elastic space. This system has self-oscillating properties, as a result of which oscillations of the crack walls can occur. Numerical analysis of equations (2) showed the oscillating solutions that can describe the self-oscillating properties of the system under consideration.

In [9], an analysis of system (1) was carried out in the case, when magma (a fluid with a high viscosity coefficient = 500 Pa*s and density = 2500 kg/m^3) flows between two elastic spaces. In this paper, the system (2) is analyzed, where the rock pressure is introduced, and water is employed as a fluid with parameters = 0.001 Pa*s and = 1000 kg/m^3. The value of the remaining parameters is taken as follows: M = 3*10^5 kg/m, A = 10^7 kg/(m*s), k = 600 MPa. The coefficient of elasticity was estimated by the formula:

$$k = \frac{\delta f}{\delta h} = \frac{1}{\delta h} L\delta p = \frac{\mu}{w\delta p} L\delta p = \frac{L\mu}{w}$$

where w – the crack extent in the z direction, μ – the shear modulus. The attached mass per meter of the wall was estimated by the formula $M = \rho_r w \cdot 1m$. For convenience, the fluid pressure in the crack is divided into two components: p – the constant part and the part that determines the pressure difference at the ends of the crack – δp. Thus, the pressure is $p_1 = p + \delta p$, $p_2 = p$.

The behavior of the system of equations (2) was investigated using numerical methods. The analysis was carried out by solving a system with various parameters, such as crack length L, pressure at the ends of the crack p_1 and p_2 and rock pressure F_r. After that, the limits of the parameters were determined, in case of which oscillations of the crack walls appeared. The characteristics of the oscillations were determined as well.

During analysis of nonlinear systems of differential equations, it is convenient to construct phase curves presenting the results of the solution appliance. Figure 2 shows phase curves in the $(h, \dot{h})$ plane for L = 5 m, p = 10 MPa, $F_r = 11$ Mpa. In Figure 2 a) the phase curve with a pressure drop δp= 3 MPa. In this case, the wall from the position corresponding to the initial data (0.001 m), will go into a static state with 0.0075 m opening, having made oscillations. In Figure 2 b) pressure drop δp = 4 MPa. Here, the crack from the position corresponding to the initial data (0.001 m) passes into the steady-state oscillation mode. The frequency of the first harmonic is 3.4 Hz. A further increase in the pressure drop leads to an increase in the frequency of the first harmonic. We note that the behavior of the numerical solutions does not depend on the initial data.

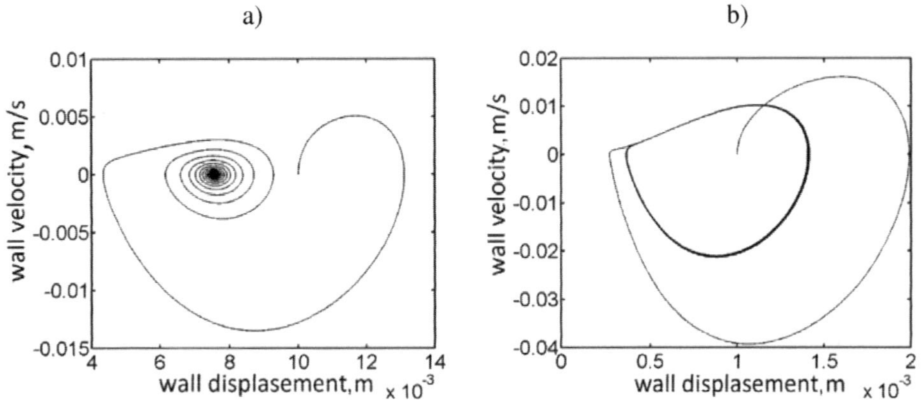

Figure 2. Phase curves in the plane $(h, \dot{h})$. Time interval for curves' calculation is 10 seconds, L=5 m, p=10 MPa, $F_r = 11$ MPa. a) – δp=3 MPa; b) – δp=4 MPa

Figure 2 demonstrates that there is a certain threshold value δp. When it is reached, steady oscillations of the crack walls are observed.

When there is a fixed static part of the pressure, the length of the crack and the rock pressure, an increase in the pressure drop at the edges of the crack leads to an increase in the oscillation frequency. For example, for L = 5 m, p = 10 MPa, $F_r = 11$, so the dependence of the frequency of the first harmonic on the pressure drop is shown in Fig. 3a). Figure 3b) shows a dependency graph of the frequency of the first harmonic on the length of the crack.

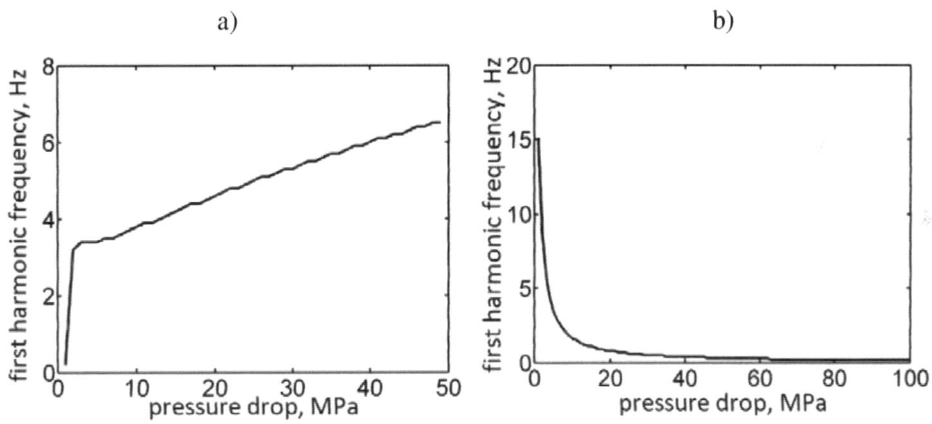

Figure 3. a) first harmonic frequency dependence on rock pressure drop if L=5 m, p=10 MPa, $F_r = 11$; b) first harmonic frequency dependence on crack length if p=10 MPa, $F_r = 11, \delta p$=5 MPa.

Figure 3 shows that if a crack length is 5 m, oscillations with a frequency of less than 4 Hz can be generated. Increase of the crack length leads to a decrease in the oscillation frequency. An increase in the pressure difference at the ends of the crack gives an increase in the frequency of the oscillations.

There are values of the parameters under which the mathematical model (2) has a solution, where the crack wall moves chaotically. For example, at p = 20 MPa, δp = 0.3 MPa, $F_r = 11$MPa, L = 5 m, this mode of motion is observed. Fig. 4 a) shows the phase curve corresponding to this case, Fig. 4 b) shows the oscillations of the crack wall depending time.

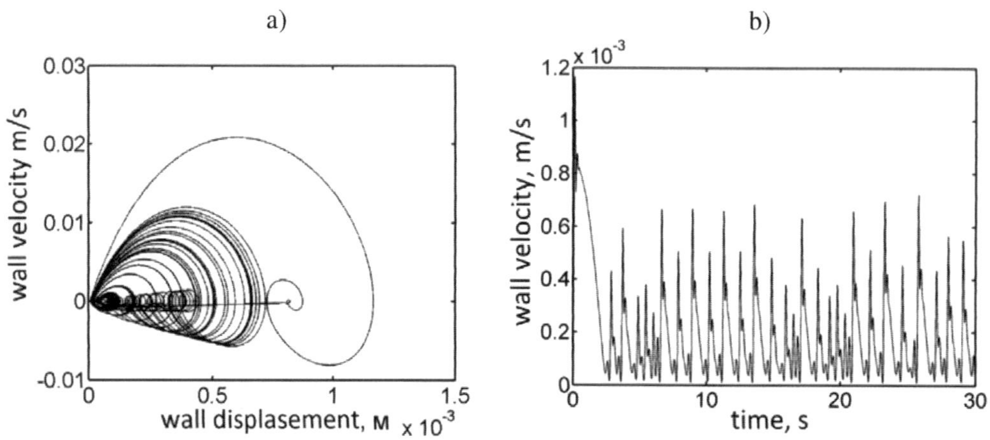

Figure 4. a) phase curve и b) wall displacement in time if p=20 MPa, δp=0.3 MPa, F_r =11 MPa, L=5 m.

4. Conclusions

The analysis of the mathematical model (2) has shown that it can describe the oscillations that occur, when there is a uniform flow of fluid through a crack in a rock. There are parameters at which low-frequency steady-state or damped oscillations are observed. The steady periodic wall motions occur, when certain threshold values of the pressure drop at the ends of the crack are reached. The characteristics of the oscillations in this case strongly depend on the size of the crack, the pressure drop of the liquid at the end of the crack and the rock pressure. In addition, model (2) has special solutions when the motion of the crack wall is chaotic.

Therefore, the model presented in this work can be used to monitor ruptures of rock through which the fluid flows. The use of the system (2) can enable the recovery of fracture and fluid flow parameters from the observed seismic radiation.

Acknowledgements

The reported study was funded by RFBR according to the research project No. 16-35-00513 мол_а

References

[1] Konstantinou KI, Schlindwein V 2003 Nature, wavefield properties and source mechanism of volcanic tremor: a review *Journal of Volcanology and Geothermal Research* Vol 119 No 1 pp 161–187

[2] Winberry JP, Anandakrishnan S., Alley R. B. 2009 Seismic observations of transient subglacial water-flow beneath MacAyeal Ice Stream, West Antarctica *Geophysical Research Letters* Vol 36 No 11

[3] Lawrence WST, Qamar A. 1979 Hydraulic transients: A seismic source in volcanoes and glaciers *Science* Vol 203 No 4381 pp 654–656

[4] Röösli C et al. 2014 Sustained seismic tremors and icequakes detected in the ablation zone of the Greenland ice sheet *Journal of Glaciology* Vol 60 No 221 pp 563–575

[5] Bame D, Fehler M 1986 Observations of long period earthquakes accompanying hydraulic fracturing *Geophysical Research Letters* Vol 13 No 2 pp 149–152

[6] Tary JB, Baan M, Eaton DW 2014 Interpretation of resonance frequencies recorded during hydraulic fracturing treatments *Journal of Geophysical Research: Solid Earth* Vol 119 No 2

IOP Conf. Series: Earth and Environmental Science **134** (2018) 012003 doi:10.1088/1755-1315/134/1/012003

pp 1295–1315

[7] Kurlenya MV, Serdyukov AS, Azarov AV, Nikitin AA 2015 Numerical modeling of wavefields of microseismic events in underground mining *Journal of Mining Science* Vol 54 No 4 pp 689–695

[8] Serdyukov SV, Azarov AV, Dergach PA, Duchkov AA 2015 Equipment for microseismic monitoring of geodynamic processes in underground hard mineral mining *Journal of Mining Science* Vol 51 No 3 pp 634–640

[9] Julian BR 1994 Volcanic tremor: nonlinear excitation by fluid flow *Journal of Geophysical Research: Solid Earth* Vol 99 No B6 pp 11859–11877

[10] Balmforth NJ., Craster RV., Rust AC. 2005 Instability in flow through elastic conduits and volcanic tremor *Journal of Fluid Mechanics* T. 527 pp 353–377

[11] Corona-Romero P, Arciniega-Ceballos A, Sánchez-Sesma FJ 2012 Simulation of LP seismic signals modeling the fluid–rock dynamic interaction *Journal of Volcanology and Geothermal Research* Vol 211 pp 92–111

[12] Rust A. 2004 Flow-induced oscillations: A source mechanism for volcanic tremor? *2003 Program of Study Non-Newtonian Geophysical Fluid Dynamics* p 113

Experimental research data on stress state of salt rock mass around an underground excavation

VD Baryshnikov* and DV Baryshnikov

Chinakal Institute of Mining, Siberian Branch, Russian Academy of Sciences, Novosibirsk, Russia

E-mail: *v-baryshnikov@yandex.ru

Abstract. The paper presents the experimental stress state data obtained in surrounding salt rock mass around an excavation in Mir Mine, ALROSA. The deformation characteristics and the values of stresses in the adjacent rock mass are determined. Using the method of drilling a pair of parallel holes in a stressed area, the authors construct linear relationship for the radial displacements of the stress measurement hole boundaries under the short-term loading of the perturbing hole. The resultant elasticity moduli of rocks are comparable with the laboratory core test data. Pre-estimates of actual stresses point at the presence of a plasticity zone in the vicinity of the underground excavation. The stress state behavior at a distance from the excavation boundary disagrees with the Dinnik–Geim hypothesis.

Mir Mine, ALROSA, extracts ore reserves under the closed open pit mine bottom by top-down cut-and-fill with cemented backfill. The underground excavations are protected from flooding by brines from high-pressure Metegero-Ichersky aquifer exposed by open pit mining using the method of 'dry' deactivation of the open pit mine with the safety pillar left under the open pit bottom [1].

To control the safety crown pillar (20 m thick), Yakutniproalmaz together with the Institute Mining developed and implemented the integrated geomechanical monitoring project in Mirny Mine in 2014. The geomechanical behavior in rock mass under the pit bottom is monitored [2], and development of geomechanical processes is predicted by mathematical modeling given available information on stress state and mechanical properties of intact rock mass [3].

Enclosing rock mass under the open pit mine bottom at the depth of extraction blocks 1 and 2 (550–750 m) below the surface is composed of salt rocks (halite) and thin carbonate interbeds [4]. By laboratory core tests, elasticity modulus of halite within Charskaya suit is 14.6–19.6 GPa, limit strength of halite at the supposed measurement depth (750 m) is 24.7 MPa [5]. At the average bulk overburden rock weight of 2.61 t/m^3 [5], the vertical stress in intact rock mass is $-\gamma H = 19.6$ MPa.

Tectonically, Mir pipe is governed by interaction of north-west, meridional and north-east subvertical faults, and the extension of the faults in some areas of the ore body is reflective of activation of the fault-and-block structure at post-ore stages [4]. Modern regional stress field has not been assessed whereas these parameters as mechanical properties of rocks are required as the boundary conditions in numerical modeling. Assessment of state and properties of rocks is a complex and laborious problem, both theoretically and practically, and requires target-oriented and long-term operation. In this connection, it is planed to carry out experimental research of stress state by stages. The first stage is the approximated estimation of the intact tock mass stresses from the viewpoint of applicability of hypotheses on stress distribution (by Dinnik and Game). Stress state of intact rock mass should be estimated beyond the influence zone of open pit and underground mining.

Aiming to estimate applicability of stress measurement in parallel drill holes [6] in salt rock mass, the experimental test series the automated measurement equipment designed at the Institute of Mining [7]. The authors think it is inexpedient to assess stresses outside the salt rock mass enclosing blocks 1 and 2 because of thin carbonate interbeds where stresses behave specifically. The test location was selected adjacent rock mass of a conveyor crossway at the level –410 m (750 m below the surface), beyond the influence zone of either open pit or underground mining.

Figure 1 depicts arrangement of the holes and orientation of the measurement direction of strainmeter (Figure 1).

The Institute of Mining has developed the method of stress measurement in parallel drill holes [6]. In this method, a measurement hole is first drilled and strainmeter is placed in it. After the strain meter readings become stable, initial stress state in rocks around the measurement hole is excited by means of drilling a parallel disturbing hole. The stresses are estimated by radial displacements caused by the parallel drilling of a disturbing hole, and deformation properties of rocks in the measurement area are determined by the change in the radial displacements under the further loading of the disturbing hole. The validity of the elastic model of rock mass behavior in stress calculation is determined by the data on rock mass response to loading of the walls of the disturbing hole.

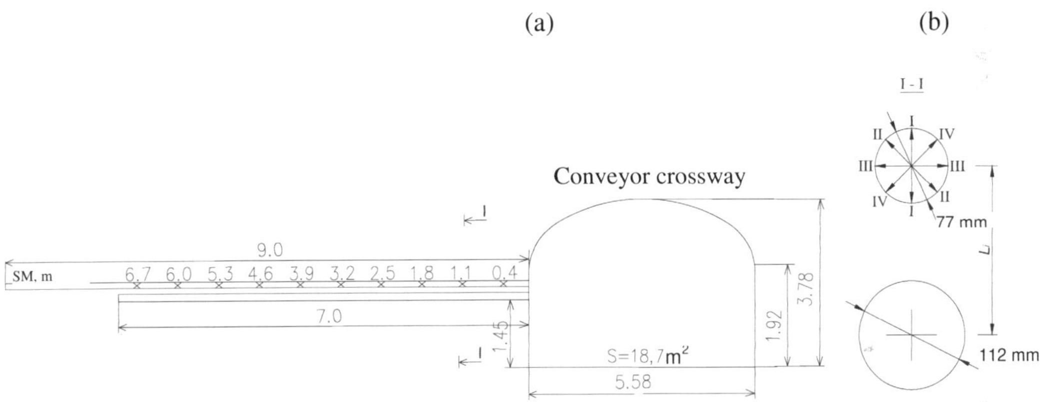

Figure 1. (a) Arrangement of measurement and disturbing holes; (b) orientation of measurement directions of strainmeter in vertical plane.

Drilling used diamond drill bits with the diameters 77 and 112 mm. Visual inspection of the walls in the measurement hole before the tests by a downhole video camera showed no defects along the whole depth of the hole (9 m).

Figure 2 shows the curves of the radial displacement in the walls of the measurement hole under disturbing hole drilling at the installation of the strainmeter (SM) 1.1 and 2.5 m away from the crossway wall. The measurement was carried when the bottom of the disturbing hole with the diameter of 120 mm was 30 mm higher and than 30–35 m lower the installation place of SM.

The analysis of the obtained results shows that:

—at the depth of 1.1 m (Figure 2a) deformation of the measurement hole walls starts when the disturbing hole bottom is 20 cm higher SM place and stops when the bottom is 30 cm lower it (the measurement accuracy 1 μm);

—after the termination of the disturbance drilling, there is a stable linear trend in all measurement directions of SM at the depth of 2.5 m (Figure 2b), which is strengthens with distance from the crossway wall.

Starting from the depth of 2.5 m and below, reading of the strainmeter show no stabilization even after the termination of drilling. The total radial displacement is composed of instantaneous elastic displacement due to the disturbing hole drilling and displacements due to irreversible deformation of salt (creep).

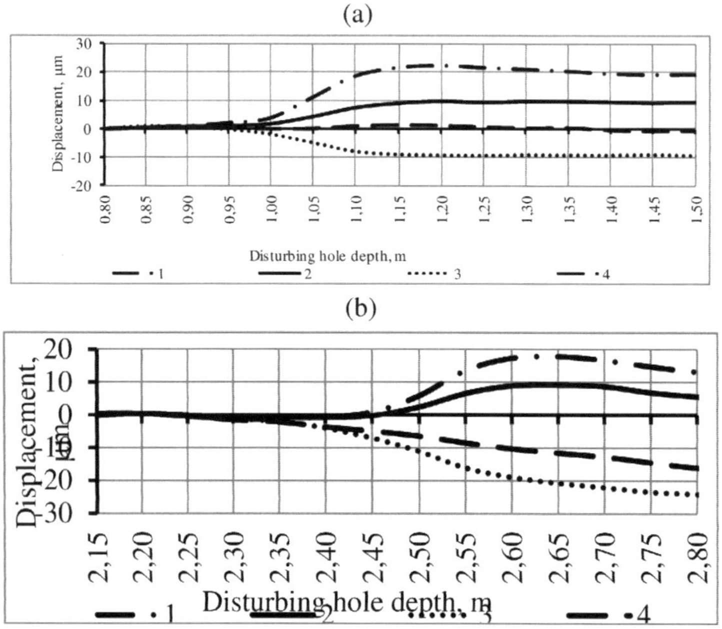

Figure 2. Displacements in measurement hole walls under disturbance drilling along the directions of SM beams at (a) 1.1 m and (b) 2.5 m.

The calculations in the parallel hole drilling method is based on the elastic model; for this reason, an approach was proposed to processing of data obtained farther than 2.5 m from the crossway wall. In each measurement direction of SM beams, the overall recorded displacements are subtracted by the displacements determined from the displacement trends after the drilling termination within the time corresponding to the time of the disturbance drilling behind the zone of SM installation (3 min). The resultant difference is assumed the elastic displacements induced by the disturbance drilling and used in the calculations of the quasi-principal stresses.

For the assessment of stresses by the radial displacements of the measurement hole walls under the disturbing hole drilling, the modulus of elastic deformation of salt, $\sigma_{æ} = 4(1-\nu^2)$, is determined by means of step-by-step loading of the disturbing hole after the termination of drilling. It is also becomes possible to validate applicability of the elastic model of the rock mass behavior.

Figure 3 illustrates the results of loading by the data of SM at 1.1 and 6.7 m. The analysis of the deformation in all 4 directions of SM beams points at the linear relation between pressure and displacement (elasticity of rocks) under short-term load and at insignificant residual displacements under unloading, not higher than the measurement error. The difference in the displacements in SM beams 2 and 4 at the depth of 6.7 m is reflective of deviation of the holes from coaxiality. Mutual position of the holes along the depth was checked by tacheometry. With regard to the obtained data, $\sigma_{æ}$ was assessed (Table 1).

The values of the elasticity modulus agree by depths. At the same time, these values of the elasticity modulus (E_e = 14.6–20.0 GPa) are comparable with the core test results (14.6–19.6 GPa).

Table 2 and Figure 4 present the values and orientations of the quasi-principal stresses in the vertical plane at different distances from the crossway wall.

The analysis of the obtained results shows the unloaded (plastic zone) 3 m away from the crossway wall; the values of stresses in this zones do not exceed the vertical stresses due to overburden rock weight ($\gamma H \approx 20$МПа). At the distance of 3 m from the crossway walls, the horizontal stresses exceed γH by 1.5–2 times.

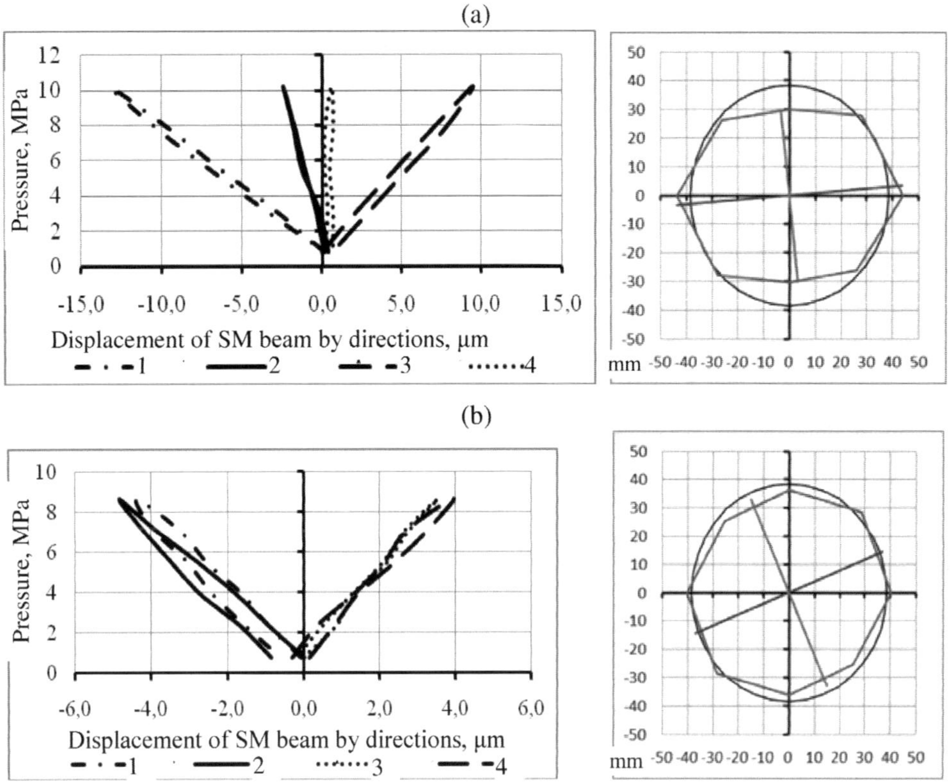

Figure 3. Radial displacements of SM beams under step-by-step loading and unloading of disturbing hole walls and orientation of principal axes of deformation in the measurement hole under the peak load at the depth of (a) 1.1 and (b) 7.6 m.

Table 1. Deformation properties of halite rock mass in the stress measurement area.

Distance to crossway wall, m	Pressure, MPa	Unloading branch		Unloading branch	
		$\sigma_{æ}$, GPa	E_d**, GPa	$\sigma_{æ}$, GPa	E_e **, GPa
0.4	8.6	4.0 (3.7÷4.4)	14.6	4.0 (3.6÷4.4)	14.6
1.1	10.2	4.6 (4.2÷5.1)	16.7	4.7 (4.3÷5.3)	17.1
1.8	10.2	4.9 (4.5÷5.3)	17.8	5.1 (4.5÷5.8)	18.6
2.5	10.2	5.3 (4.8÷5.8)	19.3	5.3 (4.7÷5.9)	19.3
3.2	10.2	5.3 (4.9÷5.8)	19.3	5.2 (4.6÷6.0)	18.9
3.9	10.2	5.3 (4.8÷5.9)	19.3	5.5 (4.8÷6.3)	20.0
4.6	10.2	5.5 (5.1÷6.0)	20.0	5.3 (4.9÷5.7)	19.3
5.3	10.2	5.2 (4.7÷5.7)	18.9	5.1 (4.8÷5.4)	18.6
6.0	10.2	5.2 (4.8÷5.6)	18.9	5.3 (4.9÷5.7)	19.3
6.7	10.2	5.1 (4.8÷5.5)	18.6	5.4 (5.2÷5.7)	19.7

* Confidence interval 90%; ** E_d—deformation modulus; E_e—elasticity modulus; $E = 3.64\sigma_{æ}$ at $\nu = 0.3$

Table 2. Values of stresses in vertical plane.

SM, m	Stresses, MPa			Quasi-principal stresses, MPa		Orientation of σ_2, deg**
	σ_v	σ_h	τ_{sh}	σ_1	σ_2	
1.10	−7.3 (0.5)	−6.8 (1.6)	0.4 (0.3)	−6.5	−7.5	66
1.80	−12.9 (1.1)	−12.7 (3.4)	1.2 (0.5)	−11.6	−14.0	56
2.50	−17.7 (0.5)	−22.0 (1.6)	1.5 (0.2)	−17.2	−22.5	28
3.20	−19.0 (1.0)	−34.2 (3.3)	1.8 (0.5)	−18.8	−34.4	19
3.90	−20.7 (1.2)	−32.8 (3.9)	0.3 (0.6)	−20.6	−32.8	16
4.60	−23.3 (2.9)	−50.3 (10.0)	1.9 (1.4)	−23.2	−50.5	20
5.30	−21.8 (1.4)	−58.2 (5.0)	0.5 (0.7)	−21.8	−58.2	18
6.00	−24.5 (1.2)	−53.5 (4.4)	5.3 (0.6)	−23.6	−54.5	29
6.70	−23.6 (2.2)	−54.9 (8.5)	3.1 (1.0)	−23.4	−55.2	28

σ_v, σ_h, τ_{sh}—vertical, horizontal and shearing stresses; *standard deviation; **positive value of the angle is assumed from the horizon counterclockwise.

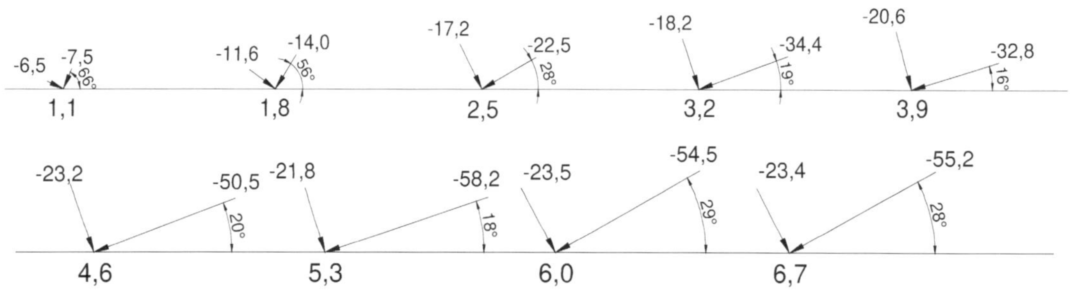

Figure 4. Orientations and values of the quasi-principal stresses (MPa) in the vertical plane.

The accuracy and reliability of the stress measurement using the method of parallel hole drilling at deeper levels in rocks can be improved by changing geometry of the experiment (first of all, spacing of the holes), which reduces deformation in rock around the measurement hole and, as a consequence, abates development of rheological processes.

Conclusion

The method of stress measurement by parallel hole drilling ensures reliable assessment of stresses in salt rock mass, which do not exceed the uniaxial compression strength of rocks. Stabilization of readings of strainmeter before and after disturbance drilling is the evidence of absence of inelastic deformation of the measurement hole walls.

The preliminary estimation of the values of the quasi-principal stresses shows the presence of a plasticity zone in rock mass around a tunnel. Beyond this zone, the subhorizontal stresses exceed the vertical stress due to overlying rock weight by 1.5–2.9 times.

The linear dependence of the radial displacements at different distances from the measurement hole walls under step-by-step loading of the disturbing wall, nearly without residual deformation under unloading, points at the elastic behavior of salt rocks under the short-term relief, and the values of the elastic modulus are compared with the laboratory test data.

References

[1] Genzel GN, Voropaev BP, Yakushenko MV, Zelberg AS, Kramskov NP, and Lobanov VV 2002 Deactivation of Mir open pit mine and flooding protection of underground mine during transition period *Proc. Sci. Conf. Mirny–2001* Moscow: Ruda Metally (in Russian)
[2] Baryshnikov VD, Fedyanin AS, Pul EK and Baryshnikov DV 2017 Geomechanical monitoring of open-pit bottom reserves in Mir Mine, ALROSA: Results *J. Min. Sci.* Vol 53 No 1 pp 34–42
[3] Baryshnikov VD, Baryshnikov DV, Gakhova LN and Kachalsky VG 2014 Practical experience of geomechanical monitoring in underground mineral mining *J. Min. Sci.* Vol 50 No 5 pp 855–864
[4] Kolganov VF, Akimov AN and Drozdov AV 2013 *Ground Conditions of Primary Diamond Deposits in Yakutia* Mirny: Mirninsk Tipograf (in Russian)
[5] Konovalenko VYa 2012 *Reference Book on Physical Properties of Rocks at Diamond Deposits in Yakutia* Novosibirsk (in Russian)
[6] Kurlenya MV, Baryshnikov VD, Popov SN et al 1981 Method to determine stresses and strains in rocks *Otrkyt. Izobret.* No 40
[7] Baryshnikov VD and Kachalsky VG 2010 Automation instrumentation to measure rock mass stresses in parallel-drilled holes J. Min. Sci. Vol 46 No 3 pp 338–342

Stress–strain state of adjacent rock mass under slice mining of steeply dipping ore bodies

VD Baryshnikov* and LN Gakhova

Chinakal Institute of Mining, Siberian Branch, Russian Academy of Sciences, Novosibirsk, Russia

E-mail: *vbar@misd.ru

Abstract. Under analysis is the stress state of rock mass surrounding stopes in the initial cutting layer displaced in plan relative to the above-lying extracted layer in the overcut rock mass. The authors determine the boundaries of the post-limiting deformation zones during stoping advance using the Mohr–Coulomb criterion. The sequence of stoping to ensure better support conditions is proposed.

Internatsionalnaya kimberlite pipe in the form of a subvertical ore body with an oval cross section is developed by ALROSA Company using the underground method of descending horizontal slicing with the layers 4–5 m high. The room-and-pillar system involves stopes 5–6 m wide with cemented backfilling [1]. Aimed to ensure the mine design productivity, the ore body is divided by cut layers into working sub-levels with a height of 30–45 m.

The geomechanics of the cut layers where stability of stopes is difficult to ensure though it is of crucial importance is analyzed in [1] in terms of a vertical ore body. Numerical modeling of stress state of rocks in combination with the in situ monitoring of deformation processes in surrounding rock mass around underground excavations made it possible to substantiate the cut layer flow chart from the center sideways with the separation of the layer into two or three panels.

The problems on stresses in rocks during cutting and backfilling were solved using the method of boundary integral equations [2, 3]. The initial parameters on natural stresses were assumed in accordance with the field data for Internatsionalny Mine: $\sigma_z^\infty = -\gamma H$, $\sigma_x^\infty = \sigma_y^\infty = -\lambda\gamma H$ at $\lambda = 0.7$, where σ_z^∞—vertical stress, $\sigma_x^\infty, \sigma_y^\infty$—horizontal stresses; λ—coefficient of lateral earth pressure; γ—bulk weight of rocks; H—mining depth. The calculated data interpretation involved the components of stress tensor and shear stresses σ_s which allowed using the Mohr–Coulomb criterion for the analysis [4]:

$$\sigma_s = \frac{\sigma_1 - \sigma_3}{2\cos\varphi} + \frac{\sigma_1 + \sigma_3}{2}\tan\varphi,$$

where $\sigma_1 > \sigma_2 > \sigma_3$—principal stresses; φ—angle of internal friction.

By the visual observations and from the numerical calculation results, it was found that failure took places in adjacent rock mass in the zones where $\sigma_s > 4$ MPa [2, 5, 6]. For this reason, $\sigma_s^C = 4$ MPa was assumed as the critical value.

As mining goes to deeper levels, the morphology of the ore body changes: its inclination, shape of the cross section. Figure 1 depicts the analytical data on the geomechanical behavior of rock mass in a cut layer at a depth of 1211 m below surface (actual -811 m) where the north–south displacement relative to the previous cut layer reached 18–20 m (Figure 1a), which considerably changed stress redistribution in the pillar between the layers. Given there are no displacement of the layers in plan view, the zones of the increased stresses σ_s form in the roof in the most stopes in the center of the cut layer, while in the displaced layer the increases stresses are generated in the roof of the stopes in the southern part of the layer (Figure 1).

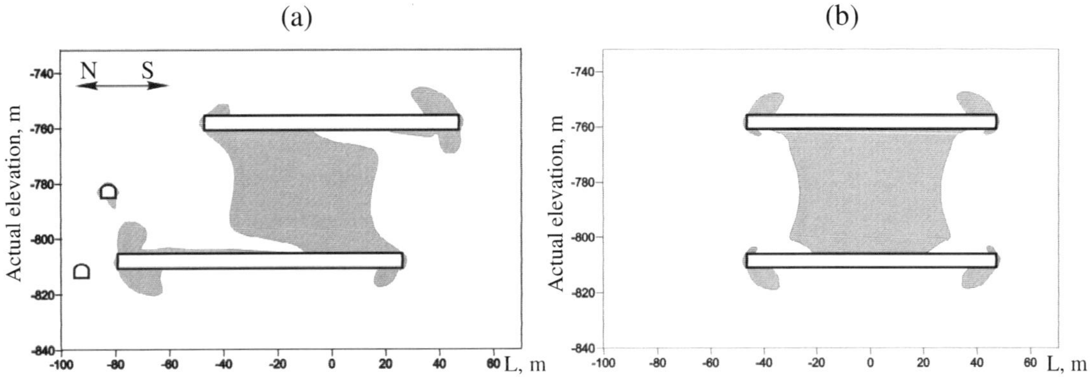

Figure 1. Distribution of σ_s (MPa) in the cut layer (elevation view): (a) actual; (b) it is assumed that the bottom layer is not displaced relative to the earlier cut top layer.

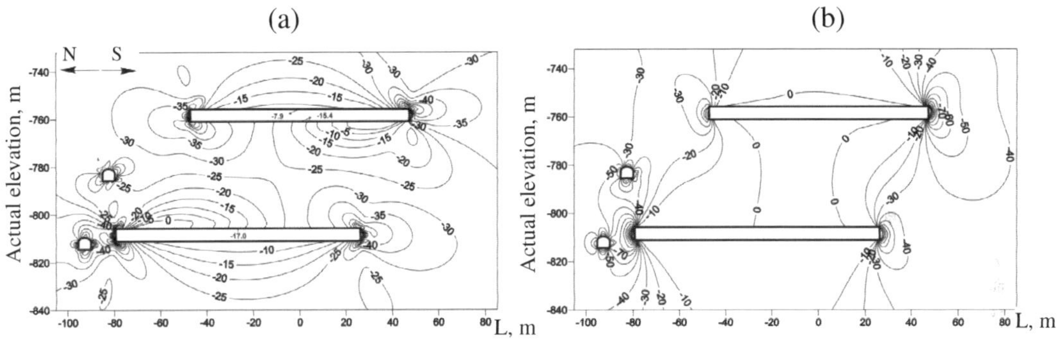

Figure 2. Distribution of horizontal (a) and vertical (b) stresses (MPa) after completion of works in the cut layer (elevation view).

Stress state in the pillar between the cut layers is characterized by concentration of horizontal stresses in the roof in the south of the cut layer and by stress relaxation in the north part; regarding the earlier cut layer, stress concentration is observed in the bottom in the north and stress relaxation takes place in the bottom in the south (Figure 2a). At the same time, the pillar is free from vertical stresses except for the ends of the layers (Figure 2b).

The analysis of the scenario of separating a cut layer into three panels similarly to [1] nearby the earlier cut and backfilled stopes (Figure 3) shows that:

—extraction of a panel in the north part (wedge-out) of the ore body will be carried out under relatively comfortable conditions in case that the first stope is cut in the near-contact zone and further mining is conducted in the directions from the center towards the extracted and backfilled chamber;

–it is indifferent whether stoping in the south part of the panel is carried out back-to-back the backfilled stopes or not: anyhow, zones of the increased stresses σ_s forms in the roof, which requires obligatory roof support;

—advance of stoping in the central panel in the line of the ore body wedge-out part induces the zone the increased stresses σ_s in the pillar in the mined-out and backfilled section of the layer, and this

IOP Conf. Series: Earth and Environmental Science **134** (2018) 012005 doi:10.1088/1755-1315/134/1/012005

zone merges with the earlier formed zone in the bottom of the above-lying layer. In the roof of the next stopes σ_s are under the critical value.

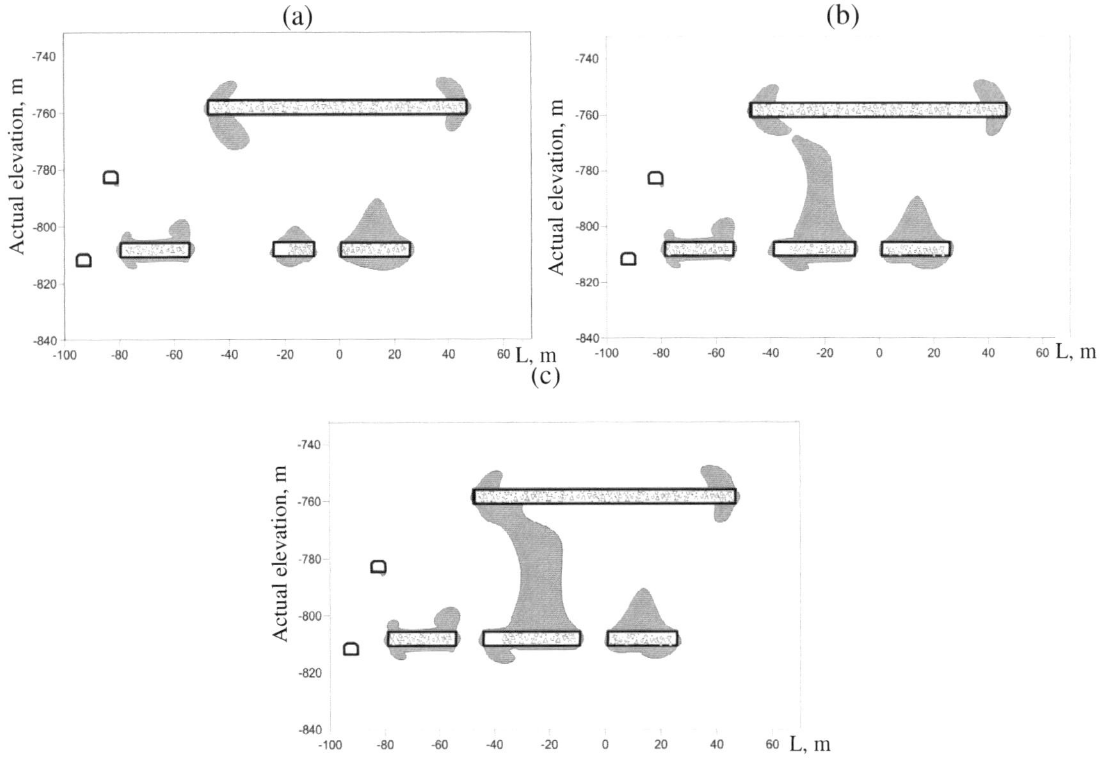

Figure 3. Inelastic strain zones during subdivision of the cut layer into three extraction panels. The end panels are mined-out, stoping in the central panel is advanced in the direction of the wedge-out part of the layer: (a)–(c)—stages of work in the central panel.

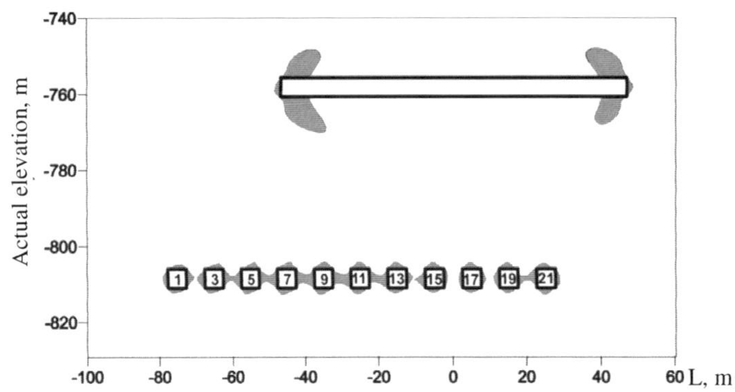

Figure 4. Post-limit deformation zones in the course of cutting the layer by stopes arranged next but one.

When the cut layer is developed by stopes arranged next but one (Figure 4), the most unfavorable conditions are in the center of the layer (5–15): σ_s higher than the critical value form in the pillars between the stopes and in their roofs at a distance to 2 m.

Conclusion

Features of stress distribution are determined in the surrounding rock mass of stopes in the cut layer made in the overmined rock mass under change in the ore body morphology. Using the Mohr–

Coulomb criterion, the authors have assessed the post-limiting deformation zones in ore body at different stages of mining operations.

When stopes in the cut layer are made back-to-back in separating the layer into three panels, it is recommended to start stoping from the ore and rock contact toward the center in the overmined (south) part of the layer and from the center toward the wedge-out end (north) in the center of the ore body.

In the two-stage stoping, for the purpose of better stability of stopes and to reduce cost of support, it is recommended to mine out ore-and-rock interface zones using the first stage stoping. Furthermore, the first stage stopes in the center of the ore body need supporting.

References

[1] *Temporary Process Guidelines on Slice Mining with Solidifying Backfill in Internatsionalny Mine* Mirny: Yakutniproalmaz 2004 (in Russian)

[2] Baryshnikov VD, Baryshnikov DV, Gakhova LN 2015 Experimental-analytical estimation of undermined safety pillar *Proc. 15th Int. Multidiscipl. GeoConference SGEM 2015* Albena: STEF92 Technology Ltd pp 285–289

[3] Baryshnikov V., Gakhova LN 2009 Geomechanical conditions of kimberlite extraction in terms of Internatsionalnaya kimberlite pipe *J. Min. Sci.* Vol 45 No 2 pp 137–145

[4] Bulychev NS 1989 *Mechanics of Underground Structures* Moscow: Nedra (in Russian)

[5] Baryshnikov VD, Gakhova LN 2015 Selection of efficient flow sheets for initial cutting in top-down mining in Internatsionalnaya pipe *J. Min. Sci.* Vol 51 No 6 pp 1165–1172

[6] Kurlenya MV, Baryshnikov VD, Gakhova LN 2012 Experimental and analytical method for assessing stability of stopes *J. Min. Sci.* Vol 48 No 4 pp 609–615

IOP Conf. Series: Earth and Environmental Science **134** (2018) 012006 doi:10.1088/1755-1315/134/1/012006

Upgrading the multi-component borehole strainmeter for the accuracy improvement of stress measurement in rocks

VD Baryshnikov[1]*, VG Kachalsky[1,2] and DV Baryshnikov[1]

[1]Chinakal Institute of Mining, Siberian Branch, Russian Academy of Sciences, Novosibirsk, Russia
[2]Novosibirsk State Technical University, Novosibirsk, Russia

E-mail: *vbar@misd.ru

Abstract. The trials of the automated stress measurement equipment have revealed the necessity to upgrade its component represented by the borehole strainmeter. Application of the digital microprocessing methods for signal measurement and conversion has allowed elimination of negative impacts and enhanced efficiency and reliability of stress–strain control in rocks.

Assessment of geomechanical behavior of rocks to substantiate safe mining flowsheets is supported by various techniques, including integrated geomechanical researches of stress state and mechanical properties of rocks. A systemic approach is required to obtaining reliable information on:

—initial stress field and its variation with depth;

—structural features and mechanical properties of ore body and enclosing rock mass;

—displacement of rocks and concentration zones of stresses;

–structural elements of mine, mining sequence and support of development entries.

As one of the method of stress state assessment in rock mass, the method of parallel drilling consists in measuring small deformations in walls of a measurement hole in the course of parallel drilling of a disturbing hole further subjected to loading by uniform pressure [1, 2]. For the implementation of this method, the Institute of Mining has developed the theoretical framework [3] and a prototype model of the automated measurement equipment set composed of the downhole strainmeter, communication unit, manual hydraulic jack and a loading device [4]. The downhole strainmeter meant for measuring radial displacements of hole walls consists of sensitive strain sensors attached or glued to steal beams; deformation of the beams is sent tot the sensors and converted into signals by a measurement channel. The full-scale trial of the equipment has shown that it needs upgrading to ensure better accuracy and reliability of measurement in case of very small radial displacements.

Aimed to increase measurement accuracy, the analysis of measurement circuits of strainmeters with different channels of amplification of signals has revealed a number of shortcomings:

—half-bridge connections in the resistive strain sensors needs that temperature stabilizes, which takes much more time for the experimentation;

—half-bridge connection of the sensors and beams incompletely uses capacities of transformation of deformation to electrical signal;

—when deformation of the beams is small, weal signals and their wire transfer to a recording station adds extra error to measurement because of the influence of deformation of wires and ambient temperature;

—stress measurement errors due to inaccurate orientation of measurement directions (diametrical beams) can be high.

These limitations of strainmeters result in considerable interference in measurement channels. The only way to handling this problem successfully is the use of well-balanced full-bridge measurement circuits on each beam and reduction in spacing of sensors and signal amplifiers. The block-diagram of the modified experimental facility (Figure 1) involves the strainmeter to be installed and advanced in a hole, communication unit and a tablet computer with bluetooth pairing.

The engineering problem is improvement of stress measurement reliability in rocks by means of more precise orientation of measurement directions of strainmeter in a measurement hole relative vertical and horizontal planes, and through more accurate determination of radial displacements along the measurement directions of strainmeter, control of measurement of displacements in the measuring hole and increased accuracy of stress state determination later on. The real-time information allows prompt and reliable estimation of the test quality on-spot. The design includes full-bridge connection circuits of strain sensors pair-wise arranged on two elastic beams and balanced during assembling. Balancing adjustment of bridges is carried of pre-deformed beams with the strainmeter installed in a calibration cylinder with the diameter conformable with the diameter of drill bit intended for the measurement hole drilling. This approach enables preserving balance of bridge circuits during measurement and allows maximum compensation of temperature drift. The measurement circuit of the multicomponent strainmeter includes bridges of resistive strain sensors, multichannel signal amplifiers, 8-channel commutator, 24 bit AD converter and a microprocessor (Figure 2).

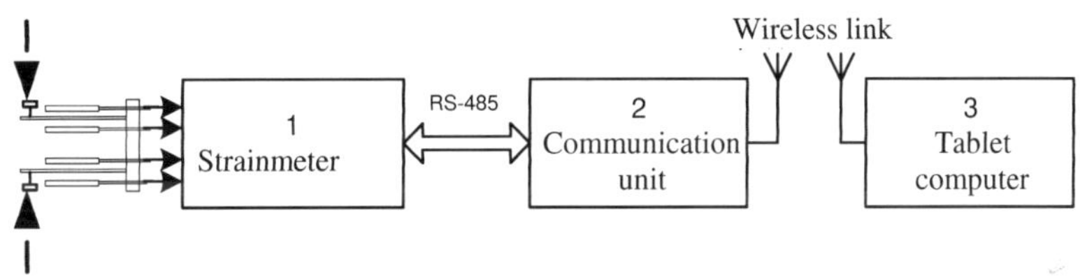

Figure 1. Block-diagram of upgraded experimental equipment for stress measurement in rocks by the method of parallel drilling.

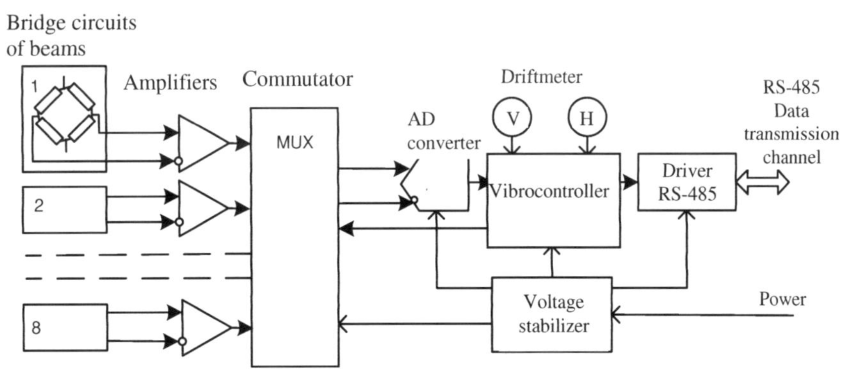

Figure 2. Microprocessor circuit to tale signals from strain sensors and driftmeters: V—vertical driftmeter (incline of the hole); H—horizontal driftmeter (orientation of measurement directions relative to horizontal plane).

The position of the strainmeter is controlled by two solid driftmeters connected, similarly to the commutator and DC converter, via the intra-circuit digital 12 C channel, to the microprocessor. All electronic devices in Figure 2 are arranged in the sealed housing of the strainmeter. The data and power connections are ensured by the communication unit (refer to Figure 1) and connected to the strainmeter by information channel RS-485 and supply.

The implemented upgrading enhances reliability of stress state control in rock mass owing to promptness and accuracy of data obtained in real time and directly at the measurement site and serving the basis for the decision-making on experimentation adjustment. Moreover, the modifications enhance efficiency of the automate data recording and processing system in terms of stress state determination in rock mass owing to the improved accuracy of measuring radial displacements of the measuring hole wall and considerably shortened time of the experimentation.

The updated software consists of two parts. The microprocessor software (Figure 3) enables serial polling of bridge circuits of the resistive strain sensor via the commutator, actuation of transformation by AD converter, waiting for interrupt by availability of data from each channel, taking of digital data on angles of position from two driftmeters, formation of data base on deformation per each of 8 strain bridges and two driftmeter and RS-485 channel transfer of the data on demand of tablet computer software.

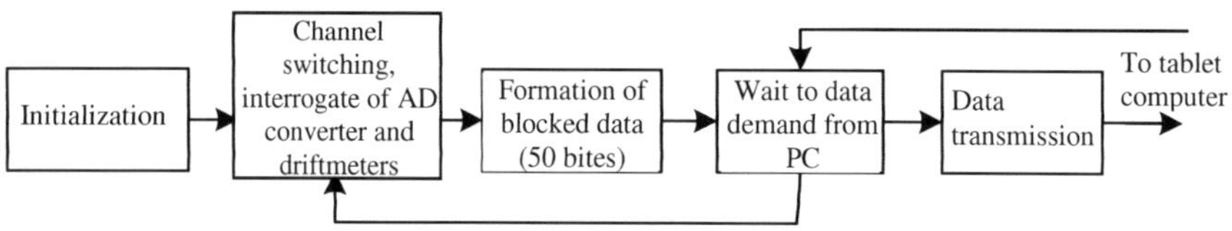

Figure 3. Block-diagram of software modules to support operation of microprocessor system of strainmeter.

Conclusion

The implemented upgrading of electronics and software allows:

—considerable reduction in temperature drift in measurement circuits of strain sensors;

—increase in strain measurement sensitivity and, thus, two times higher signal intensity owing to full-bridge connections of strain sensors and beams;

—elimination of influence exerted by cables on signals from strain sensors due to digital data communication between strainmeter and communication unit;

—improved precision of orientation of strainmeter in measurement hole by two driftmeters;

—real-time and prompt control over experimentation;

—express-analysis and comparison with the earlier data on radial displacement measurements in hole walls;

—operation decision-making on expediency of the continuing stress measurement deeper in the hole based on on-spot estimation of test data and comparison with the earlier experimental results.

References

[1] Kurlenya MV, Baryshnikov VD, Popov SN et al 1981 Method to determine stress sate in rocks, *Otkryt. Izobret.* No 40

[2] Baryshnikov VD, Kurlenya VM, Popov SN et al 1982 Method of in situ determination of the elastic properties of rocks in parallel holes *J. Min. Sci.* Vol 18 No 1 pp 69–71

[3] Kurlenya VM and Popov SN 1993 *Theoretical Basis for the Determination of Stresses in Rocks* Moscow: Nauka (in Russian)

[4] Baryshnikov VD and Kachalsky VG 2010 Automated instrumentation to measure rock mass stresses in parallel-drilled holes J. Min. Sci. Vol 46 No 3 pp 338–343

Pulsation-nature stresses on flat convergent walls of slot-type hopper under granular medium flow

AP Bobryakov, SV Klishin*, VP Kosykh and AF Revuzhenko*****

Chinakal Institute of Mining, Siberian Branch, Russian Academy of Sciences, Novosibirsk, Russia

E-mail: *sv.klishin@gmail.com, **v-kosykh@yandex.ru, ***revuzhenko@yandex.ru

Abstract. The authors investigate numerically the flow of granular materials in the converging radial channels by solving 3D problem using the finite element method. The pulsation nature of normal and shear stresses on the channel walls is demonstrated. The calculations are compared with the lab-scale experimental data.

1. Introduction

Flow of granular materials in converging channels is involved in many processing in mining: discharge from hoppers and bunkers, in ore passes etc. Direct investigation of this process in full-scale conditions faces some technical difficulties. So, numerical and experimental modeling on a laboratory scale takes on the key role. In this regard, the method of discrete elements acquires increasingly higher popularity in recent years. Laboratory-scale modeling consists of a few steps. The first step is the analysis of similarity criteria, the second step is the modeling and experimentation. The resultant outcome is the kinematics of deformation and failure and the stress distribution. There is a wide range of methods to investigate kinematics, starting from cross-hatching of material surface and finishing with X-ray imaging and various alternatives of PIV (Particle Image Velocimetry) [1, 2].

The issues connected with the stress measurements appear to be more difficult to solve. The experts in this area of research know well that it is a very intricate problem to measure stresses on a model (a fortiori in situ). Neither universal and common procedure nor equipment has been developed. This paper puts forward a variant of the specified problem solution.

2. Discrete element modeling

First, let us discuss issues of theoretical analysis. It is now popular to investigate kinematics using different variants of simulation models of flow [3, 4]. In this case, a cell is assigned, from which a material flows out. Then, certain probabilities of filling the vacant cell by the material from the neighbor cells are set. As a result, based on the very simple rules, it is possible to obtain rather complex patterns of flow, which, in some measure, simulate a real process (a sort of Conway's Game of Life). However, this approach is unsuitable in terms of stresses. It is required to use closed models that account for stresses. An adequate and simulation-like model is the discrete element model [5–7]. This method is an alternative in principle to classical methods based on conventional continuum mechanics and is widely used this time to study various modes of flow of granular materials.

Figure 1 shows an intermediate result of the 3D calculation by DEM. The figure demonstrates the kinematic pattern of deformation of a flowing medium and distribution of horizontal velocities of discrete elements at fixed timing. The analysis of the number of discrete elements required for the construction of continuum model equivalent to the assumed discrete element model was performed by the authors in [7]. It was shown that the particles to be included in the numerical experiment varied from $3{\cdot}10^4$ to $3{\cdot}10^9$. The continuum model is beyond the scope of this paper but the total number of particles in the laboratory experiment is estimated as $57{\cdot}10^6$. The numerical model includes 350 000 particles. Apparently, it is impossible to parallel a laboratory and a numerical experiment at the modern stage of computational technologies. The advantage of a numerical model is the possibility to calculate coordinates, trajectories and displacements of particles, as well as forces and stresses at each time step, which is problematic in the in situ measurement. Figure 2 shows epures of the normal forces F_n and tangential forces F_t applied by granular material on a side wall of a hopper at a fixed moment of time. It is seem that the stress distribution is nonstationary as in the experiments described below in this paper.

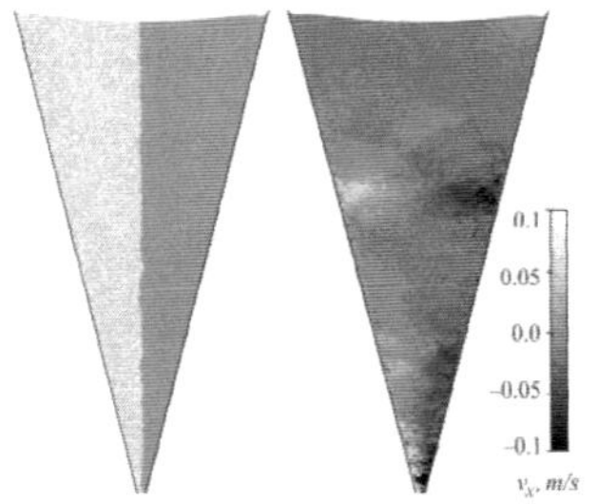

Figure 1. Kinematic pattern of deformation and distribution of horizontal PPV v_x in granular material at fixed time obtained from numerical calculation using discrete element method.

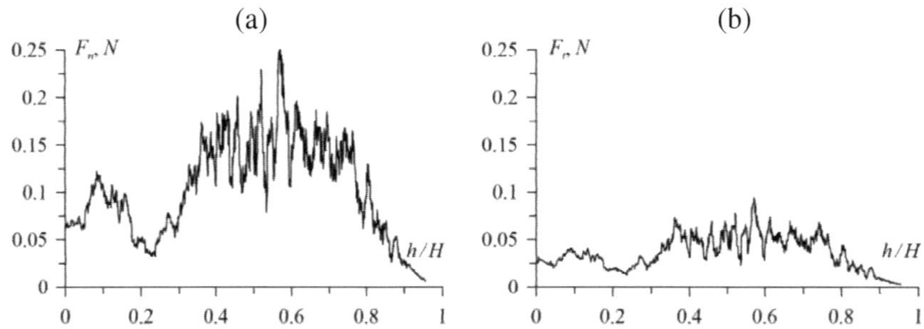

Figure 2. Distribution of the (a) normal and (b) tangential forces applied by granular material on hopper wall at fixed time moment. The coordinate origin fist with the discharge opening.

3. Stress measurement

Granular media possess elastoplastic properties and internal friction. The latter property is reflected by the stiffness of sensors. Correct measurement is only possible with the sensors possessing maximum stiffness.

On the other hand, the increase in the stiffness of elastic elements of sensors lowers their sensitivity down to a level comparable with the noise of measurement equipment. Thus, a problem of stress measurement in granular medium is design of sensors of sufficient stiffness.

Some designs of normal stress sensors possessing sufficient stiffness and sensitivity are known from literature. For example, hydraulic and mechanical pressure transducers are described in [8, 9]. The other designs are presented in [10–13]. General deficiency of the known sensor designs is the size and complexity. Such sensors are not always suitable for the analysis of properties of granular materials and soil under low stresses, for instance, on a laboratory scale. In this study, the authors used the original sensors possessing admissible sensitivity for the investigation of stress state of granular material in the laboratory conditions.

The sensitive element of a normal stress sensor was a spring steel beam rigidly attached to a steel frame on two sides. As a strain transducer, a strain gauge rosette was glued on the inner side of the beam. The size of the rosette was selected so that strain gauges on the opposite sides extended and constricted pairwise to ensure the maximum sensitivity of the sensor.

The temperature error of the sensor was decreased by making the strain gauges of a material having the temperature compensation factor conformable with the heat expansion of the beam.

The shear stress sensor [14] determines pulling forces under shear of two plates pressed to each other and placed in the granular medium. The sensor (Figure 3) has a body composed of cover 1 and bottom 2 which are simultaneously the force-transmitting elements. For pre-tensed strain gauges 3 are arranged between the cover and bottom so that one their ends are glued in the center of the cover 1 and their other ends are glued at the opposite edges of the bottom 2. Friction between the bottom and cover is reduced by means of steel balls 4 with a diameter of 0.2 mm placed between the cover and bottom. Sealing of the sensor is ensured by the thin elastic shell 5 through which contacts 6 are let out to be connected to the measurement equipment.

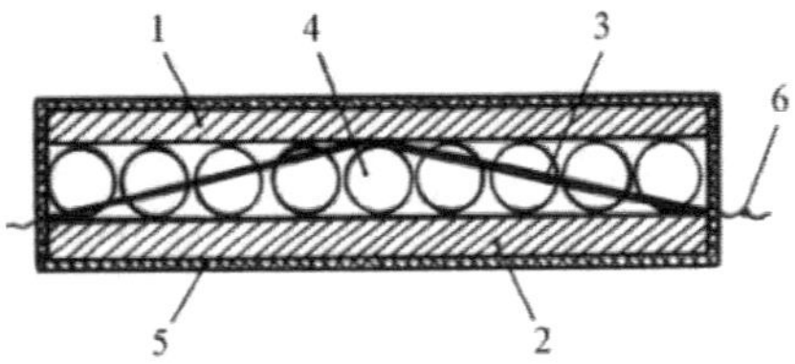

Figure 3. Shear stress sensor design 1—cover; 2—bottom; 3—strain gauges; 4—steel balls; 5—elastic shell; 6—contacts of the strain gauges.

This design totally eliminates the influence of normal stresses on the measurement of hear stresses, and heat compensation is reached thanks to the close-spaced location of all strain gauges.

The sensors are connected to LTR-212-M1 modules of crate data acquisition system L-Card. The crate data acquisition system allows amplification, digitalization and synchronous recording of signals.

4. Flow of granular materials in radial channels

Figure 4 offers a general view of the experimental slot-type hopper with the flat convergent walls 1 and the flap gate 2 of the outlet slot. On the inner surface of the transparent wall 3, a thin tube 4 filled with a colored granular material is attached. When involved in motion, the colored particles generate a marker stream to visualize the trajectory of the flow.

The stress measurement is carried out by four shear stress sensors and four normal stress sensors. On the visible wall of the hopper, sensors $\tau_1\sigma_1$ and $\tau_2\sigma_2$ are arranged at the top and the bottom, respectively; sensors $\tau_3\sigma_3$ and $\tau_4\sigma_4$ are placed at the top and bottom of the opposite wall. Thus, $\tau_1\sigma_1$ and $\tau_3\sigma_3$, as well as sensors $\tau_2\sigma_2$ and $\tau_4\sigma_4$ are arranged on the opposite walls of the hopper at the height of 235 and 140 mm, respectively.

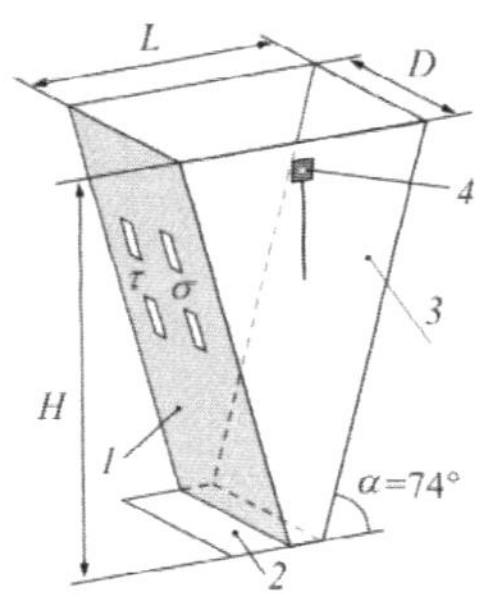

Figure 4. Slot-type hopper, general view: 1—convergent walls; 2—discharge hole gate; 3—transparent front wall; 4—source nozzle of marker stream. Hopper height $H = 310$ mm, hopper width $D = 52$ mm, upper boundary length $L = 182$ mm.

The tests involved quarts sand with particles 0.3 and 1.5 mm in size. The outlet hole size was 1.5×52 mm for particles of 0.3 mm and 3×52 mm for particles of 1.5 mm.

Regarding steady-state conditions of discharge, it is known that in a granular material, owing to pores that are nonuniformly spread in the material volume, there is always a scatter in the mechanical properties, which results in bad reproducibility of test results. This, first of all, relates the material fill conditions since fill in the mode of 'rain' or 'jet' can offer both dense and loose packing of particles. It is possible to improve structural uniformity and to stabilize mechanical parameters of a medium by subjecting it to low-amplitude alternating shears [15, 16]. In the presented studies, aimed to stabilize the experimental result on granular material discharge from hopper, it was provided that the material occurred in stationary conditions during flowing. After filling the hopper and opening the discharge hole gate, the material was continuously fed in so that the hopper was permanently filled. The feed-in was carried out through a continuous dosing unit (not shown in Figure 4). The constant weight of the material ensured stable average pressure on side walls of the hopper and the continuous discharge velocity.

The tests show that the discharge velocity of quartz sand 1.5 mm in size with the feed-in from the dosing unit is constantly 14.7 g/s, while without the feed-in the material discharge velocity grows with the height of the material column in the hopper. In the end, the average velocity makes 11.43 g/s.

Adherence to the condition of continuity was checked by simultaneous shutting the gate and the dosing unit with the subsequent weighing of the material in the hopper at the fixed moments of time. The average weight of seven samples was 2059 g at the average deviation of not higher than 6%.

So, the introduction of a stabilizing continuous-action dosing device with the adjustable capacity has made it possible to reach almost steady-state condition of the granular medium in the hopper. The average bulk density of sand is 1.4 g/cm^3 and porosity is 45%.

Figure 5 shows characteristic epures of normal and shear stresses (Figures 5a and 5b, respectively) measured by the bottom sensors $\tau_2\sigma_2$ and $\tau_4\sigma_4$ (see Figure 4) during discharge of quartz sand particles 0.3 mm in size.

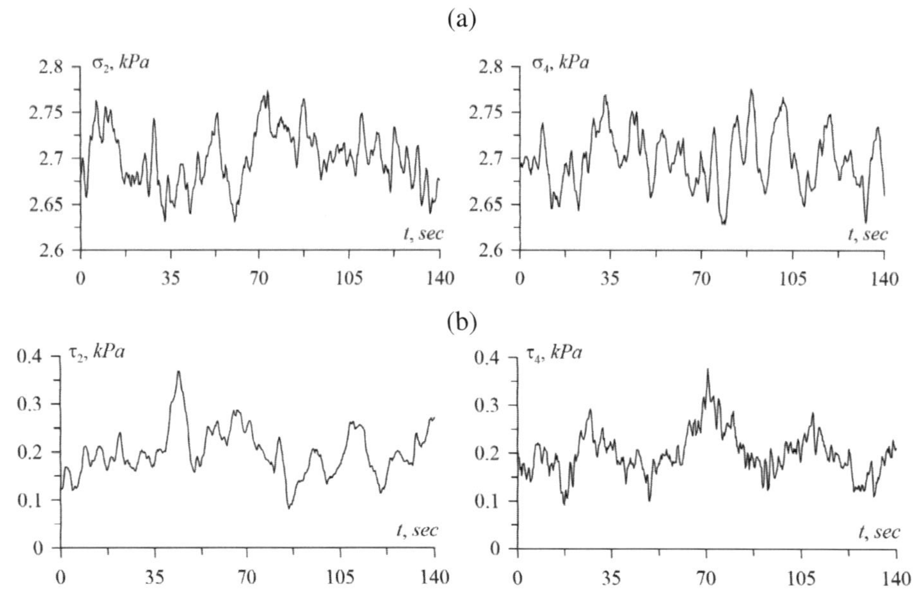

Figure 5. Diagrams of (a) normal and (b) shear stresses measured under flow of quartz particles of 0.3 mm by the bottom sensors $\tau_2\sigma_2$ and $\tau_4\sigma_4$.

The curves display the pulsation nature of the stresses during the discharge. The maximum amplitude of the normal stresses is not more than 5% of the static pressure while the amplitude of the shear stresses is 70% of the average value.

It is seen that the external stationary conditions do not make the stresses stationary. Accordingly, the flow is not stationary, either. This is evident in the kinematic patterns of the flow, too.

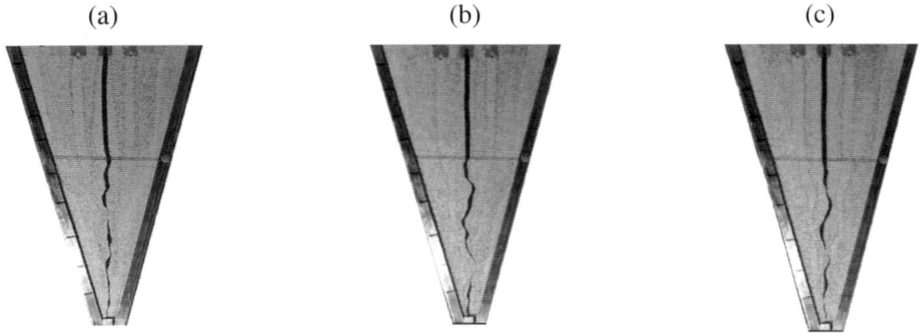

Figure 6. Deformation of the colored particle flow at the fixed time moments.

The flow is associated with the complex processes of plane shearing, localization and clustering of strains, generation of new flows and the change in their direction [17]. All that induces redistribution of stresses, generation of new stresses and their re-orientation. As an illustration, Figure 6 shows the pictures of the marker stream deformed by the flow of quartz sand with particles 0.3 mm in size. The pictures are made at the lag of 10 s. It is seen that the flow toward the discharge hole is nontrivial. At a distance of 120 mm from the hopper top, a new structure is formed, and the marker stream begins bending and stretching on either side of the direction of gravity alternatively. The new flow trajectory is characterized by the change of direction in time.

The change in the length of the straight-line portion of the marker stream versus the flow time is depicted in Figure 7. It is seen that the flow is uniform and straight-line in this portion in a relatively long distance making 1/3 of the hopper height *H*. In Figure 6 the level beneath which the change in the stream direction starts is marked by the horizontal line.

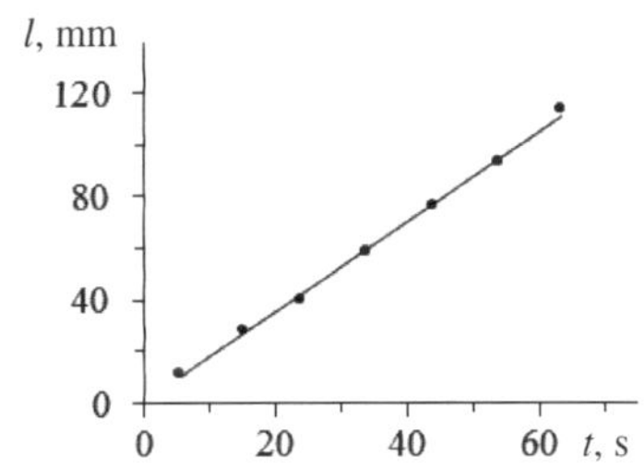

Figure 7. Change in the length of the straight-line portion of the marker stream versus time.

Repeatability of the straight-line portion length and the related flow time are illustrated in Figure 8 by the data of seven identical tests with formation of a straight marker stream of maximum length *AB*.

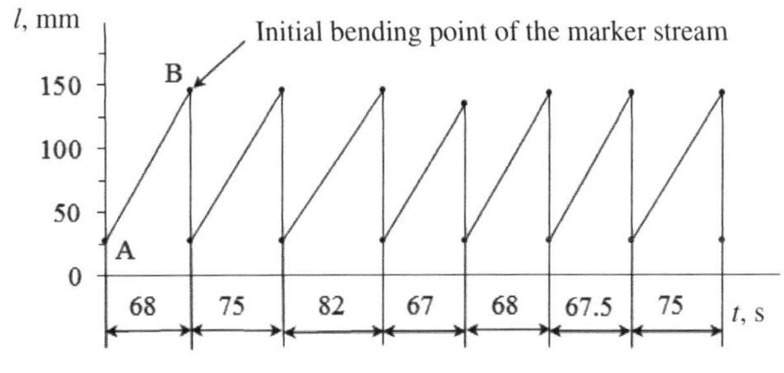

Figure 8. Formation of straight-line portion of marker stream in seven tests.

The average velocity of the marker stream in the tests was 1.69 at the standard deviation of 0.05 mm/s and maximum deviation of 6.5%.

5. Conclusions

1. The three-dimensional variant of DEM is an adequate approach top calculating gravity flows of granular media.
2. The lab-scale experiments using original stress sensors display the pulsation nature of stresses.
3. The comparison of the experimental and calculated parameters of flow shows their good agreement.

References

[1] Hall SA, Muir Wood D, Ibraim E and Viggiani G 2010 Localised deformation patterning in 2D granular materials revealed by digital image correlation *Granular Matter* Vol 12 Issue 1 pp 1–14
[2] Słomiński C, Niedostatkiewicz M and Tejchman J 2006 Deformation measurements in granular bodies using a Particle Image Velocimetry technique *Archives of Hydro-Engineering and Environmental Mechanics* Vol 53 No 1 pp 71–94
[3] Osinov VA 1993 Model of a discrete stochastic medium in the deformation and flow of friable materials *Journal of Mining Science* Vol 28 Issue 5 pp 433–440
[4] Lavrikov SV and Revuzhenko APh 2000 Stochastic models in problems of the local deformation of flowing media in radial channels *Journal of Mining Science* Vol 36 Issue 1 pp 8–16
[5] Lavrikov SV and Revuzhenko AF 2016 Numerical modeling of elastic energy accumulation and release in structurally heterogeneous geomaterials *Journal of Mining Science* Vol 52 Issue 4 pp 632–637
[6] Klishin SV and Revuzhenko AF 2017 3D Discrete element approach to the problem on abutment pressure in a gently dipping coal seam *IOP Conf. Series: Earth and Environmental Science* 84 (2017) 012003.
[7] Klishin SV and Revuzhenko AF 2015 Numerical analysis of plastic behavior of granular media exposed to deformation under broken path loading *Journal of Mining Science* Vol 51 Issue 5 pp 951–956
[8] Baranov DS and Bobylev LM 1980 Device for measuring soil stresses by Baranov–Bobylev. Author's Certificate No 724949 *Byull. Izobret.* No 12
[9] Kramadzhyan AA 1970 Sensor for measuring pressure in granular materials Underground *Ore Mining: Collection of Papers* NG Dubynin (Ed) Novosibirsk (in Russian)
[10] Ignatev EA 1992 Magnetoelastic pressure sensor. Author's Certificate No. 1486817 *Byull. Izobret.* No 42
[11] Kulikov AK and Kulikov KK 1988 Device for measuring normal and tangential components of pressure. Author's Certificate No 1408259 *Byull. Izobret.* No 25
[12] Asch G et al 2010 *Les capteursen instrumentation industrielle* 7th edition
[13] Rykov GV and Skobeev AM 1978 *Measurement of Stresses in Soils under Short-Term Loads* Moscow: Nauka (in Russian)
[14] Bobryakov AP, Revuzhenko AF and Kosykh VP 1989 Sensor for measuring shear stresses. Author's Certificate No 1485046 *Byull. Izobret.* No 5
[15] Bobryakov AP and Revuzhenko AF 1982 Uniform displacement of a granular material. Dilatancy *Journal of Mining Science* Vol 18 Issue 5 pp 373–379
[16] Bobryakov AP 2008 Influence of weak shakes on a statically stressed granular medium. Dilatancy *Journal of Mining Science* Vol 44 Issue 2 pp 115–122
[17] Revuzhenko AF, Stazhevskii SB and Shemyakin EI 1979 The asymmetry of plastic flow in a converging axisymmetric channels *Doklady Akademii Nauk SSSR* Vol 246 No 3

IOP Conf. Series: Earth and Environmental Science **134** (2018) 012008 doi:10.1088/1755-1315/134/1/012008

Selection of basic data for numerical modeling of rock mass stress state at Mirny Mining and Processing Works, Alrosa Group of Companies

IB Bokiy[1], OV Zoteev[1], VV Pul[1] and EK Pul[2]

[1]Yakutniproalmaz Institute, Republic of Sakha (Yakutia), Russia
[2]Mirny Mining and Processing Works, Republic of Sakha (Yakutia), Russia
BokiyIB@alrosa.ru

E-mail:

Abstract. The influence of structural features on the strength and elasticity modulus is studied in rock mass in the area of Mirny Mining and Processing Works. The authors make recommendations on the values of physical properties of rocks.

Analysis and prediction of geomechanical behavior of rock mass in the course of mineral mining uses the method of mathematical modeling, as rule.

In the numerical modeling and solution of geotechnical problems, the input data are governed by many parameters: mineralogy and crystal structure of rock mass, presence of weakening planes and weakening intensity, orientation of fractures, weakening nature, hydrogeology, etc. these parameters mostly condition strength and deformation characteristics of rocks, load-beating capacity and stability of exposed elements of mine system, and dimensions of the latter [1]. So, it is important to correctly select data for the numerical modeling.

The most complicated and least theoretically validated problem in geomechanics is studying strength and deformation characteristics of rocks. The current estimation procedures for the mechanical properties of rocks involve:

—in situ measurement of strength and deformation characteristics by means of full-scale testing of large specimens commensurable with the dimensions of structural elements of mining systems;

—determination of physical parameters from the inverse problem solution given the known nature of deformation, generated stresses and measured strains, based on the available analytical solutions or numerical modeling;

—inclusion of the rock mass structure influence by means of so-called structural weakening coefficients that can be determined from direct tests or by calculation.

The simplest way of transition from the mechanical properties in core tests to the properties of rock mass is allowed by the introduction of the structural weakening coefficient (Constructions Norms and Regulations SNiP I-94-80) which is determined as a number of joints per running meter. This coefficient ranges from 1 (solid intact rock mass) to 0.1 (more than 10 fractures per running meter).

The most popular relations to find the structural weakening coefficient are described below [2].

As recommended by Slesarev [3], compression strength, tension strength and cohesion of rock mass can be given by:

$$[\sigma^{rm}{}_{com}] = K_1[\sigma^0{}_{com}];\ [\sigma^{rm}{}_t] = K_2[\sigma^{rm}{}_t];\ C^{rm} = K_3C^{inv},$$

where K_1 = 0.3–0.35; K_2 = 0 at dense jointing, K_2 = 0.01–0.1 at closed joint, K_2 = 0.01–0.05 at intensive micro-jointing; K_3 = 0.6–0.9 in rock mass with a network of micro-joints, K_3 = 0.05–0.25 I rock mass with closed contact fractures, K_3 = 0.01–0.02 in rock mass with weakening bedding planes with slickensides.

According to Fisenko [4] the structural weakening coefficient K_{sw} for the estimation of rock cohesion and ultimate compression strength is

$$K_{sw} = 1/(1+a*\ln(H/d)),$$

where a—coefficient governed by rock specimen strength and jointing, grows from 1 to 7 as strength of specimens increases from 1 to 100 MPa; H—average size of exposure; d—average size of elementary block in rock mass. The size of elementary blocks is determined by the spacing of joints.

The deformation modulus can be estimated using procedures by Ruppeneyt [5]:

$$E_M = E_0/(1+5N\eta/8);\ E_M = 1.6(1+1/k)\beta E_0;\ \eta = \delta/(\beta d),$$

where E_M and E_0—deformation moduli of rock mass and rock specimen, respectively; N—number of joints in a system; δ—opening of joints, mm; d—average distance between the edges of neighbor joints; β—relative area of hard rock contacts (Rupenneyt recommended $\beta = 3*10^{-4}$); k—fracture porosity which is a ratio of volume of pores (fractures) to total volume of material. The first equality is valid in rock mass containing no more than 4 systems of joints; the second equality is for the case of chaotic jointing when rock mass behaves as granular medium.

Aside from the structural weakening coefficient, it is common to use the coefficient of long-term strength. Assessment of long-term strength in rocks has currently insufficient theoretical backup and is a complicated problem.

A feature of stress state studies is the impossibility to measure actual stresses directly in a solid medium. The stress state is examined indirectly, through measurement of other values connected with stresses by different relations. For this reason, an apparatus to convert measured values to stresses and the model of a medium included in this apparatus are the key points in the theory of measurements.

In mines of Mirny Mining and Processing Works, stresses are measured using the method of parallel drilling (by the Institute of Mining, Siberian Branch, RAS), borehole slotter and the approach based on "memory" of rocks possessing property of creep: when such rocks are subjected to long-term loading, structural rupture results in development of both elastic and plastic strains [6].

Nonuniformity of deformation characteristics influences initial stress state of rock mass. The data of in situ measurement of initial stresses are widely scattered within the same deposit both in point observations and in averaged values obtained in different areas of the deposit. Such scatter cannot be explained only by experimental imprecision or by data processing errors. The most significant factor of such pattern is the nonuniformity of the stress field conditioned by the rock mass structure, faulting and different elastic characteristics of rocks composing the deposit.

There are many procedures available for the processing of experimental results. The traditional method is the averaging of stresses determined in different areas but at the same depth in rocks; then, the averaged values are used to plot stress–depth curves. These curves are used later on as the boundary conditions in calculation of stress–strain state of enclosing rock mass around underground excavations and to determine factor of safety of underground structures.

In solving problems in geomechanics, the differentiation approach should be used in setting boundary conditions. When an underground excavation has such dimensions that its influence zone occurs within a uniform rock mass (layer), it is possible to take in situ stresses measured in this layer. Otherwise, to make the stress state calculations more accurate, it is necessary to refine boundary conditions by solving some problems on nonuniform rock mass.

For the reliable determination of the initial stress field parameters, point measurements should be carried out on a grid based on variation in properties of rocks and characteristics of jointing.

Deposits developed by Mirny Mining and Processing Works have complex geology. Ore bodies, enclosing rock mass and backfill have different strength and deformation characteristics. Enclosing rock mass is presented by alternating rocks of medium strength (limestone, dolomite) and rocks of high plasticity (rock salt) with inclusions of dolerite sill.

In situ tests on stress determination at actual elevation of –365 in Mir Mine revealed anisotropy of deformation characteristics of dolomite.

The lateral earth pressure coefficient varies from Dinnik's value to 1.2.

An emphasis should be laid on influence of brines on the strength of natural rocks and backfill. The effect of highly mineralized water of Metegero-Ichersky aquifer of strength of kimberlite, enclosing rock mass and backfill depending on length of exposure is in detail described in [6].

Table 1. Strength characteristics of rock suites at Mir kimberlite deposit (determined in tests of samples).

Suite	Occurrence depth, m	Thickness, m	Bulk weight $\gamma*10^2$, MN/m^3	Uniaxial compression strength σ_{com}, MPa	Cohesion C, MPa	Internal friction angle φ, deg
Ilginskaya	0–170	170	2.34–2.68	4.3–41.5	2.1–11.7	23–40
Verkholenskaya	170–315	145	2.40–2.53	8.3–22.4	2.4–8.9	23–41
Metegerskaya	315–430	115	2.40–2.65	17.6–24.0	4.4–7.0	27–35
Icherskaya	430–520	90	2.42–2.75	29.0	7.8–10.5	28–36
Dolerite	520–555	35	2.74–3.00	–	30.0	26
Charskaya (–187.4/–210.3// –495.6/–513.6)	555–855	300	2.55–2.63	24.0–37.5	9.3–10.6	35–36
Olekminskaya (–495.6/–513.6// –681.7/–701.6)	855–1045	190	2.65–2.79	39.0–69.0	11.2–21.4	36–38
Tolbachanskaya	1045	–	2.72–2.75	54.0–60.5	14.8–17.1	36–37

Table 2. Strength and deformation characteristics of enclosing rock mass around Mir kimberlite pipe (determined in tests of samples).

Rock	Young's modulus E, GPa	Poisson's ratio μ	Density γ, t/m^3	Compression strength, MPa	Tension strength, MPa	Internal friction angle φ, deg	Cohesion C, MPa
Dolomite, anhydrite	30.0	0.24	2.5	35	3	35	10
Limestone	40.0	0.28	2.6	51	4.8	37	11
Sill of dolerite	55.0	0.22	2.8	52	6	33	13
Halogen-carbonate strata	12.0	0.30	2.1	22	2	38	5
Kimberlite	21.0	0.26	2.45	31	3.3	35	9
Main backfill	1.0	0.18	1.7	4	0.4	30	0.9
First cut layer backfill	2.0	0.20	1.9	6	0.5	30	1.2

Table 3. Strength and deformation characteristics of enclosing rock mass around Mir kimberlite pipe in stress–strain state calculations.

No.	Suite	Rocks	Deformation modulus E, GPa	Poisson's ratio, ν	Bulk weight ρ, t/m^3	Cohesion, C, MPa	Internal friction angle , φ, deg
1	Metegerskaya (act. elev. 57//–54 m) Icherskaya (act. elev. –54//–130 m)	Dolomite, limestone, chalky clay, gypsum anhydrite	7.2	0.25	2.59	4.35	31
2	Kharyalakhdskaya (act. elev. –130//–161 m)	Gypsum anhydrite and dolomite	12	0.35	2.64	4.95	35
3	Charskaya (act. elev. –161//–507 m)	Rock salt interlaid with dolomite and anhydrite	8.5	0.33	2.15	2.5	37
4	Olekminskaya (act. elev. –507//–703 m) Tolbachanskaya (act. elev. –703//–883 m)	Dolomitic limestone, halite, dolomite, anhydritic dolomite	13.3	0.22	2.69	5.1	36
5	–	Kimberlite	7.7	0.25	2.45	2.58	35
6	–	Backfill	0.5	0.18	1.7	0.6	30

It is also evident that monolithic rock masses have higher deformation modulus and are natural stress raisors; the value f stress concentration depends on the size of discontinuities caused by deformation, ratio of elastic characteristics and other factors. Jointed areas, contact zones and backfill feature lower actual stresses and relaxation zone.

Thus, solution of the specific practical problems in geomechanics requires particular care in setting boundary conditions. The geomechanical modeling of rock mass calls for specific attention to the selection of data on strength and deformation characteristics of rocks.

The recommended values of physical properties for the first-stage calculation in geomechanical modeling are compiled in Tables 1–3.

A geomechanical model of a specific geotechnical situation is refined by the data of geomechanical monitoring. At this stage, the initial information and generalized and adjusted for the numerical modeling.

References

[1] Turchaninov IA, Iofis MA, and Kasparyan EV 1989 *Fundamentals of Rock Mechanics* Saint-Petersburg: Nedra (in Russian)

[2] Rylnikova MV and Zoteev OV 2003 *Geomechanics* Moscow: Ruda Metally (in Russian)

[3] Slesarev VD 1948 *Rock Mass Mechanics and Mine Support* Moscow: Ugletekhizdat (in Russian)

[4] *Methodical Instructions on Determination of Slope for Pitwalls, Benches and Dumps in Open Pit Mines under Operation and Construction* Leningrad: VNIMI 1972 (in Russian)

[5] Ruppeneyt KV and Liberman YM 1960 *Introduction to Rock Mechanics* Moscow: Gosgortekhizdat (in Russian)

[6] Kozeev AA, Izakson VU and Zvonarev NK 1995 *Thermo-Geomechanics of Diamond Deposits* Novosibirsk (in Russian)

[7] Baldayev ME and Bokiy IB 2014 Influence of highly mineralized water on strength properties of rock mass and backfill in Mir pipe *Proc. Int. Conf.: Science and Innovative Developments for the North* Moscow: Pero pp 35–36 (in Russian)

IOP Conf. Series: Earth and Environmental Science **134** (2018) 012009 doi:10.1088/1755-1315/134/1/012009

Geomechanical monitoring of rock mass during underground mining of Mir Pipe

AS Chaadaev and IV Zyryanov

Yakutniproalmaz Institute, Republic of Sakha (Yakutia), Russia

E-mail: ZyryanovIV@alrosa.ru

Abstract. The implementation of geomechanical monitoring in Mir Mine is discussed. The proposed integrated system of geomechanical monitoring makes it possible to improve the reliability of control over initiation and development of displacements in rock mass.

Safe mining needs implementation of special measures in hazardous production zones [1]. Mir Mine extracts ore reserves under the open pit mine bottom, under an aquifer, with the protection represented by a safety crown pillar.

Safety of mining is based on prediction and ground control in enclosing rock mass, ore body and backfill, using a hydro-geomechanical monitoring project, covering detection of water-permeable cracks and determination of strains and displacements at rock mass and ore pipe interface, in the safety ore ore crown and in the zone of artificial crown formation in case of the downward mining.

An integral part of the geomechanical monitoring is getting initial information on mechanical properties and stress state of rock mass under the open pit mine bottom, on their change in the course of mining based on the observations over undermined rock mass subsidence and damage control in hazardous contact zones. The obtained information is used to adjust parameters of numerical stress state modeling which is later on corrected using the in situ monitoring data towards enhanced reliability of prediction of state of the safety crown pillar.

Moreover, the hydro-geomechanical monitoring project includes analyses of the structure, damage and stress state of ore body, enclosing rock mass and backfill; studies into effect of brines and ambient medium on the physical properties of rocks; recommendations of safety of underground mine in case of water level in the pit exceeds the project level and water inflows in underground mine intensify; computer modeling of the monitoring object and investigation of hazardous geomechanical and hydrogeological processes in the crown pillar; development of quantitative geomechanical, geophysical and hydrogeological criteria to predict geodynamic processes in the crown pillar, ore body, enclosing rock mass and backfill.

Since 1957 till 2001 Mir kimberlite pipe had been developed by the open pit method down to a depth of 525 m, with the exposure of Metegero-Ichersky aquifer, under water inflow rate of 1200 m^3/h. The further extraction below the actual elevation -190 m involves the underground method under protection of 'dry' conservation system, which means filling the open pit mine with coarse rock material, creation of rock-and film shielding and deep-level water discharge at the actual elevation of -141 m. Conservation of the underground ore body is implemented by means of filling the open pit with three layers of rocks with a total thickness of 50 m. the bottom layer to 30 m thick is composed of coarse dolerite and is characterized as the hurricane-resistant layer. This layer is overlaid with a loamy cushion covered by an impermeable film 2 mm thick. The film is covered from the top by clayey soil.

Mining phase I in Mir Mine in block 1 (–210–310 actual elevation) was started in 2009. Extraction is carried out under the dead-storage open pit Mir in accordance with the project developed by the Yakutniproalmaz Institute (2008) in the top-down sequence. The current method is cut and fill.

For the protection of the mine under the open pit bottom (-190 m actual elevation), a crown pillar 20 m is left; the size of the pillar is approved by the the Federal Service for Ecological, Technological and Nuclear Supervision of Russia (Decree No. 13-13/3819 as of Dec 29, 2008). The reserves in the safety crown are categorized as temporally suspended. Moreover, the mine is topped with 3 safety layers each 5 m thick, i.e. overall thickness of the crown is 35 m.

The mine safety is determined by the criterion of mine water discharge which is ratio of the maximum output of mine discharge to the maximum expected water inflow.

Water protection system used the 'dry' conservation scheme. To this effect, on the pit bottom, a dolerite break stone layer is placed and then overlaid with an anti-seepage screen (-159 m actual elevation). Brines and atmospheric precipitations are discharged from the open pit via a tunnel (bottom elevation -159 m). The water inflow rate in the pit is 1200 m^3/h. To catch water inflow through the anti-seepage screen and through the pit walls, 3 catch galleries are driven at the actual elevation of -201 m, and fans of drainage holes are drill from the tunnels. At the same elevation of -210 m, auxiliary pumping stations (APS) is installed to discharge water from the mine into the open pit, while district pump stations at the actual elevations of -310 and -690 m pump inflow water to the main water discharge installation at the actual elevation of -210 m.

APS 1 is equipped with 5 pumps to pump out water from drainage and slightly inclined catch holes to the main water drainage system. APS 2 is equipped with 7 pumps (2 pumps TsNS-300/180, 2 pumps TsNS-300/120, 2 pumps 1-D-630/90 and 1 pump DE-500/63 to pump out water to the main water drainage system and the discharge gallery.

All levels of block 1 are serviced by the brine discharge system (catch facilities, sumps, by-pass holes) to redirect water to the level at the actual elevation of -327 m equipped with 4 pumps NTsSK-180/212. The pumping station pumps out water to the main water discharge system (MWDS 1 and MWDS 2 at the actual elevation of -201 m and, then, to the surface.

In 2015 the main water discharge system at -210 m was redesigned for the general water inflow rate of 1240 m^3/h.

MWDS 1 is equipped with 5 unit pumps AENTsSK(K)-180/700 and MWDS 2—with 5 unit pumps AENTsSK(K)-400/600.

In the first quarter of 2017, water discharge rate in the mine and open pit is 1200–1250 m^3/h. Some brines at the flow rate of 790–84 m^3/h are pumped out by RITs pumps fed from the pit and by APS 2 via holes 12 and 7Sh to the water discharge gallery. Pumps discharge water at the flow rate of 450–460 m^3/h to MWDS. Water flow through block 1 is 100 m^3/h regardless process water.

The drainage system operates efficiently and eliminates hydrodynamic pressure on the roof of the ore crown pillar.

Since 2011 safety of mining is ensured by the integrated hydrogeomechanical monitoring in accordance with the geotechnical monitoring project Technical Upgrading. Mir Mine. Integrated Hydro-Geomechanical Monitoring of Mining Phase I (Yakutniproalmaz Institute, Mirny, 2015) by the Geotechnical Monitoring Service of the Mirny Mining and Processing Plant jointly with the Institute of Mining, SB RAS, Research Institute of Design and Technology and NOVOTEK Center for Science and Technology.

The hydro-geomechanical monitoring project involves a set of instrumental methods to obtain initial data on deformation of enclosing rock mass, undermined ore body and backfill, including surveying lines, deep-seated plugs and directional survey. The mining-induced damage and cracking is controlled visually, using downhole endoscope and sets of perimeter and deep-seated plugs. Furthermore, the project provides for the geophysical investigation of properties and condition of the crown pillar left under the open pit mine bottom.

Decided by the management of Mirny Mining and Processing Plant and approved by ALROSA, drivage of three subparallel openings at the level of -210 m along the long axis of the horizontal position of Mir Pipe (so-called trident) with the control over hydro-geomechanical behavior of the ore crown allows reliable information on the behavior of the pillar depending on the preparatory works and actual mining in block 1. It has become possible to organize a stationary non-undermined hydrogeological and geomechanical monitoring network directly at the zone of the ore crown pillar.

The geomechanical monitoring network at the top of block 1 and in the crown pillar is aimed at:

—assessment of displacements and deformations in space and time at the bottom of the crown pillar (actual elevation –215÷ –230 m);

—plotting of the displacement trough along and across the strike of the ore body;

—determination of possible lamination zones in the ore body (actual elevation –215÷ –230 m);

—control of subsidence and horizontal movement in the backfill in the initial cut layer 4 during to-down mining in block 1;

—refinement and control of sub-backfill height in the in the initial cut layer 4 in the area of observation stations in the course of undermining of the backfill layer;

—estimation of mechanical properties of ore body and backfill in the area of observation stations;

—analysis of the measurement data in connection with the theoretical calculations upon the change in the geotechnical situation in the block and operational application of the results to the estimation of the geomechanical behavior of the ore crown pillar;

—adjustment of the geomechanical model by the obtained data for the enhanced reliability of stress state prediction in the structural elements of the mining system.

Displacements were measured by the Geotechnical Monitoring Service of Mirny Mining and Processing Plant, and the checking measurements at the deep-seated plugs and directional survey stations—by the specialists from the Institute of Mining, SB RAS.

The instrumental observation data are monthly sent to the Yakutniproalmaz Institute for the quarterly reports on the mining safety by the data of processing and analysis of the in situ monitoring of the ore body movement by the stakes of the survey lines installed in the water discharge systems at the actual elevation of –214 ÷ –219 m and in the backfill drift at the level of –260 m, deformation marks in the mine openings along the rock mass and ore contact at the level of –210 m, direct survey stations at the level of –272 m and deep-seated plugs at the levels of –265 and –272 m.

The specialists of the Yakutniproalmaz Institute and the outside contractors assess stress state in the mine using the methods of borehole slotter and parallel drilling.

The Geotechnical Monitoring Service of Mirny Mining and Processing Plant implement rock mass quality classification by the rating indexes RMR-89 (Bieniawski) and Q (Barton).

The seismic monitoring in Mir Mine uses the methods of seismoacoustics by system Prognoz-ADS. This system implements continuous monitoring of rock mass and mine structures and predicts hazardous events due to rock pressure with regard to geo-cryology [2].

The geoacoustic control allows assessment of rock mass conditions in the crown ore pillar under the influence of mining at the lower lying levels and the external load from the watered rock mass in the conditions of 'dry' conservation in the permafrost zone.

The proposed system of integrated geomechanical monitoring allows reliable control over the processes of displacements in the crown pillar and in its backfilled bottom layer.

During the analysis of displacements at the stakes of the survey lines by the method of interpolation, zoning of rock mass at the actual elevation of –210 m has been performed (Figure 1).

The convergence of the roof and floor in the catch galleries is determined using the floor stakes. The analysis of the data obtained in the observation cycle in the first quarter of 2017 (Figure 2) has allowed determining the subsidence in the north-east of the rock mass, at the actual elevation of –210 , (stakes 5–7 in the south, stakes 1–3 in the center), which is confirmed by the seismoacoustic monitoring (Figure 3).

The integrated analysis enables detecting a probable deformation zone (Figure 3) and shearing in the rock mass (shear angle > $5*10^{-3}$ deg).

As mining is advanced to deeper levels, ALROSA system of control is persistently updated and refined. The location of the new observation points is chosen based on the previous measurement auditing.

IOP Conf. Series: Earth and Environmental Science **134** (2018) 012009 doi:10.1088/1755-1315/134/1/012009

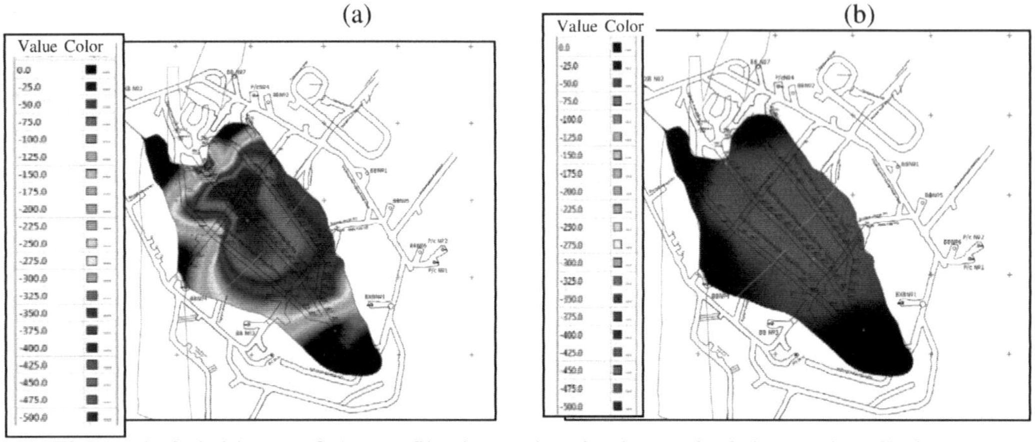

Figure 1. Subsidence of the profile line stakes in the roof of the catch galleries at the actual elevation of –210 m: (a) May 2014–March 2017, (b) first quarter of 2017.

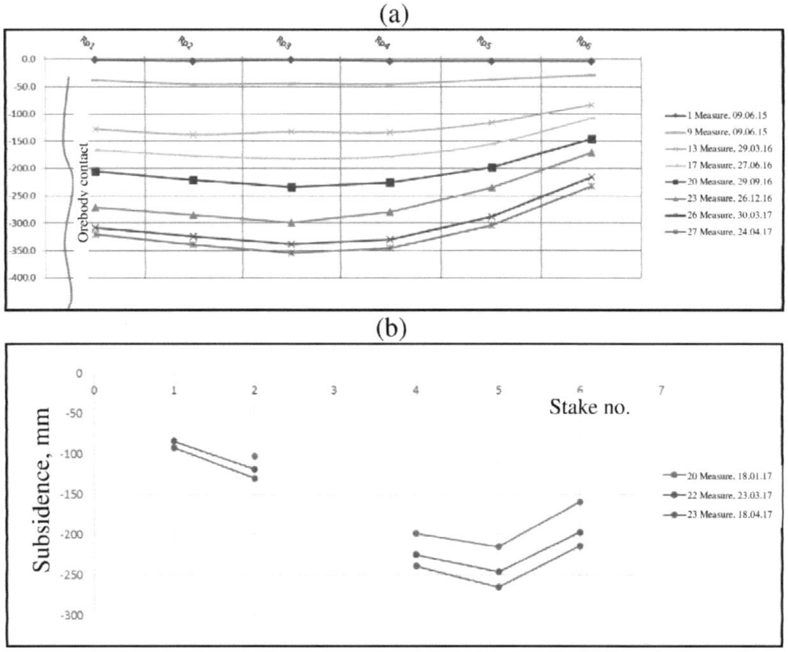

Figure 2. Subsidence of the profile line stakes in the catch galleries (a) Center and (b) South.

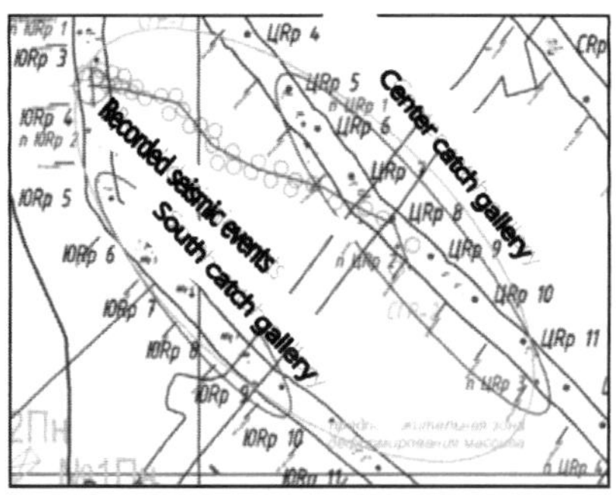

Figure 3. Probable deformation zone at the actual elevation of –210 m.

In 2014 ALROSA management has decided on the initiation of the geomechanical service in the mines. The key tasks of the service are the control of the conformity of mining processes with the approved projects, detection of violations and deviations, and determination of the influence of the latter on the alteration of the rock mass stress state in the mining zones with a view to the risk assessment of hazardous events.

References

[1] Kramskov NP et al 2007 *Regulations on Safe Mining Sequence and Control in Hazardous Areas in Mines of of ALROSA* Mirny (in Russian)
[2] Rasskazov IYu 2008 *Ground Control in Mines of the Far East Region* Moscow: Gornaya Kniga (in Russian)

The problem solution on wedge penetration in an initially anisotropic medium within the rigid-plastic scheme

AI Chanyshev[1,2]* and IM Abdulin[1]

[1]Chinakal Institute of Mining, Siberian Branch, Russian Academy of Sciences, Novosibirsk, Russia
[2]Novosibirsk State University of Economics and Management, Novosibirsk, Russia

E-mail: *a.i.chanyshev@gmail.com

Abstract. Two problems are solved in the paper: on ultimate loads in the initial stage of indentation of an absolutely rigid smooth wedge into a layer of an initially anisotropic plastic medium and in the final stage when the tool penetrates through the layer. The problems are solved with Chanyshev's constitutive relations of plasticity of the initially anisotropic medium based on use of the eigen elasticity tensors.

The process of wedge tool penetration in rocks is a subject of many researches [1–8]. Sokolovsky [1] studied wedge penetration in a rigid–plastic half-plane earlier investigated by Hill, Lee and Tupper. Some foreign researchers modeled stress state of soil under wedge introduction [2] and solved the problem on penetration of a cylinder in multi-layered soil [3]. The model of interaction between a cone and soil with regard to penetration resistance is presented in [4]. Gareeva [5] offers experimental relations for static probing by standard procedures and determines resistances on front and side faces of indenters. Relations for fining optimal parameters of a cone tool penetrating in soil are given in [6], and it is illustrated that the cone is the most rational shape.

This study deals with the penetration of a rigid smooth wedge in a layer of an initially anisotropic material (plain strain deformation). Two limit loads are determined: initial state of the edge penetration and final state when the wedge penetrates the layer and comes out from the other side (an analogous problem for isotropic medium is solved in [9].

Let in the coordinate system xOy the initial medium is deformed elastically by the Hooke law:

$$\begin{cases} \varepsilon_x = a_{11}\sigma_x - a_{12}\sigma_y, \\ \varepsilon_y = -a_{12}\sigma_x + a_{22}\sigma_y, \\ \varepsilon_{xy} = a_{33}\tau_{xy}, \end{cases} \tag{1}$$

where a_{ij}—elasticity constants characterizing ductility of the material. The choice of the Hooke law in the form of (1) with four different elasticity constant is only governed by the simplified formula of the solution; the authors will include the rest independent elastic constants a_{ij} (six!) of the plain strain deformation in the further studies.

Aiming the construct plasticity equations for the medium (1), a basis of tensors is introduced [10]:

$$T_1 = \begin{pmatrix} 1 & 0 \\ 0 & 0 \end{pmatrix},\ T_2 = \begin{pmatrix} 0 & 0 \\ 0 & 1 \end{pmatrix},\ T_3 = \frac{1}{\sqrt{2}}\begin{pmatrix} 0 & 1 \\ 1 & 0 \end{pmatrix}. \tag{2}$$

In the basis (the scalar product is determined as resultant of $\sigma_{ij}\varepsilon_{ij}$ with summing over repeated indices), the coordinates of the tensors T_σ, T_ε are, respectively given by:

$$S_1 = \sigma_x,\ S_2 = \sigma_y,\ S_3 = \sqrt{2}\tau_{xy},$$
$$\Omega_1 = \varepsilon_x,\ \Omega_2 = \varepsilon_y,\ \Omega_3 = \sqrt{2}\varepsilon_{xy}. \tag{3}$$

In the coordinates (3), the Hooke law (1) takes on matrix form:

$$\begin{pmatrix} \Omega_1 \\ \Omega_2 \\ \Omega_3 \end{pmatrix} = \begin{pmatrix} a_{11} & -a_{12} & 0 \\ -a_{12} & a_{22} & 0 \\ 0 & 0 & a_{33} \end{pmatrix}\begin{pmatrix} S_1 \\ S_2 \\ S_3 \end{pmatrix}. \tag{4}$$

For the symmetrical matrix A in (4), the eigenvalues [11] are real:

$$\lambda_1 = \frac{a_{11}+a_{22}}{2} + \sqrt{\left(\frac{a_{11}-a_{22}}{2}\right)^2 + a_{12}^2},\ \lambda_2 = \frac{a_{11}+a_{22}}{2} - \sqrt{\left(\frac{a_{11}-a_{22}}{2}\right)^2 + a_{12}^2},\ \lambda_3 = a_{33}. \tag{5}$$

The eigenvectors, pursuant to the symmetry of the matrix A, are mutually orthogonal:

$$\vec{b}_1 = (\cos\beta, -\sin\beta, 0),\ \vec{b}_2 = (\sin\beta, \cos\beta, 0),\ \vec{b}_3 = (0, 0, 1), \tag{6}$$

where $\operatorname{tg} 2\beta = 2a_{12}/(a_{11}-a_{22})$ and the constants a_{ij} are determined by (1).

Taking into account (5), (6), (2), the eigen tensors are:

$$\tilde{T}_1 = \begin{pmatrix} \cos\beta & 0 \\ 0 & -\sin\beta \end{pmatrix},\ \tilde{T}_2 = \begin{pmatrix} \sin\beta & 0 \\ 0 & \cos\beta \end{pmatrix},\ \tilde{T}_3 = \frac{1}{\sqrt{2}}\begin{pmatrix} 0 & 1 \\ 1 & 0 \end{pmatrix}. \tag{7}$$

Then, thee stress tensor T_σ is expanded by the basis (7). As a result, we have the coordinates of the tensor T_σ in the new basis:

$$\tilde{S}_1 = \sigma_x\cos\beta - \sigma_y\sin\beta,\ \tilde{S}_2 = \sigma_x\sin\beta + \sigma_y\cos\beta,\ \tilde{S}_3 = \sqrt{2}\tau_{xy}. \tag{8}$$

For the coordinates $\tilde{S}_i$, $\tilde{\Omega}_i$ in the basisв (7), there exists proportional elastic relations:

$$\tilde{\Omega}_1 = \lambda_1\tilde{S}_1,\ \tilde{\Omega}_2 = \lambda_2\tilde{S}_2,\ \tilde{\Omega}_3 = \lambda_3\tilde{S}_3, \tag{9}$$

where λ_i—the eigenvalues (5). Apparently, Poisson's ratio is absent, i.e. a force directed along a tensor $\tilde{T}_i$ causes deformation only along this direction.

The ductilities λ_1, λ_2, λ_3 for the initially isotropic medium are distributed so that $\lambda_1 = \lambda_3 > \lambda_2$ and plasticity develops in a plane intersecting the tensors $\tilde{T}_1$, $\tilde{T}_3$ where there are the highest ductilities. In our case of (1), considering (9) and the hypothesis that λ_1, λ_2, λ_3 are distributed as $\lambda_1 > \lambda_3 > \lambda_2$, we assume that plasticity in the anisotropic medium (1) also develops in the plane $\tilde{T}_1$, $\tilde{T}_3$ where ductilities are higher than in the line of $\tilde{T}_2$.

The plasticity criteria in the case of $\lambda_1 > \lambda_3 > \lambda_2$ are given by:

$$\left|\tilde{S}_1\right| = \left|\sigma_x \cos\beta - \sigma_y \sin\beta\right| = K_1,\ \left|\tilde{S}_3\right| = \sqrt{2}\left|\tau_{xy}\right| = K_3, \tag{10}$$

where K_1, K_3—constants which are the elastic limits in the medium (1) in the lines of the tensors $\tilde{T}_1$, $\tilde{T}_3$ (in the plane $\tilde{S}_1$, $\tilde{S}_3$ the plasticity conditions is a rectangle with the sides (10)). The idea that the eigen tensors of elasticity are the eigen tensors of plasticity, creep and failure belongs to Anvar Chanyshev [12].

Suing the common scheme [13], we analyze the system of equilibrium equations:

$$\frac{\partial \sigma_x}{\partial x} + \frac{\partial \tau_{xy}}{\partial y} = 0,\ \frac{\partial \tau_{xy}}{\partial x} + \frac{\partial \sigma_y}{\partial y} = 0 \tag{11}$$

together with the plasticity conditions (10). Using (11) and the plasticity conditions (10) allows finding the axes in xOy as the straight lines:

$$\frac{dy}{dx} = \pm\sqrt{\text{ctg}\beta} \tag{12}$$

(the angle β is determined from (6)) and the equilibrium equations along them:

$$d\sigma_y + d\tau_{xy}\frac{dy}{dx} = 0.$$

Combined with (10), this yields the stress distribution along the lines (12):

$$\begin{cases} \dfrac{dy}{dx} = +\sqrt{\text{ctg}\beta},\ \sigma_y + \tau_{xy}\sqrt{\text{ctg}\beta} = \text{const},\ \sigma_x = \sigma_y\ \text{tg}\beta \pm \dfrac{K_1}{\cos\beta}, \\ \dfrac{dy}{dx} = -\sqrt{\text{ctg}\beta},\ \sigma_y - \tau_{xy}\sqrt{\text{ctg}\beta} = \text{const},\ \sigma_x = \sigma_y\ \text{tg}\beta \pm \dfrac{K_1}{\cos\beta}. \end{cases} \tag{13}$$

Using the second plasticity condition (10) with (11) produces the stress distributions below:

$$\sigma_x = f(y),\ \sigma_y = \varphi(x),\ \tau_{xy} = \pm\frac{K_3}{\sqrt{2}}, \tag{14}$$

where f, φ—arbitrary functions of the appropriate coordinates x, y.

In layered anisotropic medium, the state (14) conforms with the plastic shearing of the layers relative one another. The formulas (13) describe plastic deformation of the layers.

Finally, we use (13), (14) in solving a problem on penetration of a perfectly rigid smooth wedge in a layer composed of the material (1).

The first illustration is given in Figure 1. A half-plane xOy has a recess *BCD* where a wedge with a nose angle 2γ is placed. It is required to find the limit load on the wedge faces *BC* and *CD* such that material starts yielding along the slide lines (12) subject to the conditions (13).

Denoting the angle between the characteristics (12) and the axis x as α brings the case of a sharp wedge when $\pi/2 - \gamma > \alpha$ as in Figure 1 or a blunted wedge when $\pi/2 - \gamma < \alpha$. The latter variant will be studied later on.

Let us analyze the straight lines parallel to *CE* with the equations $dy/dx = \sqrt{\text{ctg}\beta}$. Along them, $\sigma_y + \tau_{xy}\sqrt{\text{ctg}\beta} = 0$ (because of $\sigma_y = \tau_{xy} = 0$ at the boundary *DE*), and, thus, in these lines, $\sigma_y = -\tau_{xy}\sqrt{\text{ctg}\beta}$, $\sigma_x = \sigma_y\ \text{tg}\beta \pm \dfrac{K_1}{\cos\beta}$.

IOP Conf. Series: Earth and Environmental Science **134** (2018) 012010 doi:10.1088/1755-1315/134/1/012010

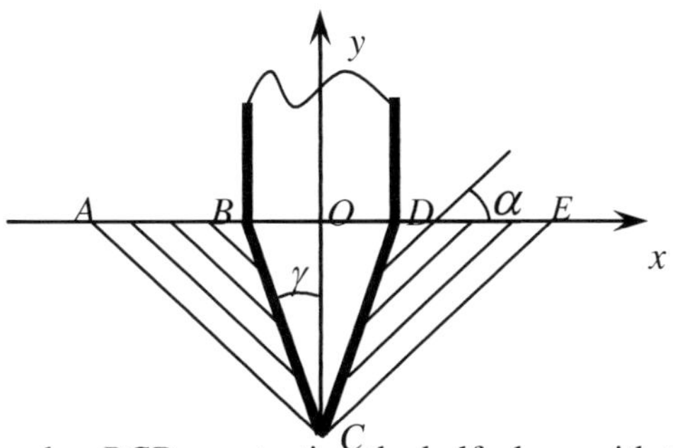

Figure 1. The wedge *BCD* penetrating the half-plane with the boundary $y = 0$.

The calculation starts on the face *CD*. The shear stress is zero *CD* (as the wedge is assumed perfectly smooth). Plot an external normal to *CD*. It follows from Figure 1 that:

$$\vec{n} = (-\cos\gamma, \sin\gamma), \ \vec{t} = (\sin\gamma, \cos\gamma), \tag{15}$$

where $\vec{t}$—unit vector of a tangent. Considering (15), we find the stress vector $\vec{p}$ on *CD*:

$$\vec{p} = (-\sigma_x \cos\gamma + \tau_{xy} \sin\gamma)\vec{i} + (-\tau_{xy} \cos\gamma + \sigma_y \sin\gamma)\vec{j}, \tag{16}$$

where $\vec{i}$, $\vec{j}$–unitary vectors in xOy. Using (16), (15), determine the shear and normal stresses on *CD*:

$$\begin{aligned} p_t &= \vec{p}\cdot\vec{t} = -\tau_{xy}\cos 2\gamma - \frac{\sigma_x - \sigma_y}{2}\sin 2\gamma = 0, \\ p_n &= \vec{p}\cdot\vec{n} = \sigma_x \cos^2\gamma + \sigma_y \sin^2\gamma - \tau_{xy}\sin 2\gamma = 0. \end{aligned} \tag{17}$$

From the first equation in (17) and the conditions $\sigma_y = -\tau_{xy}\sqrt{\text{ctg}\beta}$, $\sigma_x = -\tau_{xy}\sqrt{\text{tg}\beta} - K_1/\cos\beta$, we determine the stress τ_{xy} on *CD* as:

$$\tau_{xy} = \frac{K_1 \sin 2\gamma}{\cos\beta[2\cos 2\gamma + \sin 2\gamma(\sqrt{\text{ctg}\beta} - \sqrt{\text{tg}\beta})]}. \tag{18}$$

The second equation (17) yields the load p_n:

$$p_n = -\frac{K_1}{\cos\beta}\left[\cos^2\gamma + \frac{\sin 2\gamma(\sqrt{\text{tg}\beta}\cos^2\gamma + \sin 2\gamma + \sqrt{\text{ctg}\beta}\sin^2\gamma)}{2\cos 2\gamma + \sin 2\gamma(\sqrt{\text{ctg}\beta} - \sqrt{\text{tg}\beta})}\right]. \tag{19}$$

With the known load on *CD*, we find the force

$$\vec{F}_n = p_n \frac{h}{\cos\gamma}\vec{n} \tag{20}$$

applied by the wedge to the anisotropic body (*h*—varied penetration depth of the wedge). Projection of (20) on the axis *y* and change of the sign in the expression produces the relation to determine the resistance force of the medium to the wedge penetration:

$$2F_y = -2p_n h \ \text{tg}\gamma = 2|p_n| \, h \ \text{tg}\gamma. \tag{21}$$

The relation (21) corresponds to the entry of the wedge tip in the layer, i.e. it is valid when $0 \le h \le h_0$ (Figure 2a), where h_0—height of the conical part of the tool. Considering the entry of the tool in the layer (Figs. 2b and 2c), it is evident that the formulas (19)–(21) hold true upon the assumption that the sides of the tool are free from friction. At the same time, the formulas should assume $h = h_0$, i.e. along $h_0 \le h \le H$ (*H*—layer thickness), the penetration resistance is constant: $2F_y = 2|p_n|\, h_0 \,\mathrm{tg}\gamma$. After that, there comes the stage when the wedge tip comes out of the layer (Figure 2d). The calculations show that the resistance is still equal to (21) with the variable *h*: $0 \le h \le h_0$.

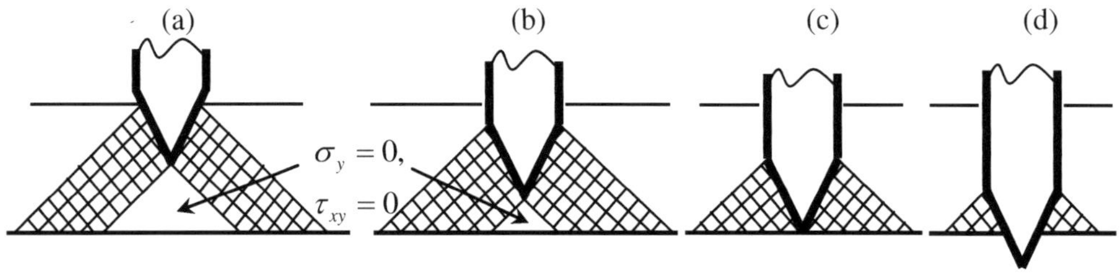

Figure 2. Stages of penetration of the perfectly rigid wedge in the anisotropic body layer: (a) $0 \le h < h_0$; (b) $h_0 \le h < H$; (c) $h = H$; (d) $h \ge H$: h—penetration depth; h_0—height of conical part of the wedge; H—layer thickness.

The calculations are performed for the maximum possible thick layer breakable by a tool. The tool approaches the body at the velocity v_0. The tool mass is a sum of the mass of the rectangle with a length *L* and width $2h_0 \,\mathrm{tg}\gamma$ and the mass of the conical head with an area $h_0^2 \,\mathrm{tg}\gamma$. The tip penetration velocity is denoted by v_1, the head velocity at the other side of the layer—v_2, and the head velocity outside the layer—v_3. In the limit case, $v_3 = 0$. Then:

$$\frac{mv_1^2}{2} - \frac{mv_0^2}{2} = -\frac{ch_0^2}{2}, \tag{22}$$

$$\frac{mv_2^2}{2} - \frac{mv_1^2}{2} = -ch_0(H - h_0), \tag{23}$$

$$\frac{mv_3^2}{2} - \frac{mv_2^2}{2} = -\frac{ch_0^2}{2}, \tag{24}$$

where $c = 2|p_n|\, h_0 \,\mathrm{tg}\gamma$. Summing up (22)–(24) gives in the limit case $v_3 = 0$ the relation of the maximum possible penetration depth:

$$H = \frac{mv_0^2}{2ch_0}. \tag{25}$$

Inserting the mass $m = \left(2L\, h_0 \,\mathrm{tg}\gamma + h_0^2 \,\mathrm{tg}\gamma\right)\rho g$, density ρ and free fall acceleration g in equation (25) finally yields:

$$H = \frac{\rho g}{4|p_n|}\left(\frac{2L}{h_0} + 1\right)v_0^2. \tag{26}$$

The implications of (26) are quite evident: the penetration depth is higher at the higher initial velocity v_0; the penetration depth is lower with the higher rigidity of the medium in the line of the normal drawn to the wedge; the penetration depth is higher with the higher ratio L/h_0 (bulkier tail of the tool); the penetration depth is higher with the higher density of the wedge. The most interesting is the dependence of the medium resistance p_n on the medium properties (angle β) and the wedge nose angle γ. This study includes calculations of the resistance versus the listed parameters.

Conclusion

In the framework of a rigid–plastic body, the authors have constructed the mathematical model of penetration of a rigid tool in an initially anisotropic medium. The maximum possible penetration depth has been determined. The dependence of the penetration depth on geometry of the wedge and on the anisotropic parameters of the medium has been analyzed.

References

[1] Sokolovsky VV 1969 *The Theory of Plasticity* Moscow: Vyssh Shkola (in Russian)
[2] Ling-Yu Xu, Fei Cai, Guo-Xin Wang and Keizo Ugai 2013 Nonlinear analysis of laterally loaded single piles in sand using modified strain wedge model *Computers and Geotechnics* No 51 pp 60–71
[3] Seo H and Prezzi M 2006 Analytical solutions for a vertically loaded pile in multi-layered soil *Geomechanics and Geoengineering: An International Journal* pp 1–10
[4] Tekeste MZ, Raper RL and Tollner EW 2007 Finite element analysis of cone penetration in soil for prediction of hardpan location *American Society of Agricultural and Biological Engineers* ISSN 0001-2351 Vol 50 pp 23–31
[5] Gareeva NB 2003 Development of methods for designing foundations of foundations for digital models of soil massifs on the basis of sounding data *Author's Abstract Diss. Doct. Tech. Sci.* Ufa (in Russian)
[6] Ponomarenko YuI 2002 Improving the efficiency of the pile foundations in sealed soils *Author's Abstract Diss. Doct. Tech. Sci.* Omsk (in Russian)
[7] Kotov VL and Linnik EYu 2014 Numerical calculation of the optimal shape of the body of revolution during motion with a constant speed in a soil medium *Vychislit. Mekh. Sploshn. Sred* Vol 7 No 2 pp 142–150
[8] Afanasyev AI and Fedoseev AP 2014 Contact stresses in an elliptical working tool of a machine for destruction of oversized material Fundamental. Issled. No 9 pp 497–500
[9] Davydov DV and Myasnikin YuM 2009 On introduction of bodies into a rigid-plastic medium, *Vestn. VGU, Ser. Fiz. Mat. Nauk* No 1 pp 94–100
[10] Novozhilov VV 1963 On the forms of connection between stresses and strains in initially isotropic inelastic bodies (geometrical standpoint) *Prikl. Matem. Mekh.* Vol 27 No 5 pp 794–812
[11] Korn GA and Korn TM 1973 *Mathematical Handbook for Scientists and Engineers* Moscow: Nauka (in Russian)
[12] Chanyshev AI 1984 Plasticity of anisotropic media *J. Appl. Mech. Tech. Phys.* Vol 25 No 2 pp 311–314
[13] Kachanov LM 1956 The Fundamentals of the Theory of Plasticity Moscow: Gostekhizdat (in Russian)

IOP Conf. Series: Earth and Environmental Science **134** (2018) 012011 doi:10.1088/1755-1315/134/1/012011

Inclusion of inhomogeneous deformation and strength characteristics in the problem on zonal disintegration of rocks

AI Chanyshev[1,2]* and OE Belousova[1]

[1]Chinakal Institute of Mining, Siberian Branch, Russian Academy of Sciences, Novosibirsk, Russia
[2]Novosibirsk State University of Economics and Management, Novosibirsk, Russia

E-mail: *a.i.chanyshev@gmail.com

Abstract. The authors determine stress and deformation in a heterogeneous rock mass at the preset displacement and Cauchy stress vector at the boundary of an underground excavation. The influence of coordinates on Young's modulus, shear modulus and ultimate strength is shown. It is found that regions of tension and compression alternate at the excavation boundary—i.e. zonal rock disintegration phenomenon is observed.

1. Introduction

In solid mechanics and, specifically, in rock mechanics, preferable formulation of boundary value problems are: setting the Cauchy stress vectors at the known boundaries of a body, setting displacement vectors at the boundaries and setting the Cauchy stress vectors at some boundaries and displacement vector at the other boundaries (combined problem) [1–3]. In elasticity, the theorems of existence and uniqueness of solution are proved [4–5]. This study addresses such formulation of a boundary value problem when the Cauchy stress vector and the displacement vector are set at the same boundary. The problem includes the simplest boundary in the form of a circumference and the preset boundary conditions are independent of coordinates.

2. The problem of the theory of elasticity for rock mass surrounding a tunnel

Let there be a tunnel with a radius a (Figure 1) with the preset boundary conditions given by:

$$\begin{aligned} u_r &= -u_0, \quad u_\theta = 0, \\ \sigma_r &= -p_0, \quad \tau_{r\theta} = 0. \end{aligned} \tag{1}$$

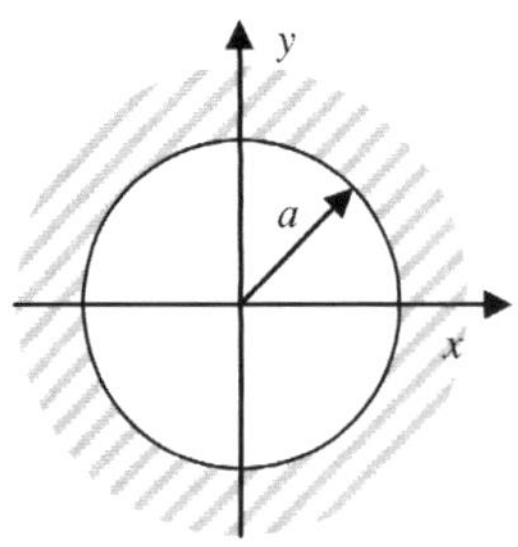

Figure 1. Tunnel with a radius a and the Cauchy stress vector and displacement vector preset as the boundary conditions.

Rock mass around the tunnel is in elastic state. It is required to find distribution of stresses, strains and displacement around the tunnel.

The problem is solved by integration of a second-order differential equation under the boundary conditions (1). The differential equation in terms of stresses is given by:

$$\frac{d\sigma_r}{dr} + \frac{\sigma_r - \sigma_\theta}{r} = 0, \tag{2}$$

the Hooke low for the plain strain:

$$\begin{cases} \sigma_r - \sigma_\theta = 2\mu(\varepsilon_r - \varepsilon_\theta), \\ \sigma_r + \sigma_\theta = 2k(\varepsilon_r + \varepsilon_\theta), \end{cases} \tag{3}$$

$$2\mu = E/(1-\nu),\ 2k = E/(1-2\nu)(1+\nu).$$

The Cauchy relations:

$$\varepsilon_r = \frac{du_r}{dr},\ \varepsilon_\theta = \frac{u_r}{r}. \tag{4}$$

Placement of (4) in (3), (3) in (2) brings a second-order differential equation in terms of radial displacement u_r:

$$\frac{d^2u_r}{dr^2} + \frac{1}{r}\frac{du_r}{dr} - \frac{u_r}{r^2} = 0. \tag{5}$$

The unknown is Eq. (5). Its solving needs setting the function ($u_r = -u_0$) and the normal derivative (coordinate r) at the boundary; the coordinate derivative r is:

$$\frac{du_r}{dr} = \frac{\sigma_r}{k+\mu} - \frac{k-\mu}{k+\mu}\varepsilon_\theta. \tag{6}$$

In view of (1), the condition (6) is re-written as:

$$\left.\frac{du_r}{dr}\right|_{r=a} = -\frac{p}{k+\mu} + \frac{k-\mu}{k+\mu}\frac{u_0}{a}. \tag{7}$$

The general solution of (5) is presented as a sum $u_r = C_1 r + \frac{C_2}{r}$, where C_1, C_2—arbitrary constants.

From the initial conditions (1), (7), the constants:

$$\begin{gathered} C_1 = -\frac{1}{2(k+\mu)}\left(p + 2\mu\frac{u_0}{a}\right), \\ C_2 = -au_0 - C_1 a^2. \end{gathered} \tag{8}$$

Then:

$$\begin{cases} u_r = C_1 r\left(1 - \frac{a^2}{r^2}\right) - \frac{au_0}{r}, \\ \varepsilon_r = -\frac{p}{2(k+\mu)}\left(1 + \frac{a^2}{r^2}\right) - \frac{\mu u_0}{(k+\mu)a} + \frac{k}{k+\mu}\frac{u_0}{a}\frac{a^2}{r^2}, \\ \sigma_r = -\frac{p}{k+\mu}\left(k + \mu\frac{a^2}{r^2}\right) - \frac{2k\mu}{(k+\mu)}\frac{u_0}{a}\left(1 - \frac{a^2}{r^2}\right). \end{cases} \tag{9}$$

It follows from (8), (9) that when $r \geq a$ the displacement u_r is negative. At the same time, the strain ε_r can always be positive, e.g. at $p = 0$, $k = \infty$. This fact seems unusual as under longitudinal and transversal compression of rock mass along the axis r the strain ε_r is positive, i.e. tensile. Actually, there is nothing unusual as the strain $\varepsilon_\theta = \dfrac{u_r}{r}$ is negative (for u_r is negative) while for an incompressible material $\varepsilon_r + \varepsilon_\theta = 0$, thus, $\varepsilon_r = -\varepsilon_\theta$.

To sum up this Section, it follows from (1), (7) that setting the stress vector at the tunnel boundary is equal to setting the normal derivative of the function of displacement at this boundary.

The equation (5) yields the formula to determine σ_r at 'infinity':

$$\sigma_r\Big|_{r\to\infty} = -\frac{pk}{k+\mu} - \frac{2k\,\mu}{k+\mu}\frac{u_0}{a}$$

(as $k \to \infty$ $\sigma_r\Big|_{r\to\infty} = -p - 2\mu u_0/a$).

Near the tunnel, in the radial direction, always tension takes place, which is the source of the zonal disintegration phenomenon [6, 7].

3. Problem solving with due regard to inhomogeneous distribution of shear modulus

The Cauchy problem is formulated so that to take into account inhomogeneous distribution of shear modulus around a cylindrical tunnel. Let a rock mass around a tunnel with a radius r_0 have inhomogeneous distribution of shear modulus:

$$2\mu = f\left(\frac{r}{r_0}\right), \tag{10}$$

where f —arbitrary function of the radial coordinate.

The solution range is $r \geq r_0$. We have the second-order linear differential equation with the variable coefficients to determine u ($u = u_r$):

$$\frac{d^2u}{dr^2} + \frac{d\mu}{(k+\mu)dr}\left(\frac{du}{dr} - \frac{u}{r}\right) + \frac{1}{r}\frac{du}{dr} - \frac{u}{r^2} = 0\,. \tag{11}$$

We introduce (11) with a logarithmic coordinate ξ and a function $P(\xi)$ given by:

$$\xi = \ln\left(\frac{r}{r_0}\right), \quad P(\xi) = \frac{d}{d\xi}\ln(k+\mu)\,. \tag{12}$$

In this case, from (11) we have:

$$\frac{d^2u}{d\xi^2} + P(\xi)\left(\frac{du}{d\xi} - u\right) - u = 0\,. \tag{13}$$

Reduce the order of the differential equation (11) by introducing a new function:

$$v = \frac{du}{d\xi} - u\,. \tag{14}$$

From (13) with regard to 914), we have the system of the first-order linear differential equations for finding the functions v and u:

$$\begin{cases} \dfrac{dv}{d\xi} + (1 + P(\xi)v = 0, \\ \dfrac{du}{d\xi} - u = v. \end{cases} \tag{15}$$

The system (15) is solved analytically. First, from (13), it is found that:

$$v = v_0 e^{-\int_0^\infty [1+P(\xi)]d\xi},$$

where v_0—the value of v at the boundary $r = r_0$. Later on, from the second equation of (13), we find the function u:

$$u = u_0 e^{\xi} + v_0 \left\{ \int_0^{\xi} e^{\int_0^{\xi}[1+P(\xi)]d\xi} e^{-\xi} d\xi \right\} e^{\xi}, \tag{16}$$

where u_0—the value of the displacement u at the boundary $r = r_0$.

It is noteworthy that (16) is a general solution of Eq. (13), where u_0, v_0—constants. Let us analyze the case when the shear modulus 2μ in rock mass around a tunnel changes by the law:

$$k + \mu = (k + \mu_0)(1 + \aleph \sin(2\pi n \xi)); \tag{17}$$

here, n defines the number of 'blocks' in the range of the coordinate ξ from 0 to 1 (the radius r varies from r_0 to $r = e r_0$, where $e \approx 2.718$). Substitution of (17) in (12) yields $P(\xi) = \dfrac{2\pi \aleph \cos(2\pi n \xi)}{1 + \aleph \sin(2\pi n \xi)}$. Having integrated the first equation (15), we have:

$$v = v_0 \frac{e^{-\xi}}{1 + \aleph \sin(2\pi n \xi)}. \tag{18}$$

Finding $u(\xi)$ needs using (16), i.e. taking integral of the function $e^{-2\xi} / (1 + \aleph \sin(2\pi n \xi))$. Having that the parameter $\aleph$ in (17) changes from 0 to 1 and $|\sin(2\pi n \xi)| \le 1$, we expand the function $1/(1 + \aleph \sin(2\pi n \xi))$ in an infinitely decreasing progression. In this case:

$$\frac{e^{-2\xi}}{1 + \aleph \sin(2\pi n \xi)} \approx e^{-2\xi}(1 - \aleph \sin(2\pi n \xi) + \aleph^2 \sin^2(2\pi n \xi) - \aleph^3 \sin^3(2\pi n \xi) + \ldots).$$

The unknown integral is presented as a sum of integrals: $\int e^{-2\xi} d\zeta$, $\int e^{-2\xi} \sin(2\pi n \xi) d\zeta$, $\int e^{-2\xi} \sin^2(2\pi n \xi) d\zeta, \ldots$ with the coefficients $1, -\aleph, \aleph^2, -\aleph^3, \ldots$. Each integral is calculated using the formula of integration by parts [8].

Then, we find the connection between the constants u_0 and v_0 in (16). It is assume that the tunnel boundary is stress-free, i.e. $\sigma_r|_{\xi=0} = 0$. In this case:

$$r_0 \sigma_r = (k + \mu_0)\frac{du}{d\xi} + (k - \mu_0)u = 0, \quad u = u_0, \quad \left(\frac{du}{d\xi}\right)\Bigg|_{\xi=0} = u_0 + v_0. \tag{19}$$

In the end, we have: $(k+\mu_0)[u_0+v_0]+(k-\mu_0)u_0=0$, where $v_0=-\frac{2k\,u_0}{k+\mu_0}=-\frac{u_0}{1-\nu}$, ν— Poisson's ratio. The final expression for $u(\xi)$ is given by:

$$\frac{u(\xi)}{u_0}=e^{\xi}\left(1-\frac{1}{1-\nu}\int_0^{\xi}e^{-2\xi}(1-\aleph\sin(2\pi n\xi)+\aleph^2\sin^2(2\pi n\xi)-\aleph^3\sin^3(2\pi n\xi)+...)d\xi\right). \quad (20)$$

Figure 2a shows the curve of the relative displacement $u(\xi)/u_0$ and the logarithmic coordinate ξ: up to the value $\xi\approx 0.8$ the derivative $u(\xi)/u_0$ of the coordinate ξ is negative, i.e. the material experiences tension in the line of the radial coordinate in the range of ξ from 0 to 0.8. Later on, after $\xi\approx 0.8$, he derivative becomes positive, which means compression of the material in this domain ($\xi\geq 0.8$).

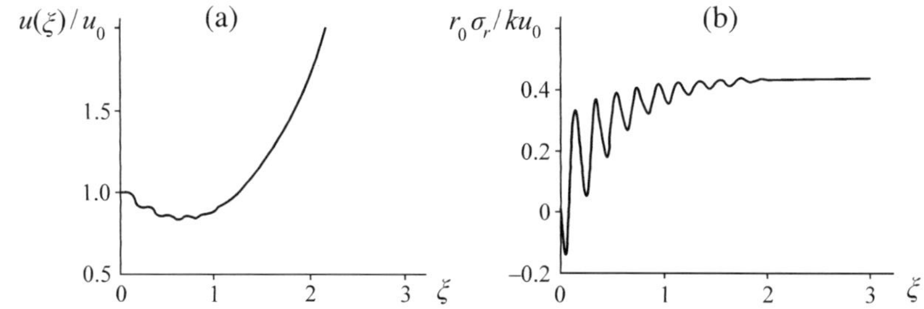

Figure 2. (a) Relative displacement $u(\xi)/u_0$ versus logarithmic coordinate ξ at $\aleph=0.5$, $\nu=0.3$, and $n=5$; (b) radial stress $r_0\sigma_r/ku_0$ versus logarithmic coordinate ξ at $\aleph=0.5$, $\nu=0.3$, $n=5$.

On the interval $\xi\in[0;\ 0.8]$, the derivative can be both positive and negative, i.e. at the tunnel boundary, the zones of tensile and compressive strains alternate—this is a zonal disintegration of rocks around the tunnel [6].

With the known displacements, the radial stress is found using the Hook law. In this case, the expression for the radial stress will take on form:

$$\frac{r_0\sigma_r}{u_0k}=e^{-\xi}\left(\frac{\frac{du}{u_0}}{d\xi}+\frac{u}{u_0}\right)+(2(1-\nu)(1+\aleph\sin 2\pi n\xi)-1)\left(\frac{\frac{du}{u_0}}{d\xi}-\frac{u}{u_0}\right)e^{-\xi}. \quad (21)$$

Below in the article, the numerical solution of Eq. (15) is presented. The second-order differential scheme is given by:

$$\frac{u_{\xi+2}-2u_{\xi+1}+u_{\xi}}{h^2}+P(\xi)\frac{u_{\xi+2}-u_{\xi}}{2h}-P(\xi)u_{\xi+1}-u_{\xi+1}=0. \quad (22)$$

The initial conditions are described by (19).

Figure 3 depicts the relative displacement $u(\xi)/u_0$ versus the logarithmic coordinate ξ at $\aleph=0.5$, $\nu=0.3$, and the number of blocks is $n=5$ (similarly to Figure 2a). The solid line is the analytical solution using (19), the points show the numerical solution using (22). Figure 3 illustrates total coincidence of the analytical and numerical solutions.

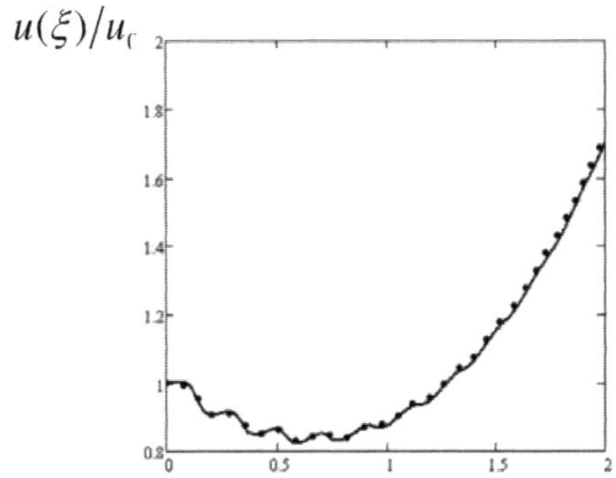

Figure 3. Relative displacement $u(\xi)/u_0$ versus logarithmic coordinate ξ at $\aleph = 0.5$, $\nu = 0.3$, and $n = 5$. Solid line–analytical solution (19); points—numerical solution (22).

In the same manner, the problem on the Young modulus versus radius is solved. In this case, instead of (11), we have:

$$\frac{d^2u}{dr^2} + \frac{1}{r}\frac{du}{dr} - \frac{u}{r^2} + \frac{1}{E}\frac{dE}{dr}\left(\frac{du}{dr} + \frac{\nu}{1-\nu}\frac{u}{r}\right) = 0 . \tag{23}$$

The Cauchy conditions:

$$u\big|_{r=a} = -u_0 , \quad \sigma_r\big|_{r=a} = -p_0 = \frac{(1-\nu)E}{(1+\nu)(1-2\nu)}\left(\frac{du}{dr} + \frac{\nu}{1-\nu}\frac{u}{r}\right)\Bigg|_{r=a} . \tag{24}$$

The problem (23), (24) is solved numerically.

4. Conclusion

The problem on stress state in rock mass around a cylindrical tunnel with the Cauchy conditions and inhomogeneous distribution of shear modulus has been analytically and numerically solved.

It is shown that zones of tensile and compressive strains alternate in adjacent rock mass, which the zonal disintegration phenomenon in rocks.

References

[1] Kamke E 1971 *Manual on Ordinary Differential Equations* Moscow: Nauka (Russian translation from German)
[2] Naymark MA 1969 *Linear Differential Operators* Moscow: Nauka (in Russian)
[3] Kalitkin NN 1978 *Numerical Methods* Moscow: Nauka (in Russian)
[4] Tikhonov AN, Vasilieva AD and Sveshnikov AG 2005 *Higher Mathematics and Mathematical Physics Course. Differential Equations* Moscow: Nauka (in Russian)
[5] Hartman P 1964 *Ordinary Differential Equations* New Your: Wiley
[6] Oparin VN 1997 Scale factor of the phenomenon of zonal disintegration of rocks and stratification of the lunar interior from seismic data *J. Min. Sci.* Vol 33 No 6 pp 497–507
[7] Kurlenya MV and Oparin VN 1996 Scale factor of phenomenon of zonal disintegration of rock, and canonical series of atomic and ionic radii *J. Min. Sci.* Vol 32 No 2 pp 81–90
[8] Korn GA and Korn TM 1978 *Mathematical Handbook for Scientists and Engineers* Moscow: Nauka (in Russian)

IOP Conf. Series: Earth and Environmental Science **134** (2018) 012012 doi:10.1088/1755-1315/134/1/012012

Inclined indentation of smooth wedge in rock mass

AI Chanyshev[1,2]*, GM Podyminogin[1] and OA Lukyashko[1]

[1]Chinakal Institute of Mining, Siberian Branch, Russian Academy of Sciences, Novosibirsk, Russia
[2]Novosibirsk State University of Economics and Management, Novosibirsk, Russia

E-mail: *a.i.chanyshev@gmail.com

Abstract. The article focuses on the inclined rigid wedge indentation into a rigid-plastic half-plane of rocks with the Mohr–Coulomb-Mohr plasticity. The limiting loads on different sides of the wedge are determined versus the internal friction angle, cohesion and wedge angle. It is shown that when the force is applied along the symmetry axis of the wedge, the zone of plasticity is formed only on one wedge side. In order to form the plasticity zone on both sides of the wedge, it is necessary to apply the force asymmetrically relative to the wedge symmetry axis. An engineering solution for the asymmetrical case implementation is suggested.

The problem on destruction of a material by a wedge is the basic in mining [1–5]. It has been solved in elasticity [6–7] and in plasticity [8–10]. The most commonly used approach is rigid–plastic body when elastic strains are neglected and fully plastic strains are the framework. A wedge can be indented along a straight line and at an angle. The straight-line indentation is when the wedge faces have the same angles with the surface of a body the wedge penetrates. The inclined indentation when the inclines of the wedge faces relative to the body are different on all sides. Both direct and inclined inundation processes have been studied [2–3 and 8–11]. The objective was to determine forces that make the wedge penetrate in the body.

Unlike the earlier research, the present paper authors aim to determine loading conditions when a wedge moves in the preset direction. In this case, it is required to control forces and moments applied to the wedge. Let us discuss the case of plain strain deformation of rock mass under perfectly rigid interaction between a nondeformable wedge and a perfectly smooth surface. These requirements are traditional and can be withdrawn by adding the mathematical model with the contact friction and considering more complex deformation state. We take this simplest example to understand the essence of the problem.

So, we analyze the problem on the inclined indentation of a smooth wedge with a nose angle 2γ and the axis of symmetry inclined at the abscissa axis at a constant angle α (Figure 1).

It is assumed that the half-lane material is a medium under limit equilibrium by the Mohr–Coulomb condition:

$$\max_{n}\{\tau_n + \mathrm{tg}\,\psi \cdot \sigma_n\} = K\,, \tag{1}$$

where ψ —internal friction angle; K —cohesion coefficient; n —normal to an arbitrarily oriented plane in the coordinate xO_yz. In the principal axes of the stress tensor, (1) is rewritten as:

$$\frac{\sigma_1-\sigma_3}{2\cos\psi}+\mathrm{tg}\,\psi\cdot\frac{\sigma_1+\sigma_3}{2}=K, \tag{2}$$

where σ_1, σ_2, σ_3—principal stresses of the tensor T_σ. The medium equilibrium equations are:

$$\partial\sigma_x/\partial x+\partial\tau_{xy}/\partial y=0,\ \partial\tau_{xy}/\partial x+\partial\sigma_y/\partial y=0. \tag{3}$$

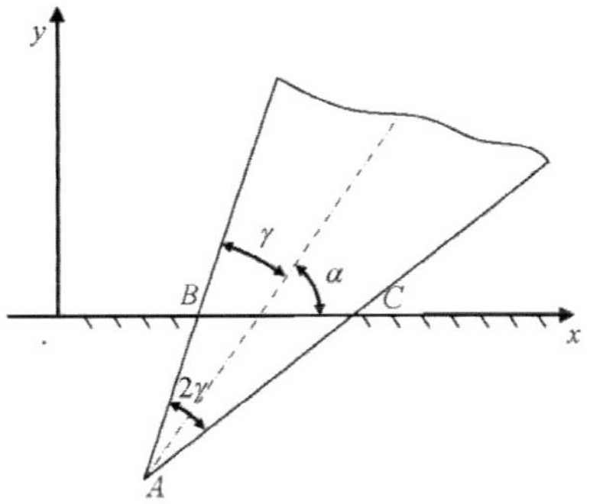

Figure 1. Schematic indentation of wedge with the nose angle 2γ in the rigid–plastic half-plane.

After substitutions $(\sigma_x-\sigma_y)/2=T\cos 2\Theta$, $\tau_{xy}=T\sin 2\Theta$, $(\sigma_x+\sigma_y)/2=\sigma$, where $T=(\sigma_1-\sigma_3)/2$, $\sigma=(\sigma_1+\sigma_3)/2$, $\mathrm{tg}2\Theta=2\tau_{xy}/(\sigma_x-\sigma_y)$, $\sigma_1\ge\sigma_2\ge\sigma_3$, Θ—angle between the first direction for T_σ and the x axis, the stresses $\sigma_x,\sigma_y,\tau_{xy}$ are related to two independent values $\sigma=\sigma(x,y)$, $\Theta=\Theta(x,y)$:

$$\begin{cases}\sigma_x=\sigma(1-\sin\psi\cdot\cos 2\Theta)+K\cdot\cos\psi\cdot\cos 2\Theta,\\ \sigma_y=\sigma(1+\sin\psi\cdot\cos 2\Theta)-K\cdot\cos\psi\cdot\cos 2\Theta,\\ \tau_{xy}=-\sigma\cdot\sin\psi\cdot\sin 2\Theta+K\cdot\cos\psi\cdot\sin 2\Theta.\end{cases} \tag{4}$$

Substituting (4) in (3) results in two equilibrium equations:

$$d\sigma[\sin 2\Theta\cdot dy+(\sin\psi+\cos 2\Theta)\cdot dx]+2\cdot(\sigma\sin\psi-K\cos\psi)\cdot d\Theta\cdot dy=0 \tag{5}$$

along two coordinate lines with the equations:

$$(dy/dx)_1=\mathrm{tg}(\Theta+\pi/4+\psi/2),\ (dy/dx)_2=\mathrm{tg}(\Theta-\pi/4-\psi/2). \tag{6}$$

It follows from (6) that that the characteristics (lines) are arranged symmetrically relative to the first principal direction.

In order to determine limiting loads on the right-hand and left-hand faces of the wedge (Figure 1), we take first the right-hand face AC depicted in Figure 2 with a segment CE showing the free boundary. At the boundary CE, the normal stress $\sigma_y=0$ and the shear stress $\tau_{xy}=0$. Consequently, the normal to the boundary CE and the tangent (axis X) to this boundary are the principal directions. Inasmuch as the segment CE undergoes compression under the wedge penetration, $\sigma_y=\sigma_1=0$, $\sigma_x=\sigma_3<0$, which means that the first principal; direction is perpendicular to the axis X . In other words, the angle $\Theta=\pi/2$ and the characteristics are the line with the equations:

$$dy=\mathrm{tg}(3\pi/4+\psi/2)dx,\ dy=\mathrm{tg}(\pi/4-\psi/2)dx. \tag{7}$$

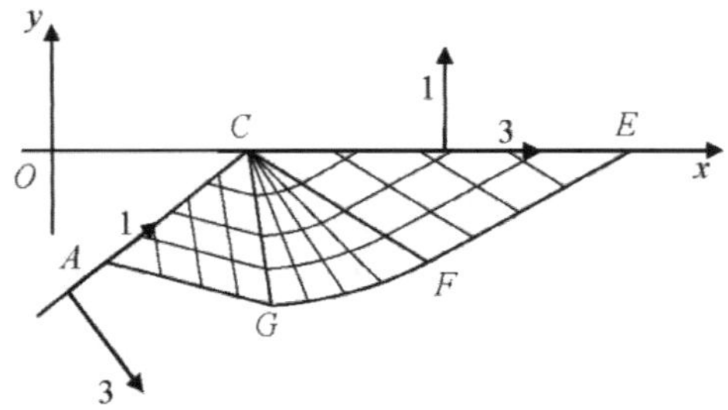

Figure 2. Determination of limit load on the right-hand face of the wedge indenter: 1, 3—directions of the principal stresses.

It is assumed that the angle Θ is constant in the triangle CEF, i.e. the lines (7) are the straight lines from CE to AC and are described by the equations:

$$y = \mathrm{tg}(\pi/4 - \psi/2) \cdot x + \mathrm{const}.$$

Inasmuch as (2) should be valid in the whole triangle CEF, under $\sigma_1 = 0$, $\sigma_x = \sigma_3$, we obtain $\sigma_3 = -2K \cdot \cos\psi/(1 - \sin\psi)$; i.e. $\sigma = (\sigma_1 + \sigma_3)/2$ in CEF is:

$$\sigma = -K \cdot \cos\psi/(1 - \sin\psi). \tag{8}$$

In the same manner, we find stress state in the triangle ACG (Figure 2). In this case, the principal directions are the normal to AC and the tangent. The face is inclined to the axis X at an angle $\alpha - \gamma$. It is assumed that the pressure long the normal to AC (negative) is higher in absolute value than along AC, i.e. the first principal direction in T_σ is oriented along AC. In other words, Θ is equal to $\alpha - \gamma$. The pressure applied to the face AC is denoted by $p_n = \sigma_3$. Let us express p_n in terms σ. From the limit equilibrium condition (2), we have:

$$\sigma_3 = \sigma(1 + \sin\psi) - K\cos\psi. \tag{9}$$

This means that finding $p_n = \sigma_3$ requires calculating σ on AC. As previously, Θ is assumed constant in the triangle ACG, which implies uniform stress state.

We suppose that the fields ACG and CEF are connected by a centered field with the polar coordinate system with the pole C. Meanwhile:

$$x = \rho \cdot \cos\varphi,\ y = \rho \cdot \sin\varphi, \tag{10}$$

where the φ is related with Θ:

$$\varphi = \Theta - 3\pi/4 + \psi/2. \tag{11}$$

When $\Theta = \pi/2$ the angle φ equals $-\pi/4 + \psi/2$, when $\Theta = \gamma - \alpha$ $\varphi = \Theta - 3\pi/4 + \psi/2 + \gamma - \alpha$.

$$dx = d\rho \cdot \cos\varphi - \rho \cdot \sin\varphi \cdot d\varphi,\ dy = d\rho \cdot \sin\varphi + \rho \cdot \cos\varphi \cdot d\varphi, \tag{12}$$

where dx, $d\rho$, $d\varphi$, dy—differentials of functions. Placement of (12) in (7) brings the coordinate lines:

$$d\varphi = 0,\ \rho = \rho_0 \cdot e^{\mathrm{tg}\psi \cdot (\varphi - \varphi_0)},$$

where $\varphi_0 = -\pi/4 + \psi/2$. At $\varphi = -3\pi/4 + \psi/2 + \gamma - \alpha$ we have $\rho = \rho_0 \cdot e^{-\mathrm{tg}\psi \cdot (\pi/2 - \gamma + \alpha)}$, which means that ρ for this value of φ is less than ρ_0 for $\varphi = \varphi_0 = -\pi/4 + \psi/2$.

Then, we analyze the relation in the characteristics (5). We have $d\varphi = d\Theta$, $dy/dx = \mathrm{tg}(\Theta - \pi/4 - \psi/2) = \mathrm{tg}(\varphi - \psi + \pi/2) = -\mathrm{ctg}(\varphi - \psi)$. Substitution of this value in (5), after transformations, allows a relation:

$$\frac{d\sigma}{\sigma \cdot \sin\psi - K \cdot \cos\psi} + \frac{2d\varphi}{\cos\psi} = 0;$$

the latter is integrated and yields:

$$\sigma \cdot \sin\psi - K \cdot \cos\psi = (\sigma \cdot \sin\psi - K \cdot \cos\psi) \cdot e^{-2\mathrm{tg}\psi \cdot (\varphi - \varphi_0)}.$$

The integration constant is found from the condition that at $\varphi = \varphi_0 = -\pi/4 + \psi/2$ the value of σ coincides with the value of σ from (8). In this case, at $\varphi = -3\pi/4 + \psi/2 + \gamma - \alpha$, we have:

$$\sigma \cdot \sin\psi - K \cdot \cos\psi = -\frac{K \cdot \cos\psi}{1 - \sin\psi} \cdot e^{2\mathrm{tg}\psi \cdot (\pi/2 - \gamma + \alpha)}$$

or

$$\sigma = \frac{K \cdot \cos\psi}{\sin\psi} \cdot \left[1 - \frac{e^{2\mathrm{tg}\psi \cdot (\pi/2 - \gamma + \alpha)}}{1 - \sin\psi}\right].$$

Using (9), find σ_3 at AC:

$$\sigma_3 = \frac{K \cdot \cos\psi}{\sin\psi} \cdot \left[1 - \frac{1 + \sin\psi}{1 - \sin\psi} \cdot e^{2\mathrm{tg}\psi(\pi/2 - \gamma + \alpha)}\right]. \quad (13)$$

Multiplying (13) by the length of *AC* gives the force required to be applied to AC to obtain stress fields as in Figure 2.

For another thing, in the analysis of the stress field to the left of the line *AB* in Figure 1, it is assumed that it is inclined to the axis x at an angle $\alpha + \gamma$ less than $\pi/2$ (otherwise we have the situation des cribbed above). In so doing, we obtain two stress fields *HDB* and *ADB* (Figure 3).

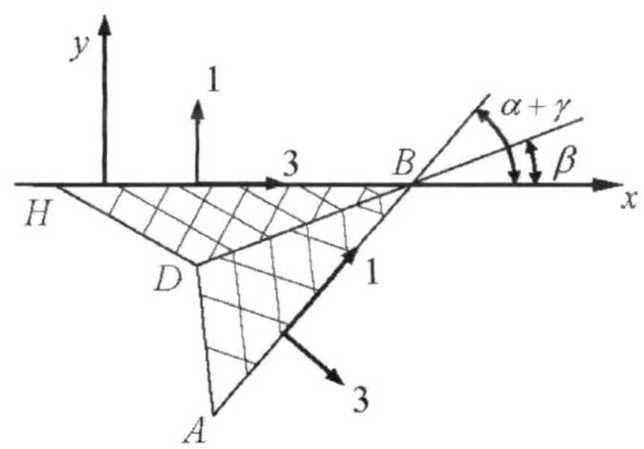

Figure 3. Stress fields HDB and ADB separated by a straight line with the unknown angular coefficient $\mathrm{tg}\beta$.

The stress state is assumed constant (uniform) in the triangle HDB:

$$\tau_{xy} = 0,\ \sigma_y = \sigma_1 = 0,\ \sigma_x = \sigma_3,\ (\sigma_1 - \sigma_3)/2\cos\psi + \mathrm{tg}\psi \cdot (\sigma_1 + \sigma_3)/2 = K, \quad (14)$$

$$\sigma_x = \sigma_3 = -2K\cos\psi/(1 - \sin\psi).$$

The same computations are made for the triangle ADB. In this case,

$$\Theta = \alpha + \gamma,\ \sigma_x = \sigma + T\cos 2\gamma,\ \sigma_y = \sigma - T\cos 2\gamma,\ \tau_{xy} = T\sin 2\gamma, \tag{15}$$

where $T = (\sigma_1 - \sigma_3)/2$, $\sigma = (\sigma_1 + \sigma_3)/2$. By the Mohr–Coulomb condition, $T = K\cos\psi - \sigma\sin\psi$. Therefore, instead of (15), we have:

$$\begin{cases} \tau_{xy} = K\cos\psi\sin 2\gamma - \sigma\sin\psi\sin 2\gamma, \\ \sigma_x = K\cos\psi\cos 2\gamma + \sigma(1 - \sin\psi\cos 2\gamma), \\ \sigma_y = -K\cos\psi\cos 2\gamma + \sigma(1 + \sin\psi\cos 2\gamma). \end{cases} \tag{16}$$

Figure 3 shows the straight line with the angular coefficient $\mathrm{tg}\beta$ separating the fields (14) and (14). To connect the fields, we have the condition of continuity of the Cauchy vector when intersecting this line (third Newton law). Introduce the vector of the normal n to DB: $n = (\sin\beta - \cos\beta)$ and obtain:

$$\begin{cases} \sigma_x^+ \sin\beta - \tau_{xy}^+ \cos\beta = \sigma_x^- \sin\beta - \tau_{xy}^- \cos\beta, \\ \tau_{xy}^+ \sin\beta - \sigma_y^+ \cos\beta = \tau_{xy}^- \sin\beta - \sigma_y^- \cos\beta. \end{cases} \tag{17}$$

If the stress fields to the left and right of DB in Figure 3 are given the plus and minus signs, respectively, we have:

$$\begin{cases} \left(\sigma_x^+ - \sigma_x^-\right)\sin\beta + \tau_{xy}^- \cos\beta = 0, \\ -\tau_{xy}^- \sin\beta + \left(\sigma_y^- - \sigma_y^+\right)\cos\beta = 0. \end{cases} \tag{18}$$

The system (18) is a uniform system of two equations to determine two unknown values: $\sin\beta$ and $\cos\beta$. In order that solution exists, it necessary that the determinant of the system (18) is zero. With the zero determinant, the stress σ in ADB in Figure 3 is given by:

$$\frac{\sigma}{K} = -\frac{\cos^2\psi + \sin\psi 2\cos^2\gamma}{\cos\psi(1 - \sin\psi)} \pm \frac{2\cos\gamma\sqrt{1 - \sin^2\gamma\sin^2\psi}}{\cos\psi(1 - \sin\psi)}, \tag{19}$$

and the angle β is found from (18), e.g.:

$$\mathrm{tg}\beta = \left(\sigma_y^- - \sigma_y^+\right)/\tau_{xy}^- . \tag{20}$$

The calculation of σ_3 at AB use the Mohr–Coulomb condition:

$$\sigma_3 = \sigma(1 + \sin\psi) - K\cos\psi . \tag{21}$$

Comments. It is apparent that the limit loads on the right-hand size (13) and left-hand side (21) of the wedge are different. Furthermore, the forces are different (as the stresses are multiplied by different areas). This fact means that under the force applied to the wedge along the wedge axis of symmetry, the material where the limit force is lower (on the left-hand side of the wedge in Figure 1) will 'yield' first. Naturally, to reach the plastic deformation of the material on the left and right of the wedge, it is required to apply the force not along the axis of symmetry of the wedge but at a certain angle connected with the formulas (13), (21) and the lengths of the faces AB and AC of the wedge in Figure 1. One of the possible diagrams of nonsymmetrical application of force along the wedge axis is shown in Figure 4.

Conclusion

1. Limit loads on a wedge indented in rock mass at an angle have been determined.

2. The authors have suggested the scheme of nonsymmetrical loading of a wedge during its inclined indentation.

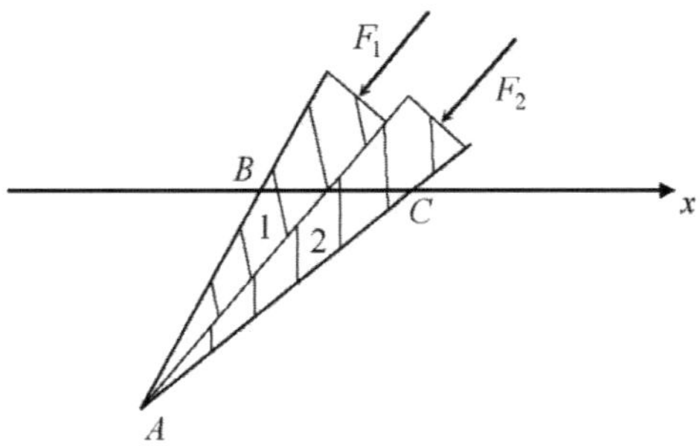

Figure 4. Different forces F_1 and F_2 applied to 'constituents' 1 and 2 of a wedge.

References

[1] Hill R 1953 *The Mathematical Theory of Plasticity* Oxford: Clarendom Press
[2] Vorontsov AL, Sultanzade NM, and Albagachiev AYu 2008 Mathematical theory of cutting. State-of-the-art of the issue in the foreign science *Tekhnol. Metallov* No 9 pp. 18–26
[3] Davydov DV and Myasnyankin YuM 2009 Indentation of bodies in rigid-plastic environment *Vestn. Voronezh. Gos. Univer., Ser. Fizika, Matem.* No 1 pp 94–100
[4] Sokolovsky VV 1969 *Theory of Plasticity* Moscow: Vyssh Shkola (in Russian)
[5] Kachanov LM 1969 *Basics of the Theory of Creep* Moscow: Nauka (in Russian)
[6] Ivlev DD 1957 Pressing of a thin blade into a plastic medium *Izv. AN SSSR, OTN* Vol 10 pp 89–93
[7] Ivlev DD and Maksimova LA 2000 Pressing an indenter into an ideal rigid-plastic strip *Izv. RAN. Mekh. Tverd. Tela* No 3 pp 131–136
[8] Hill R, Lee EH and Tupper SJ 1947 The theory of wedge indentation in ductile materials *Proc. Roy. Soc. L. Ser. A* Vol 188.
[9] Anisimov AN and Khromov AI 2007 Pressing a wedge ion a half-space under the Mohr–Coulomb yield condition *Vestn. SGTU, Ser. Fiz. Mat. Nauki* No 1(14) pp 44–49
[10] Chanyshev AI and Abdulin IM 2012 Dependence of limiting loads on elastic parameters of anisotropic medium in the problem on toll indentation into initially anisotropic medium. Part II: Problem on wedge *Interexpo GEO-Siber* Novosobirsk: SGUGiT Vol 2 No 1 pp 171–176 (in Russian)
[11] Babakov VA and Karimov IM 1978 Pressing a flat stamp into a plastic medium *J. Min. Sci.* No 1 pp 30–35

IOP Conf. Series: Earth and Environmental Science **134** (2018) 012013 doi:10.1088/1755-1315/134/1/012013

Application of synthetic aperture radars for the ground displacement monitoring in mineral mining areas

VV Dobrynchenko*, IS Kokorinand LV Shebalkova

Novosibirsk State Technical University,
Novosibirsk, Russia

E-mail: *rector@nstu.ru

Abstract. The authors discuss applicability of synthesized aperture radars to monitorthe ground surface displacement in mineral mining areas in terms of a synthesized-aperture interferometric radar. The operation principle of the interferometric method is demonstrated on studies of the ground surface displacements in areas of oil and gas reservoirs. The advantages of the synthetic aperture radar are substantiated.

A number of merits of synthesized aperture radars (SAR) and advanced SAR-based techniques for making radiolocation images of Earth surface (subsurface) provide the grounds to consider the feasibility to employ them to monitor mineral deposits. As for SAR merits, first, of prime importance is indiscriminateness of these systems to brightness of a study area and immunity to weather conditions in a survey zone along with rather high spatial resolution. High penetration capacity of radio waves allows overcoming the screen effect of vegetation mantle, to detect subsurface formations and to observe geophysical processes [1, 2].

In the present research the feasibility to apply SAR to monitor mineral deposits is considered in terms of satellite SAR interferometry because of its competitive advantages over the surface techniques:

–a single image covers area up to 100 km^2;

–precision of displacement evaluation is within centimeter scale;

–feasibility to obtain and to analyze archive images.

In view of the above it is explicit that SAR interferometry is applicable to survey displacements of natural and technogenic objects. Survey of displacements in mine fields, sedimentations over mines and tunnels, monitoring of bridge stability and overhead roads are important to provide life safety, and to minimize risks of negative effects on technogenic objects.

Satellite interferometry is widely used in development of oil and gas deposits in the Russian Federation. Consideringclimateandgeographicpeculiaritiesofsurveyterritoriesit should be kept in mind that SAR data processing is a complicated procedure.

SARinterferometryitselfrepresentsconstructionofdigitalreliefmodels (DRM) and enables to monitor Earth surface dynamics for a certain period.

Interferometry procedure provides a pair of images of the same Earth surface area [3], made from two close locally parallel orbits (Figure 1). The second image is made by the same satellite or its tandem couple (this process was realized in the period of simultaneous operation of satellites ERS-1 and ERS-2). Two images are used to compute an integrated interferogram representing also 2D matrix, where each element is product of the first image signal and a complex conjugate signal of the second image. Thus, a phase of each element of a complex interferogram is equal to phase difference of two images [3]. To introduce corrections described in details below requires a digital relief model which can be derived based on independent evidence or two radar images of the same area. The images must be made in periods of insignificant deformations. Provided that displacements of a reflecting surface element are absent, the phase difference is expressed through difference of respective inclined lengths to the survey area in two satellite positions r_{1G} and r_{2G} + (Figure 1):

$$\varphi_G = \frac{4\pi}{\lambda}\left(r_{2G} - r_{1G}\right), \tag{1}$$

whereλ is wave length, corresponding to carrying frequency of radar radiation.

On-board radiolocation satellite systems (hereinafter RSS) of European satellites RS-1/2 and ENVISAT have λ = 5.6 cm. For PALSAR RSS (abbreviated version of Phased Array type L-band Synthetic Aperture Radar), mounted in Japanese satellite ALOS (Advanced Land Observing Satellite) λ = 23.6 cm. Difference of inclined lengths (Figure 1) can be expressed through distance from a survey area to vertical projection of satellite position on Earth surface (length by spheroid) and from the survey location topography, namely, altitude of the area where the reflecting element locates, over spheroid. The use of data on orbit position and DRM, respectively, makes it possible to exclude interference of these dependences in the differential interferogram [1]. Thereto, the second dependence itself is the basis of the process for derivation of the digital relief models with the help of SAR-interferometry. Local altitude z can be expressed as:

$$z = H - r_{1G}\cos\theta, \tag{2}$$

where H is orbit altitude over averaged Earth surface (Figure 1).

Let the reflecting surface displaced for a period between two successive shots (vector D). Then an inclined length becomes equal to r'_{2G}. As it was proved above, because of difference in RSS positions (equivalent to presence of distance B, termed as interferometer base) the phase of interferogram depends not only on surface displacements, but on topography as well. Using DRM before deformations the differential interferometry method allows elimination of topography interference. In formula of phase difference in the case of Earth surface displacement it is possible to single out member D_G, corresponding to difference of inclined lengths related to reflector displacement, member H_G conditioned by interference of topography, atmosphere dynamics and noises is:

$$\varphi_G \frac{\lambda}{4\pi} = \left(r'_{2G} - r_{2G}\right) + \left(r_{2G} - r_{1G}\right) = D_G + H_G. \tag{3}$$

The inclined length exceeds multifold displacements, so beams r'_{2G} and r_{2G} can be assumed parallel (so called approximation of the far zone or parallel beams [4]), it means that displacement component DG can be assumed equal to projection of vector D in direction towards satellite – LOS:

$$D_G = |D|\sin\left(\theta_{inc} - \alpha_G\right). \tag{4}$$

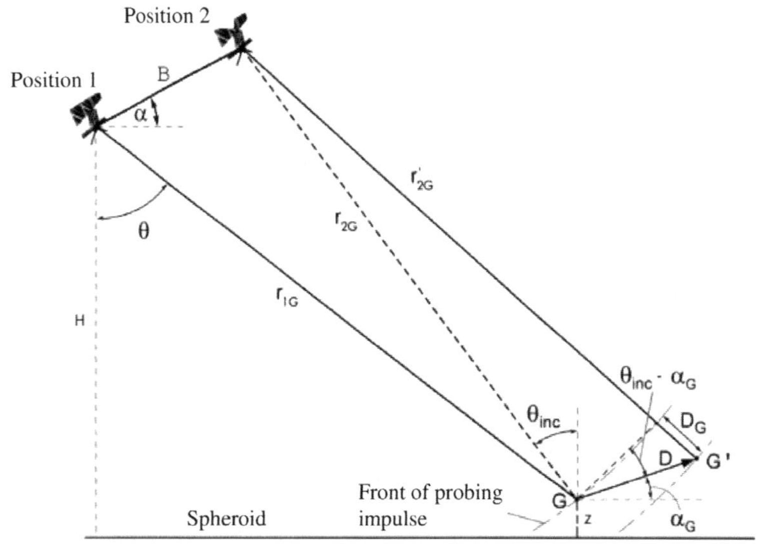

Figure 1. Geometry of interferometric imaging: r_{1G} is distance from the first satellite position to a reflector G on the Earth surface at altitude z; r_{2G} is the second satellite position to the same point given that deformation is absent; r'_{2G} is distance from the second satellite position to point G' if displacement D is present in the period between images; θ is observation angle of RSS in position 1; θ_{inc} is falling angle of probing beam RSS in position 2; B is interfometer base; H is orbit altitude over averaged Earth surface; α is angle of base inclination; α_G is incline angle of displacement vector D; D is Earth surface displacement in the period between two imagings; D_G is projection of displacement vector D on the observation line or LOS (line-of-sight).

Thus disregarding that DRM precision ranges within a decimeter at best, the differential interferogram enables to determine surface displacements comparable to wavelength of RSS, viz., in centimeters. Interference of atmosphere errors used to be weak to affect estimates of large-amplitude Earth surface displacements [5]. At average low velocities of displacements the interference of atmosphere errors is removed by special signal processing methods, briefed on below. ItisimpossibletoevaluatedirectlyaltitudesorEarthsurfacedisplacementsfrom a phase interferogram as it contains not absolute phase shift values, but values convoluted by module 2π, to be precise, lying in the range $[0, 2\pi]$. The problem of phase nonuniqueness is solved by so-called unfolding of the phase interferogram. This classic problem has been arisen long before radar interferometry development [6–8]. The problem of transition to absolute phase values is reduced to addition of a required number of full phase cycles to every phase measurement. There are different algorithms for unfolding of phase interferogram. Earth surface displacement evaluated by an interferogarm is equal to projection of the full vector of displacements in the direction to the satellite, this direction is characterized by a view angle (or beam tilt angle). Analyzing data on ascending and descending orbits, it is possible to identify two projections, but it is not enough to determine three components of displacement vector. The complete vector can be determined with the use of additional data (for example, surface or satellite geodesy) or additional presumptions.

Methods of radar satellite interferometry are employed to monitor different processes accompanied with displacements of the daylight surface, such as development of mineral deposits (oil, gas, coal, and ores).

One of pioneer attempts to use SAR-interferometry to monitor subsidence over oil reservoir was made at Lost Hills and Belridge in San Joaquin, California [1]. Analysis of dual interferograms, made by only 5 images of satellites ERS-1 and ERS-2 with interval from 35 days to 26 months provided an opportunity to single out subsidence zones of funnel shape (2–5 km 52 and 2x15 km) at subsidence velocity up to 40 mm/year. Later the results of interferometric processing on Lost Hills and Belridge oil fields, California were used not only to monitor subsidence of surface over oil fields, but also to control underground processes running in oil- and gas-bearing beds with the help of geomechanical modeling [1].

The application of differential interferometry methods (DinSAR-assessment of displacements by pairs of interferograms) has a number of restrictions related to impossibility to make pairs of interferograms with good coherence under complex conditions when Earth surface is covered with vegetation or humidity is very high, etc. The methods of stable reflectors can overcome these imperfections of interferometric methods. At present there are three well–established versions of methods of stable reflectors: PSInSAR, DePSI, and StaMPS [1].

The interferometric processing technologies used in Russia so far are primitive with actually little chance to obtain robust results. More advanced technologies, for example, method of stable reflectors (DePsi, PSInSAR, StaMPS) are scantily applied in Russia to date.

Conclusion

1. Data of SAR-interferometry make it possible to obtain also in-time variations in displacement rates along with robust evaluation of amplitudes of Earth surface displacements.

2. There are a number of factors restricting the use of radar interferometry to monitor mineral deposit areas.

3. The being of a developed infrastructure in the vicinity of survey area can be considered as a positive factor as such objects are known as stable reflectors.

From the analysis of monitoring data on mining are as it is concluded that reliability and informativity of survey evidence depends on a number of factors, including integrity of satellite data, strategy of their obtaining, processing methods, efficiency of combined interpretation of surface and satellite data. The present research is based on the latest achievements of the global science in SAR-interferometry. The solution of the research problems is of actual importance for professionals working in this field.

References

[1] Dmitriev PN 2013 New methods of processing and interpretation of radar satellite interferometry data *Dissertation Thesis* Moscow (in Russian)
[2] Zakharov AI 2012 Methods of remote sensing of the Earth by radars with synthesized aperture. *Dissertation Thesis* Fryazino (in Russian)
[3] Hanssen RF 2001 *Radar Interferometry: Data Interpretation and Error Analysis* Kluwer Academic Publishers
[4] Zebker HA 1986 *Topographic Mapping from Interferometric Synthetic Aperture Radar Observations* GeofizIssled pp 4993–4999
[5] Dinga LX 2004 Atmospheric effects on InSAR measurements in Southern China and Australia:

acomparative study *ISPRS Proc* Vol 35 Part 1 pp 70–75
[6] Prati CA, Giani M and Leuratti N 1990 2D phase unwrapping technique based on phase and absolute value information *Conference IGARSS'90 Proc* pp 2043–2046
[7] Lin Q Vesecky JF and Zebker HA 1992 New approaches in interferometric SAR data processing *IEEE Trans Geosci Remote Sensing* Vol 30 pp 560–567
[8] Hooper AA 2004 New method for measuring deformation on volcanoes and other natural terrains using InSAR persistent scatterers *Geophysical Research Letters* Vol 31 L 23611 pp 1–5

IOP Conf. Series: Earth and Environmental Science **134** (2018) 012014 doi:10.1088/1755-1315/134/1/012014

Tensile fracture of rocks under uniaxial compression

VP Efimov

Chinakal Institute of Mining, Siberian Branch, Russian Academy of Sciences, Novosibirsk, Russia

E-mail: *efimov-pedan@mail.ru

Abstract. The uniaxial compressive tests of brittle rock sample show that the samples fail in the form of columnar fracturing along the axis of the force application. The researcher attempts to relate the test data with the characteristic values of tensile strength.

The uniaxial compression often induces cleaving of rock samples along a plane coinciding with the direction of the compressive force application. Such fracture is typical for brittle rocks but is sometimes observed in plastic rocks when loading pates have sizes equal to the dimension of cross section of a sample, or when some measures are undertaken to reduce constraint on the sample faces [1, 2]. The term 'cleaving' highlight the difference of this type of fracture from shearing, which also takes place under uniaxial compression. The experiments conducted by the present paper author and by other researches [3–5] aimed to determine physical parameters of rocks that govern their failure show that the initial energy of fracture activation in hard rocks is independent of the type of applied fracturing load. This parameter is the same for the compression and tension of such rocks. This fact suggests that the compression fracture of such rocks is based on the mechanism of fracture of strained bonds by tension. In passing from the breakage of the strained bonds to the increase in the existing defects in the framework of a model of thermal fluctuations in failure, it is possible to suggest development of failure under the action of tension in the conditions of total compression. The relationship between the compression and tension strengths in Figure 1 indirectly proves the said model of failure.

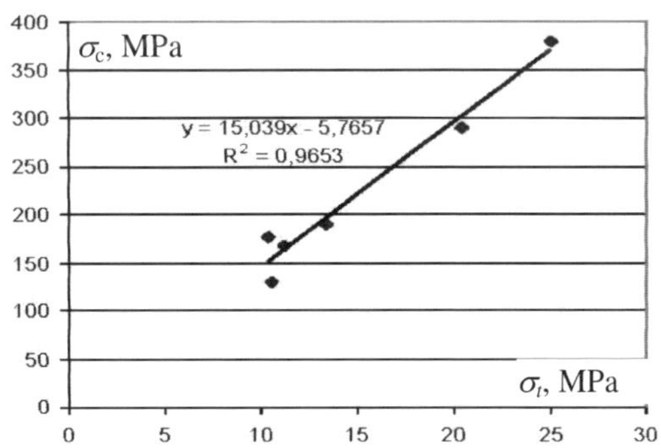

Figure 1. Compression strength versus tension strength of tested rocks.

Griffiths [6] showed that tensile stresses appeared at the tips of cracks oriented at an angle to the compression axis. Furthermore, the conclusion drawn on the basis that a solid is a medium containing differently oriented defects in the form of cracks is well known: ratio of compression and tension strengths for brittle media is equal to 8. It is worthy of mentioning that the criterion of failure was chosen the tension strength of molecular bond.

Presentation of a sample as a medium with cracks is not a single possible model to transform compressive stresses to tensile stresses. The well known solution of an elastic problem on stress distribution around a spherical pore in an infinite body subjected to compression at infinity [7] yields tensile stress at the poles of the pore, and this stress also can be a center of initiation of tensile micro-cracks leading to macro-failure. The values of the initiated tensile stresses are close to the values of the compressive stresses but are extremely localized in space. The use of the local strength criteria in the determination of tensile stresses in such geometry should result in a small difference in the values of the compression and tension strengths in samples containing such defects, which contradicts the reality. For the description of failure in nonuniform stress fields, the models involve nonlocal strength criteria [8–14] which is effective in this case. In such models, failure is not analyzed at a hazardous point of material but in some vicinity the linear dimension of which is a characteristic of the material. It is assumed that failure takes place when stresses, averaged over a certain length (or equivalent) reach the value equal to tension strength measured by standard in a uniform field. With such approach, the ratio of the compression and tension strengths depends on the characteristics of material. Let us analyze this approach in terms of compression of a sample with a single spherical pore.

The stresses $\sigma_{\theta\theta}$ and $\sigma_{\varphi\varphi}$ under uniaxial compression or tension are determined from the solution to the problem of Leon [7]:

$$\sigma_{\vartheta\vartheta} = \sigma\left[1 + \frac{4-5\nu}{14-10\nu}\frac{a^3}{r^3} + \frac{9}{14-10\nu}\frac{a^5}{r^5} + \left(-1 + 5\frac{2-5\nu}{14-10\nu}\frac{a^3}{r^3} - \frac{21}{14-10\nu}\frac{a^5}{r^5}\right)\cos^2\theta\right],$$

$$\sigma_{\varphi\varphi} = \sigma\left[-3\frac{2-5\nu}{14-10\nu}\frac{a^3}{r^3} + \frac{3}{14-10\nu}\frac{a^5}{r^5} + \left(15\frac{1-2\nu}{14-10\nu}\frac{a^3}{r^3} - \frac{15}{14-10\nu}\frac{a^5}{r^5}\right)\cos^2\theta\right], \tag{1}$$

where a—radius of the pore; r—distance from the center of the pore; θ—polar angle counted from the axis of force application; φ—azimuth direction angle; ν—Poisson's ratio. The distribution of stresses in the vicinity of hazardous points:

—at the poles along the force application line:

$$\sigma_{\vartheta\vartheta} = \sigma_{\varphi\varphi} = \frac{3\sigma}{14-10\nu}\left((3-5\nu)\frac{a^3}{r^3} - 4\frac{a^5}{r^5}\right); \tag{2}$$

—at the equator of the pore:

$$\sigma_{\vartheta\vartheta} = \sigma\left(1 + \frac{4-5\nu}{14-10\nu}\frac{a^3}{r^3} + \frac{9}{14-10\nu}\frac{a^5}{r^5}\right). \tag{3}$$

It is evident from (2) that the stresses at the poles have opposite sign to the applied force while there is the concentration of stresses at the equator. Stress distribution decreases with the growth of r.

We take the failure criterion as the integral criterion by Novozhilov:

$$\sigma_t = \frac{1}{\delta}\int_0^{\delta}\sigma_n(r)dr, \tag{4}$$

where σ_t—tension strength measured in uniform field; δ—characteristic structural constant of a material. Integrating (2) and (3) produces the ratio of the compressive and tensile strengths:

$$\frac{\sigma_c}{\sigma_t} = \frac{1}{6}\left(\frac{4(14-10\nu)(1+b)^4 + 2(4-5\nu)(2+b)(1+b)^2 + 9(4+6b+4b^2+b^3)}{(3-5\nu)(2+b)(1+b)^2 - 2(4+6b+4b^2+b^3)}\right), \tag{5}$$

where $b=\delta/a$—dimensionless parameter.

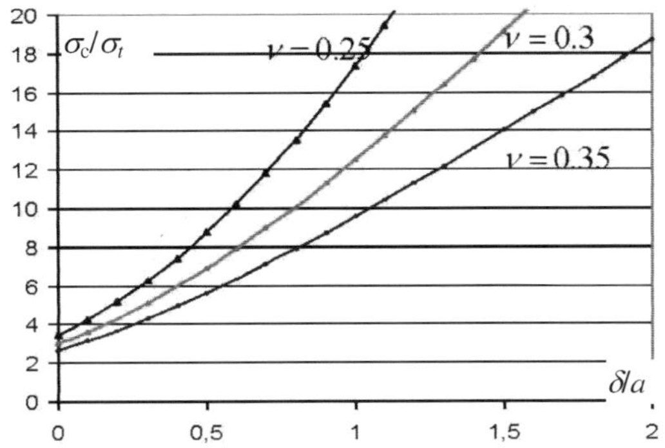

Figure 2. The ratio of compressive and tensile strength versus the ratio of structural parameter and pore radius at different values of Poisson's ratio.

Figure 2 shows the curves of this ratio and the ratio of structural parameter to pore radius. Evidently, the compression/tension strength ratio depends on the structural characteristics of materials.

In passing to a real rock, it is expected to obtain qualitatively the same results, namely, the influence of the structural features of rocks on the ratio of compression and tension strength. The works [15, 16] analyzed different models of failure under compression in materials having ordered structure some elements of which can transform compression to tension. Considering intact rock mass not as a solid medium with averaged parameter but as an aggregation of grains with different mechanical characteristics, we obtain a heterogeneous medium and transformation of the compressive stresses to tensile stresses is possible at the boundaries of these grains. In hard rocks, such granite, porosity is low, and the discussed effect of transformation of the compressive stresses to tensile stresses should not be connected with the size of pores but with the size of grains, and the hardest grains should be assumed inclusions and soft grains—filled pores. The relationship between the strength and average size of grains in such rocks is depicted in Figure 3.

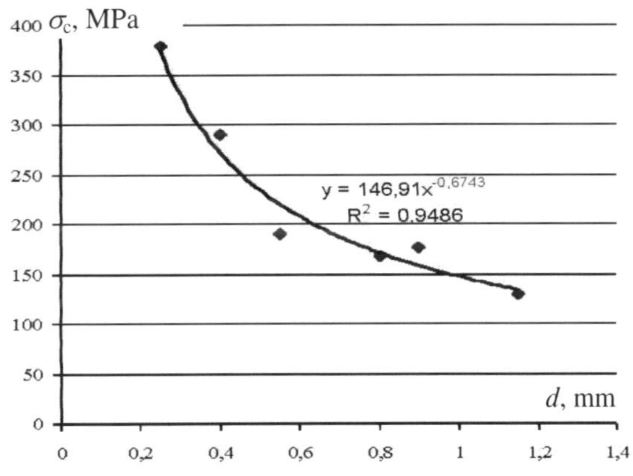

Figure 3. Compression strength versus grain size of rocks.

The curve shows the power dependence of strength on grain size in rocks. This curve is plotted for the test rocks the mechanical characteristics of which are described in the table: *d*—average grain size; K_{1c}—critical SIF of tensile cracks; σ_c—compression strength; σ_t—Brazilian test tension strength; σ_b^4—4-point bending strength; δ—structural parameter having dimension of length, determined as the ratio of critical SIF to tension strength [10]:

$$\delta = \frac{2}{\pi}\left(\frac{K_{1c}}{\sigma_t}\right)^2. \tag{6}$$

Mechanical characteristics of tested rocks.

Rock	d, mm	K_{1c}, MPa·m$^{1/2}$	σ_c, MPa	σ_t, MPa	σ_b^4, MPa	$\frac{\sigma_c}{\sigma_e}$	$\frac{\sigma_b^4}{\sigma_t}$	δ, mm	δ/d
Dolerite	0.25	1.90	379	25.0	33.4	15.20	1.34	4.3	17.20
Gabbroide	0.40	2.00	290	20.4	33.6	14.20	1.65	5.9	14.75
Gabbro-diorite	0.45	1.10	190	13.4	19.4	14.20	1.45	5.6	12.40
Granite	0.80	1.14	168	11.2	16.6	15.00	1.38	6.6	8.25
Biotite granite	0.90	0.70	177	10.4	11.2	17.00	1.08	2.9	3.20
Leucocratic granite	1.15	0.80	130	10.6	13.4	12.30	1.26	3.6	3.10

The ratio of the structural parameter to the grain size, as follows from the table, takes on value from 3 to 20 for the tested rocks, i.e. the structural parameter covers a few grains at the least. The dependence of the compressive stresses on the dimensionless parameter δ/d is shown in Figure 4. The conclusion is that the more grains are covered by the structural parameter, the stronger is rock. This conclusion is valid both in compression and tension.

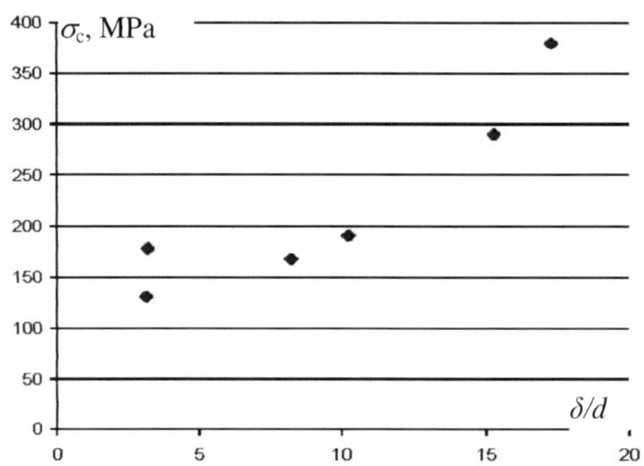

Figure 4. Compression strength versus dimensionless parameter δ/d.

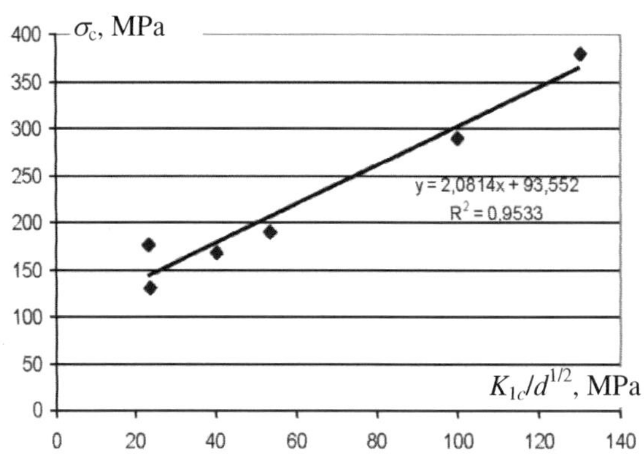

Figure 5. Curve of compression strength of rocks.

Figure 5 demonstrates, in reduced coordinates, the curve of compression strength in the tested rocks on the value which has dimension of stresses and is composed of the product of SIF for tensile cracks and inverse value of average grain size to the ½ power.

Conclusion

Fracture of rocks by compression greatly depends on the surface conditions of compressing plates. There can be three scenarios: failure along a conical or pyramidal surface, along the diagonal or cleaving along the force application line. Fracture with the formation of cones or pyramids at the contact surface due to strong constraint of materials initiates in the central part. When the constraint is weak, fracture fragments can have a clearly columnar structure. The correlation relations between compression strength, tension strength and tensile crack resistance in hard rocks are reflective of the close connection of fracture processes in these tests in case when compression initiates formation of the columnar structure. The average ratio of compression and tension strengths for the test rocks is equal to 15, and it should be adjusted according to the mechanical characteristics of grains composing rocks.

References

[1] Fairhurst C and Cook NGW 1966 The phenomenon of rock splitting parallel to the direction of maximum compression in the neighbourhood of a surface *Proc. I ISRM Congress*, Lisbon, pp 687–692
[2] Bock H (Ed) 1978 *An Introduction to Rock Mechanics* Townsville: James Cook University of North Queensland
[3] Efimov VP 2007 Investigation into the long term strength of rock under loading with a constant rate *J. Min. Sci.* Vol 43 No 6 pp 600–606
[4] Kucheryavy FI, Mikhalyuk AV and Demchenko LA 1980 Energy of activation and energy intensity of rock fracture *Ivz. Vuzov Gorny. Zh.* No 5 pp 57–63
[5] Wexler YuA 1979 Durability of rocks under compression *J. Min. Sci.* No 3 pp 71–76
[6] Griffith AA 1924 Theory of Rupture *Proc. I Int. Congr. On Applied Mechanics* Delft pp 55–63
[7] Hahn H 1985 *Elastizitatstheorie* Stuttgart (in German)
[8] Novozhilov VV 1969 To the basics of the theory of equilibrium cracks *J. Appl. Math. Mech.* Vol 33 No 5 pp 797–812
[9] Lajtai EZ 1972 Effect of tensile stress gradient on brittle fracture initiation Int. J. Rock Mech. Min. Sci. 9 pp 569–578
[10] Suknev SV and Novopashin MD 2000 Determination of local mechanical properties of materials *Dokl. RAN* Vol 373 No 1 pp 487–500
[11] Legan MA 1993 Correlation of local strength gradient criteria in a stress concentration zone with linear fracture mechanics *J. Appl. Mech. Tech. Phys.* Vol 34 No 5 pp 585–592
[12] Kharlab VD and Minin VA 1989 Strength criterion considering effect of stress gradient *Research on Mechanics of Engineering Structures and Materials: Collected Papers* Leningrad: LISI pp 53–57 (in Russian)
[13] Kornev VM 2002 Generalized sufficient strength criterion. Description of the pre-fracture zone J. Appl. Mech. Tech. Phys. Vol 43 No 5 pp 763–769
[14] Efimov VP 2013 Rock tests in nonuniform fields of tensile strength J. Appl. Mech. Tech. Phys. Vol 54 No 5 pp 857–865
[15] Goldstein RV 2003 Fracture under compression *Usp. Mekhaniki* Vol 2 pp 3–20
[16] Goldstein RV and Osipenko NM 1999 Structures in failure processes of fracture *Izv. RAN Mekh. Tverd. Tela* No 5 pp 49–71

IOP Conf. Series: Earth and Environmental Science **134** (2018) 012015 doi:10.1088/1755-1315/134/1/012015

Drilling and blasting parameters in sublevel caving in Sheregesh mine

AA Eremenko[1*], VN Filippov[1], AI Konurin[1], AP Khmelinin[1], DV Baryshnikov[1] and EA Khristolyubov[2]

[1]Chinakal Institute of Mining, Siberian Branch, Russian Academy of Sciences, Novosibirsk, Russia
[2]Gornaya Shoria Division, EVRAZRUDA, Russia

E-mail: *eremenko@ngs.ru

Abstract. The factors that influence geomechanical state of rock mass in Sheregesh Mine are determined. The authors discuss a variant of geotechnology with fan drilling. The drill-hole patterns and drilling-and-blasting parameters are presented. The revealed causes of low-quality fragmentation of rocks include the presence of closed and open fractures at different distances from drill-hole mouths, both in case of rings and fans, as well as the blocking of drill-holes with rocks.

Sheregesh iron ore deposit occurs in Gornaya Shoria within the Altai–Sayan folded zone featuring active tectonics and seismicity to Richter magnitude 7 and higher [1, 2]. The current mining depth is 800 m, the deposits is ranked in the rockburst-hazardous category [3].

Sheregesh Mine widely applies blasting technologies with changes 89 and 160 mm in diameter in mining with caving by blocks and levels with levels to 70–80 m high and blocks from 27 to 30 m wide. Ore levels are blasted in the direction of compensation space or are subjected to confined blasting. Blocks are undercut. Ore drawing involves vibration machines VDPU-4TM and scrapers.

Blasting is a process that has a highest influence on rock mass. After large-scale production and process blasts, aside from ore fragmentation, dynamic events due to rock pressure are observed, including spalling, bumps, microshocks etc of any intensity [4–6]. In connection with this, Podruslovy site was subjected to mining with sublevel caving at sublevel +255 ÷ +185 m.

Stoping starts with making a cut along the length of a block at the ore and rock interface. Slot raises in the extraction blocks are made by blasting in parallel holes drill from level +255 m. Then, ore is blasted from the cut towards the slot raise. Widening of the cut is made from drifts at level +220 m. The cut has the width not les than 3 m, its length is equal to the length of the block, the height equals the height of the sublevel +255/+220 m (35 m). While the cut is made, rocks from the caving zone above level +255 m is bypassed to level +220 m and fills the cut.

Confined blasting is performed toward the broken rocks in the cut (at the first stage) and then towards mined-out void filled by rocks. Fan of upward holes are drilled from drilling blind drifts at level +220 m. Ore is stopes is blasted by a layer per a fan. As mined-out stope is enlarged, ore blasting is made by 2–4 fans of blastholes.

Ore is discharged from stopes using load–haul–dumpers via entries of drilling blind drifts and in diagonal loading accesses driven from the haulage drifts at the level. For the safety of the haulage drifts, ore pillars are left above them. Complete ore extraction after total ore breakage in two adjacent

IOP Conf. Series: Earth and Environmental Science **134** (2018) 012015 doi:10.1088/1755-1315/134/1/012015

drilling drifts, the ore pillars left above the haulage drifts are drilled and broken in the backward direction towards the interblock ore pillars. As ore pillars are broken above the haulage drifts, broken ore is discharged from the haulage drift ends. An average block has a height of 45 m, length of 40 m and a width of 20 m.

Ore in stopes is broken by sections. At first, when blasting is performed toward the cut, a section is made by 1–2 fans of blastholes. After a few explosions in the stope, the confining space becomes wider to 12–15 m, and a blasting section can be 2–4 fans. In blasting, explosive Grammonit M21 is arranged in holes with a diameter of 89 mm. In the confined blasting by fans at the fragmentation standard of 500 mm, the powder factor ranged from 0.73 to 64 kg/m^3.

The burden in the confined blasting was 1.8 m, the spacing of ends of holes in fans was 2.3 m. The drilling-and-blasting design in a fan is described in Tables 1 and 2.

From the evidence of the analysis, specific consumption of explosive for secondary fragmentation remains high and is 0.3 t/m^3. Therefore, the studies were initiated to understand the causes of off-standard fragmentation by means of georadar surveys and visual observation in fans of blastholes.

Table 1. Drilling-and-blasting design.

Hole no.	Hole length, m	Charge length, m	Uncharged length, m	Explosive weight, kg
1	8.9	3.2	5.7	23
2	10.9	6.8	4.1	48
3	13.7	7.9	5.8	56
4	17.7	14.7	3.0	104
5	16.7	9.4	7.3	67
6	16.0	13.5	2.5	96
7	15.6	8.8	6.8	62
8	15.5	13.1	2.4	93
9	15.6	8.8	6.8	63
10	16.0	13.5	2.5	96
11	16.7	9.4	7.3	67
12	17.7	14.7	3.0	104
13	13.7	7.9	5.8	56
14	10.9	6.8	4.1	48
15	8.9	3.2	5.7	23
Total:	214.5	141.7		1006

Table 2. Blasting parameters.

No.	Description	Unit of measurement	Total
1	Broken ore volume	m^3	604
2	Hole length	m	14.3
3	Total hole drilling length	m	214.5
4	Number of holes	holes	15
5	Quantity of: Explosives Electric detonators	 kg detonators	 1006 24
6	Specific consumption of explosives for secondary fragmentation	t/m^3	0.2–0.3

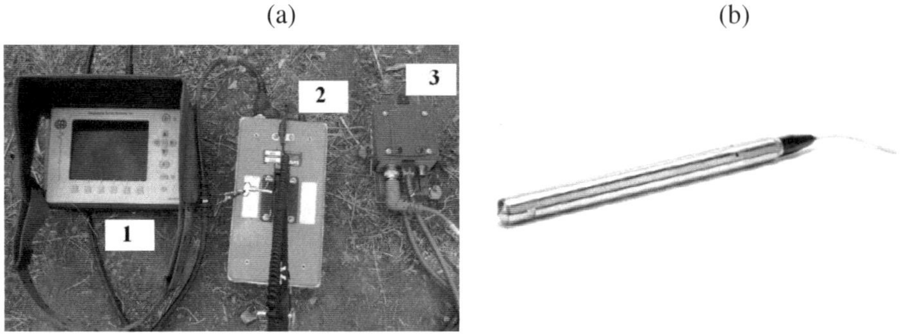

Figure 1. (a) Georadar SIR-3000 with antenna assemblies: 1—georadar; 2—antenna assembly with the central frequency of 900 MHz; 3—antenna assembly with the central frequency of 2600 MHz; (b) down-the-hole probe with the central frequency of 1000 MHz.

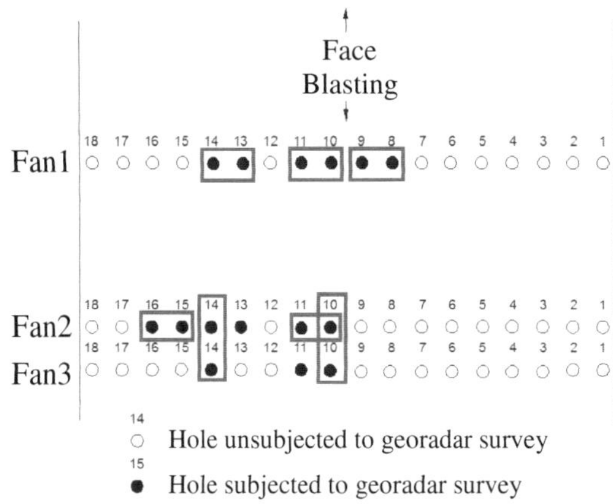

Figure 2. Drilling patterns in fans 1–3.

Georadar survey was carried out using georadar SIR-3000 with an antenna assembly having a center frequency of 2600 MH for calibration tests and using hole probes with center frequencies of 500 MHz and 1000 MHz (Figure 1). The patterns of drilling and arrangement of fans are shown in Figure 2.

The measurement probe was installed in a hole, and a metal rod was placed in the neighbor hole. The probe and the rod were simultaneously put deeper in the holes, and georadar made marks in the hole each 50 cm with a view to match the georadar survey data and the hole length together. In the radargrams, these marks are shown as white dashed lines. Upon reaching the hole bottom, the data recording stopped and the test was performed in the backward direction. After the in situ tests, the detailed analysis of the obtained radargrams was undertaken with RADAN.

Figure 3. Radargram after calibration tests of georadar on a specimen.

After the georadar calibration test on a specimen of ore, it was found that the probing signal attenuated greatly and the radargram contained numerous multiple frequencies. The presence of such frequencies can be explained by conduction currents that are the sources of secondary waves inconsistent with the reflections from the ore–air interface (Figure 3).

After calibration and adjustment of the value of capacitivity, holes 8–9, 10–11 and 13–14 in the first fan were surveyed (refer to Figure 2).

The analysis of the radargrams in the listed holes yields that the useful signal is heavily depressed. At a depth of 70–80 mm, the clear line of correlation corresponds to the reflection from the inward surface of the hole. At a depth of 280–300 mm, the line of correlation is conditioned by spurious multiple frequencies that are noise and useless. So, in the tested segment 2 m long, the hole spacing is 40–50 mm, the useful signal corresponding to the reflection from the metal rod attenuate with depth and is barely distinguishable in the radargrams.

In the second stage, the tests were carried out in the second fan after the first fan blasting was completed.

At a depth of 80–85 mm, the clear line of correlation conforms with the reflection from the inward surface of the hole as in the radargrams in the first fan. At a depth from 280 to 340 mm, the line of correlation is conditioned by spurious multiple frequencies that are noise and have no useful information.

In the segment of the first 2 m in the hole, there are correlation lines corresponding to the reflections from the metal rod. The spacing of the hole at this depth is from 45 to 60 mm. The useful signal attenuates with depth in the hole and is untraceable against noise.

In the third stage, the holes in the neighbor fans were examined, namely, in the second and the third fans.

The radargrams from the neighbor fans contain correlation lines corresponding to the reflection from the inner surface of the holes (at a depth from 80 to 85 mm) and spurious signals that are noise. The line of correlation to fit the reflection from the metal rod is undistinguishable in the radargrams, which is explained by heavy attenuation of the signal in ore by the numerous multiple re-reflections induced by spurious conduction currents.

The visual inspection exhibited locations of closed and open cracks in the hole and hole bottoms (Figure 4 and Table 3).

It has been found that after explosion, the coverage is observed at the depth from 11.1 to 14 m in fan 1, from 12.1 to 12.4 m in fan 2 and at 12.4 m in fan 3. The length of the holes is 9.4 to 20.2 m. Cracks grow in-between the holes at a depth of 5.2 and 10.4 m (holes 3 and 5), 13.4, 15.8, 16.9 and 18.7 m (holes 5 and 8) etc. Between fans 2 and 3, cracks merge at a depth of 2.8, 4.1, 8.0 and 13 m and deeper. Cracks from fan 1 seldom reach fan 2: at a depth of 10.8, 15.8, 17 and 19.9 m.

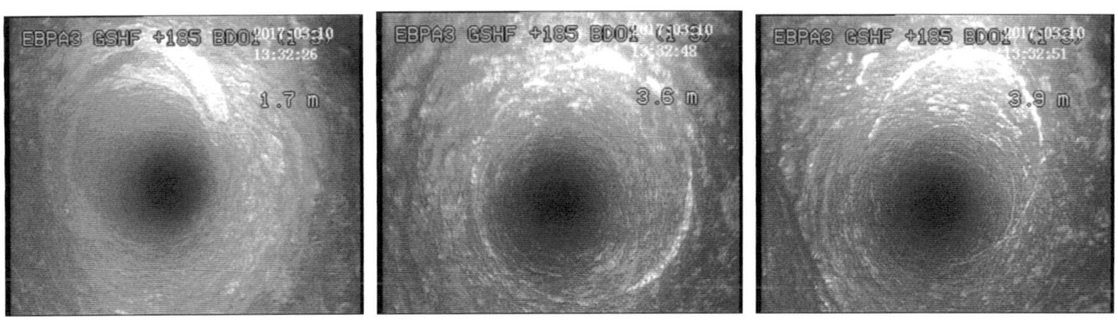

Figure 4. Images of holes and cracks in fan 1.

Table 3. Summary data on depths of holes in fans 1–3.

Fan 1	3	5	8	9	10	11	12	13	14	15	16	17	18
Depth, m	13.0	>20.2 (nd)	>19.3 (nd)	>19.4 (nd)	>13.8 (c)	11.7 (c)	11.1 (c)	10.4 (c)	11.0 (c)	11.1 (c)	15.0 (c)	124	9.4
Fan 2	3	6	–	9	–	11	–	13	–	15	–	–	–
Depth, m	13.1	>20.0 (nd)	–	>12.2 (c)	–	>12.1 (c)	–	>12.4 (c)	–	>13.6 (c)	–	–	–
Fan 3	4	6	8	–	–	11	–	–	–	15	–	–	–
Depth, m	>18.6 (nd)	>19.4 (nd	>18.4 (nd)	–	–	12.4 (c)	–	–	–	>13.2	–	–	–

Comment: nd—depth not determined; c—coverage

Conclusion

The drilling-and-blasting patterns have been design for fan drilling and blasting in the conditions of Sheregesh deposit.

The research has revealed factors of influence on the blasting quality, including drilling patterns and the presence of crack at different distances from the hole mouth.

The research proves that the georadar survey aimed to estimate expansion of hole spacing in magnetite ore is impossible. The further research will use the method of ultrasonic sounding.

Acknowledgments

This study has been supported by the Federal Agency for Scientific Organizations of Russia in the framework of Program IX.132.4, Project No. 0321-2016-0003.

References

[1] Platonova SG, Skripko VV 2003 *Earthquake in Gorny Altai: Causes, Consequences and Forecasts* Kemerovo: Kuzbass–Irbis (in Russian)

[2] Kuznetsov VA 1954 *Geotectonic Zoning of the Altai-Sayan Folded Area. Geology of Asia* Moscow: AN SSSR (in Russian)

[3] *Guidelines on Safe Mining under Condition of Rockburst Hazard in Gornaya Shoria* Novosibirsk–Novokuznetsk 2015 (in Russian)

[4] Shtirts VA, Eremenko AA, Konurin AI, Klishin IV 2015 Experience of forecasting geodynamic events under large-scale blasting *GIAB* No 7 pp 196–206

[5] Konurin AI, Eremenko AA, Filippov VN 2017 Assessment of rock mass conditions under production blasting and geodynamic events *GIAB* No 7 pp 153–160

[6] Bashkov VI, Shtirtz VA, Eremenko AA, Konurin AI 2015 Assessment of geomechanical conditions of rockburst-hazardous rock mass under large-scale blasting in Gornaya Shoria *J. Fundament. Appl. Min. Sci.* Vol 2 pp 205–2013 (in Russian)

Effect of vibro-impact exposure on intensity of geodynamic events in rock mass

AA Eremenko[1], VV Timonin[1*], AA Bespalko[2], VN Karpov[1] and VA Shtirts[3]

[1]Chinakal Institute of Mining, Siberian Branch, Russian Academy of Sciences, Novosibirsk, Russia
[2]Institute of Nondestructive Testing, Tomsk National Research Polytechnic University, Russia
[3]Tashtagol Division, EVRAZRUDA, Russia

E-mail: *timonin@misd.ru

Abstract. Effect of vibrational impacts on rockburst-hazardous iron ore body and enclosing rocks is experimentally studied using the microseismic, electromagnetic, acoustic and electrometric methods. The nature of the change in the intensity of electrical resistance, and in electromagnetic and acoustic signals in a series of impacts is determined. The variation in the amplitude dispersion of spectral components of signals at different frequencies is described.

Iron ore mining operations in Gornaya Shoria are scheduled for the transition to levels below 1000 m below ground surface in the nearest future. As a result of mining, for instance, at Tashtagol deposit, large areas of subsoil are changed quantitatively and technologically altered rock masses are generated. As a consequence of different rate of destruction, such altered rock masses contain zones where properties of rocks and, thus, geomechanical behavior differ from the initial conditions [1, 2].

In deeper level mining, it becomes important to understand geodynamics of rocks under induced factors since blasting and stoping result in restructuring of rock mass and, accordingly, in accumulation of energy of such external influences. In this connection, it is necessary to analyze rock mass behavior with regard to the nature and rate of the technological impact with the purpose of predicting and preventing rock burst toward safety of mines [3]. At the same time, it is of the current concern to study the influence exerted by long weak dynamic shock actions on stress–strain state of rock mass with subsequent reduction in the shock action intensity in the influence zone of mining operations, or at the redistribution of impacts deeper inside rock masses [4–6].

The experimental studies were carried out at Level –350 m in cross drift 8 at a depth of 800 m below the ground surface in Vostochny site of Tashtagol deposit (Figure 1). Vostochny site ore body is composed of magnetite with the admixtures of quartz, calcite, chlorite, epidote and other minerals. The horizontal thickness of the ore body is 15 to 60 m and more. The dip angle varies from 70 to 90°. B the dynamic source of P-waves is a compression vacuum percussion machine (CVPM). The specification of the machine is given the tale below, while Figure 2 shows its physical view.

The details of the machine operation in the experiments [7], and trials of CVPM machines are described in [8, 9].

IOP Conf. Series: Earth and Environmental Science **134** (2018) 012016 doi:10.1088/1755-1315/134/1/012016

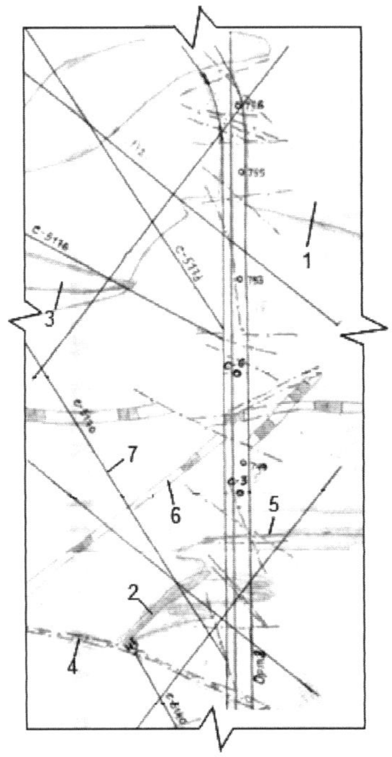

Figure 1. Arrangement of cross drift 8 at Level –350 m: 1—ore; 2—syenite; 3—skarn; 4—fault; 5—fractures; 6—dykes; 7—geological exploration wells.

At the location site of CVPM, dynamic impact recording used the methods of microseismicity and electrometry, and electromagnetic an acoustic emission. It was found that the percussion action induced shocks within the classes of energy from 2 to 2.4 in cross drifts 9 and 10 (center).

Table 1. Specifications of CVPM

Description	Characteristics
	CVPM 1 model
Power drive:	Electric
–basic	Original
–power, kW	2
–weight, kg	4
Process chamber	1
Compression and vacuum unit air lines used to generate impacts	Vacuum
Body material	Polyethylene
Elastic reverse	–
Length of power stroke of hammer, mm	1300
Inner diameter of the body, mm	200
Hammer weight, kg	40
Unit blow energy, J	to 300
Blows per minute	50–60
Control mode	Automatic
Percussion angle, deg	–90°
Weight, kg	56

Figure 2. General view of compression vacuum percussion machine CVPM.

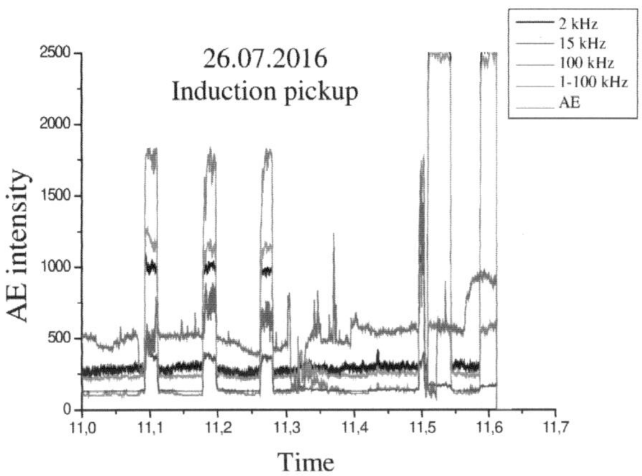

Figure 3. Variation in EME and AE under series of impacts. The first three peaks are obtained under percussive treatment of rock without sandy gravel; the last two peaks—rock with a sandy gravel layer 1 cm thick (particles to 10 mm in sizes).

Concurrently with vibro-treatment, electrometric sounding was carried out in rocks mass. The electrometric measurement at a depth of 2 m showed electrical resistance of 0.35 Ω·m before the vibro-treatment and 31 Ω·m after the treatment. The electrical resistance increased owing to "opening" of cracks. However, due to low energy of blows, no considerable increase was recorded at all measurement point (an average increase in the electrical resistance made 1.8 and 3 times).

The percussive excitation of electromagnetic signals was modeled physically. The aim of the modeling was to find variations in the intensity of EME under continuous acoustic excitation pulses generated by initiation, growth and manifestation of geodynamic events. At the level above (–280) m, neither blasting, no first or second mining was performed. The percussion machine had blow energy of 300 J and frequency 1 Hz. The electromagnetic signal recording used EM detector REMS-1 with the EM signal amplitude averaged to 1 s and REMAS-1 with the EM signal amplitude averaged to 10 ms.

EME was received using induction pickups. Figure 3 illustrates EME and AE intensity by the data of detector REMS-1 during a series of hammer blows in rocks.

Initially a series was not less than 20 blows. Specific number of blows is seen in Figure 4.

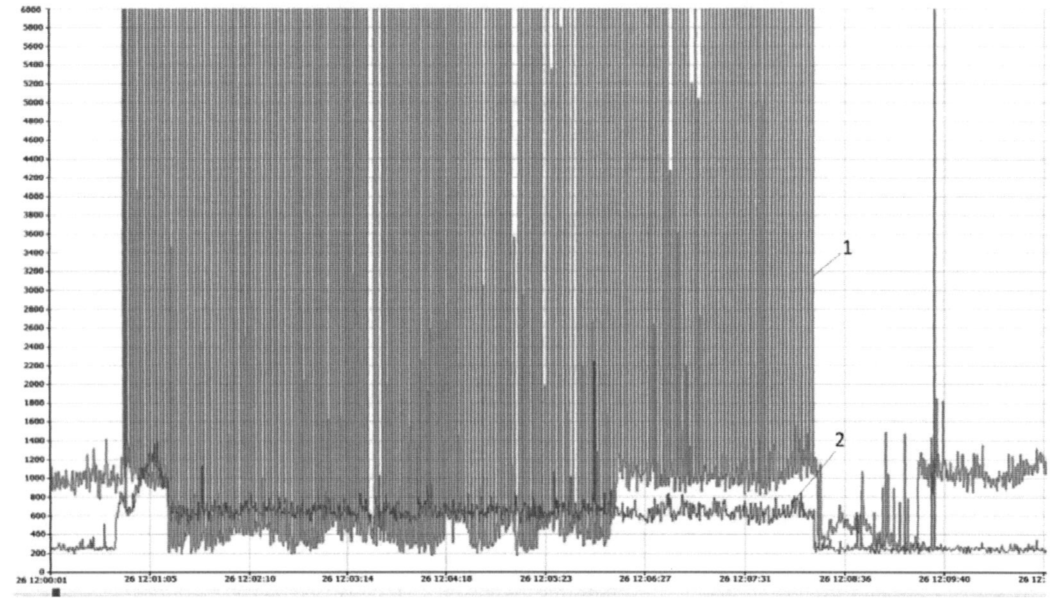

Figure 4. The last 8th expanded fragment of measured dispersion in the amplitude of spectra of electromagnetic signals in rocks at the frequencies of 1 (1) and 13 (2) kHz in the first series of hammer blows.

Analysis of the electromagnetic and acoustic emission measurements and dispersion of amplitudes of the selected spectra of electromagnetic signals generated by a series of hammer blows reveals an interesting pattern. In the beginning of the series, the amplitudes of EME and electromagnetic signals. Grows (Figure 4), then distribution of the amplitude takes place to a certain value and is sustained over the whole time period of the blow series. It is highly probable that subsequent blows (or shocks) excite new electromagnetic response in rocks and the previous blows maintain the emitting rock excited, and the level of amplitudes of electromagnetic signals lowers against the background of residual excitation field.

The dispersion of amplitudes is measured in the selected spectra of electromagnetic signals at the frequencies 1 and 13 kHz from the records obtained by detector REMAS-1 with the amplitude averaging at 10 ms.

Thus, the vibro-percussion treatment has an influence on initiation and manifestation of geodynamic events in the form of shocks with an energy class from 2 to 2.4, variation in the electrical resistance from 1 to 31 Ω·m and dispersion of amplitude of electromagnetic signals at the frequencies of 1 and 1 3 kHz with the amplitude averaging to 10 ms.

Acknowledgments
This study has been supported by the Russian Academy of Sciences, Project No. 0321-2015-0002.

References

[1] *Guidelines on Rockburst Prevention with Regard to Mineral Deposit Geodynamics* Leningrad: VNIMI 1983 (in Russian)
[2] Egorov PV et al 1999 *Rock Mass Monitoring in Mines in Gornaya Shoria* Kemerovo: KuzSTU

(in Russian)
[3] Kurlenya MV, Eremenko AA, Shrepp BV 2001 *Geomechanical Problems of Iron-Ore Mining in Siberia* Novosibirsk: Nauka (in Russian)
[4] Shtirts VA, Eremenko AA, Konurin AI, Klishin IV 2015 Experience of forecasting geodynamic events under large-scale blasting *GIAB* No 7 pp 196–206
[5] Bashkov VI, Shtirtz VA, Eremenko AA, Konurin AI 2015 Assessment of geomechanical conditions of rockburst-hazardous rock mass under large-scale blasting in Gornaya Shoria *J Fundament Appl Min Sci* Vol 2 pp 205–213 (in Russian)
[6] Konurin AI, Eremenko AA, Filippov VN 2017 Assessment of rock mass conditions under production blasting and geodynamic events *GIAB* No 7 pp 153–160
[7] https://www.youtube.com/channel/UCSB2wauPNrLMvDTpMvgHkww
[8] Timonin VV, Tkachuk AK and Karpov VN 2016 Double-action compression/vacuum–impact machine *J. Min. Sci.* Vol 52 No 5 pp 927–932 DOI: 10.1134/S1062739116041418
[9] Repin AA, Tkachuk AK, Karpov VN, Beloborodov VN, YaroslavtsevAG and Zhikin AA 2016 Engineering and analysis of independent movable compression–vacuum percussion source of P-waves in seismic survey *J. Min. Sci.* Vol 52 No 1 pp 146–152 DOI: 10.1134/S1062739116010236

IOP Conf. Series: Earth and Environmental Science **134** (2018) 012017 doi:10.1088/1755-1315/134/1/012017

Geodynamic risk magnitude as an objective indicator of rockburst prevention effectiveness (in terms of apatite mines in Khibiny)

YuV Fedotova* and VI Panin

Mining Institute, Kola Science Center, Russian Academy of Science, Apatity, Russia

E-mail: *fjulia@mail.ru

Abstract. The results of the statistical retrospective analysis of the officially recorded geodynamic events in mines of Apatit Company within the Khibiny Massif are presented. The risks and aftereffects of geodynamic events have been calculated. Under discussion are the results of three calculation variants taking into account the scale of human impact on rock mass. The analysis shows that the main damage due to geodynamic events is different-degree destruction of mine workings while the remaining aftereffects account for less than ten percent. That is, the geodynamic risk in apatite mines can be identified as technological.

1. Introduction

Among the numerous problems in the mining industry, the most acute issue is still induced seismicity—rock bursts and induced earthquakes governing geodynamic risks. This problem remains of concern in Russia, Australia, South Africa, Canada, USA, South America and in many other countries. Irrespective of ownership, the most widely used standard risk factor is individual risk characterizing number of fatal accidents (or severe injury accidents) and including probability of negative event. Different values of maximum permissible level (MPL) of individual risk in different countries show that mortality risk higher than 10^{-4} per year is inapplicable (too high) as MPL. Expert estimates based on generalization of implemented research findings, including avowed safety, state that MPL of individual risk in Russia should be within 10^{-4}–10^{-5} per year [1, 2].

Risk ratings and indexes customary in geodynamic safety are mostly qualitative. This is explained by the complex physical nature of initiation and instance of dynamic events, huge number of interfering factors and difficulty of mathematical formalization for determining quantitative parameters of danger level. Geodynamic risk, depending on probability of hazardous event, hazardous situation or hazardous process, is ranked as low, medium and high (ore negligible, admissible and inadmissible).

All that is positively applicable in estimation of geodynamic risk at rockburst-hazardous mineral deposits in Khibiny [3].

2. Object of research

Khibiny is an alkaline intrusive of central type, the natural events are seldom and have magnitudes not higher than 1–2 on a scale from 1 to 12 by Medvedev–Sponheuer–Karnik (MSK-64). Shallow induced earthquakes with magnitudes 2–4 (MSK-64) and low-energy geodynamic events recorded in the Khibiny mines are explained by effects (both natural and technical) on the actual large-scale mining.

Out of natural effects, mines are heavily affected by variation in water content of rock mass, which is seasonal (except for the cases when mines expose aquifer strata, which is independent of seasons). Temperature changes, lunar and solar cycles, gas emanation, etc. have weaker influence on mining performance. Influence of natural factors on geodynamic events depends on structural features of rock mass—the impact is stronger near faulted and oxidized zones and lower in weakly damaged zones of hard rock mass.

Technology factors also have different effect on geodynamic behavior of rock mass. Geodynamic risks differ by the scale of their aftereffects. In a greater degree, the influence of technology factors is conditioned by the scale of mining operations: in the course of apatite–nepheline mining in Khibiny, nearly 4.5 Bt of rock mass are extracted, displaced and stored partly. This is added by the influence of change in the mining technology (for example, transition from block caving to sublevel caving). In a lesser degree, the technology impact is connected with the adjustment of separate process designs (e.g. transition to another type of explosives, etc.).

A geodynamic event may end in no failure in mines but is unambiguously indicative of a hazard. It is clear that aftereffect of a geodynamic event is governed the power of the event and its source coordinates [4]. After a remote event, tremors are possible in underground excavations, and the sound and strength of the tremor depends on the energy of the geodynamic event; after a near by geodynamic event, a rock burst is possible in a mine. The risk assessment should account for influence exerted by different factors.

The expert risk estimate (in the first approximation) uses standard methods of the statistical analysis in order to find trends of change in parameters; thereupon, a decision is made on expediency of detailed elaboration and multifactor analysis.

3. Calculation procedure

For the geodynamic risk assessment in apatite mines, based on the statistical post-even analysis of records, probability of geodynamic events and degree of their aftereffects were calculated using the available data base on geodynamic events recorded in mines by the Commission on Rockbursts of Apatit Company over the period from 1981 to 2015. The assessment was implemented for seven spans of 35, 30, 25, 20, 15, 10 and 5 years with regard to the change in the geotechnical situation by the ends of these periods.

Within the whole time period mention, round 70 geodynamic events were recorded, with different-degree destruction in mines and, sometimes, equipment damage and injuries. Geodynamic events without aftereffects were also recorded.

For the geodynamic risk assessment, 54 events were picked up in the data base as their records contained all information on all parameters involved in the analysis.

The probability of a geodynamic event and its aftereffects (if any) were calculated using the formulas presented below in this section.

The geodynamic event probability $P_{\text{total}} = N_{\text{total}}/T_{\text{total}}$, where N_{total}—number of recorded events; T_{total}—period of time within which these events were recorded.

It is possible to calculate a general probability (all events irrespective of class), or a probability per a class (for rock bursts only, or for induced earthquakes only). This paper presents data calculated based on all recorded events of different classes in the specified time spans.

The excavation failure probability $P_{bw} = (N_{bw} \cdot L_{bw})/L_w$, where N_{bw}—number of recorded events resulted in failure of underground excavations; L_{bw}—failure zone length; L_w—total length of excavations.

These calculations used information from logs of specific rock bursts recorded by the rockburst prediction and prevention services of mines. On the whole, these formulas can involve prehistory data (length of damages during spalling, extensive formation of loose rocks in walls and roofs in mines). In this case, the information is taken from field books of the rockburst prediction and prevention services of mines.

The equipment damage probability $P_{be} = N_{be} / (T \cdot N_e)$, where N_{be}—number of recorded events; T—time; N_e—number of machines in excavations.

Unfortunately, not every process damage is recorded (mainly, the most severe cases). For example, when a dynamic event damages equipment but the machine leaves the place in self-propelled mode, this case remains beyond records (here, social and economic factors come forward). In this respect, this parameter is calculated very approximately (at different times, the value can increase by 10 to 30%).

The injury probability $P_{bm} = N_{bm} / (T \cdot N_m)$, where N_{bm}—number of recorded events; T—recording period of time; N_m—number of personnel in excavations.

This parameter depends on the recording process quality, too. Only permanent injuries and fatal cases are entered on rockburst logs. Thus, the calculated values of this parameter do not show the truth picture and impair the statistical estimation results.

The no-aftereffect probability $P_w = N_w / T$, where N_w—number of recorded events; T—recording period of time. The data on this parameter are seldom recorded as are identified as tremors. As a rule, it is economically unreasonable to interrupt production processes in case when no damages are revealed. Thus, these statistical calculation results also can be insufficiently representative.

The calculation procedure assumed some constraints. For instance, the probabilities of geodynamic events and aftereffects were calculated for 1 day for year, the excavation damage probabilities—for 1 m^3 of excavation for year. It was also taken into account that within 86 years-long mining at apatite–nepheline deposits in Khibiny since 1929 till 2015 by Apatit Company, tunnel driving reached 600 km log (7000 m per year on average). The analysis used data on geodynamic events recorded in all mines of Apatit Company, in one mine and on a single level.

4. Results and discussion

We present the calculated results with respect to the mining scale for there variants: 1—for all mines of Apatit Company; 2—for Kirov Mine (KM) of Apatit and 3—for a level in KM.

Variant 1 for all mines: the calculations of the event probability and risk level took into account that: for 35 years tunnel drivage totaled 245000 running meters out of which in 20% of this length (average 1400 m/yr) damage was recorded, and 20000 machines and 4000 people were there within a year.

Table 1 gives the values of the calculated probabilities for geodynamic events of a certain class for different spans of times and their aftereffect probabilities for the period from 1981 to 1985 (5 years) and from 1981 to 1990 (10 years) and so on. The time span was selected in order to trace the change in the values of the calculated parameters as mining front was advanced.

It is clear from Table 1 that the calculated probabilities have values within 10^{-9}–10^{-7}, i.e. they are very low and their risk depends on the whole period of observations and on the spans of its constituents.

Variant 2 for Kirov Mine: the calculation conditions were 164000 running meters of tunnel drivage (7000 m per year on average) out of which 20% (average 1400 m) were damaged and accommodated 5000 machines and 1000 miners within a year. The calculated results are compiled in Table 2.

It follows from Table 2 that, similarly to Variant 1, the calculated probabilities and risk levels are very small and are within 10^{-10}–10^{-7}. As against Variant 1, the range expanded towards lower values, and the values of some parameters decrease by an order of magnitude.

Variant 3 for a level in a mine: the calculations assumed that drivage made 1000 running meters on the level over the period from 1981 till 2015, and 600 machines and 250 miners were yearly present in the excavations on this level.

In Variant 3 the calculated values are within 10^{-9}–10^{-7} per year. As compared with the data obtained in the first two variants, the third variant values of probabilities and risks lie in a narrower range – 10^{-9}–10^{-8} вper year. The no-aftereffect probability in this variant is higher by an order of magnitude. Nonetheless, the values of the other parameters remain at the level of negligible risk.

Table 1. General probability of negative geodynamic events and their aftereffects in Apatit Co. mines between 1981 and 2015.

Time span T_{total}, years	Number of events N_{total}	Geodynamic event probability P_{total}	Risk of event N_{bw} per 1 m^3 of excavation per year	Excavation damage probability P_{bw}	Equipment damage probability P_{be}	Risk of event N_{be} per year	Injury rate N_{bm}, men	Risk of event P_{bm} per year	No-aftereffect event N_w	Risk of event P_w per year
35	54	2.E-08	51	2.E-07	4	4.E-09	4	2.E-08	4	3.E-04
30	49	2.E-08	43	2.E-07	2	3.E-09	3	2.E-08	3	3.E-04
25	41	3.E-08	37	2.E-07	1	2.E-09	3	3.E-08	2	2.E-04
20	34	3.E-08	31	2.E-07	1	3.E-09	3	4.E-08	2	3.E-04
15	30	5.E-08	30	3.E-07	1	5.E-09	3	7.E-08	1	2.E-04
10	23	9.E-08	20	3.E-07	1	9.E-09	0	–	0	–
5	12	2.E-07	12	3.E-07	0	–	0	–	0	–

The comparison of the values obtained in all three variant shows that, on the whole, the level of risk of a geodynamic events and its aftereffects is low in apatite mines and lies within the range of negligible individual risk values—10^{-11}–10^{-7} per year. This means that, depending on the data analysis scale (for the whole mine, or for a level in a mine), the values of individual risk change but the risk level is yet low.

Table 2. General probability of negative geodynamic events and their aftereffects in Kirov Mine of Apatit Co. between 1981 and 2015.

Time span T_{total}, years	Number of events N_{total}	Geodynamic event probability P_{total}	Risk of event N_{bw} per 1 m^3 of excavation per year	Excavation damage probability P_{bw}	Equipment damage probability P_{be}	Risk of event N_{be} per year	Injury rate N_{bm}, men	Risk of event P_{bm} per year	No-aftereffect event N_w	Risk of event P_w per year
35	21	7.E-09	18	2.E-09	3	2.E-10	3	6.E-08	2	2.E-04
30	16	7.E-09	16	2.E-09	3	4.E-10	3	8.E-08	2	2.E-04
25	13	8.E-09	13	3.E-09	2	4.E-10	3	1.E-07	1	1.E-04
20	11	1.E-08	11	4.E-09	2	8.E-10	0	–	1	1.E-04
15	6	1.E-08	6	3.E-09	2	2.E-09	0	–	0	–
10	2	8.E-09	2	3.E-09	1	3.E-09	0	–	0	–
5	1	2.E-08	1	5.E-09	1	2.E-08	0	–	0	–

Table 3. General probability of negative geodynamic events and their aftereffects on a mining level in Kirov Mine of Apatit Co. between 1981 and 2015.

Time span T_{total}, years	Number of events N_{total}	Geodynamic event probability P_{total}	Risk of event N_{bw} per 1 m^3 of excavation per year	Excavation damage probability P_{bw}	Equipment damage probability P_{be}	Risk of event N_{be} per year	Injury rate N_{bm}, men	Risk of event P_{bm} per year	No-aftereffect event N_w	Risk of event P_w per year
35	16	4.E-08	14	1.E-08	5	5.E-09	2	4.E-09	1	8.E-05
30	12	4.E-08	12	1.E-08	4	6.E-09	2	7.E-09	1	9.E-05
25	11	5.E-08	10	1.E-08	3	7.E-09	2	1.E-08	1	1.E-04
20	9	6.E-08	8	2.E-08	2	9.E-09	1	1.E-08	0	–
15	6	7.E-08	6	2.E-08	1	1.E-08	1	2.E-08	0	–
10	2	5.E-08	2	2.E-08	1	3.E-08	0	–	0	–
5	1	1.E-07	1	4.E-08	0	–	0	–	0	–

5. Conclusion

The data reported in this paper show that the main impact of geodynamic events consists of the various degree damage caused to underground excavations while the other aftereffects total less than 10 percent. This means that the geodynamic risk can be identified as technological.

Thus, geodynamic risks are common in mining in high-stress hard rock mass. The calculations of separate risks over certain time period, based on the available statistics for the conditions of mine-technical systems in apatite mines show that the individual risk level under geodynamic event and its aftereffects is very low and lies within 10^{-11}–10^{-8} per year, i.e. inside the range of negligible quantities, which is an evidence of high efficiency of rockburst countermeasures. It has been found that the highest damage is caused to mine workings, i.e. to the technology process. For this reason, the issues of the geodynamic risk control should be the top-priority goal alongside with their prediction and estimation.

References

[1] Elokhin AN 2002 *Analysis and Risk Management: Theory and Practice* Moscow: PoliMedia (in Russian)

[2] Bykov AA and Porfiryev BN 2006 Rrisk analysis, conceptions and classifications *Problemy Analiza Riska* Vol 3 No 4 pp 319–337

[3] Fedotova YuV and Panin VI 2016 Assessment of level of geodynamic safety in apatite mines *GIAB* No 4 pp 344–353

[4] Kozyrev AA, Panin VI and Fedotova YuV 2012 Safety issues at mining operations in highly stressed rock massifs *Zapis. Gorn. Inst.* Vol 198 pp 150–156

Selection of criteria to estimate deformation of kimberlites in seismic-acoustic monitoring of Mir pipe

AS Fedyanin

ALROSA, Republic of Sakha (Yakutia), Russia
E-mail: BokiyIB@alrosa.ru

Abstract. In terms of the seismo-acoustic monitoring in Mir Mine, criteria are proposed to estimate deformation of the ore body. The relationship between the linear size of fractures and the intensity of radiated energy is revealed.

1. Introduction

Geotechnical estimation is an important part of mineral mining process. The estimation should include basic physical and dynamic events in rock mass and the accepted method of mining [1].

Mines of ALROSA Company use geotechnical control systems based on the:

—estimation of regional-scale geotechnical risks inside the area of a mineral deposit, considering water bodies, and local-scale risks within a mine field;

—routine monitoring of pre-accident signs towards prompt precaution and prevention;

—adjustment of control system with regard to variation in mining technology and mine-technical conditions;

—real-time visualization of monitoring data and storage in data bases of geotechnical service and supervision office;

—improvement of control procedures and refinement of hazard criteria for each specific deposit with regard to a mining method.

On the ground of the listed principles, the Company has developed a selection algorithm for optimized control, including research and development, for each deposit. For example, the integrated monitoring of diamond-bearing kimberlite mining in Mir Mine uses seismo-acoustic and surveying control, and measurement of displacements by deep-seated survey plugs.

In practical seismometry, coordinates of seismic events are determined by the first arrivals of P- and S-waves, while mechanism and energy of events are defined by the duration and spectrum of seismo-acoustic signals [2]. However, interpretation of seismic signals allows indirect estimation of processes running in rocks, by inversion of seismograms. A tectonic shear or an explosion can be described using force dipoles, whereas it is difficult to represent failure of interchamber pillars or deformation of backfill stoping belts by equivalent forces based on seismograms.

Similar problems arise in assessment of state of kimberlite ore featuring low strength characteristics and incapable to accumulate considerable stress for the generation of rock bursts.

In this connection, it is an important task of mine safety to determine a zoning criterion to divide rock mass into stable and failure-hazardous areas.

2. Key points of monitoring system selection for Mir kimberlite pipe

The framework of the integrated monitoring system for kimberlite mining includes:

—Mir kimberlite strength is from 45 to 60 MPa, peak seismic wave velocity reaches 1700 m/s;

—the optimal choice of sensor spacing in the network depends on the signal sampling rate with regard to the wave velocity;

—failure of kimberlite on the same level in a mine may be both brittle and elastoplastic. In view of geometry and block structure of the ore body and enclosing rock mass, it is inexpedient to record initial stages of fracturing in a seismic range to 800 Hz. The reliable location of natural seismic emission which points at failure requires that the recording frequency band is 0.5–12 kHz;

—an additional estimate of deformation process intensity is drift velocity of a cluster of seismic events, which is reflective of the area of the fracture growth in rock mass;

—it is advisable to confirm coordinates of a localized deformation zone by check measurements in local sites using a portable device (rock mass sounding). Susceptibility of rock mass to rockbursting can be estimated by acoustic emission (AE) only base on recording of brittle failure events with weak energies without artificial activation of acoustic processes;

—number of AE impulses recorded within a certain time interval, density of spatial distribution of AE sources, as well as energy and amplitude of AE signal are the basic source data for the determination of a critical isoline and estimation of deformation processes.

3. Selection of monitoring system

Using the above methodological framework, it has been decided to carry out monitoring in Mir Mine based using methods of seismoacoustic and Prognoz-ADS system that detects hazardous events due to rock pressure in underground mineral mining [3].

The system includes spatially distributed piezo-electric g-meters–geophones to detect and measure impulses of natural AE. The data on seismoacoustic events (SAE) are accumulated in data base for the further analysis and interpretation. The built-in software enables detection and recording of acoustic events, and determines parameters of acoustically active zones, which allows assessment of geomechanical behavior of rocks. The measurement data are used in 2D or 3D mapping of the acoustic activity of rock mass.

BY analyzing regular patterns of distribution of AE sources, areas with high energy and density of AE are delineated and rated in terms of potential hazard [4].

The basic signs of an AE source are: concentration and localization of AE events; frequency and persistency of acoustic activity periods, velocity, volume, total energy and migration direction of acoustically active zones, distance between a source and exposed rocks, etc. These characteristics are experimentally determined at each specific deposit.

For the conditions of Mir kimberlite pipe, the optimal parameters of the network of seismoacoustic sensors, their number and frequency bands for reliable recording of events which point at failure in kimberlite have been found from in situ experiments. The resultant seismoacoustic control system covers the zone of a crown pillar below the open pit bottom with an area of 65 thou m^2. With the spacing of 120 m, the error of SAE coordinates is not more than 5 m even in the complex structure rock mass.

4. Seismoacoustic event criteria

The seismoacoustic (geoacoustic) method has a high resolution capability as it allows recording of weak elastic impulses (with an energy from 1 J and above) emitted at the early stages of deformation in rocks. In this case, it becomes possible to monitor the process of initiation of failure zones starting from the beginning, which essentially improves reliability of prediction of hazardous events due to rock pressure and offers some time for precaution decision-making and implementation.

The analytical research involves a simplified algorithm of assessment of acoustic events by the impulse recording time, and by signal duration and amplitude; later on, these data are used to determine energy and coordinates of AE source as well as basic characteristics of signals to distinguish between the natural and induced AE events.

As a result, new data on interaction between deformation processes in roc mass subjected to critical stresses and variation in acoustic field parameters have been obtained.

Earlier (since 2010) monitoring of the kimberltite pipe was based on the measured displacements of plugs installed in the roof and floor of underground excavations. The comparison of the subsidence measurements and the acoustic activity characteristics over the period from April to September 2016 proves their good agreement.

A complementary control technique was chosen surveying by plugs in roof rocks and estimation of convergence in underground excavations.

The analysis of the composite map of subsidence and acoustic emission in 2016 shows that the boundaries of different damage zones agree well with the subsidence in the center of the crown pillar.

For the visualization of the process of deformation development, the crown pillar along its height (interval of actual elevations –190/–230 m) was divided into equal vertical cells with an areal size 20×20 m. Then, inside these cells, energies of all AE events recorded in 1–20 and 20 June–10 July were summed up. Plotting of the isolines of the total energy per unit cell volume allowed tracing intensive failure at ore and rock mass interface over the period from June to July 2016. This situation was expectable and connected with the accepted mining technology, difference in strength characteristics of ore body and rocks and high dislocation and damage of the ore body. Large-block undamaged rock mass is in the stress concentration zone adjoining the ore and rock mass interface.

It is worthy of mentioning that that approach provided a qualitative picture of deformation even in the regions beyond the coverage of the earlier net of plugs 'tied up' with underground excavations.

Aiming to determine safety criterion, it was decided to estimate interrelation between AE energy and sizes of induced joints. The subvertical monotonous drift of AE signal coordinates in the crown pillars was considered formation of failure plane or fracturing plane.

The revealed regularity in the change of actual elevations of AE coordinates connected with the level of emitted energy points at the initiation of potentially hazardous failure processes in kimberlite rock mass.

From the analysis of characteristics of AE signals recorded in the crown pillar under the open pit mine bottom in 2016, the linear fracture size *Lm* has been related with the AE energy intensity *E*(*J*) in kimberlite rock mass. This relation at the approximation factor 0.92 can be represented by the exponential and polynomial functions:

$$E(J) = 654.66 * e^{0.351 * Lm},$$

$$E(J) = 159.82 * Lm^2 - 450.09 * Lm + 1358.2.$$

In this manner, on the ground of the revealed regularities, it is possible to detect signals of potentially hazardous events and to model failure planes in rock masses. Localization of fracturing zones is based on the direct monitoring carried out by geotechnical service in mines using the methods of visual control by portable seismoacoustic devices, which enables more accurate delineation of hazardous zones and estimation of their effect on mining safety.

5. Conclusion

1. For the conditions of Mir kimberlite pipe, for the first time, the criteria and procedure have been developed for the detection of hazardous deformation in rocks using the geoacoustic method in the event of insufficient data on failure (brittle or elastoplastic).

2. Regarding the safety criterion, the interrelation has been revealed between the linear size of growing fractures and the intensity of emission energy in kimberlite rock mass.

3. Using the proposed package of predictive indicators of critical state by parameters and behavior of seismoacoustic emission in rocks enables the early-stage zoning of rock mass by the degree of damage and stress state estimation as well as later detection of failure stages with differentiation between natural and induced signals.

References

[1] Bagdasaryan AG, Luishov BG, Rodionov VN and Fedyanin AS 2008 Detection of feature of a rupture structure in walls of an open pit in terms of the Muruntau open pit *J. Min. Sci.* Vol 44 No 1 pp 73–81

[2] Rasskazov IYu 2008 *Ground Control in Mines in the Russian Far East Region* Moscow: Gornaya Kniga (in Russian)
[3] Gilyarov VL, Damaskinskaya EE, Kadomtsev AG and Rasskazov IYu 2014 Analysis of statistical parameters of geoacoustic monitoring data fort the Antey uranium deposit *J. Min. Sci.* Vol 50 No 3 pp 443–447
[4] Kuksenko VS 2009 Changes in structure of natural heterogenous materials under deformation *Gorny Zh.* Vol 45 no 4 pp 355–358

IOP Conf. Series: Earth and Environmental Science **134** (2018) 012019 doi:10.1088/1755-1315/134/1/012019

Validation of choice and determination of geotechnology parameters with regard to stress–strain state of rocks

AM Freidin, SA Neverov*, AA Neverov and AI Konurin

Chinakal Institute of Mining, Siberian Branch, Russian Academy of Sciences, Novosibirsk, Russia

E-mail: *nsa_nsk@mail.ru

Abstract. The paper illustrates efficiency and reliability of types of rock mass stress state conditioned by geological and structural features of rocks in design, selection and validation of geotechnology parameters. The authors of the paper present calculation of stresses in rock mass under sublevel stoping depending on the type of the geosphere and on the depth of the ore body occurrence.

One of the possible ways to deal with the problems connected with great depth is the conceptualization of rock mass by the type of stress state versus the geological structure.

The identified types of geological–tectonic structures are related with the stress state of rock mass (see the table below). The joint analysis of these two factors makes in possible to substantiate boundary conditions to solve practical problems of geomechanics and to select mining technologies in this bases.

Aiming to test methodology of geomechanical modeling (with regard to stress state conditioned by geological–tectonic structure) [1, 2], mining systems are divided into classes subject to model representation of state of rocks at different depths (table).

The review of geotechnologies practiced in more than 90 mines of the world has allowed some conclusions to be drawn.

In relatively simple geological and geotechnical conditions, all known systems of mining are applicable [3].

The main outcome of the implemented generalization is the depth-related overlap of different mining system with the formation of cross-effect areas (table). This cross-effect areas point at the capability of safe operation of alternative geotechnologies and combination of different ground control method, which expands the range of application of inexpensive systems of mining at deep levels. In this case, the choice of a mining system is governed equally by the type of stress state of rock mass, and regional and local features of a deposit, for instance, geomorphology, seismic activity, deep faulting etc.

The proposed geomechanical model of geomedia improve estimation of geotechnical situation in mines and favor more correct mine design and planning.

Validation of key solutions on deeper level mineral mining, as well as their reliability and certainty reduces to finding rational geotechnologies in the sphere of influence of geomechanical models of geomedia represented by different tectonic types of rock mass and their comparison with the practical examples of deep ore body mining.

IOP Conf. Series: Earth and Environmental Science **134** (2018) 012019 doi:10.1088/1755-1315/134/1/012019

Classes of mining system per stress state of rock mass and mining depth

Stress state, correlation coefficient C_c	Geological–tectonic structure	Stress state versus depth	Depth-related mining system
1. Geodynamic $300 \le H \le 1300$ $1.6 \le \lambda \le 4.0$ $C_c > 0.87$	Stable basal complexes and shelves of platforms. Mobile and seismically active folded mountainous systems	$\sigma_{H\max} = \alpha\lambda^{\kappa}\ln(\gamma H) - \delta \approx \sigma_1$ $\sigma_{h\min} = \frac{(\sigma_{H\max} + \sigma_v)}{2} \approx \sigma_2$ $\sigma_v \approx \gamma H \approx \sigma_3$	to 0.5 km—caving, sublevel drifting below 0.5 km—backfilling
2. Tectonic $300 \le H \le 2000$ $1.2 \le \lambda \le 2.1$ $C_c > 0.88$	Mobile and seismically active folded mountainous systems, geosynclinals folded mobile belts and overthrust folding and faulting	$\sigma_{H\max} = 2.8e^{\kappa\lambda}\gamma H^{0.7} \approx \sigma_1$ или $\sigma_{H\max} \approx \psi\lambda^{\kappa}\gamma H + \theta \approx \sigma_1$ $\sigma_{h\min} = (0.6 \text{ ч } 0.85)\sigma_{H\max} \approx \sigma_2$ $\sigma_v \approx \gamma H \approx \sigma_3$	to 0.5 km—with natural support from 0.5 to 1.0 km—ore caving, sublevel drifts below 1.0 km—cut-and-fill and combined methods
3. Geostatic $200 \le H \le 5000$ $0.8 \le \lambda \le 1.3$ $C_c > 0.90$	Young mobile platforms, fault troughs, island arcs, overthrust folding and faulting nontectonic faulting	$\sigma_{H\max} = \lambda\gamma H \approx \sigma_1 \approx \sigma_2$ $\sigma_{h\min} = (0.8 \text{ ч } 1.0)\sigma_{H\max} \approx \sigma_3$ $\sigma_v \approx \gamma H \approx \sigma_1 \approx \sigma_2$ $\sigma_1 \approx \sigma_2 \approx \sigma_3 \approx \gamma H$	to 0.8 km—with natural support from 0.81 to 1.5 km—ore caving, sublevel drifts below 1.2 km—cut-and-fill methods under 2 km—physico-technical and physicochemical methods
4. Gravity $100 \le H \le 5000$ $\lambda = \frac{\nu}{1-\nu}$, $C_c > 0.92$	Sedimentary covers, mobile shelves and platforms. Caledonian structures in the form of overlapped folds and faults	$\sigma_{H\max} \approx \lambda\gamma H \approx \sigma_2$ $\sigma_{h\min} \approx \lambda\gamma H \approx \sigma_3$ $\sigma_v \approx \gamma H \approx \sigma_1$	to 0.8 km—with natural support and caving from 0.81 to 1.5 km—with caving below 1.2 km—with backfill and combined methods under 20 km—physicochemical methods

*γ—bulk weight of overlying rocks, MN/m^3; H—occurrence depth, m; λ—lateral earth pressure coefficient; ν—Poisson's ratio; κ—empirical coefficient: for very hard rocks $\kappa = 0.12$–0.30, for medium hard and fair rocks $\kappa = 0.08$–0.12; α, δ, ψ and θ—empirical coefficient of proportionality: $\alpha \approx 32$–37, $\delta \approx 65$–80, $\psi \approx 0.8$–0.9, $\theta \approx 5$–10.

With a view to validating reliable determination of geotechnology design based on stress state conditioned by a type of geological–tectonic structure of rock mass, case studies have been carried out to find application area and limit depth for combination of room-and-pillar and caving method in two variants (Figures 1 and 2) and sublevel caving technologies (Figure 3).

For the corrected comparison of the calculated results and actual performance of real mine, the international experience of gently dipping ore body mining using room-and-pillar method has been generalized as this method is close to the discussed geotechnologies [4] and sublevel caving systems.

At the present time, mining depth has reached 1000 m below the ground surface and deeper, and the methods of room-and-pillar and caving technologies yet remain to be studied more comprehensively.

The international mining practice has many examples of room-and-pillar operation in unfavorable mine-technical conditions as well as geotechnologies with caving at the depths below 700–100 m. Mines Konrad (Germany), Denison and New-Quirk (Canada) first extracted gently dipping ore bodies to 12 m thick using the room-and-pillar method [4].

Reliability of parameters and application fields of the mining systems with regard to stress state of rock mass was verified by comparing the research findings with the actual data of mines operating under different geomechanical conditions and at various depths.

The check calculations have been performed to assess stability of structural elements in room-and-pillar method in terms of Konrad mine (Germany) for the depth of 1000 m and in sublevel caving under different geological conditions of an iron ore body.

Figures 1 and 2 show areas of probable damage of rocks in the structural elements of a combined mining system at the structural weakening coefficient C_c 0.5 and 0.8, while Figure 3 depicts tension zones in the structural elements of sublevel caving technology.

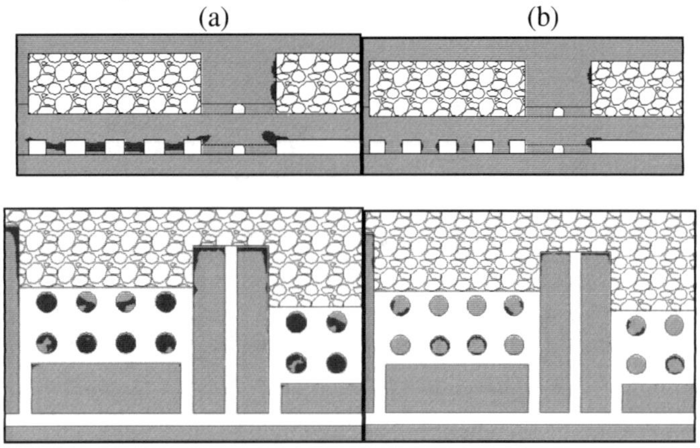

Figure 1. Prediction of rock damage zones in combination of room-and-pillar and sublevel caving technologies: (a) $C_c = 0.5$; (b) $C_c = 0.8$; ●—probable damage zones.

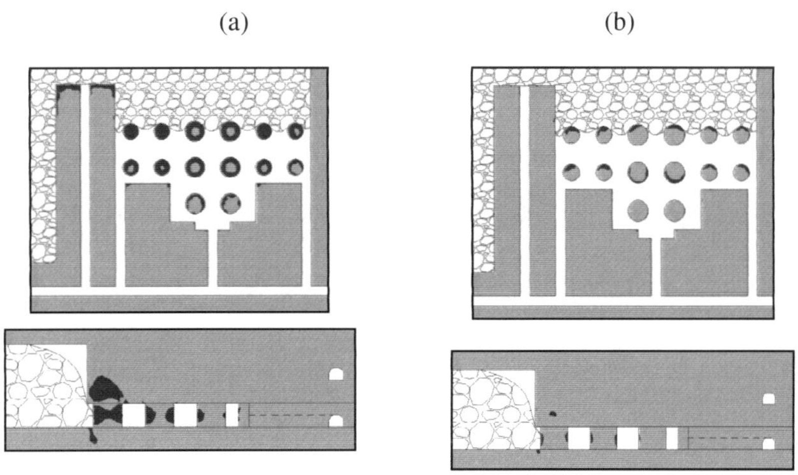

Figure 2. Prediction of post-limit deformation zones in room-and-pillar mining with periodic extraction of ore reserves from pillar and roof caving: (a) $C_c = 0.5$; (b) $C_c = 0.8$.

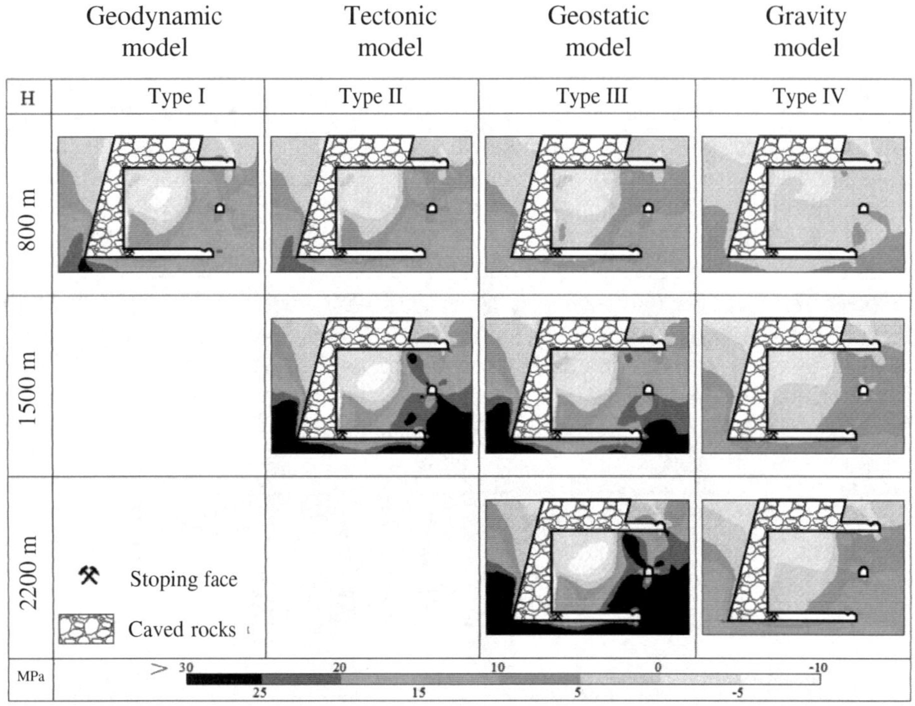

Figure 3. Distribution of minor principal stresses (tensile stress zones).

The outcome of the studies [5] is given in brief below.

The discussed combination geotechnologies with room-and-pillar and caving in the conditions of the gravity model of a geomedium have the limit application depth not deeper than 1000 m in good and fair quality rock mass. Below the depth of 1000 m, these mining systems are only applicable after obligatory relaxation of rock mass from high stresses.

The comparative estimation undertaken shows good consistency between the calculated and actual data (in terms of Konrad mine in Germany). According to the available information, the mine transited from the classical room-and-pillar to the combination of this method with caving and performed safely at the depth of 1000 m.

The prediction estimate of application depths for the sublevel caving method satisfactorily agrees with the actual data from operating mine, namely:

—mines Artem and Irtysh, Easter Kazakhstan—geostatic stress state, $H = 800$ m (no observable changes in rocks);

—mines Creighton, Sudbury, Canada—tectonic and geostatic stress states, $H = 1300–1800$ m (mining in stable and undamaged areas);

—mine Lucky Friday, USA—gravity stress state, $H = 1710$ m (local application);

—mine Kiruna, Sweden—tectonic stress state, $H = 1300$ m (applied without limitation and observable changes in rock mass);

—Mine Nikolaevsky, Vostok-2 deposit, Russian Far East—tectonic stress state, $H = 900$ m (local application);

—Lenin mine, Krivbass, Russia—gravity stress state, H below 1000 m (satisfactory correlation with the undertaken research findings in terms of combination of systems with caving and backfilling).

Conclusion

Selection and validation of mining systems for deep ore bodies should be based on the prediction of type and value of rock mass stresses at the stage of mine design and planning. The analysis of geological conditions performed by the authors on the database on 90 ore deposits in Russia and in the world has shown that the existing geological structure of ore deposits correlates with the definite type of stress state of enclosing rock mass with regard to the ore occurrence depth.

References

[1] Neverov SA 2012 Types of orebodies on the basis of the occurrence depth and stress state. Part I: Modern concept of the stress state versus depth *J. Min. Sci.* Vol 48 No 2 pp 249–259

[2] Neverov SA 2012 Types of orebodies on the basis of the occurrence depth and stress state. Part II: Orebody tectonotypes and geomedium models *J. Min. Sci.* Vol 48 No 3 pp 421–428

[3] Tapsiev AP, Freidin AM, Uskov VA, Filippov PA, Neverov AA and Neverov SA 2014 Resource-saving geotechnolgies for thick gently dipping complex ore deposits in the Norilsk Region *J. Min. Sci.* Vol 50 No 5 pp 904–913

[4] Freidin AM, Neverov SA et al 2016 Geomechanical assessment of geotechnology at a project stage of underground ore mining *Gorny Zh.* DOI: 10.17580/gzh.2016.02.08

[5] Neverov AA, Vasichev SYu and Neverov SA 2013 Experience of gently dipping orebody mining and basic trends in development of the promising mining systems *Proc. All-Russian Conf. Miners' Offspring–2013* Novosibirsk: IGD SO RAN (in Russian)

Modeling stress–strain state of rock mass under mining of complex-shape extraction pillar

VN Fryanov* and LD Pavlova**

Siberian State Industrial University, Novokuznetsk, Russia

E-mail: *zzz338@rdtc.ru, **ld_pavlova@mail.ru

Abstract. Based on the results of numerical modeling of stresses and strains in rock mass, geomechanical parameters of development workings adjacent to coal face operation area are provided for multi-entry preparation and extraction of flat seams with production faces of variable length. The negative effects on the geomechanical situation during the transition from the longwall to shortwall mining in a fully mechanized extraction face are found.

1. Introduction

Efficient fully mechanized mining is possible in gently dipping coal mine fields 3–4 k in size along the dip and strike. Extraction panels should be single-wing and rectangular. In practice, in the presence f synclinal and anticlinal folds, disjunctive dislocations or different purpose coal pillars, a mine project can assumed complex-shaped extraction panels. Figure 1 depicts a variant of preparation and extraction in a panel with disjunctive dislocations revealed during driving of ventilation drift 01. When a fully mechanized mining face is advanced from the flank shafts to the central shafts, longwall will have varied length.

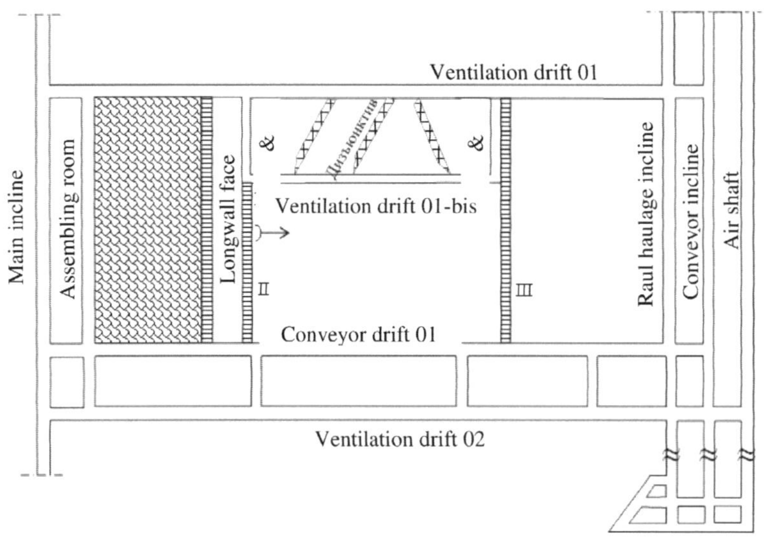

Figure 1. Operation in complex-shaped extraction panel in the influence zone of disjunctive dislocations: I, II, III—locations of working face.

Under such conditions, it becomes necessary to support systems of parallel longwalls at a mining depth of 200–800 m by coal pillars with a width of 20–60 m. The width of pillars is chosen based on specific geological and geotechnical conditions, including the length of longwall. In the transition from shortwall to longwall fully mechanized mining and vice versa, it is required to determine parameters of stable coal pillars and support designs for entry ways. Justification of the width of stable coal pillars and reliable support design for roadways outlining an extraction panel, it is required to estimate stresses in coal and enclosing rock mass.

During fully mechanized longwalling, it is required to shorten the fully mechanized mining system by the length of cross split 20–21 by disassembling some sections of powered support. After extracting coal in a shortwall between cross splits 1 and 2, the length of the fully mechanized system is increased again, and coal development past cross split 2 is continued by a longwall [1].

The geomechanical assessment of safe mining in case of a complex shape extraction panel in Figure 1 involves a few engineering problems to be solved:

—validation of an algorithm to select support design and ensure stability of cross splits in the abatement pressure zones as the fully mechanized longwall system approaches them;

—determination of input data to calculate geometry for a safe entry of the fully mechanized mining system in cross split 1;

—determination of input data for the support design in ventilation drift 01-bis, conveyor drift 01 and ventilation drift 02.

The available practical guides ensure incomplete solution of the set problems.

So, the necessity to model stress state in coal-and-rock mass in case of complex shape of longwalls and to design support for roadway outlining mined-out area in case of bench mining is conditioned by:

—formation of stress raisors at the corner edges of the coal bed when the width of the coal pillar between the work face and cross split 1 is shortened and dynamic events due to rock pressure are possible;

—unavailability of quantitative relations for characteristics of abatement pressure in rock mass around roadways in the varied length sections of longwalls;

—difficulty of validating parameters of entry of the fully mechanized mining system to a disassembling room and ensuring its stability during removal of powered support units.

2. Model and methods

Geometrical model of extraction panel includes such characteristics of work face position (Figure 1) as:

—position I—before entry in the dissembling room (cross split 1) for the transition from longwall to shortwall fully mechanized mining;

—position II—shortwall mining;

—position III—after addition of powered support units in cross split 2 and transition to longwall mining.

The process objects where it is necessary to assess vertical stresses is the pillar between the fully mechanized longwall face and cross split 1, cross split 1, as well as ventilation and conveyor drifts (Figure 1).

The mathematical model of coal-and-rock mass stress state is based in the constitutive relations of the theory of deformation solid in the form of a boundary value problem [2, 3] and the finite element method [4, 5].

Aimed to reveal mechanisms of vertical stress distribution in coal bed and roof rocks in a complex-shaped longwall, the computational experimentation has been carried out using the authorial package of problem-oriented programs [6, 7].

The source data of modeling were: extraction panel 800 m long, work face length 300 m in the longwall and 200 m in the shortwall, development entries 5 m wide, mining depth of 600 m, coal bed thickness 2 m, immediate roof 7 m thick, main roof thickness 30 m, coal strength 10 MPa, roof rock strength 30–60 MPa.

3. Results and discussion

In the modeling, the width of the pillar between the fully mechanized face and cross split 1 was shortened from 50 to 0 m at step of 10 m. Figure 2 shows the stress isolines in the longwall when the fully mechanized face is spaced at 10 m from cross split 1.

The analysis of the distribution of stresses, displacements and coal disintegration factor shows that:

—with the narrower pillar between the fully mechanized mining system and the cross split, the most hazardous zone is the juncture of ventilation drift 01-bis and cross split 1; the vertical stress concentration factor is 4.6 in this zone, and the vertical stresses are much higher than the coal compression strength;

—the maximum of the abatement pressure epure is at the edge of the coal bed when the coal pillar is 50 m wide and drifts to the center as the width of the pillar is reduced, which initiates concentration of energy that can release in the form of a rock burst;

—the presence of the disjunctive dislocation with high methane content between cross splits 1 and 2 an outburst of coal, rocks and gas is possible in the juncture zone of ventilation drift 01-bis and cross split 1;

—the diagrams of disintegration coefficient in coal and roof rocks show that when the coal pillar is 5 m wide, coal fails in the pillar and in the sides of cross split 1 while the roof of the cross split disintegrates into blocks under the action of the horizontal and shearing stresses.

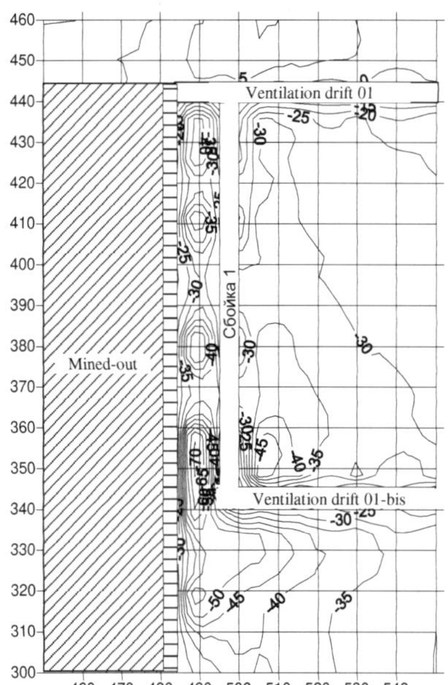

Figure 2. Distribution of vertical stresses on approach of longwall face to cross split 1.

Based on the determined distribution of stresses in coal and rock mass and using the estimated probability of gas-dynamic events, ejection of coal from sidewalls and roof rock falls, the developed engineering solutions recommend making a disassembling room instead of cross split 1 as the fully mechanized mining system is advanced.

After transition the fully mechanized system from longwall to shortwall mining, the extraction situation has normalized (Figure 3).

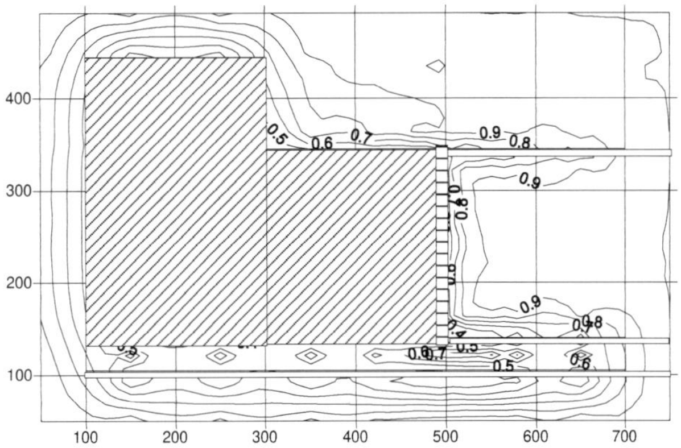

Figure 3. Distribution of coal disintegration coefficient in shortwall 200 m long.

The concentration coefficient of vertical stresses at the edge of the coal bed in the middle of the shortwall has halved as compared to the same coefficient in the longwall and made 2.2. The negative effect exerted by the bench at the juncture of cross split 1 and ventilation drift 01-bis on stability of the latter was felt as the fully mechanized mining system moved away from the bench to the extent of 0–0.2*H*, where *H*—mining depth. In the shortwall the residual/initial strength ratio of coal became 0.8–1.0, which means reduction in probability of sloughing of edge coal.

The process of transition of the fully mechanized mining system from the longwall to the shortwall brought no essential complications to the geomechanical situation (Figure 4). The units of the powered support are assembled in cross split 2 and ensure stability of roof rocks is the assembling room. At the first stage of longwall mining, overhanging rocks of the bench at the juncture of driven ventilation drift 01-bis and cross split 1 induce local increase in the vertical stresses in the edge coal; however, as the face is advanced and the roof rocks are caved, the stresses become uniform longwise the face. Further longwall mining causes no geomechanical complications (Figure 4).

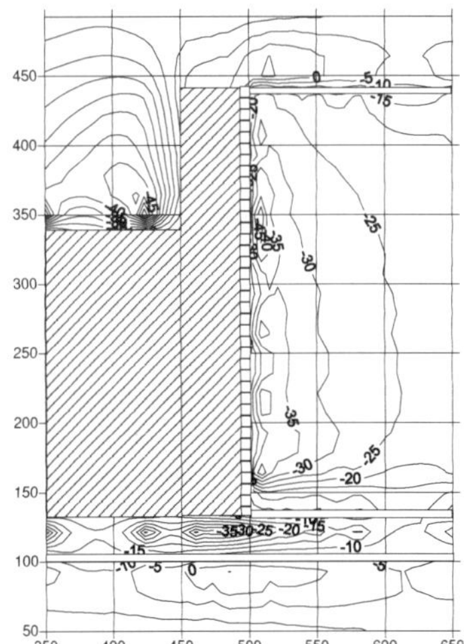

Figure 4. Distribution of vertical stresses in transition from shortwall to longwall mining.

Based on the research findings, the concentration coefficient of vertical stresses in coal pillar between conveyor drift 01 and ventilation drift 02 is given by:

$$k = 0.25 + 4.00\frac{l}{H},$$

where k—concentration coefficient of vertical stresses; l—length of longwall, m; H—depth of mining, m.

This formula is recommended for determining an estimated depth of mining, normal displacements of roof and sidewall rocks in roadways and a width of pillars in conformity with the guidelines provided by VNIMI.

4. Conclusion

The authors have found that the transition of the fully mechanized system face from longwall to shortwall mining can affect geomechanical situation in an extraction panel in the form of: increased stress zone formed at junctures of rooms when the width of the coal pillar between the working face and the disassembling room is reduced; probability of gas dynamic events in the coal pillar and at the coal bench where energy raisor is generated; failure of roof rocks and sidewalls in the disassembling room when the coal pillar width is under 5 m.

It is recommended to transit between longwall and shortwall mining within an extraction panel by means of cutting a disassembling room, roof rock support and sequential disassembling of mining equipment.

Acknowledgements

The studies have been supported by the Ministry of Education and Science of the Russian Federation in the framework of the Federal Targeted Program on R&D in Priority Areas of Advancement in the Science and Technology of Russia for 2014–2020, Topic: Efficient Robotic Longwall Top Coal Caving Technology, Unique Identifier RFMEF160417X0173.

References

[1] Anufriev VE, Kalinin SI et al 2006 *Geomechanical State of Adjacent Rock Mass of a Disassembly Chamber* Kemerovo: InstUglaUglekhim SB RAS RANK (in Russian)

[2] Tsvetkov AB and Pavlova LD 2013 Mathematical modeling of the stress-strain state of geomass under integrated action of gravitational and geotectonic forces *Nauch.-Tekh. Vestn. Povolzhie*, No 6 pp 485–488

[3] Tsvetkov AB, Pavlova LD and FryanovVN 2016 Construction of the approximant of complete diagram for rock deformation *IOP Conference Series: Earth and Environmental Science* Vol 45 (012009) (Available at: http://iopscience.iop.org/issue/1755-1315/45/1)

[4] Kornev ES, Pavlova LD and Fryanov VN 2014 *Numerical Modeling of Geomechanical Processes during Shortwall Mining of Coal Seams* Prague: Vědeckovydavatelskécentrum Sociosféra-CZ

[5] Kornev ES, Pavlova LD and Fryanov VN 2015 Numerical modeling of geomechanical state of a geomass in the bilateral extraction of short coal pillars *Vestn. KuzGTU* No 2 (108) pp 20–26

[6] Kornev ES, Pavlova LD and Fryanov VN 2013 Development of a package of problem-oriented programs for modeling geomechanical processes by the finite element method *Vestn. KuzGTU* No 2 pp 65–69

[7] Kornev ES and Pavlova LD 2015 Application of software for finite element analysis for the solution of three-dimensional problems of mining geomechanics *Proc. Int. Conf.: High Technologies for Mineral Mining and Use* Novokuznetsk pp 126–129 (in Russian)

IOP Conf. Series: Earth and Environmental Science **134** (2018) 012021 doi:10.1088/1755-1315/134/1/012021

Temperature-induced stresses in reinforced concrete structures

LN Gakhova[1,2]

[1]Chinakal Institute of Mining, Siberian Branch, Russian Academy of Sciences, Novosibirsk, Russia
[2]Sayano-Shushensk Division, Northern (Arctic) Federal University, Russia

E-mail: gahoval@mail.ru

Abstract. The author estimates stress state in steel-and-concrete power conduit arranged at the concrete downstream face with regard to temperature and combination of loads. It is shown that the maximal tensile stresses are formed under minimum temperature in winter at the conduit points remote from the downstream face.

In case of the heat exchange in rock mass under conditions of high-amplitude seasonal variations in air temperatures, in deep level mines in the North, during operation of large concrete structures and in other similar cases, it is necessary to take into account temperature effects on stress–strain state of geomaterials.

Regarding waterworks, the most critical structures of their design are steel-and-concrete pressure conduits and spiral chambers of hydropower units. Reliable and safe operation of these critical structures is one of the top-priority objectives both at the stage of design and during service period. An obligatory element of serviceability and safety check in high-pressure waterworks and their structural elements is the stress–strain state modeling for the whole structure and its most critical elements.

This paper reports the 3D numerical modeling of stress state of the steel-and-concrete power conduit of Sayano-Shushenskaya hydropower plant subjected to thermal forces. The computation used bundled software for solving problems on thermal elasticity of block structured rock masses [1]. In the stress state calculation for a steel-and-concrete power conduit, the blocks are assumed the conduit layers.

A steel-and-concrete conduit of circular section was modeled by a multi-layer ring (Figure 1) composed of an inner steel shell structure and an external reinforced concrete shell structure with two layers of reinforcing steel (equivalent to two rows of hoop reinforcement). The thickness of each of the reinforcing rings was accepted based on the condition that cross section area of a ring was equal to cross section area of rebars in the corresponding reinforcement row.

The consequence of thermal effect on a block structured rock mass is the redistribution of stresses inside and in-between the blocks. The dynamics of temperature field in a block is taken into account with the help of the equation of transient heat conduction. The system of differential equations of the unbound quasistatic thermoelastic problem without the sources of heat and bulk forces is given by [1]:

$$\mu^{[n]}\Delta u^{[n]} + (\lambda^{[n]} + \mu^{[n]})\,\mathrm{grad\,div}\,u^{[n]} = \gamma^{[n]}\,\mathrm{grad}\,\theta^{[n]}, \tag{1}$$

$$\Delta\theta^{[n]} = \beta \frac{\partial \theta^{[n]}}{\partial t}, \tag{2}$$

where $u^{[n]}(X, t)$—vector of displacements at point X (x, y) at time t; $\theta^{[n]}$ $(X, t) = T(X, t) - T_0$—temperature counted from the initial state; $\lambda^{[n]}$, $\mu^{[n]}$—Lamé constants; $\gamma^{[n]} = \alpha^{[n]}(3\lambda^{[n]} + 2\mu^{[n]})$; $\alpha^{[n]}$—coefficient of linear thermal expansion; Δ—Laplace operator; $\beta = 1/æ^{[n]}$, where $æ^{[n]}$—thermal diffusivity.

(a) (b)

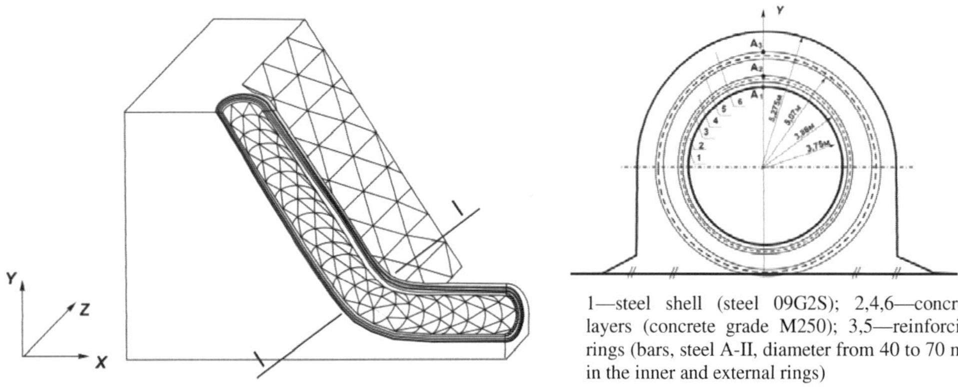

1—steel shell (steel 09G2S); 2,4,6—concrete layers (concrete grade M250); 3,5—reinforcing rings (bars, steel A-II, diameter from 40 to 70 mm in the inner and external rings)

Figure 1. Computational scheme: *a*—bottom of the conduit; *b*—cross section I–I.

It is assumed that the temperature difference is such that the properties of rocks remain unaltered and, consequently, the stresses and strains are related by the Duhamel–Najman relation [1].

The temperature field is determined regardless of the field of displacement by solving Eq. (2) with the boundary and initial conditions:

$$\theta^{[n]}(X,0) = 0, \quad X \in G^{[n]} U \Gamma^{[n]}, \tag{3}$$

$$\frac{\partial \theta^{[n]}}{\partial n(Y)} + h^{[n]}(Y,t)\,\theta^{[n]}(Y,t) = h^{[n]}(Y,t)\,\overline{\theta}^{[n]}(Y,t), \quad Y \in \Gamma^{[n]}, t > 0, \tag{4}$$

$$p_i(Y,t) = f_i(Y,t), \quad Y \in \Gamma^{[n]}, t > 0, \quad i = 1, 2, \tag{5}$$

$h^{[n]}$ (Y, t)—heat exchange coefficient; $\overline{\theta}^{[n]}$ (Y, t)—ambient temperature or the neighbor block temperature; $\partial\theta^{[n]} / \partial n(Y)$—derivative with respect to the outside normal to a block. The obtained value of the temperature is used to solve Eq. (1) where the time t is considered as a parameter.

The unbound quasistatic thermoelastic problem is reduced to sequential solution of the system of integral equations for finding the densities φi, $i = 1, 2, 3$. The values of the stresses and strains are determined by integrating over the boundary (for the densities $\varphi 1$, $\varphi 2$ obtained from the system of integral equations of isothermic elasticity) or over the boundary and time (for $\varphi 3$ from the integral equation of thermal potential of simple fiber [1, 2]).

At the interfaces of adjacent media (solids, e.g. neighbor blocks $G^{[n]}$ and $G^{[m]}$)m the conjugation conditions are set for heat flows and temperatures [4]: $q_{S^+_{n,m}} = q_{S^-_{n,m}}$; $t_{S^+_{n,m}} = t_{S^-_{n,m}}$, where $q_{S^+_{n,m}}$, $t_{S^+_{n,m}}$—respectively, heat flow and temperature of a boundary element $S^+_{n,m}$ at its contact with the block $G^{[n]}$, and

$q_{S^-_{n,m}}$ and $t_{S^-_{n,m}}$ — heat flow and temperature of a boundary element $S^-_{n,m}$ at its contact with the neighbor block $G^{[m]}$.

The conditions of conjugation of heat flows at the interface of the immobile medium (solid) and a fluid (water, air) are set in terms of the coefficients of heat transfer α_1 W/(m^2·deg) and heat exchange $h_1 = \alpha_1/k$ (1/m), where k—heat conduction coefficient, W/(m·deg).

The problem is implemented using the method of regularization by Tikhonov. For practical problems taking much time to be solved, algorithms are based on the concept of parallel computing [1, 2].

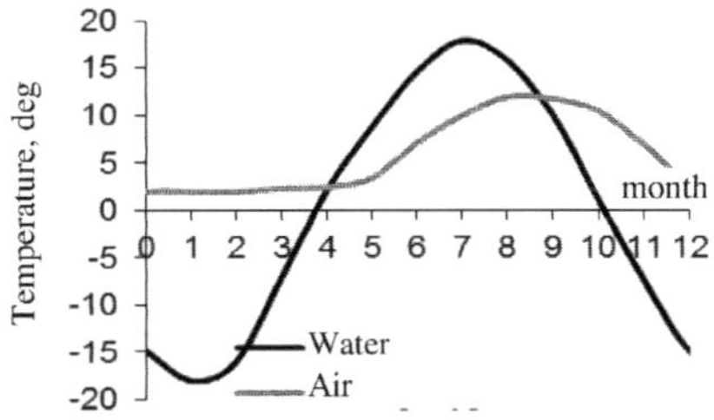

Figure 2. Annual cycle of water temperature at the water intake and air temperature in the area of Sayano-Shushenskaya HPP.

The stresses and strains were analyzed in structural elements in the middle of the bottom section of the steel-and-concrete conduit (Figure 1) under seasonal variation in the water temperature at the water intake and in the air temperature (Figure 2). The steel-and-concrete conduit of circular section was modeled by a multi-layer ring composed of an inner steel shell structure and an external reinforced concrete shell structure with two layers of reinforcing steel (equivalent to two rows of hoop reinforcement). The thickness of each of the reinforcing rings was accepted based on the condition that cross section area of a ring was equal to cross section area of rebars in the corresponding reinforcement row.

It is assumed that steel, concrete and material the reinforcing rings are made of are the linearly deformable materials. Table 1 presents the source data for the stress calculation of the steel-and-concrete power conduit. Given the conjugation of heat flows at the solid and fluid (water, air) interface [3]), the heat transfer coefficients are β_1 = 6 W/(m^2·°C) for air and β_2 = 600 W/(m^2·°C) for water; the heat exchange coefficient are $h_1 = \beta_1/\lambda_b$ = 3.45 1/m for air–concrete system and $h_2 = \beta_2/\lambda_b$ = 17.14 1/m for water–steel system.

The calculations accounted for fracture formation of concrete [1, 3, 6].

Table 1. Physical properties of structural elements of steel-and-concrete conduits.

Physical property	Steel shell	Concrete	Reinforcing ring material
Elasticity modulus E, MPa	$2.1\cdot10^5$	$2.9\cdot10^4$	$2.1\cdot10^5$
Density ρ, kg/m^3	7794	2400	7784
Coefficients:			
Poisson's ratio ν	0.28	0.17	0.25
Linear thermal expansion α, 1/deg	$1.2\cdot10^{-5}$	$0.95\cdot10^{-5}$	$1.1\cdot10^{-5}$
Thermal conduction k, W/(m·deg)	35	1	30
Thermal diffusivity æ , m^2/s	$7.32\cdot10^{-6}$	$0.54\cdot10^{-6}$	$0.84\cdot10^{-6}$
Heat transfer α_1, W/(m^2·deg)	3.51	0.87	2.1

During operation of conduits, their stress state is conditioned by: stress state of the bottom edge of the dam, process stresses generated during construction, cyclical hydrostatic loads, seasonal variation in the temperature of ambient air and water in the conduit. Since the bottom edged of the dame is compressed in the longitudinal direction (along the conduit), it is assumable that the steel shell structure and the reinforcement of the conduit are also axially compressed and inactive under the action of hydrostatic pressure in the conduit [2–5]. The circumferencial stresses in the steel shell and reinforcement are generated under the hydrostatic load and seasonal temperature variation in air and water.

The tensile (circumferencial) stresses have the determining influence on the strength of the power conduit constructs. Table 2 and Figure 3 give the maximum values of σ_1 for the main load combination at the level of head water (LHW) of 539 m at the minimal and maximal temperatures of air and water in the conduit.

The circumferencial stresses are nonuniformly spread along the conduit perimeter: maximal tensile stresses are formed in the steel shell structure at the most remote points from the bottom edge of the dam while minimal tensile stresses concentrate nearby the bottom edge (see Figure 3). In summer, σ_1 in the upper part of the shell lowers by 10% (Table 2).

Table 2. Maximal principal tensile stresses σ_1 (MPa) in structural elements of the power conduit under the main combination of loads.

Structural element	σ_1, MPa	
	$T^{air}_{min} = -17.3°C$; $t_w = 2°C$	$T^{air}_{max} = +17.9°C$; $t_w = 12.5°C$
Steel shell	132.1	119.5
Inner reinforcing ring	80.0	57.1
External reinforcing ring	98.2	71.4

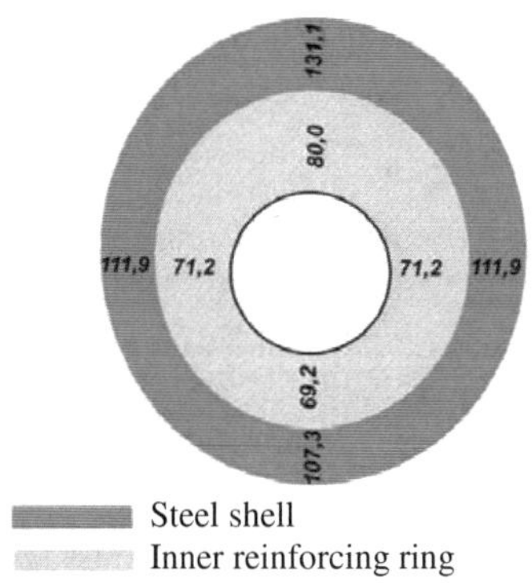

Figure 3. Epure of principal stress σ_1 (MPa) in the steel shell structure and inner reinforcing ring in the straight section of the conduit (Figure 1b) at LHW = 539 m, $T^{air}_{min} = -17.3°C$, $T^{w}_{min} = 2°C$.

The same distribution of σ_1 is observed in the inner reinforcing ring: the tensile stresses in the sides are not higher than 88% of the maximal values (σ_1 reaches 80 MPa in the top portion of the reinforcing ring); lowest values of the tensile stresses are recorded at the conduit attachment with the bottom edge of the dam. The tensile stresses σ_1 in the external reinforcing ring are higher than in the inner ring by 15–20%.

In cold season when the temperature of ambient air and water lowers (see Figure 2), the tensile stresses in concrete and in reinforcement nearby the outside boundary of the conduit grow. In the steel shell structure, the tensile stresses increase insignificantly. In summer, due to heating of the external concrete ring and the external shell structure, the tensile stresses reduce in the external reinforcement:

σ_1 nearby the outside boundary of the conduit decrease significantly, from 98.2 MPa at T^{air}_{min} = –17.3°C to 71.4 MPa at T^{air}_{max} = +17.9°C. The rise in the water temperature in the conduit results in the reduction in the tensile stresses in the steel shell structure.

Sum up, the efficiency of the solution of the quasistatic thermoelastic problems by the method of singular integral equations for piecewise-homogenous media has been illustrated in terms of the stress state calculation for a multi-layered steel-and-concrete power conduit of the real waterworks. The results of the numerical calculations prove their applicability in the analysis of thermal stress state in a composite material with regard to the temperature variation in adjacent media (rock mass, ambient air and water in the conduit). The calculated stresses in the structural elements of the hydropower plant conduit under thermal effects are quantitatively and qualitatively similar to the in situ stress measurements [4, 5].

References

[1] Gakhova LN 2010 Temperature-induced stresses in blocks-structure rock mass *Proc. Sci. Conf. on Fundamental Problems of Geoenvironment Formation under Industrial Impact* Novosibirsk: IGD SO RAN Vol 1 pp 117–121 (in Russian)

[2] Gakhova LN 2012 Stresses in Reinforced-Concrete Sandwich Structures, Considering Fracturing in the Concrete *Proc. 2nd Russia–China Conf. on Nonlinear Geomechanical-Geodynamic Processes in Deep Mining* Novosibirsk: IGD SO RAN pp 316–320 (in Russian)

[3] Kutateladze SS 1990 *Heat Transmission and Hydrodynamic Resistance* Moscow: Energoatomizdat (in Russian)

[4] Gakhova LN, Kuznetsova YA 2016 Stressed state in composite reinforced concrete turbine penstocks in a high-head dam structure Power Technology and Engineering Springer New York Consultants Bureau Vol 49 No 5 pp 319–323

[5] Permyakova LS, Rasskazchikov VA, Ulyashinsky VA, Epifanov AP, Kuznetsova YuA 2008 Stress state of the pressure line elements in the turbine installations of Sayano-Shushenskaya hydroelectric power station *Gidrotekhnich. Stroit.* No 11 pp 11–18

Calculation of craters resulting from impact rupture of rock mass using pulse hydrodynamic problem formulation

LV Gorodilov* and TB Rasputina

Chinakal Institute of Mining, Siberian Branch, Russian Academy of Sciences, Novosibirsk, Russia

E-mail: *gor@misd.ru

Abstract. A liquid–solid hydrodynamic model is used to determine shapes and sizes of craters generated by impact rupture of rocks. Near the impact location, rock is modeled by an ideal incompressible liquid, in the distance—by an absolute solid. The calculated data are compared with the experimental results obtained under impact treatment of marble by a wedge-shaped tool.

Impact rupture of rocks is carried out by jumper bits or rotary–percussion drills. Despite numerous theoretical and experimental studies, the problem connected with the quantitative description and prediction of impact fracture remains yet to be solved as the present paper authors think. One of approaches to solving the problem may be maximum simplification of analytical models which describe the processes using minimum set of input parameters but at practically reasonable accuracy.

In this study, the theoretical determination of sizes of craters made by a wedge tool in soil used a liquid–solid hydrodynamic model earlier applied to blasting craters [1–3]. This model assumes that under impact rupture of soil, strength of soil near the impact tool as opposed to inertia forces can be neglected. In the distance from the tool, soil is assumed hard. The border between the zones of flowing and immovable soil is the flow line at which the critical velocity V = const.

The problem is formulated as finding an unknown section Γ_1 of the boundary Γ (Figure 1) so that the function φ satisfies Laplace's equation in the domain Ω:

$$\Delta\varphi = 0, \quad (1)$$

and the boundary conditions at Γ:

$$\partial\varphi / \partial s = V \quad \text{at} \quad \Gamma_1, \quad (2)$$

$$\partial\varphi / \partial n = 0 \quad \text{at} \quad \Gamma_1, \quad (3)$$

$$\varphi = 0 \quad \text{at} \quad \Gamma_2, \quad (4)$$

$$\varphi = -\varphi_0 \quad \text{at} \quad \Gamma_3. \quad (5)$$

By symmetry of the crater relative to its axis, Figure 1 only shows the right-hand part. Here, X, Y—coordinate axes; Γ_1—boundary of the crater in soil; Γ_2—free surface; Γ_3—boundary of pulse source (edge of the tool); r_0–half-width of the tool; φ_0—potential at the tool boundary; s and n—angular position and outward normal drawn to the boundary.

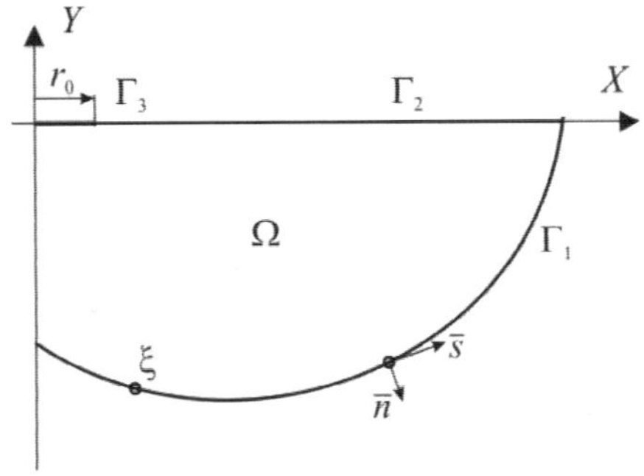

Figure 1. Analytical pulse hydrodynamic model to determine profile of crater.

Transition to dimensionless parameters in (1)–(5) enables understanding the the size of the crater depends on one dimensionless parameter:

$$\overline{\varphi}_0 = \frac{\varphi_0}{r_0 V}. \tag{6}$$

Inasmuch as a part of the boundary of unknown and is to be determined, we used the iteration algorithm [3–5] and boundary element method. The initial position of the boundary Γ_1 was set arbitrarily.

The solution of (1)–(5) of Laplace's equation used the equation [6]:

$$c(\xi)\varphi(\xi) + \int_\Gamma \varphi(\eta) q'(\xi,\eta) d\Gamma(\eta) = \int q(\eta)\varphi'(\xi,\eta) d\Gamma(\eta), \tag{7}$$

where $c(\xi)$—constant and equals γ/π (γ—solid angle at which the boundary Γ from the point ξ); $\varphi(\eta)$ and $q'(\eta)$—potential and derivate of the potential with respect to the outward normal to Ω at the point $\eta \in \Gamma$, respectively; $\varphi'(\xi, \eta)$—solution of Laplace's equation; $q' = \partial\varphi' / \partial n$.

For the plane symmetry used in the calculations, the boundary Γ is a boundary of a cylinder. Numerical solution of Eq. (7) used the method by Krylov–Bogolyubov [7] which consists in the substitution of integral equations for algebraic equations. The boundary of the domain Ω was split into approximately equal linear sections—boundary elements, φ and $\partial\varphi/\partial n$ at each element were assumed constant and related to the middle points of the sections—nodal points. Equation (7) was written in digital form for each boundary element and a system of linear algebraic equations was obtained as a result.

By solving the system, the values of the potential at the nodal points were calculated together with the velocity along Γ_1 at the intersection of j-th and (j+1)-th elements from the formula:

$$v_j = 0.5(\varphi_{j+1} - \varphi_j)/(l_{j+1} + l_j) = 0, \tag{8}$$

where φ_j, φ_{j+1} and l_{j+1}, l_j—potentials at j-th and (j+1)-th elements and the lengths, respectively. Then, the intersection points of the boundary elements were shifter along the normal drawn to them, and a new curve Γ_1 was plotted. In case that the solution failed to satisfy (2) at a preset accuracy (deviation of the velocity along the boundary from the critical value was higher than the preset limit), the next iteration was performed.

The shift of Γ_1 on each iteration was fulfilled in two stages. First, the boundary element next the symmetry axis was fixed, and the other elements were shifted, and the velocity v_j on the elements was compared with v_1. Then, comparing the velocity on the first section, v_1, and the critical velocity V, the whole boundary was either expanded or contracted proportionally. For better convergence, the shape of the boundary was regularized. In more details, the procedure is described in [8].

Figure 2a shows the calculated shape of the crater and the approximate profile of the craters in the experiments [9], in Figure 2 there is a picture of the experimental crater.

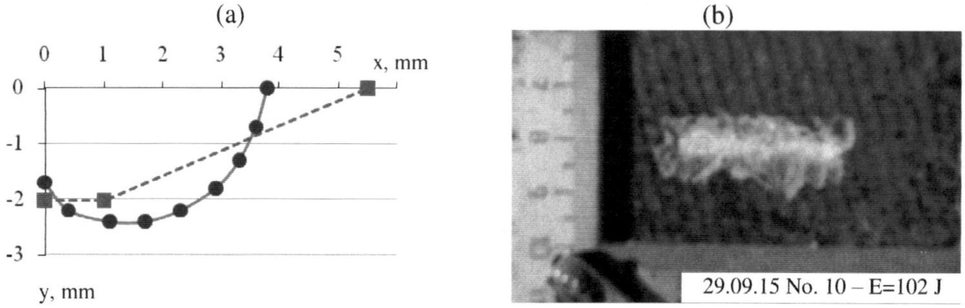

Figure 2. (a) Calculated and experimental craters and (b) picture of a crater after impact by a wedge tool: ●—calculation, ■—experiment.

One of the major issues of calculating craters in the pulse hydrodynamic model is the selection of values of potentials on the surface of pulse sources (in the case under discussion, this is governed by the energy of the wedge tool) and critical velocities governing properties of the medium. Precise theoretical determination of these values is impossible due to abstract nature of the model. A certain relative estimate of the potential φ_0 in case of a wedge tool can be the value proportional to the tool energy density per unit surface area interacting with the tool edge:

$$\varphi_0 = \kappa \cdot \frac{E}{r_0 l},$$

where E—energy of the wedge; l—longitudinal size of the wedge (perpendicular to the plane XY); κ—proportionality coefficient.

In the experiments in [9], kinetic energy transmitted by a hammer to a wedge varied from 15 to 125 J. The front edge surface area of the wedge was $2{\cdot}30 = 60$ mm^2. The critical velocity was assumed as 1 m/s. The proportionality coefficient for the good correlation of the calculation and test data was chosen as $\kappa = 0.8{\cdot}10^{-8}$ m^2 s/kg.

Figure 2 shows the calculated curves of the crater depths $H0$ in the axis of symmetry, crater radii and wedge energy, the corresponding experimental data and the trend of the experimental results (second-order polynomial curve). The analysis of the plots (refer to Figure 2a) shows that theoretical and experimental depths agree well while the data on crater radii scatter considerably. This is a reflection of a feature of the pulse hydrodynamic model which cannot account for the brittle failure—split-off in walls of the crater, which results in the considerably higher R. In principle, the result can be corrected by introducing dependence of the critical velocity V on the potential φ.

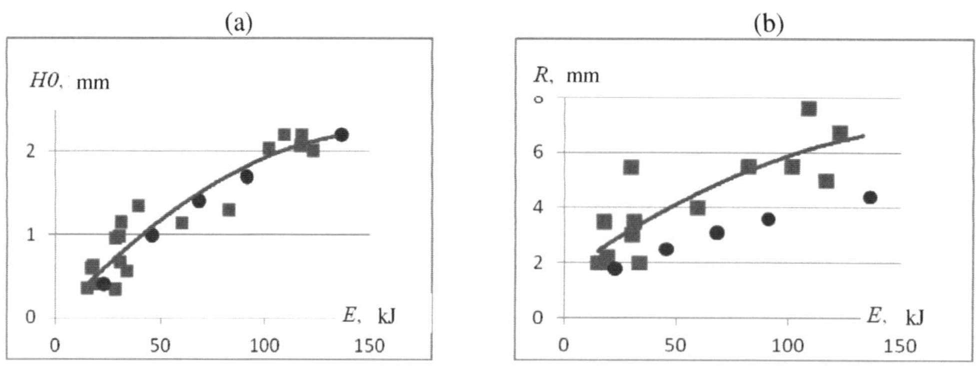

Figure 3. (a) The calculated depth and (b) radius of crater versus the energy of the wedge tool: ●—calculation, ■—experiment, ——experimental data trend.

IOP Conf. Series: Earth and Environmental Science **134** (2018) 012022 doi:10.1088/1755-1315/134/1/012022

In this manner, within the framework of the pulse hydrodynamic model, the authors have calculated the profile of the craters generated on the half-space surface by an impact wedge. The comparison with the experimental results obtained on a marble block shows that the calculated radii of craters are much smaller. The analytical model can be improved by the introduction of the critical velocity–potential dependence. By selection of the proportionality coefficient between the wedge energy and the potential at the boundary of the pulse source, a good agreement has been reached between the calculation and experiment results on the depth of the craters (scatter within 10–15 %).

The described analytical model and the results may be recommended as an express-method to determine the size of craters formed in solid material under impact loading. Simplicity and low computational efforts make the model adaptable to the models for the assessment of efficiency of impact systems.

References

[1] Kuznetsov VM 1960 Shape of a crater after explosion on the ground surface *Prikl. Mekh. Tekh. Fiz.* No 3 pp 152–156

[2] Ilinsky NB and Potashov VA 1986 *Boundary-Value Problems of Explosion Theory* Kazan: Kazan. Univer. (in Russian)

[3] Sher EN, Gorodilov LV and Chernikov AG 1997 Loosening cone of an annular blasthole charge in soft ground *J. Min. Sci.* Vol 33 No 5 pp 422–426

[4] Krasnov VK 1984 *Solution of Axisymmetric Explosion Problems Using the Method of Boundary Integral Equations* Moscow: VINITI (in Russian)

[5] Gorodilov LV 1996 Numerical solution of the problem of ejection explosion in a two-layer medium in a pulse hydrodynamic formulation *J. Appl. Mech. Tech. Phys.* Vol 37 No 3 pp 357–363

[6] Brebbia CA, Telles JCF and Wrobel LC 1984 *Boundary Elements Techniques* Springer-Verlag Berlin Heidelberg

[7] Kantorovich LV and Krylov VI 1952 Approximate Methods of Higher Analysis Moscow–Leningrad: Gostekhteoretizdat (in Russian)

[8] Gorodilov LV 1993 Mechanical action of underwater pressure charges in soils and rocks *Theses of Dissertation for Candidate of Physical and Mathematical Sciences* Novosibirsk: IGD SO RAN (in Russian)

[9] Gorodilov LV, Efimov VP and Kudryavtsev VG 2016 Experimental research of marble block fracture under action of hammer head–impact tool percussion system *J. Fundament. Appl. Min. Sci.* Vol 3 pp 44–50 (in Russian)

IOP Conf. Series: Earth and Environmental Science **134** (2018) 012023 doi:10.1088/1755-1315/134/1/012023

Effect of water on fracture of rocks under diametral compression

VA Kalachev[1], DV Zaitsev[1], AN Kochanov[2], YuA Kostandov[3] and PE Panfilov[1]*

[1]Ural Federal University, Yekaterinburg, Russia
[2]Academcian Melnikov Institute of Integrated Mineral Resources Development–IPKON, Russian Academy of Sciences, Moscow, Russia
[3]Crimean Federal University, Simferopol, Russia

E-mail: *peter_panfilov@mail.ru

Abstract. The researchers study the effect of water on the strength of small-size specimens made of metamorphic rocks (carbonaceous quartzite, serpentine) and an analog of sedimentary rocks (sand-and-cement mixture, or artificial sandstone) exposed to diametral compression. It is found that the specimens exhibit the same brittle deformation behavior after water storage for a day. All specimens show reduced strength while quartzite and serpentine feature a decrease in deformability down to failure. Microscopic fracturing is viscoelastic and independent of water. It is suggested that the influence of water on the deformation behavior of the specimens can be explained by the Rebinder effect.

Water environment has a considerable influence on mechanical properties of rocks. For instance, water penetrated in a material changes its elastic and strength characteristics [1–3]. These changes should be taken into account in geomechanical assessment-based grounding of mining, e.g. in the stability designs of mines and underground structures. Such calculations need data on mechanical characteristics of rocks [4]. It is difficult to carry out physical tests of large specimens of rock, including the requirement of pre-set shaping. For instance, for the Brazilian tests, specimens need to be shaped as disks with a diameter of 10–15 cm, and it is difficult achieve evenness and parallelism of the face surfaces of such disks. Furthermore, such large-scale experimentation needs expensive testing machines. With smaller size specimens, the manufacturing procedure is much simpler, the test are essentially less expensive, and different-scale metallography of the specimens becomes possible without the loss in their integrity [5]. This study focuses on the influence exerted by water on the mechanical properties and fracturing behavior in small-size specimens of rocks in indirect tensile tests (Brazilian testing).

It was chosen to test metamorphic rocks (coaly quartzite, serpentinite) and a sedimentary rock analog (artificial sandstone or cement-and-sand mixture) [6]. The specimens of quartzite and serpentinite were shaped as disks with a diameter of 5 mm and a thickness of 3 mm; the artificial sandstone disk had a diameter of 11 mm and was 5 mm thick. The rough workpieces to make specimens were drilled from parallel planes under water with a hollow diamond drill; the planes had a thickness not less than 5 diameters of a future specimen. The cylindrical rough workpieces were cut by a small-size diamond saw 45 mm in diameter and 0.1 mm thick under continuous spraying by water. Then, the side surfaces of the specimens were polished up to flatness on a tool-grinding diamond wheel. The indirect tensile tests (Brazilian tests) were carried out under room temperature on Shimadzu AG-50K XD tension test frame (crosshead speed of 0.1 mm/min). The testing was terminated when the stress–strain curve acquired a dog-leg reflective of initiation of a critical fracture in a specimen. For each type of rocks, two groups of 10 specimens each were tested. The first group was the specimens in the initial state, and the second group was the specimens held in tape water for a day. The test data processing used materials testing software Trapezium of Shimadzu. The surfaces of the specimens before and after the tests were examined using Epson Perfection V750 Pro scanner (augmentation ×10) and MIM-8M metallographic microscope with Canon high-resolution camera (augmentation ×100). Using the microscopic images and Adobe Photoshop, topograms of the fractured surfaces of the specimens were plotted. In the topograms, the shape, length, width and number of fractures were determined and statistically processed using a standard program. The array data characteristics were chose the arithmetical average and standard deviation.

The stress–strain curves were plotted for each type of rocks. The curves of initial specimens and specimens held in water exhibited the same behavior. Unexceptionally, deformation endured by the specimens down to failure was of the order of 1%. It can be concluded that the specimens subjected to the tensile load behaved as a brittle material irrespective of being held in water or not. The only difference lied in the quantitative characteristics of such parameters as ultimate stress limit and deformation to failure. The data on the mechanical properties of the test rocks are given in Table 1. Evidently, the strength characteristics of the test rocks after the day-long water storage reduced.

Table 1. Mechanical properties of model materials.

Index	Coaly quartzite		Serpentinite		Artificial sandstone	
	air	water	air	water	air	water
σ, MPa	32	25	23	18	6	4
δ, %	1.3	0.9	1.3	0.9	0.6	0.6

After the tests, the untreated quartzite specimens came apart while the untreated specimens of serpentinite and sandstone preserved integrity. This suggests two competing mechanisms of stress relaxation in a material subjected to tensile loading: initiation–growth of fractures and deformation (reversible and irreversible). In serpentinite, which like all rocks is incapable for macroscopic irreversible deformation, the mechanism to compete with failure will be reversible or elastic deformation. In sandstone, due to high porosity, the contribution of irreversible deformation to stress relaxation cannot be neglected as some pores can collapse under loading. In quartzite, the main relaxation mechanism is fracturing while the part of deformation is insignificant.

Although the ultimate strength and deformation to failure decreased in the specimens after water storage, only a half of the quartzite specimens came apart after the tests whereas the specimens made of serpentine and sandstone preserved integrity. This means that the both mechanisms of stress relaxation, i.e. fracturing and deformation, are active in serpentinite and sandstone, while in quartzite the initiation and growth of fractures is competed by the deformation mechanism.

The microscopic studies of the fractures showed that their tips were sharp irrespective of the type of rock a specimen was made of, and the trajectories of the fractures were oriented along the load application axis (Figs. 1a and 1c, Figs. 2a and 2c). The length of the main fracture was different in the treated and untreated specimens. In the dry specimens, the length of the main fracture was comparable with the specimen diameter; in the specimens after water storage, it was smaller (Table 2). The microscopic examination revealed that the main fracture was composed of fine lens-shaped fractures tending to coalescence (Figs. 1b and 1dm Figs. 2b and 2d). Most fractures had blunt tips though there were sharp tip fractures. Fractures in the dry specimens were 2–3 times wider than in the wet specimens (Table 2). An inclusion on the pathway of the main fracture could deflect the fracture from the initial trajectory.

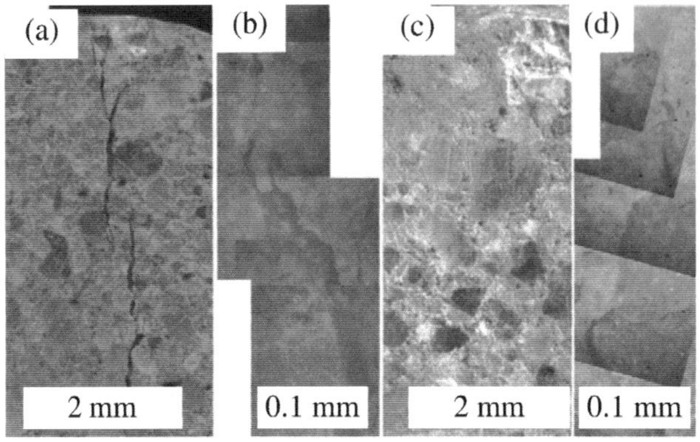

Figure 1. Failure of initial sandstone specimen and after water storage on a macroscale (a), (c) and on a microscale (b), (d).

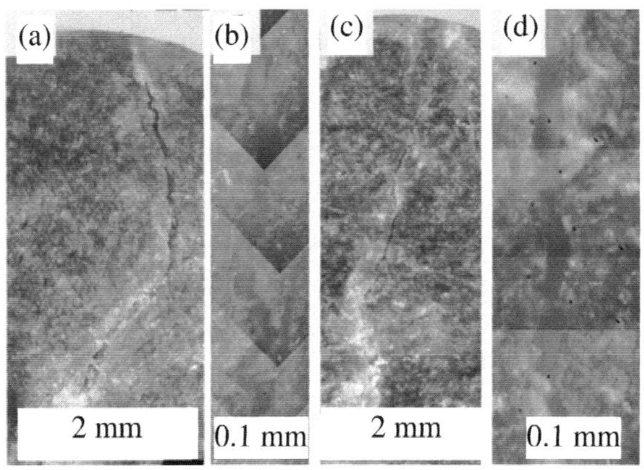

Figure 2. Failure of initial serpentinite specimen and after water storage on a macroscale (a), (c) and on a microscale (b), (d).

Table 2. Fracturing of model materials.

Specimen		Coaly quartzite		Serpentinite		Artificial sandstone	
		air	water	air	water	air	water
Diameter, mm		6	6	6	6	10	10
Total length of fractures, mm		3.98±0.12	4.18±0.04	3.86±0.16	3.42± 0.07	8.20±0.24	7.21±0.07
Number of fractures		2	6	3	4	3	14
Fracture length, mm	Left side	2.09±0.01	0.30±0.03	1.92±0.07	0.48±0.03	2.17±0.15	0.56±0.04
	Center	–	1.72±0.04	0.88±0.24	1.79±0.08	5.33±0.27	0.07±0.02
	Right side	1.89±0.15	0.16±0.01	0.43±0.09	0.13±0.04	0.70±0.08	1.17±0.09
Fracture width, mm	Left side	0.03±0.02	0.01±0.00	0.04±0.01	0.03±0.01	0.04±0.04	0.02±0.01
	Center	–	0.01±0.00	0.03±0.01	0.04±0.01	0.10±0.04	0.01±0.00
	Right side	0.03±0.01	0.01±0.00	0.03±0.01	0.02±0.01	0.02±0.01	0.02±0.01

The macro-level analysis of deformation behavior of the specimens proves the brittle behavior of rocks under tensile loading irrespective of the rock type. The presence of water unalters the brittle deformation behavior but reduces the strength characteristics of the specimens. The data on fractures on the macroscale conform with the inference on the brittle failure of the specimens: the main fractures have sharp tips and straight profile. The analysis of the microscale failure yields that the main fracture results from the coalescence of lens-shaped fractures with blunt tips, and the width of the fractures is governed by the presence of water in the specimen. The failure mode described above suggests two mechanisms of elastic energy relaxation in the specimens: fracture growth and deformation. Consequently, the microscale mode of failure can be characterized as viscoplasticity. The reduction in the ultimate strength limit and deformation to failure, as well as the decrease in the width of fractures under action of water is the effect of Rebinder [7].

It is noteworthy that according to the test data on water-saturate granite specimens [8], the mode of failure changes during deformation. In case of dry specimens, failure follows the two-stage mechanism: disperse accumulation of defects and localization and growth of a single failure source, as a rule; in water-saturated specimens, chaotic failure and high-rate damage embrace the whole volume of the material.

Conclusion

The studies have shown that the test rock materials under tensile loading exhibit brittle behavior on macroscale and viscoelastic behavior on microscale. The influence of water on deformation behavior of specimens can be explained by the Rebinder effect. The mechanism of the water effect on strength needs further analysis.

References

[1] Adushkin VV and Turuntaev SB 2005 *Mining-Induced Processes in Earth Crust (Hazards and Catastrophes)* Moscow: INEK (in Russian)
[2] Sobolev GA 1993 *Fundamentals of Earthquake Prediction* Moscow: Nauka (in Russian)
[3] Scholz CH, Syke LR and Agarwal YP 1973 Earthquake prediction: A physical basis *Science* Vol 181 pp 803–809
[4] Melnikova NV, Rzhevsky VV and Protodyakonov MM (Eds) 1975 *Handbook (Cadastre) of Physical Properties of Rocks* Moscow: Nedra (in Russian)
[5] Zaytsev DV, Kochanov AN, Panteleev IA and Panfilov PE 2017 Influence of scale effect in strength tests of rock samples *Izv. RAN Ser. Fizich.* Vol 81 No 3 pp 366–369
[6] Vernon RH 2004 *A Practical Guide to Rock Microstructure* Cambridge University Press
[7] Rebinder PA 1979 *Selectals. Surface Phenomena in Disperse Systems. Physicochemical Mechanics* Moscow: Nauka (in Russian)
[8] Kuksenko VS, Damaskinskaya EE and Kadomtsev AG 2011 Fracture of granite under various strain conditions *Fiz. Zemli* No 10 pp 25–31

Importance of early adjustment of rotary-percussion drilling tool to mineral mining conditions

VN Karpov,* VV Timonin**

Chinakal Institute of Mining SB RAS,
Novosibirsk, Russia

E-mail: *karpov@misd.ru, **timonin@misd.ru,

Abstract. The authors discuss the technical and economic criteria for the evaluationof efficient application of rotary-percussion drilling tools. The determinants of efficiency of downhole pneumatic drilling hammers are identified. The authors set conditions of high-velocity hole-making and find maximum indices of long-term performance for downhole pneumatic rotary-percussion drilling tools.

The application of the modern high-tech *main active members* of rotary-percussion drilling rigs, namely, down-hole percussion machines equipped with rock-cutting tools is a promising condition to lower mining costs in production of ore and construction materials when the rotary-percussion process is employed to make holes. The use of modern high-automated drilling complexes allows prompt auxiliary operations and optimal adjustment of operation parameters through the control panel, but does not eliminate the need in modification of the machinery and their adaptation to mining conditions of a mineral deposit [2, 3]

The term "main active or working member" is introduced by researchers working for IM, SB RAS, who were the first to introduce the full-scale rotary percussion drilling by downhole drilling tools as an ore-breaking process with deep blasthole charges in Russian mines in 1954–1955 [1].

Research reports on application of different versions of active members of rotary-percussion drilling machines, actually, air hammers and their rock-cutting tools are not systemized and unified. As a rule, available publications on new versions of down-hole percussion machines is restricted with such specifications as average mechanical drilling velocity, rock strength by Protodiakov's scale, rated energy-carrier pressure, production rate per a shift and general description of a mineral deposit, perhaps, in terms of the full-scale production experience and acting standards based on rock drillability classification [4, 5]. The reports disregard physical-mechanical properties of a rock mass in view to the recommended rated drilling parameters and complete auxiliary devices sets in order to gain rational technical and economic hole-making parameters with account for mining-geological and technical conditions of a specific mineral deposit. The lack of so-called efficient drilling pattern and "route chart", which could guarantee technical and economic indices of rational application of main active rotary-percussive drilling members under conditions of a specific mineral deposit constitutes one of acute challenges in domestic drilling practice in quarries and mines [6].

Let consider the technical and economic criteria for evaluation of rational application scope of main drilling rig executives on the down-hole air hammers as an example to make deep blast holes. The production cost of one running meter drilling of a blast hole in combination with the drilling capacity per a shift basically depends on intensity of rock breakage, viz., mechanical drilling velocity V_m, route rate of drilling V_r and sinking per a trip L_s considering the effects of natural, technical and technological factors.

Moreover, it is next to impossible to find the maximum feasible productivity of down-hole air hammers based on average mechanical drilling velocity and sinking length per a run, because the parameters are under strong effects of variable natural or technogenic factors. In this connection one of parameters of the maximum drilling productivity can be the maximum feasible velocity of sinking $V_{r\,max}$ at optimal time of mechanical drilling t_{opt} per a run under rated operation parameters of rock cutting.

It is important that established correlation relations between physical-mechanical properties of rocks and average penetration speed $V_{av\text{-}p}$ of pneumatic drilling is of restricted scientific and practical value, as a velocity can be calculated based on different time parameters possible during a route run. This peculiarity is typical of most research works, where empirical relationships are derived based on average mechanical drilling velocity which does not match the maximum drill velocity $V_{r\,max}$, viz. $V_{av\text{-}p}= L/t$, where $t \neq t_{opt}$, $L \neq L_{opt}$. This feature is specific for earlier mentioned standards based on classification of rocks by their drillability, where an average penetration speed of drilling is presented with no justification of sinking length per a run L provided that $V_{r\,max}$ is gained per a run.

As practice shows the normalization of drilling operations is an actual and rather acute issue at quarries and mines. Performance norms used to be set without proper account for qualitative parameters of drilling. To be precise, the well known mine machines NKR-100M are not equipped with devices to measure operation parameters of drilling. A drill operator quantitatively feels working pressure, forward force, rotation velocity, in brief, responsibility for hole-making norms lies on a drill operator, his motivation, his skills. This leads to excess- or understated production norms. The human factor remains a constitutive factor in evaluation of productivity norms for a drill crew at a certain drill section, no matter whether there is the detailed information on physical-mechanical properties of a rock mass, and a high-tech powerful pneumatic hammer. Figure 1 demonstrates a general view of the first characteristic sinking of a blasthole by two drill operators by the same air hammer under the rest identical conditions, excluding a choice of operation parameters capable to affect a character of rock failure.

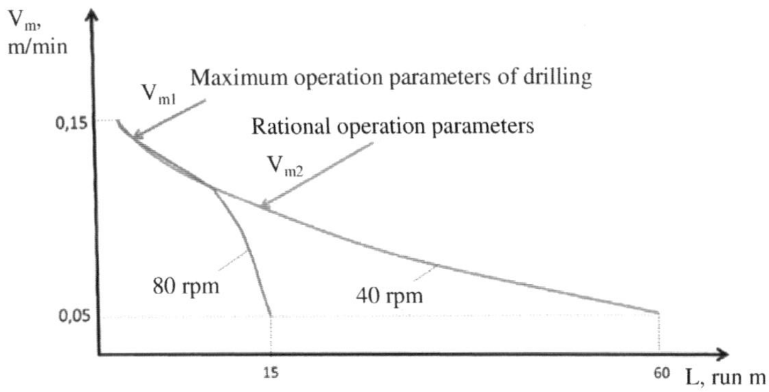

Fugure 1. Relationships of mechanical-economic parameters of drilling versus a choice of maximum and rational operation parameters of blasthole making.

In [7] the researchers established that in rotary-percussion hole-making under the constant nominal pressure of energy carrier (air) there is a wide zone of tool rotation angles between percussion loads within which productivity of a drilling machine is maximum and practically the same. Thereto, the zones have boundary parameters in regard to a rock mass, but drill bits have differing resistance. This problem becomes evident in making deep blastholes since the first sinking run.

In hard and abrasive rocks the pressure of energy carrier is low within 0.5 MPa and indentors of drill bits are underloaded relative to the rated parameters because of low capacity parameters of production air hammers, and the surface failure is realized and then it transfers to a fatigue wear as a result of indentor wear-out [3]. Increase in rotations causes increase in angle between impact loads, and this leads to drastic wear of hard drill-bit alloy, as the reduced effects of impacts per every rotation of the tool and enhanced cutting effect like in rotary drilling. The hard alloy of drill bit is subjected to drastic wear because it is not intended for this drilling process [8, 9].

This situation results in greater number of runs: up to 6 runsper 50 running meters in making deep blastholes, thus boosting total drilling costs because of excess energy-carrier consumption by a drill rig making multiple runs. The typical case of the maximum rotation frequency of the rig (80 rpm) in drilling an upper-layer blasthole semi-ring of 105 mm in diameter is the use of P105PM air hammer in NKR100MA rig, Starooskolsky Mechanical Plant Co., Russia, on weakly mineralized quartzites in Gubkina Mine. The field tests of Starooskolsky machines with electric motor of MO-5 swivel head in NKR-100 modified versions justified that they do not have variants of column rotation variations other than design ones from 1.25 Hz and higher [10 –13]. This state of things does not enable to realize rational principles of rotary-percussion hole drilling by downhole air hammers, as the principles are based on provision of the maximum durability of drill bits at the minimum power-intensity of drilling in executing runs in blastholes [14 – 16]. That is the reason why the known global companies Mitsubishi (Japan) and Rock More Internationale (USA, Austria) proposed their services to solve the problem and developed their own tool for domestic down-hole machines, but did not manage to find solution to this problem.

The penetration speed of drilling can not serve a constitutive factor in evaluation of performance of downhole air hammers and its maximum indices, in particular, as a choice of operation conditions providing high V_m can promote excessive wear of drill bits and incidents (jamming of drill strings, and even its loss, breakdown of main parts of the rig, etc.). Herewith, with increasing depth of a blasthole a high penetration speed becomes less beneficial than a footage per a run because the duration of round-trip operation tends to grow with increase in blasthole depth. Alternatively, increase in footage per a bit run L sharply reduces a number of round-trip operations cuts down specific operation costs per 1 run m of drilled holes. Therefore, it is reasonable to consider the choice of optimal modes of drilling providing the minimal energy-intensity of rock cutting at the maximum durability of the rock-cutting tools when researchers develop and master the methods for evaluation of performance of the principal executive members with armament considering the mechanical properties of rocks. This issue is to be discussed in details in other publications by the present authors.

Let consider specific operation costs C, the general formula to calculate production cost of 1 run m of drilled holes is:

$$C = \frac{C_{dr}\cdot(t_{dr}+t_{auh})+C_b+C_{ah}\cdot K_{hw}}{L}, \text{ as } \frac{L}{t_{dr+t_{aux}}} = V_p, \text{ expression is } C = \frac{C_{dr}}{V_p} + \frac{C_b+C_{ah}\cdot K_{hw}}{L}$$

where C_{dr} is working cost of a drill rig, RUB/h; C_b is cost of bit of air hammer, RUB; C_{ah} is cost of air hammer, RUB; L is run length per a bit of air hammer to drill a hole in a rock mass, run m; K_{hw} is coefficient of air hammer depreciation, calculated from ratio L/L_{gl}; L_{gl} is guaranteed life of air hammer, run m.

It is important to state that in some Russian publications the drilling cost for an air hammer is calculated disregarding the cost of a downhole air hammer, namely, index C_b is considered and index C_{ah} is disregarded,the reliability of such calculations is low and the cost of 1 running meter of a drilled hole is under-evaluated. The adoption of calculation formulas for rotary hole-making process is explicit. As a rule, the wear life of downhole tools is few times higher than for bits $L_{gl} > L$; as is the case with the cost. For example, the cost of modern high-pressure air hammers of Atlas Copco Co. with the rate of application 0.23 u/1000 run m. is 6–8 times higher than the cost of a bit of the same producer (according to data of ZAO "UGMKRugormash-Voronezh") [17]. Respectively, it is possible to obtain reliable results in calculations of production costs provided that coefficient K_{hw} is used to specify the costs of application of downhole tools. Unfortunately, there are cases in the actual practice when rates of application for air hammers are higher or comparable to rates of their rock-cutting tool at $L_{gl} \leq L$, $K_{hw} \geq 1$ [6].

Figure 2 demonstrates the model of diagram for technical and economic assessment of hole-making efficiency by downhole percussion machines on rotary-percussion drilling machines developed at Chinakal Institute of Mining SB RAS.

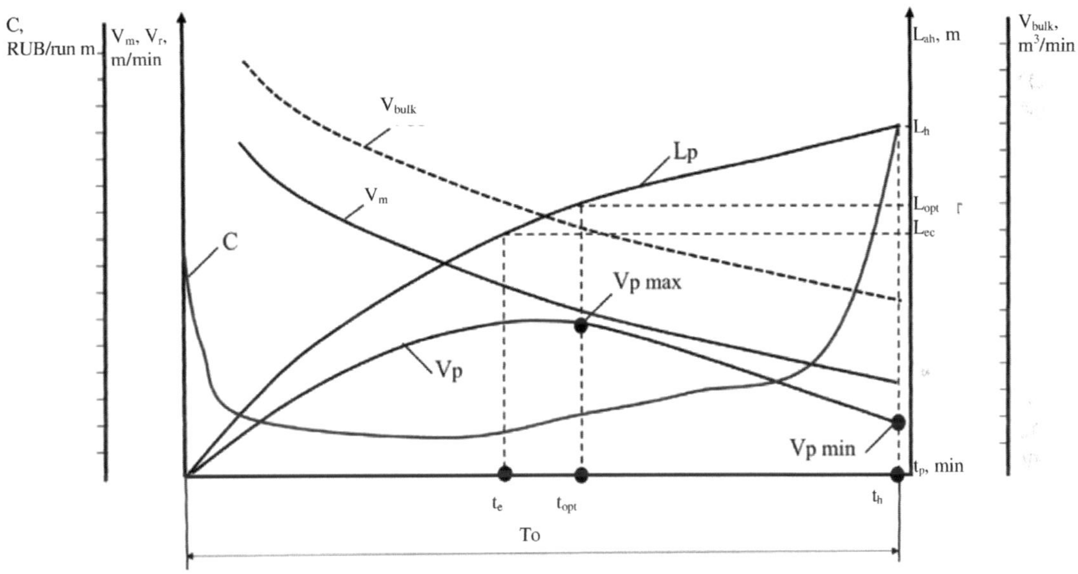

Figure 2. Model of diagram plotted based on the full-scale research data to evaluate the technical-economic efficiency of blasthole-making by downhole air hammers on the rotary-percussion drilling rigs in underground conditions.

The diagram represents variations in penetration V_p, run V_r and bulk V_{bulk} speeds of drilling; the run length executed by the air hammer L_{ah}, and working cost C of one running meter of drilling under preset drilling conditions for operation time T_o per a hole. Intervals for calculation of proposed parameters and plotting of relationships do not exceed length of drilling strings used in operation. In the diagram t_e – economic penetration time for a downhole air hammer; $t_p = t_e$, where $L_{ah} = L_{er}$, L_{er} – economic run of air hammer in a drilled hole; top–optimal penetration time for downhole air hammer $t_p = t_{op}$, where $L_{ah} = L_{op}$, L_{op} – optimal penetration of executive member into a geomedium; t_h – drilling time to failure for air hammer or to a depth preset by drilling-blasting pattern $t_p = t_h$, where $L_{ah} = L_{hr}$, L_{hr} – run of air hammer to failure or to a wanted depth according to drilling-blasting pattern or to above 50 % decline in drilling productivity because of drill bit wear or network pressure fall which makes drilling hazardous or uneconomic.

The most unfavorable combinations of drilling parameters reach critical values at $V_r = V_{r\ min}$, $t_p = t_h$, $L_{ah} = L_h$, at the maximum costs, curve C (Figure 2). It is important to emphasize that parameter C is always the highest in the case of accident-free termination of the run, as dismounting of the drilling string is an energy-intensive operation and in most depends on the drill rig automation.

The model of the technical-economic diagram to evaluate efficiency of pneumatic percussion drilling under full-scale production conditions is elaborated based on theoretical experience of Russian prominent researchers in rotary drilling of large-depth production and geological holes [18].

It is essential that Russian researchers engaged in investigation into rotary-percussion drilling with the use of downhole percussion machines did not pay proper attention to the bit run speed in making deep blast holes of 50–80 m, in particular. Drilling by light-weight rotary-percussion machines in essence represents a physical model of boring of deep oil, geological and other-type holes by applying stationary drilling complexes withlarge number of rods in a drill string and in a less degree by geological rock mass diversity. Field research of technical-economic characteristics of drilling machines revealed that the bit run speed in making lower blasthole semi-ring at a routine failure-free mode repeats the trend of penetration rate practically from the beginning to the end of the run. This means that the rig crewneglects a forced cleanout of a hole in the course of drilling by every drill rod and performs the hole cleanout only before extention of a flight. This state of things and the lack of manometers capable to assess loss of metering characteristics of downhole tools, as well as a fall in an ascending-air-flow speed during the hole cleaning lead to frequent break-downs and even loss of downhole tools and complete drill strings [6]. In production drilling in oil and gas fields such situation is impossible and prohibitive, as the cost of a lost borehole of several thousand meters in length amounts to millions roubles.

The model of the diagram for technical-economic evaluation of efficiency of downhole air hammers is reasonable to apply during runs of a drilled hole in making deep blastholes. The proposed model is also applicable in making deep oil, gas, and exploration holes by rotary-percussion drilling. The equipment assessment in terms of the mine geology permits to make rational choice, adaptation of downhole tools, facilities and drilling conditions to specific local rock characteristics, to work out adaptive measures for years in advance at the planning stage with the aim to solve technological problems specified by mining systems and boring-blasting patterns. The high-class grade guaranteed by domestic and foreign producers of the respective equipment is of prime importance as well. The application of the new-proposed model in full-scale production conditions is reported in [10–12].

Let consider the modern approaches to adaptation of downhole tools at open pit mines. When drilling blastholes by the modern high-tech air hammers [19–21] at consumption rate 0.05–0.5 run m per 1000 run m (foreign data) the service life of the equipment is not specified. For example, in the case study the Table 1 contains drilling specifications with a wide range of working air pressure on rocks of 203 MPa in strength with the use of DM45HP machine (Sweden) and air hammer QL65QM equipped with a bit 6-3/4 Q6FE (171.4 mm in diameter) according to database of Mining Solutions Co., a distributor of Atlas Copco, but the Table 1 does not contain such parameters as energy of a single impact A_{ah} for air hammer, applicable for the cited operation pressure range and capacity N_{ah}. These very parameters would be helpful in calculations of the single impact energy applied to hole bottom in order to break a rock mass. It is known that in rotary-percussion drilling the drilling thrust is required to provide the permanent contact of an air hammer and rocks. The drilling thrust is selected with regard to minimum parameters to eliminate vibration effects generated by an actuated air hammer. Respective penetration speed resulted from a dynamic effect on a rock mass mainly depends on a single impact energy of an air hammer executing mechanical failure of rocks, cutting structure of a bit, impact frequency, rod rotation, hole cleaning from slime and last but not least physicomechanical properties of rocks.

Table 1. Specifications of hole making by DM45HP drill rig with QL65QM downhole hammer

Air pressure	MPa	2.4	2.3	2.2	2.1	2
Air flow rate	m^3/min	27.3	25.8	24.4	23.0	21.6
Drilling speed	m/h	46.0	44.3	42.7	41.1	39.3
Impact frequency	imp/min	1876	1844	1813	1783	1751
Rotation speed	rpm	40	39	37	36	34
Thrust	kgs	1308	1262	1222	1181	1137
Blow rate	m/min	2626	2482	2341	2212	2086
Service life of an air hammer	m	9853	11029	12400	13852	15769
Service life of a bit	m	1793	1726	1664	1601	1532

The lack of the above parameters can be easily explained. The single impact energy of the first and last runs (on the achievement of the critical service life level for an air hammer) can differ greatly in drilling with the use of the same drill bit model in the failure-free period in a certain drilling area with a set of physicomechanical properties of local rock mass and other identical conditions. One of reasons for difference can be catastrophic wear of basic parts of an air hammer. As a result there are two absolutely different tracks of drilled holes passed by a new drill bit and a new air hammer and a track passed by a worn-out air hammer with low energy parameters, a rock failure pattern varied from the bulk- to fatigue- or surface-type rock failure.

In [22] it is reported that the cumulative cross-section areas of gaps exert negative influence on both impact energy and impact frequency. Reduction in these parameters lowers capacity of the machine and drilling productivity as a whole. Researchers of IM SB RAS focus on this problem in development, design and investigation into high-pressure air hammers under both laboratory and production conditions. In [23] the full-scale research data on the problem under consideration are presented for PV-170M air hammer. The similar situation is observed if two machines of the same type-size line with differing energy parameters and the identical drill-bit models are used under the same operation conditions. Failure energy requirements and respective costs of 1 run m drilling are different and described in detail in [10–12].

The users should know optimal intermediate parameters between the first and conventionally "last" run of a drilled hole. It is essential for timely replacement of principal parts of air hammer or a worn-out downhole tool for a new one. Respectively, air hammers of the same type-size line and identical cutting structures under rest identical conditions can have quite different energy parameters after the equal service life intervals. In other words air hammers used to have their own specific parameters.

Figure 3 demonstrates relationships of service life parameters for QL65QM air hammer and 6-3/4Q6FE drill bits versus working pressure of compressed air at σca = 203 MPa.

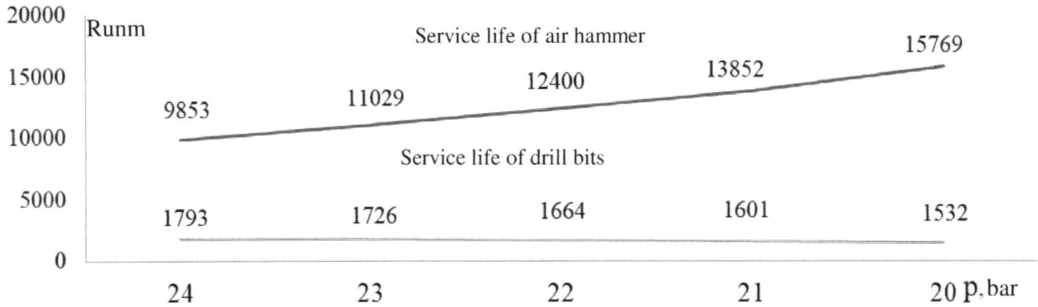

Figure 3. Relationships of service life parameters for QL65QM air hammer and 6-3/4Q6FE drill bits versus working pressure of compressed air at σ_{ca} = 203 MPa.

The present paper is intended to potential users of Atlas Copco products. There are issues to solve, let say, how many repair sets per a downhole machine in cost equal to 50% of the air hammer cost (8 - 10 thousand EURO per a unit of the mentioned type-size line) are required to a user, provided that according to producer's certificate the machine is capable to make holes of at least 5000 run m in length before it reaches the final service life parameters. Nobody can guarantee that a repair set is capable to maintain the design capacity of the machine without reduction in drill penetration speed, excessive consumption of drill bits, increased maintenance costs of compressors and as a consequence of increased costof 1 run m of drilled holes.

Assessment of technical-economic performance of downhole air hammers at open pit mines has specific features. Higher drilling performance, limited number of drill strings, as a rule no more than three, minute time of auxiliary operations in the course of a run (string removal velocity being 0.5–09 m/s) leads to a large number of runs of drilled holes per a shift. Inasmuch as it makes sense to assess the technical-economic performance based on run meterage before the replacement of a rock-breaking tool of a downhole air hammer. In theworld practice of drilling at open pit mines the decline of drilling performance by 30% because of wear-out of hard alloy of drill bits is an attribute to replace a rock-cutting tool from the economic point of view. At mines this parameter can be higher in drilling of large-length blastholes, as it is not reasonable to replace a drill bit in last 5-10 meter interval of the total 60 m length of a blasthole to be made. Dismounting of a drill string is energy-intensive operation. Thereto, the decline in performance is often due to not abrasive wear-out of a drill bit, but due to a fall of pressure in mine network (in tool input), or leakage of energy carrier during a run of a drill string to a downhole machine or through watered levels, variance in mining-geological conditions. In making large-length blastholes by downhole tools the critical fall in drill penetration speed is equal to 70% of rated parameters. Moreover, this fall occurs in making the second half of a blasthole length with downhole machines and a tool with the minimum service period, namely, new tools.

In Figure 4 the technical-economic diagram for open mining processes was worked out at IM SB RAS. In the diagram the research data are on drilling performance of PV170M downhole air hammer to make blastholes of 172 mm in diameter and 18 m in length at SWDB165 machine at the rated air pressure p = 1.2 MPa in Borok quarry, Novosibirsk.

IOP Conf. Series: Earth and Environmental Science **134** (2018) 012024 doi:10.1088/1755-1315/134/1/012024

Drilling was performed without reconditioning of the rock-cutting drill-bit insertions. The estimated parameters in this diagram are similar to the previous one, but considering the specific character of drilling operation, viz., large number of runs of a drilled hole per a shift, etc., the given data on technical-economic parameters were plotted for every collared hole, for operative drilling duration, for T_o per every shift. Within a shift the actual operative drilling time amounted to 91.6% of a shift time, excluding personal needs, and preparatory operations. T_o is a precise time of drilling operation at the quarry under the RF normative acts. Thus exclusively operative hole-drilling time per a shift before replacement of a drill bit is reported in the diagram..

It is explicit in Figure 4 that it is reasonable to replace a drill bit according the design drilling specifications at the quarry and its mechanical properties in the drilling area when $L_p = L_{opt}$ after 14 holes are made; their total length being 252 run m, the time of cost-effective drilling tэis gained at area $L_p = L_e$. In this case it would make sense to use a tool grinding facility for drill bits and to substantiate expediency of its acquisition for future research projects or to test other procedures for improvement of drilling performance. Assessment of cost-performance ratio on pilot units in terms of producer's specifications under full-scale production conditions makes sense before decision making on purchase of new equipment.

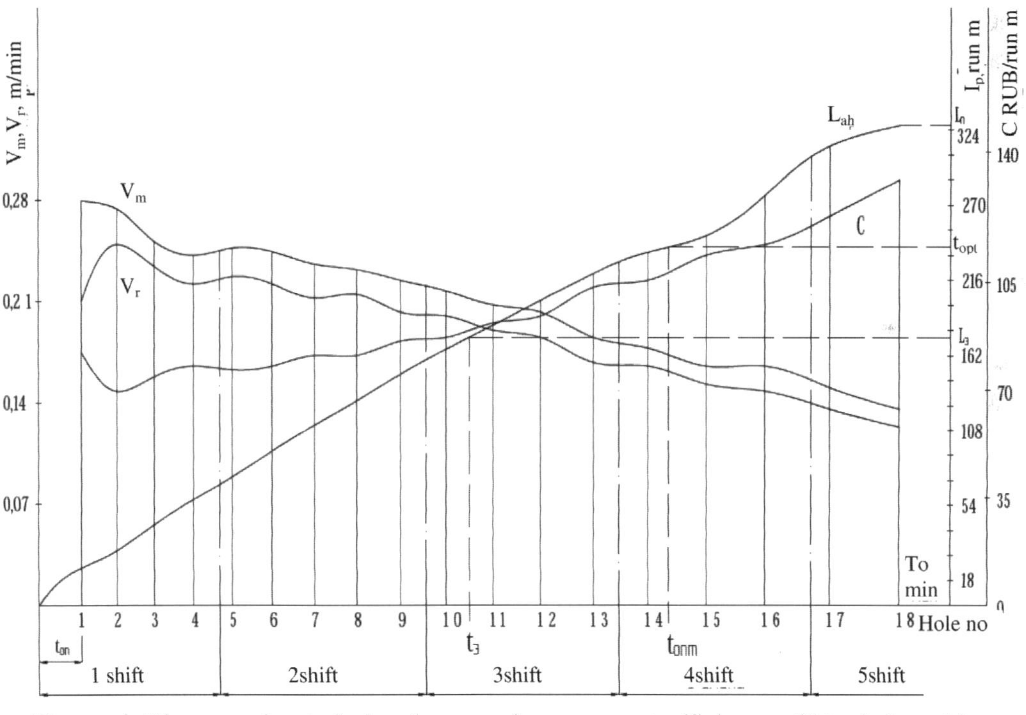

Figure 4. Diagram of technical and economic assessment efficiency of blasthole making by downhole air hammers in open pit mines.

Investigations in terms of the same mineral deposit are preferable to assess the service life parameters of serial and experimental downhole air hammers including replacement for new machines or individual important parts in order to gain optimal capacity specifications. In the new-proposed approach investigation into runs of drilled holes followed with further replacement of bits of downhole air hammer (or in the design replacement intervals) makes it feasible to establish "the last final" run, in other words, the range of non-failure operation life when the downhole hammer should be replaced for a new one. The above diagram of technical-economic efficiency of air hammers in openpit mines can be helpful in underground mines to make short-length blastholes by high-power rotary-percussion drill machines (high-pressure air hammers (2–3.5 MPa), hydroperforators, hydropercussion units).

Researchers working for IM SB RAS elaborated the procedure for evaluation of performance of rotary-percussion drilling machines under production conditions in terms of physical-and-mechanical properties and drillability with account for environment, technical, technological factors affecting the drilling capability. In the procedure evaluation of technical-and-economic parameters by the new-proposed diagrams is one of the basic but not single analytical criterion of evaluation. The detailed description is reported in [10–12] and later publications by authors.

Available methods for determination of rational application scope for air hammers and drilling bit cutting structure under production conditions imply industrial tests intended to determine experimentally the drilling penetration speed, run length, or total run of air hammer and bits in rocks with differing physical-and-mechanical properties, disregarding proper studies of variations in technical-economic indices in the course of hole-making. In specific tests the researchers focused on essential number of runs (30–60) with the minimal penetration per a run no more than 2-3 meters. The test data are used to make approximate calculations of production rate per a shift, cost of 1 running meter hole and to establish the optimal application scope for air hammers and rock-cutting tools [18, 24, 25]. In the case under consideration a large number of short runs leads to overestimated values of drillability and underestimation of actual cost of 1 run m of made hole. The similar state of things is observed in laboratory tests which results should be treated as preliminary evidence requiring correction in terms of environment, technical and technological factors, influencing the penetration speed under conditions of a certain mineral deposit [2].

Conclusion

Adaptation of controllable drilling parameters: drilling pattern, capacity of an air hammer, and others in terms of uncontrollable parameters of a mineral deposit: rock properties, geological and structural specific features makes it feasible to improve drilling productivity in rotary-percussion hole-making with application of downhole percussion machines, thereto the maximum service life of the main active members is gained.

References

[1] Yemelyanov PM, Yesin NN, Zinovyev AA, Semenov LI and Suksov GI 1965 *Cars for Well-Drilling by Submersible Hammers in Underground Conditions* Novosibirsk: Academy of Sciences of the USSR (in Russain)

[2] Oparin VN, Timonin VV and Karpov VN 2017 Quantitative estimate of rotary–percussion drilling efficiency in rocks *Journal of Mining Science* Vol 52 No 6 pp 1100–1111 DOI: 10.1134/S1062739116061637

[3] Lifentsov AS 2010 Power dependences of submersible pneumatic impact tools and drill bits for drilling of explosive wells Scientific and technical collection: *Development of Ore Fields* Issue 93 Kryvyi Rih: KTU (in Russain)

[4] *Uniform Performance Standards and Time for Underground Clearing, Tunnelling and Cut Mining Operations* Part I, approved by committee of the USSR on work and social problems and the All-Union Central Council of Trade Unions. The resolution No. 326/20-93 of December 31, 1982 (in Russain)

[5] *The Resolution of Ministry of Labor of the Russian Federation* dated 4/21/1993 No 8 «About

the approval of the Integrated standards of time for tunnelling and cut mining operations and mines and in geological exploration» (in Russain)

[6] Eremenko VA, Karpov VN, Timonin VV, Shakhtorin IO and Barnov NG 2015 Basic trends in development of drilling equipment for ore mining with block caving method. *Journal of Mining Science* Vol 51 No 6 pp 1113–1125 DOI: 10.1134/S106273911506037X

[7] Begagoen IA, Dyadura AG and Bazhal AI 1972 *Drilling Machines* Moscow: Nedra (in Russian)

[8] Hartman HL 1959. Basic studies of percussion drilling American Institute of Mining, *Metallurgical and Petroleum Engineers* Vol.214 pp 68–75

[9] Hard sintered alloys Brands

[10] Karpov VN 2016 Assessment testing procedure for downhole pneumatic hammers under production conditions *J. Fundament Appl. Min. Sci.* Vol 2 No 3 pp 74–80

[11] Timonin VV and Karpov VN 2016 Assessment of rock failure process under percussion-rotary drilling *J. Fundament Appl. Min. Sci.* Vol 3 No 2 pp 172–176

[12] https://www.youtube.com/watch?v=3C0AgNkm20g IM SB RAS

[13] http://www.somz.ru/products/good/1419196834.html UGMK Rudgormash-Voronezh

[14] Trofimov SN and Lagunova YuA 2011 Drill rigs for open mining of NIPIGormash Gorn. Oborud. *Electromekhanika* No 10 pp 13–18

[15] Fox B et al 2011 *Blasthole Drilling in Open Pit Mining* Garland Texas USA: Atlas Copco Drilling Solutions LLC

[16] *A – Z of DTH drilling* 2016 Halco Rock Tools

[17] https://rgmural.ru/files/content/info/other/konkurentosposobnost-burovogo-oborudovaniia-rgm.pdf (in Russain)

[18] Vozdvizhensky BI, Melnichuk IP and Peshalov YuA 1973 *Physicomechanical Properties of Rocks and their Influence on Efficiency of Drilling* Moscow: Nedra pp 188–191 (in Russain)

[19] Eremenko VA, Karpov VN, Filatov AP and Kotlyarov AA 2014 Improvement of system of development with ore crushing on compressed medium in the time of mastering of bump hazardous deposits *Gornyi Zhurnal* Issue 1 pp 50-55 (in Russian)

[20] Kondratenko AS, Konurin AI, Timonin VV, Denisova EV and Khmelinin AP 2016 Different-purpose longhole directional drilling with the drilling trajectore control Proceedings of the *International Multidisciplinary Scientific GeoConference SGEM* Vol 2 pp 327–332, DOI: 10.5593/SGEM2016/ B12/S03.043

[21] Kondratenko AS, Timonin VV and Patutin AV 2016 Prospects for directional drilling in hard rocks *Journal of Mining Science* Vol 52 No 1 pp 129–134 DOI: 10.1134/S1062739116010212

[22] Repin AA, Smolyanitsky BN, Alekseev SE, Popelyukh AI, Timonin VV and Karpov VN 2014 Downhole high-pressure air hammers for open pit mining *Journal of Mining Science* Vol 50 Issue 5 pp 929–937 DOI: 10.1134/S1062739114050123

[23] Repin AA, Alekseev SE, Timonin VV and Karpov VN 2015 Analysis of the compressed air distribution in down-the-hole hammer drills in Collected Papers: *Miner's week - 2015 reports of the XXIII international scientific symposium* pp 475–482

[24] Shadrina A and Saruev L 2014 Exploration and determination of the principles of rotary-percussive underground slimhole drilling *International Journal of Mining Science and Technology* No 24(2) pp 245–249

[25] Shadrina A, Kabanova T, Krets V and Saruev L 2014 A Study of Specific Fracture Energy at Percussion Drilling *Proc.XVIII International Scientific Symposium in Honour of Academician M. A. Usov: IOP Conf. Series: Earth and Environmental Science* 21

IOP Conf. Series: Earth and Environmental Science **134** (2018) 012025 doi:10.1088/1755-1315/134/1/012025

Modeling complex state of block-layered rock mass with hierarchic inclusions

OA Khachay[1]* and OYu Khachay[2]

[1]Bulashevich Institute of Geophysics, Ural Branch, Russian Academy of Sciences, Yekaterinburg, Russia
[2]Ural Federal University, Yekaterinburg, Russia

E-mail: *olgakhachay@yandex.ru

Abstract. Iterative algorithms are constructed for 2D modeling of diffraction of sound and linearly polarized S-wave on an inclusion having a hierarchical elastic structure in a J-th layer of an N-layer elastic medium. Under analysis are the cases when the inclusion of each rank and the enclosing medium have different densities and the same elastic parameters and when they have the same densities and different elastic parameters. The space–frequency representation of the wave field distribution uses the methods of the integral and integral–differential equations.

1. Introduction

Using 3D electromagnetic induction space-and-time monitoring, the authors [1, 2] illustrated applicability of the hierarchical discrete medium model to describing structure of rock mass composed of various substances [3]. In the framework of a specific modification of the model, two hierarchical levels were analyzed. Disintegration zones [4] around a tunnel occur asymmetrically at the tunnel roof and floor, which may be an evidence of the nonequilibrium state of the medium. These zones occur discretely, i.e. they are absent in certain intervals in rock mass surrounding tunnels. The maximum change in rock mass exposed to mining impact is the change in the morphology of spatial layout of these zones. The obtained results make it possible to conclude that a rock mass is a multi-rank hierarchical structure. The rock mass dynamics, structure and self-organization can be studied using the flatbed multi-level electromagnetic induction method with a controllable source and the appropriate data processing and interpretation procedure to reveal disintegration zones–indicators of rock mass stability.

Currently the results of theoretical modeling of electromagnetic and seismic field in a layered medium with inclusions of hierarchical structure are commonly used in construction of new systems for integrated monitoring of complex structure media [5]. The modeling algorithm for 3D and 2D nonuniformities in electromagnetic and seismic cases, respectively, are constructed [6–8]. It is shown that as the hierarchical pattern of a medium becomes more complex, the degree of spatial nonlinearity in seismic and electromagnetic fields grows, which agrees with the detailed monitoring tests in rockburst-hazardous Tashtagol and North-Ural Bauxite Mines. The developed theory demonstrates complication of the process of complexing methods using electromagnetic and seismic fields to study response of medium with a hierarchical structure. This problem is inseparable from the formulation and solving of an inverse problem of propagating electromagnetic and seismic fields in such complex media. The problem of the inverse problem algorithm using the equations of the theoretical inverse problem for the 2D Helmholtz equation is discussed in [9, 10]. The explicit equation of the theoretical inverse problem are written for the cases of the electromagnetic field dissipation (polarization E and H) and the linearly polarized elastic wave dissipation in a layered conductive medium and in an elastic medium having a hierarchical conductive or elastic inclusion, which are the framework for delineating non-coaxial inclusions of the first rank of the hierarchical structure. Evidently, the inverse problem solution uses the input data of monitoring adjusted to hierarchical structure of a medium. On the other hand, in a more complex medium, each wave field brings information on the internal structure and condition of the medium, and the interpretation of the seismic and electromagnetic fields should be carried out separately. This study addresses the issue of complexing of different-type geophysical fields propagating in hierarchically structured media.

2. Modeling sound diffraction on 2D hierarchical nonuniformity of anomalous density in an N-layer elastic medium

The work [11] describes the modeling algorithm for the sound diffraction in on 2D elastic hierarchical inclusion in a J-th layer of an N-layer medium. In this case, $k_{1ji}^2 = \omega^2(\sigma_{ji} / \lambda_{ji})$ —P-wave number, where *ji* indicate properties inside the nonuniformity while *ja*—outside the nonuniformity, λ—Lame constant, σ—density; ω—angular rate, $\vec{u}$ = gradφ—vector of displacements, φ^0 —potential of the normal seismic field in the layered medium without nonuniformity: $\varphi_{ji}^0 = \varphi_{ja}^0$. It is assumed that elastic characteristics of the hierarchical inclusion are similar for all ranks l and enclosing layer, while its density differs at all ranks from the density of the enclosing medium; then the system of equations [3] is rewritten as:

$$\begin{aligned}
&\frac{(k_{1ji}^2 - k_{1j}^2)}{2\pi} \iint_{Scl} \varphi_l(M) G_{Sp,j}(M, M^0) d\tau_M + \frac{\sigma_{ja}}{\sigma_{ji}} \varphi_{l-1}^0(M^0) - \\
&- \frac{(\sigma_{ja} - \sigma_{jil})}{\sigma_{jil} 2\pi} \oint_{Cl} G_{Sp,j} \frac{\partial \varphi_l}{\partial n} dc = \varphi(M^0), M^0 \in S_{Cl}, \\
&\frac{\sigma_{jil}(k_{1jil}^2 - k_{1j}^2)}{\sigma(M^0) 2\pi} \iint_{Scl} \varphi_l(M) G_{Sp,j}(M, M^0) d\tau_M + \varphi_{l-1}^0(M^0) - \\
&- \frac{(\sigma_{ja} - \sigma_{jil})}{\sigma(M^0) 2\pi} \oint_{Cl} G_{Sp,j} \frac{\partial \varphi_l}{\partial n} dc = \varphi_l(M^0), M^0 \notin S_{Cl},
\end{aligned} \tag{1}$$

where $k_{1ji}^2 = \omega^2(\sigma_{ji} / \lambda_{ji})$ —P-wave number, where *ji* indicate properties inside the nonuniformity while *ja*—outside the nonuniformity, $l = 1 \ldots L$—hierarchical level number, φ_l^0— potential of the normal seismic field in the layered medium when there is no nonuniformity of the previous rank: $\varphi_l^0 = \varphi_{l-1}$ when $l = 2 \ldots L$ and $\varphi_l^0 = \varphi^0$ when $l = 1$, which conforms with the corresponding expression [11].

Given that the transition to the next hierarchical level causes no changes in the two-dimensionality index and only the geometry of cross-sections of the nested structures is changed, it is possible to describe the iterative process of acoustic field modeling (the case of formation of P-wave only). The iterative process relates with the modeling the displacement vector during transition to the next hierarchical level. Inside each hierarchical level, the integral-differential equation and integral-differential representation are are calculated using by the algorithm (1). When the local nonuniformity structure divides into a number of nonuniformities at a certain hierarchical level, the integrals in the in (1) are taken around all non-uniformities.

3. Modeling diffraction of elastic S-wave on hierarchical nonuniformity of anomalous density in an *N*-layer elastic medium

Similarly to (1) modeling of an elastic S-wave in an *N*-layer medium having 2D structural hierarchy of arbitrary morphology section is written using the integral relations from [11]

$$\frac{(k_{2jil}^2-k_{2j}^2)}{2\pi}\iint_{SCl} u_{xl}(M)G_{Ss,j}(M,M^0)d\tau_M + u_{x(l-1)}^0(M^0)=u_{xl}(M^0),\ M^0\in S_{Cl},$$
$$\frac{\mu_{jil}(k_{2jil}^2-k_{2j}^2)}{\mu(M^0)2\pi}\iint_{SCl} u_{xl}(M)G_{Ss,j}(M,M^0)d\tau_M + u_{x(l-1)}^0(M^0)=u_{xl}(M^0),\ M^0\notin S_{Cl}, \tag{2}$$

where $G_{Ss,j}(M,M^0)$—function of the seismic field source, coincides with the Green function written in [11] for the corresponding problem, $k_{2jil}^2=\omega^2(\sigma_{jil}/\mu_{jil})$ $\mu_{jil}=\mu_{ja}$—S-wave number; u_{xl}—component of the displacement vector; u_{xl}^0—component of the displacement vector of seismic field in the layered medium when there is no nonuniformity of the previous rank: $u_{xl}^0=u_{x(l-1)}$ when $l=2...L$ and $u_{xl}^0=u_x^0$ when $l=1$, which agrees with the corresponding expression for the normal field in [11]. It is noteworthy that the structure of the equations (1) conforms with the general case when the hierarchical nonuniformity has both density and elastic characteristics different from the enclosing layer at all ranks. The only difference of the problem is the wave number values. This, the S-wave-related response is more sensitive to the nonuniform density domain in a medium. This should be taken into account in the analysis of a complex-structure geo-medium.

4. Modeling sound diffraction on 2D hierarchical nonuniformity of anomalous stress state in an *N*-layer elastic medium

The work [11] describes the modeling algorithm for the sound diffraction in on 2D elastic hierarchical inclusion in a *J*-th layer of an *N*-layer medium. $G_{Sp,j}(M,M^0)$—function of seismic field source, the boundary problem for this function is formulated in [12], $k_{1ji}^2=\omega^2(\sigma_{ji}/\lambda_{ji})$—P-wave number. Let the density of the hierarchical inclusion at all ranks *l* be the same as the density of the enclosing layer, while the elastic characteristics of the hierarchical inclusion differ at all ranks from the elastic characteristics of the enclosing medium; then the system of equations (3) is rewritten as:

$$\frac{(k_{1jil}^2-k_{1j}^2)}{2\pi}\iint_{SCl} \varphi_l(M)G_{Sp,j}(M,M^0)d\tau_M + \varphi_{l-1}^0(M^0)=\varphi_l(M^0), M^0\in S_{Cl},$$
$$\frac{\sigma_{jil}(k_{1jil}^2-k_{1j}^2)}{\sigma(M^0)2\pi}\iint_{SCl} \varphi_l(M)G_{Sp,j}(M,M^0)d\tau_M + \varphi_{l-1}^0(M^0)=\varphi_l(M^0), M^0\notin S_{Cl}. \tag{3}$$

The notation is the same as in the system of equations (1).

5. Modeling diffraction of elastic S-wave on hierarchical nonuniformity of anomalous stress state in an *N*-layer elastic medium

Similarly to (3) the algorithm for modeling an elastic S-wave propagation in an *N*-layer medium with a two-dimensional hierarchical structure of an arbitrary morphology section is written using the relations from [11]:

$$
\begin{aligned}
&\frac{(k_{2jil}^2-k_{2j}^2)}{2\pi}\iint_{SCl}u_{xl}(M)G_{Ss,j}(M,M^0)d\tau_M+\frac{\mu_{ja}}{\mu_{jil}}u_{x(l-1)}^0(M^0)+\\
&+\frac{(\mu_{ja}-\mu_{jil})}{\mu_{jil}2\pi}\oint_{Cl}u_{xl}(M)\frac{\partial G_{Ss,j}}{\partial n}dc=u_{xl}(M^0),\ M^0\in S_{Cl},\\
&\frac{\mu_{jil}(k_{2jil}^2-k_{2j}^2)}{\mu(M^0)2\pi}\iint_{SCl}u_{xl}(M)G_{Ss,j}(M,M^0)d\tau_M+u_{x(l-1)}^0(M^0)+\\
&+\frac{(\mu_{ja}-\mu_{jil})}{\mu(M^0)2\pi}\oint_{Cl}u_{xl}(M)\frac{\partial G_{Ss,j}}{\partial n}dc=u_{xl}(M^0),\ M^0\in S_{Cl}.
\end{aligned}
\tag{4}
$$

The notation is the same as in the system of equations (2). The structure of the equations (4) coincides with the general case when both the elastic characteristics and density of the hierarchical inclusion differ at all ranks from the elastic characteristics and density of the enclosing layer. The difference of this problem is the wave number values. In this manner, the P-wave-related response of the medium is more sensitive to the elastic nonuniformity inclusion. This should be taken into account in the analysis of a complex-structure geo-medium.

Conclusion

The comparison of the expressions (1) and (2), (3) and (4) allows some conclusions to be drawn. In the integrated seismic-gravitational model disregarding anomalous stress state inside the inclusion, the analysis of the anomalous acoustic effect based on the P-wave propagation data shows its higher sensitivity relative to the shape of an inclusion as compared with the acoustic effect of S-wave propagation. It follows from the presented expressions that elastic characteristics of an enclosing medium cannot be neglected in the seismic model as these characteristics have influence on the values of anomalous density during interpretation. Otherwise the gravitational model will not reflect material constitution of a medium under analysis. In the anomalously stress state modeling without regard to anomalous influence of density nonuniformity inside the inclusions, the analysis of the anomalous acoustic effect using the S-wave propagation data shows its higher sensitivity as compared to the acoustic effect of P-wave propagation. However, the density of enclosing medium should not be disregarded in the seismic model as the density values influence the interpretation of values of the anomalous elastic characteristics generating the anomalous stress state. Otherwise, the geomechanical model will not reflect adequately stress state of the analyzed medium.

References

[1] Khachay OA 2004 On the study of structure and state of a geological heterogeneous nonstationary medium in the framework of a discrete hierarchical model *Ross. Geofiz. Zh.* No 33–34 pp 32–37

[2] Khachay OA, Novgorodova EN and Khachay OYu 2003 New technique for detecting zones of disintegration in the near-development space of rock mass *GIAB* No 11 pp 26–29

[3] Khachay OA 2007 Investigation of the development of instability in the rocks mass using the method of active electromagnetic monitoring *Fiz. Zemli* No 4 pp 65–70

[4] Shemyakin EI et al 1986 Effect of zonal disintegration of rocks around underground excavations *DAN SSSR* Vol 289 no 5 pp 1088–1084

[5] Khachay OA and Khachay OYu 2017 New monitoring systems to evaluate mine seismicity, taking into account the hierarchical heterogeneity of rock mass *Monitoring. Nauka Tekhnol.* No 1(30) pp 55–61

[6] Khachay OA, Khachay OYu and Khachay AYu 2016 New methods of geoinformatics for the integration of seismic and gravitational fields in hierarchical environments *Geoinformatika* No 3 pp 25–29

[7] Khachay OA, Khachay OYu and Khachay AYu 2015 New methods of geoinformatic for wave fields monitoring in hierarchical environments *Geoinformatika* No 3 pp 45–51
[8] Khachay OA and Khachay AYu 2016 Simulation of seismic field propagation in a layer-block elastic medium with hierarchical plastic inclusions *GIAB* No 12 pp 318–326
[9] Khachay OA and Khachay AYu 2016 Defining of the 2-d surface of the anomaly stressed hierarchical object located in the layered blocked geological medium using the data of acoustic monitoring *GIAB* No 4 pp 354–356
[10] Khachay OA and Khachay AYu 2015 Determination of the surface of fluid-saturated porous inclusion in a hierarchical layered block environment according to electromagnetic monitoring data *GIAB* No 4 pp. 150–154
[11] Khachay OA and Khachay AYu 2013 Modeling of electromagnetic and seismic fields in hierarchically inhomogeneous media *Vestn. YuUGRU, Ser. Vychislit. Matem. Inform.* Vol 2 No 2 pp 49–56
[12] Khachay OA and Khachay AYu 2011 On the integration of seismic and electromagnetic active methods for mapping and monitoring the state of two-dimensional heterogeneities in an N-layer medium *Vestn. YuUGRU, Ser. Komp. Tekhnol., Upravl., Radioelektron.* No. 2(219) pp 49–56

Catastrophe theory—one of the basic components in the analysis of the seismic response of rock mass to explosions

OA Khachay[1]* and OYu Khachay[2]

[1]Bulashevich Institute of Geophysics, Ural Branch, Russian Academy of Sciences, Yekaterinburg, Russia
[2]Ural Federal University, Yekaterinburg, Russia

E-mail: *olgakhachay@yandex.ru

Abstract. It is shown that the dynamic process of mining can be controlled using the catastrophe theory. The control parameters can be values of blasting energy and locations of explosions relative to an area under study or operation. The kinematic and dynamic parameters of the deformation waves, as well as the structural features of rock mass through which these waves pass act as internal parameters. The use of the analysis methods for short-term and medium-term forecast of rock mass condition with the control parameters only is insufficient in the presence of sharp heterogeneity. However, the joint use of qualitative recommendations of the catastrophe theory and spatial–temporal data of changes in the internal parameters of rock mass will allow accident prevention in the course of mining.

1. Introduction

The research into rock mass conditions is continued using the developed algorithm for processing data of detailed seismicity catalogue [1], providing supplemental information essential for predicting hazardous events in deep-level ore mining [2] with blasting. Furthermore, in spotlight is the process of initiation of resonant energy release, including the blasting impact and weak series energy responses in the form of tremors contributing to the resonant release. The message-bearing signs of initiation of a high-energy dynamic event are used: response time after an impact and volume of a rockburst source in the form of ranking of distance from the expected focus. The multimedia seismological information is of importance in prediction of hazardous events in ore mines.

The theoretical results on the causes of chaotization in nonlinear dissipative dynamic systems are compared with the data of processing detailed mine seismicity catalogue—seismic responses of rockburst-hazardous rock mass to blasting—by the method of phase diagrams. Theoretically, the common cause of the chaotization and stochastization in a dynamic system is the loss of stability, which can be recorded as exponential gapping between close-spaced phase trajectory, combined with the common boundedness and compaction. This result agrees with the phase diagrams plotted based on the mine seismicity catalogue data. [3, 4]. Currently, it is of significance that physics and mathematics start interacting, and needs of physics influence mathematical methods while mathematics has influence on physical knowledge. However, there are some physical problems for which the linear mathematics apparatus is either insufficient or inapplicable [5]. For all-around coverage of the diverse phenomena in acoustics and mechanics, the mathematical apparatus of linear differential equations is absolutely deficient and leaves aside the events which are the most specific and interesting. The fact is that the differential equations which describe these events are knowingly nonlinear. Accordingly, we talk of 'nonlinear' systems. The foundations of the mathematical apparatus relevant to individual problems and the whole cycle of nonlinear problems were laid by Poincaré and Lyapunov [6, 7].

In this study, the mathematical approaches of the catastrophe theory are applied to analyzing mine seismicity catalogue data on rock mass stability and instability under dynamic impact. One of the first researchers focusing on the loss of quasi-static equilibrium in rock mass using the catastrophe theory was Odintsev [8]. The important theoretical results were obtained in the sphere of estimation and prediction of dynamic conditions in rock mass under mechanical effect, and in assessment of post-dynamic condition, which is of specific significance in work planning in mine after rock burst. The application of these theoretical results needs stress monitoring at rock mass and mine interface and inside the rock mass, which is laborious. The way out is geophysical monitoring of wave fields containing information both on structure and stresses. One of such data bases is mine seismicity catalogue.

2. Review of the catastrophe theory methods for instability study in linear dynamic systems

The first studies on the catastrophe theory were published approximately in 1979 and told that this theory gives a universal tool to analyze all jump-type transitions, ruptures and qualitative changes [9]. According to Arnold [9], the sources of the catastrophe theory are Whitney' theory of smooth mapping and theory of bifurcation in dynamical systems by Poincaré and Andronov. Bifurcation means partition and describes qualitative restructuring of objects upon the change of parameters the objects depend on. Catastrophes are the jump-type changes as a spontaneous response of a system to smooth change in external conditions. The evolutionary process describing the system response to the applied effect is mathematically described by a vector field in a phase space [10]. The point in the phase space defines condition of the system, and the vector at this point denotes the rate of change in this condition. The points at which vectors turn to zero are named the balance positions. In course of time, vibration may arise in the system and equilibrium becomes unstable. On a phase plane, the stationary vibrations are represented by a closed curve called a limit cycle. The evolution of a dynamical system was studied in [9] with regard to time variations in internal and governing parameters. Generally, the application flowchart of the catastrophe theory starts with the assumption that the process under study is described using a certain number of governing and internal parameters. The equilibrium states of the process form a surface of some or other number of measurements in this space. The mapping of the surface of equilibriums onto plane of governing parameters has singularities called the singularities of general positions. The theory related with these singularities predicts geometry of 'catastrophes,' i.e. jumps between states under the change in the governing parameters. One of the most well-known singularities is Whitney's umbrella. After the loss of stability of equilibrium, two steady-state conditions can be observed. One is an oscillating periodic mode. This type of instability is called soft instability as the stabilizing oscillating mode, in case of small post-criticality, differs insignificantly from the equilibrium state. The second mode is connected with some features: before a steady-state mode loses stability, the domain of attraction of this mode becomes very small, and random disturbances throw the system from this domain before it vanishes. This type of loss in stability is called rough instability, and the system jumps from the stationary mode to another mode of motion. The latter can be another steady-state stationary mode, or steady-state vibrations, or event more complex motion. The steady-state modes of motion are attractors. The modes differing from equilibrium and strictly periodic oscillations are called strange attractors and connected with the problem of turbulence [10].

3. Analysis of the data of active mine seismicity monitoring using the catastrophe theory flowchart

The main objective of monitoring is detecting pre-critical condition of rock mass under mining with blasting. The pre-critical condition is defined as the response energy not higher than 10^4 J. The response energy in the range 10^5–10^7 J means the critical condition of rock mass. The response energy of 10^8–10^9 J and higher defines post-critical or catastrophic condition.

On Nov 25, 2007, in Tashtagol Mine cross-cut 29 at a depth of 714 m, rock burst with the energy of $8.14 \cdot 10^8$ J took place. Let us trace initiation of this event from the viewpoint of rock mass instability and connection with the effect and locations of blasting. The rock mass volume under study is represented in the coordinates *OX* (25–31 tunnel ≈ 240 m), *OY* (length of cross-cuts between fringe drift and air way is ≈ 240 m) and *OZ* (mine levels –140, –425 m, ≈ 590–875 m).

Below we analyze the morphology of time of rock mass responses using the data from the mine seismicity catalog since June 4, 2006 till November 25, 2007.

4 June–13 August 2006. The loss in equilibrium after four impacts takes places in the form of oscillations; the first blast is carried out inside the object under study but it has the least effect out of the four blasts, and the maximum energy response is within the pre-critical range.

20 August–17 September 2006. Classical mode of instability after two blasts. There is a response in the form of almost uniform time distribution of energy passing to oscillating process; Sep 3 blast energy exceeded 10^8 J, while Sep 17 blast energy exceeded 10^6 J thought the total energy of response reduces and is not higher than 10^3 J.

24 September–14 October 2006. The response has different morphology, the oscillating process changes to jump-type release follows by oscillation with the background level of soft instability. The next responses occur beyond the domain under study and behave as low-amplitude energy oscillations. On Oct 14, 2006, the seismic response morphology is the classical soft instability with the attenuation of oscillating energy release.

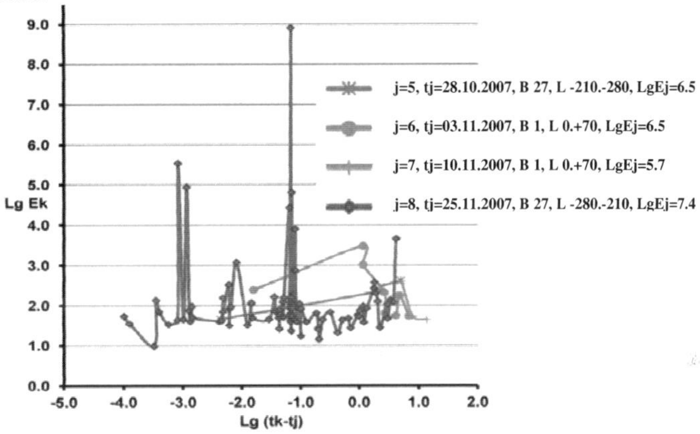

Figure 1. Morphology of energy *Ek* (J) of rock mass response to blasting in time from Oct 28 to Nov 25, 2007: *tk*—response time; *tj*—impact date and time (in days); *Ej*—disturbance energy (J), B—block; L—level

21 October–12 November 2006. Blasting is carried out inside the study domain but the morphology of soft instability is more pronounced after the blast on Nov 12. Immediately after the blast, oscillating energy release begins, then goes jump followed by oscillation and again jump with an amplitude more than 10^4 J. the Background energy oscillations exceed 10^2 J.

19 November–17 December 2006. Outside the study domain, a blast with the energy higher than 10^9 J is carried out, and the rock mass responds by two jump-type energy releases with the energy of the jumps on Nov 12, 2006. The next blast in the study domain on Nov 26, 2006 causes the soft instability response on the energy background of the second period of the analysis. The next blast on Dec 3, 2006 induces higher response with oscillations and jumps as in the third period of the analysis.

24 December 2006–21 January 2007. The blast inside the study domain (nearby the place of the future rock burst) with an energy of 10^8 J causes sharp though critical instability in the rock mass, two jumps with an energy under 10^5 J and three jumps with an energy of 10^4–10^3 J. The oscillations are more intensive. The other blasts in the distance from the study domain over the period from Dec 30, 2006 to Jan 14, 2007, sustain oscillations induced by the blast on Dec 24. The blast on Jan 21 with the energy lower than 10^5 J at a long distance from the study domain does not prevent jump-type energy release reflective of prolongation of stress relaxation period in the system.

25 March–8 April 2007. Blasting is conducted far from the study domain. The rock mass response is random and not oscillating, the energy is mostly not higher than 10^4 J, except for the response to the blast on Apr 1, 2007 when the energy jumps over 10^4 J.

6–20 May 2007. Basting in the north of the rock mass, than, on May 13 simultaneous blasting in the north and inside the study domain (in the cross-cut where the catastrophic rock burst occurred later on), rock mass response is a jump with an energy of 10^5 J.

13 June–15 September 2007. The study domain mostly responds to the outside blasts, the response to the internal blasting is jumps of energy to 10^5 J without preliminary oscillating.

30 September–21 October 2007. Blasting only in the study domain nearby the rock burst cross-cut. The energy distribution morphology shows the low-energy dependence on time and very short period of oscillations.

It is seen in the figure above that blasting between October 28 and November 10 induces no oscillating energy response in rock mass. The blasting on Nov 25, 2007, in block 27 causes a complex dynamic catastrophe process. There are three foreshocks, one aftershock, four pre-critical and one post-critical jump of energy. All these events are accompanied by low-energy oscillations.

4. Conclusion

Mining is a dynamic process which is controllable using the recommendations of the catastrophe theory. The governing parameters of the process can be blast energy and blast locations relative to the rock mass domain subjected to the analysis. The internal parameters are the kinematic and dynamic characteristics of deformation waves [11, 12] and the rock mass structure where the waves propagate [13]. It is insufficient to make short-term and long-term prediction of sharply nonuniform rock mass conditions sing only the governing parameters. On the other hand, the joint application of the qualitative recommendations of the catastrophe theory and the data on space and time of change in internal parameters will make it possible to prevent catastrophes in mines.

References

[1] Khachay OA and Khachay OYu 2014 Comparison peculiarities of synergetic properties of rockburst-hazardous rock mass, defined from the seismic and induction electromagnetic monitoring data *Monitoring. Nauka Tekhnol.* No 3 pp 50–55

[2] Khachay OA, Khachay OYu, Klimko VK and Shipeev OV 2015 Informative precursors of high-energy dynamic events using data of seismological mine monitoring *GIAB* No 4 pp 155–216

[3] Naimark YuI 1978 *Dynamic Systems and Controllable Processes* Moscow: Nauka (in Russian)

[4] Khachay OA and Khachay OYu 2015 Nonlinear dynamic modeling rock mass response to strong impacts *Geofiz. Issled.* Vol 16 No 2 pp 19–26 UT RSCI:23569010

[5] Andronov AA, Vitt AA and Khaikin SE 1981 *Theory of Oscillations* Moscow, Nauka (in Russian)

[6] Poincaré H 1971–1974 *Collected Works* Moscow: Nauka (in Russian)

[7] Lyapunov AM 1954–1959 *Collected Works* Moscow–Leningrad: AN SSSR (in Russian)

[8] Odintsev VN 1995 Analytical prediction of the dynamic effects of mine pressure J. Min. Sci. Vol 31 No 4 pp 245–255

[9] Arnold VI 1990 *Catastrophe Theory* Moscow: Nauka (in Russian)

[10] Klimontovitch YuL 2007 *Turbulent Motion and Chaotic Structure* Moscow: Kom. Kniga (in Russian)

[11] Khachay OA, Khachay OYu and Klimko VK 2013 Dynamic characteristics of slow deformation waves as a rock mass response to blasting *GIAB* No 5 pp 208–214

[12] Khachay OA, Khachay OYu and Shipeev OV 2013 Study of the hierarchic structure of dynamic characteristics of slow deformation waves *GIAB* No 5 pp 215–222

[13] Khachay OA, Khachay OYu, Klimko VK and Shipeev OV 2016 Study the non linear dynamic state of rock mass using the seismic catalogue data obtained in deep mines under anthropogenic influence *GIAB* No 4 pp 217–230

On the calculation of air flow rates to ventilate closed-type stations in subway with the double-track tunnel

LA Kiyanitsa

Chinakal Institute of Mining, Siberian Branch, Russian Academy of Science, Novosibirsk, Russia

E-mail: lavrentij.kijanitza@yandex.ru

Abstract. Metro is not only the most promising kind of public transport but also an important part of infrastructure in a modern city. As a place where large groups of people gather, subway is to ensure the required air exchange at the passenger platforms of the stations. The air flow rate for airing the stations is also determined based on the required temperature, humidity and MAC of gases. The present study estimates the required air flow rate at the passenger platform of the closed-type subway station with the double-track tunnel given the standard air temperature, humidity and gas concentration, as well as based on the condition of the specified air flow feed and air changes per hour. The article proposes the scheme of air recirculation from the double-track tunnel to the station.

1. Introduction

Microclimate in subway underground structures is provided by tunnel and station ventilation systems. Choice of equipment and operation mode parameters depends on air flow rates calculated in terms of heat-humidity balance in main line tunnels and at Metro stations. Modern subway construction trends imply erection of closed-type stations and double-track tunnels. Case studies are construction of double-track tunnel in Frunze radius in Saint Petersburg and Kozhukhov line in Moscow. Of actual significance is the development of scientific-methodological fundamentals of design of closed-type subway with double-track tunnels under severe continental climate conditions, as the experience in design of similar subways is not available so far.

The basic specific feature of closed-type stations alternatively to other types is aerodynamic isolation of passenger platforms from track sections and line tunnels with no respective heat and humidity exchange between these subway sections. This explains existence of separate aeration modes for closed-type stations and double-track tunnels [1].

The airflow performance of a ventilation system is selected based on heat and humidity balance in passenger platforms and track sections. Microclimate parameters at stations and tunnels are normalized by the Russian Federation acting acts [2, 3].

To calculate a ventilation air flow rate at passenger platforms it is necessary to calculate amount of released heat, and humidity, and repugnant substances and to derive an equation of heat, humidity balance and hazardous substance balance in order to determine a required assimilating capacity of the fresh air flow.

2. Calculation methods

The principle components of the station heat balance are passengers and personnel, escalators, heat exchange between a station and a track tunnel, light systems, heat exchange between peripheral soil mass and a station, subway stationary equipment. In the case of the lack of a permanent water flow the main sources of moisture and CO_2 at a station are passengers and personnel. The basic components of heat balance in a tunnel and a track section are trains, passengers, lighting systems, cable-channels, heat exchange between a station and a track section, stationary equipment, combines trailer-secondary substations (if waste air is removed from a station), heat exchange between subway infrastructure and a peripheral soil mass. The principal source of moisture and CO_2 in a tunnel is passengers.

Heat generation (average per an hour) q_n, kW, for trains is determined from formula [4]:

$$q_n = \frac{2 \cdot \left(T_1 - T_2 \pm \sum A_G\right) \cdot N_{t/d}}{24 \cdot 3600 \cdot 1000}, \tag{1}$$

where T_1 is kinetic energy of a train at the initial slow-downing stage, J, T_2 is kinetic energy at the final braking stage, $T_2 = 0$ J; ΣA_G is job of gravitation forces in the given section, it is assumed that $\Sigma A_G = 0$; 2 is a magnitude of twoness of traffic; $N_{t/d}$ is a number of train couples per a day.

Heat flow q_{pc}, kW, from a track section to passenger platforms by heat transmission through a partition can be calculated from formula [5]:

$$q_{pc} = \frac{F}{1000} \cdot \frac{\left(t_{\text{ts}} - t_{\text{st}}\right)}{\dfrac{1}{\alpha_{\text{t-p}}} + \sum_{i=1}^{n} \dfrac{\delta_n}{\lambda_n} + \dfrac{1}{\alpha_{\text{s-p}}}}, \tag{2}$$

where F is area of a partition wall, m^2; t_{ts} is temperature in a track section, °C; t_{st} is air temperature at a station, °C; $\alpha_{t\text{-}p}$ is air heat transfer factor in a track section to a partition wall, W/(m$^2\cdot$°C); $\alpha_{s\text{-}p}$ – is air heat transfer factor in a station to a partition wall, W/(m$^2\cdot$°C); δ_n is thickness of the i-th wall layer, m; λ_n is heat conductivity factor of the i-the layer of the wall, W/m·°C; n is number of wall layers, u.

The numerical modeling of heat transition between a track section and a station through open gateway of a track section enabled to establish that warm air mainly passed trough an upper one third of gateway opening. A warm air flow through open gateways of a track section q_{og}, kW was calculated from formula [5]:

$$q_{\text{og}} = \vartheta \cdot \frac{1}{3} h \cdot b \cdot \rho_{\text{a}} \cdot c_{\text{a}} \cdot \left(t_{\text{ts}} - t_{\text{st}}\right) \cdot n_{\text{c}} \cdot n_d \cdot N_{t/d} \frac{2 \cdot \tau_{st}}{24 \cdot 1000 \cdot 3600}, \tag{3}$$

where ϑ is air velocity in the upper one third of gate opening, m/s, determined experimentally; h and $\cdot b$ are dimensions of gate opening, m; ρ_a is air density, kg/m^3; c_a is isobar heat capacity of air, J/(kg·°C); t_{ts} is temperature in a track section, °C; t_{st} is temperature at a station, °C; n_c is number of cars in a train, u.; n_d number of door openings per a car, u.; τ_{st} is stopping time at a station, J is assumed 30 s; 2 is a magnitude of twoness of train traffic.

Heat generation of escalators, light systems and power substations are calculated under procedures, described in [6]. Heat generation of stationary equipment q_{eq}, kW, is determined from formula [7, 8]:

$$q_{\text{eq}} = \frac{P_{\text{m}} \cdot (1 - \eta)}{\eta} \cdot K, \tag{4}$$

where P_m is rated motor capacity of a facility, kW; η is engine efficiency; K is activity ratio.

Heat flow from station platform and rooms to a soil mass per a year is of reverse character. Herewith heat loss of a station in a soil mass in the first year of exploitation differs from heat loss at a routine operation mode. Heat loss of passenger platform and a track section of a station in a soil mass in the initial operation period $q^{pp}{}_{init}$, W is calculated under procedure described in [6].

Specific heat loss at a routine operation mode in passenger platform and a track section $q^{pp}{}_{ro}$, W/m$^{2\cdot}$ are calculated from:

$$q_{\text{ro}}^{\text{pp}}(\tau) = B(h) + A(h) \cdot \cos(0.0172 \cdot \tau + \omega(h)), \tag{5}$$

where $A(h)$ amplitude of specific heat flow fluctuations, W/m^2; $B(h)$ is average annual specific heat flow, W/m^2; $\omega(h)$ is initial phase of fluctuation of specific heat flow, rad; 0.0172 is angular factor, 1/day. Coefficients $A(h)$, $B(h)$, $\omega(h)$ are determined in terms of station depth h, m.

Heat loss in a peripheral soil mass of a running meter of double-track tunnel is calculated under the procedure [9].

Heat generation q_{pas}, W, moisture production W_{pas}, kg/h, and CO_2 emission B_{pas}, kg/h, on the part of passengers and personnel are calculated by formulas:

–passenger platform of the station:

$$q^{st}{}_{pas} = q_{pas} \cdot n_{pas}\,;\; W^{st}{}_{pas} = w_{pas} \cdot n_{pas}\,;\; B^{st}{}_{pas} = b_{pas} \cdot n_{pas}\,, \tag{6}$$

where q_{pas}, w_{pas}, b_{pas} are total heat generation by a passenger calculated by [6] and n_{pas} – number of passengers at a platform;

–two-track tunnel and a track section:

$$q^{tt}_{pas} = q_{pas} \cdot n_{pas} \cdot n_t\,;\; W^{tt}_{pas} = w_{pas} \cdot n_{pas} \cdot n_t\,;\; B^{tt}_{pas} = b_{pas} \cdot n_{pas} \cdot n_t\,, \tag{7}$$

where q_{pas}, w_{pas}, b_{pas} are total heat generation by a passenger calculated by [6] and n_{pas} is number of passengers in a train; n_t is number of trains per a track section, calculated by formula:

$$n_t = \frac{2 \cdot N_{t/d} \cdot \left(\dfrac{l - l_m - l_p}{\vartheta} + \tau_c + \tau_m + \tau_p \right)}{24 \cdot 3600} \le 2\,, \tag{8}$$

where $N_{t/d}$ is number of train pairs per a day; l is track span length, m; l_{sd}, l_{acc} are length of slowdown and acceleration sections, respectively, m; ϑ is operating speed of train motion, m/s; τ_s, τ_{sd}, τ_{acc} rated time of stopping, slowdown, and acceleration of a train, s.

Rated heat amount q_{vent}, kW, humidity W_{vent}, kg/h, and CO_2 B_{vent}, kg/h, intended for removal by station ventilation systems and calculated by formulas:

$$\sum_{i=1}^{k} q_i = q_{vent}\,;\; \sum_{i=1}^{k} W_i = W_{vent}\,;\; \sum_{i=1}^{k} B_i = B_{vent}\,, \tag{9}$$

where i is number of a toxic source; k is quantity of toxic sources, q_i, W_i, B_i amount of toxic emission of the i-th source;

The air flow rate of overall-exchange ventilation and air conditioning systems is computed under the procedure described in [10] to provide sanitary-hygienic conditions, fire-and-explosion safety and to eliminate condensate generation. The air flow rate should be evaluated separately for warm, cold, and transitional periods of the year in terms of assimilation of heat and moisture generation and mass of harmful substance emission, approving the maximum values of the resultant air flow rate data.

2. Results

Cumulative average hourly heat emission/loss q, kW, and production of moisture W and CO_2 B, kg/h for passenger platforms of the closed-type station (Gusinobrodskaya station, Novosibirsk Metro), track section and double-track tunnel (l = 1990 m) (Table 1).

The design air flow rate was calculated for: (a) track section and main track line tunnel, (b) passenger platform and rooms of the station. It is found that in public places of the closed-type station the maximum air flow rate is evaluated based on the rated air exchange condition, in a track section and main track line tunnel the maximum air flow rate is calculated considering the conditions of assimilation of heat production. For a double-track tunnel and a track section the calculated ventilation air flow rate amounted to (a) 78.93 m^3/s in warm season; (b) 47.24 m^3/s in transitional season, and (c) 20.22 m^3/s in cold season. The minimum fresh air flow rate for a double-track tunnel is 11.33 m^3/s. Calculated ventilation air flow rates for passenger platform of the closed-type station and recirculating air flow from a track section to public rooms are cited in Table 2.

Table 1. Cumulative average hourly heat emission, production of moisture and CO_2

Period of a year		Passenger platform			Tunnel and track tunnel sections		
		Warm period	Transitional period	Cold period	Warm period	Transitional period	Cold period
The first operation year	q, kW	–396.49	–64.49	–59.24	299.06	963.25	961.94
	W, kg/h	42.81	25.56	18.72	240.3	134.84	53.4
	B, kg/h	13.86	13.86	13.86	60.07	60.07	60.07
Routine operation year	q, kW	–45.47	–49.75	–95.43	947.7	963.64	1091.64
	W, kg/h	42.81	25.56	18.72	240.3	134.84	53.4
	B, kg/h	13.86	13.86	13.86	60.07	60.07	60.07

Table 2. Calculated ventilation air flow rate for public rooms of the closed-type station and a share of recirculating tunnel air

Period of a year	The first year of operation			Routine operation mode		
	Warm[1]	Transitional	Cold	Warm	Transitional	Cold
Calculated ventilation air flow rate L, m^3/s	9.94	9.94	9.94	9.94	9.94	9.94
The minimum fresh air flow rate L^{min}_{fa}, m^3/s	2.57	2.57	2.57	2.57	2.57	2.57
Fresh air flow rate L_{fa}, m^3/s	6.17	2.60	3.98	6.71	2.57	4.29
Share of recirculating tunnel air r, %	37.9	73.8	60.0	32.5	74.1	56.8

[1] – additional heat supply is required.

Recirculation share r, %, is determined from:

$$r = \frac{L - L^{\min}_{\text{fa}}}{L}, \tag{10}$$

where L is calculated air flow rate, m^3/s; $L^{\min}_{fa}$ is the minimum required fresh air flow rate, m^3/s.

To compensate heat deficit in passenger rooms of the closed-type stations (Table 1) the design of separate ventilation systems is proposed for the closed-type stations and a two-track tunnel with recirculation of tunnel air in passenger rooms of the station. This design enables to lower load on air-heating ventilation systems in winter and transitional periods (Figure 1) and to save heat energy up to 43% in the cold period of the year.

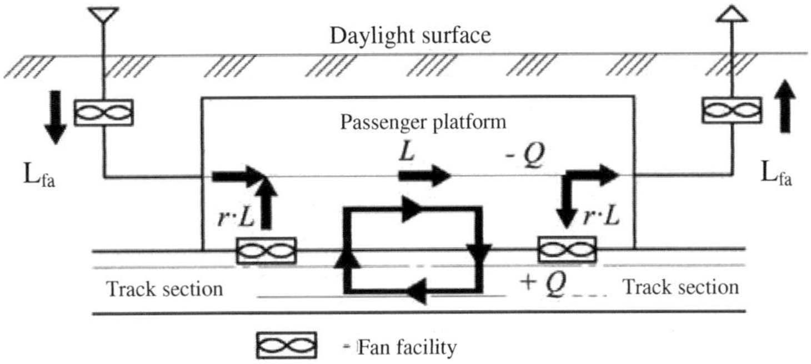

Figure 1. Schematic of the separate ventilation system of the closed-type station with recirculation of tunnel air in passenger platform of the station.

3. Conclusions

1. Average hourly heat generation, amount of moisture and CO_2 emissions designed to be assimilated by ventilation systems in passenger rooms of the closed-type station and two-track tunnel are evaluated;

2. The design air flow rate for ventilation of two-track tunnel and fresh air flow rate plus recirculating tunnel air for station ventilation are computed;

3. The new-proposed design of separate ventilation systems for the closed-type station and two-track tunnel with recirculation of tunnel air to passenger rooms of the station makes it possible to lower heat load on the ventilation air-heating systems and to save up to 43% of heat energy in the cold period of the year.

References

[1] Kiyanitsa LA and Lugin IV 2016 Heat and humidity balance as influencing design parameter of selective ventilation in subway with closed-type stations of *J. Fundament App. Min. Sci.* Vol 3 No 2 pp 86–91 (in Russian)

[2] *RF Construction Regulations SP 120.13330.2012. Subways* 2013 Moskow Ministry of Regional Development (in Russian)

[3] *RF Construction Regulations SP 2.5.1337-03 Sanitary rules for subways approved by the chief sanotory inspector of Russia* 2010 Moskow: Ministry of Health (in Russian)

[4] Krasyuk AM 2006 *Tunnel Ventilation of Subways* Novosibirsk: Nauka (in Russian)

[5] Kiyanitsa LA 2017 Modeling of heat exchange between passenger areas and double-track tunnel at a closed-type metro station *Technical Regulation in Transport Construction* Vol 21 No 1 URL: trts.esrae.ru/40-240 (in Russian)

[6] Tsodikov VYa 1975 *Ventilation and Heating Subways* Moskow: Nedra (in Russian)

[7] Ligerman II 1963 *Ventilation of Electric Machines of Industrial Enterprises* Moscov-Leningrad: Gosenergoizdat, (in Russian)

[8] *Regulation Code RD 22.18-355-89 Method for Determining Heat Emissions of Electrical Equipment* 1989 Rostov-na-Donu: GiproNIIStroiMash (in Russian)

[9] Lugin IV and Alferova EL 2016 Influence of the annual cyclicity of the change in the heat flux to the ground on the calculated heat balance of the double-track tunnel of the subway *Interekspo Geo-Sibir', 2016* No 3 part 2 pp 191–196 (in Russian)

[10] *RF Construction Rules 60.13330.2012 2013: Heating, Ventilation and Air Conditioning* approved by Ministry of Regional Development, dated Jan. 1, 2013

Physical modeling of axisymmetric hydrofracturing by plastic material injection in elastic medium

I.V. Kolykhalov

Chinakal Institute of Mining, Siberian Branch, Russian Academy of Sciences, Novosibirsk, Russia

E-mail:ikolykhalov@mail.ru

Abstract. The article describes the experimental and numerical investigation of hydraulic fracture propagation under injection of a plastic material near the free surface and the surface loaded by a die block to simulate the effect of an open fracture in the course of the multiple hydrofracturing. The experimental and calculated data are compared.

1. Introduction

Hydrofracturing is a technological process designed to disintegrate rocks. In hydrofracturing a viscous fluid or a plastic material is pumped into a natural or artificial fracture thus inducing growth of a fracture up to a wanted size. The process under consideration provides a better opportunity to extract a mineral deposit in a safe and ecofriendly manner without mine construction. The directed hydrofracturing became popular in mining of tight roof, coal seam degasing, sulfide ore leaching, rock plugging, blast-free extraction of precious crystal-bearing rock blocks, and production of construction stone [1, 2]. The latest novel application of hydrofracturing is intensification of oil and gas production from low-permeability reservoirs [3, 4]. Concept of the process runs as follows: a series of fractures is initiated perpendicularly to a borehole with a preset interval in order to increase maximally an area of hydrocarbons intake. The maximum performance of the process is gained when induced fractures are of disk-type and distance between them is equal to radius of drainage area. This condition is not always fulfilled, as induced fractures can be curved in shape and intercrossing. Geometry of induced fractures is appreciably influenced by presence of earlier artificial or natural fracture, free surfaces, fluid pumping modes, and properties of a working fluid. Theoretical studies on hydrofracture development nearby earlier available fractures are reported in [5, 6], where the algorithm for calculation of a path of successive evolution of co-axial fractures is developed for an axisymmetric statement of the problem. In the present paper the researcher models propagation of an axisymmetric fracture in hydrofracturing with a plastic material in the vicinity of earlier available open fracture and free surface. The aim of the present studies is to check compliance of experimental fracture parameters with numerical calculated results.

2. Experimental procedure

Physical modeling was carried out on blocks made of polymethylacrylate (organic glass) of 230×160×106 mm and 170×170×36 mm in size. An available open hydrofracture was modeled by

IOP Conf. Series: Earth and Environmental Science **134** (2018) 012028 doi:10.1088/1755-1315/134/1/012028

surface loading with a round ellipsoid-shape print press with displaced center $U = 0.1$ mm (small semi-axis) and diameter of 100 mm (large semi-axis). The print was pressed and fixed with a preset force with bolts of 6 mm in diameter on the maximum-size side of the test specimen (Figure 1a).

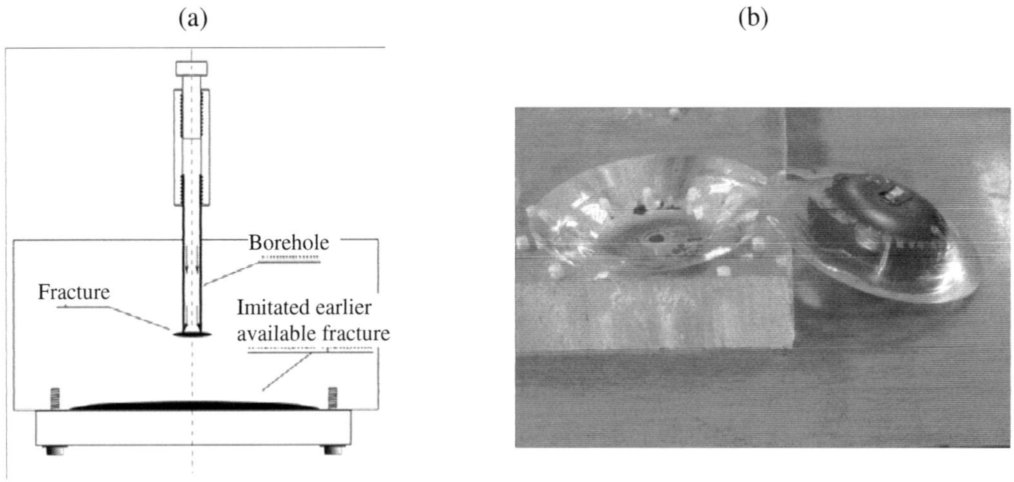

Figure 1. Schematic of experiments on (a) development of closely located hydrofractures and (b) external view of an induced fracture shape.

To imitate a production well, a hole of 10 mm in diameter was made on the opposite side along axis of the available fracture. A narrow incipient axisymmetric fracture imitating a transverse well was cut in the bottom of the hole in distance h from the available fracture. The fracture diameter with account for well's diameter was 14 mm. Next, a proppant agent was fed by means of a screw auger to the incipient fracture at flowrate of 0.5–0.7 cm^3/min. The fracture developed at the near-equilibrium mode up to the moment until projection of its radius on a plane of the available fracture reached 50–60 mm. The proppant agent was plasticine OST 6-15-1525-86 applied at temperature 24°C. A series of tests was carried out in different distances h between a growing fracture and the available one (Figure 1b). To study the effect of free surface on parameters of growing fracture the tests were performed under the described procedure with no print press loading of the test specimen in different distances to free surface.

3. Numerical calculations

Numerical softwares developed by discontinuous component methods were used to calculate fracture growth [7]. Computation algorithm is described in details in [5, 6]. The experimental problem statement was: three co-axial parallel disk-type fractures locate normally to axis z in an elastic space infinitely compressed under principal stresses (Figure 2). The first fracture has radius Q and boundary conditions on its edges $\sigma_n = 0$, $\tau_s = 0$. This fracture imitates stress-free surface of a block. The second and third fractures have radii R_0 and r and represent the earlier available and incipient fractures, respectively. Distance H is between free surface and the available fracture and h is between the incipient and available fractures. Boundary conditions on edges of the available fracture are prescribed by ellipsoid opening equal to $2U_0$ in its center. Pressure in the incipient fracture is assumed $p(s,t)$, where s is coordinate along the length of the fracture line in meridian cross-section, t is time.

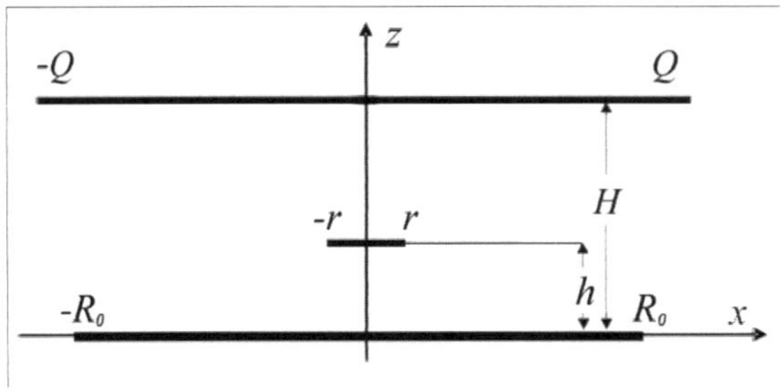

Figure 2. Initial position of fractures in cross-section plane $y = 0$

Development of the fracture was computed in near-equilibrium mode. The working fluid flow is described by the law of the plastic material deformation. The solution to material motion along the fracture was obtained by AM Linkov's hypothesis on proportionality of effective normal stresses acting on fracture edges relative to their normal displacements [8]. Considering this hypothesis the set of equations of the plastic material flow is:

$$p(s,t) = \frac{\pi E}{4(1-\upsilon^2)} \cdot \frac{U(s,t)}{R(t)},$$
$$\frac{\partial p(s,t)}{\partial s} = -\frac{\tau_0}{U(s,t)}, \qquad (1)$$
$$q_0 = \frac{dV}{dR_\gamma} \cdot \dot{R}_\gamma(t),$$

where E is elasticity modulus of a medium; υ is Poisson ratio; $U(s,t)$ is semi-opening of a fracture; R is length of fracture line in meridian section; R_γ is a length of a fracture section filled with a plastic material; τ_0 is yield limit of a plastic material; V is fracture volume; q_0 is plastic material consumption.

From first two equations (1) for calculation of $p(s,t)$ we obtain differential equation:

$$\frac{\partial p(s,t)}{\partial s} = -\frac{\pi \tau_0 E}{4(1-\nu^2)} \frac{1}{p(s,t) R(t)}, \qquad (2)$$

which solution should meet condition $p(0,t) = P_0(t)$. $P_0(t)$ is pressure in the center of the fracture where the equilibrium fracture growth condition is fulfilled; $dW/dS = 2\gamma$ is Griffith's energy fracture criterion, where dW/dS is intensity of liberated elastic energy; 2γ is failure viscosity. In terms of boundary conditions in the center of the fracture the analytical solution is:

$$p(s,t) = \sqrt{P_0^2 - \frac{\pi \tau_0 E}{2(1-\nu^2)} \cdot \frac{s}{R(t)}}. \qquad .(3)$$

In numerical calculations the following parameters were used: $E = 3.3 \cdot 10^3$ MPa, $\upsilon = 0.33$, $R_0 = 50$ mm, $2U_0 = 0.2$ mm, failure viscosity $2\gamma = 2.7 \cdot 10^{-4}$ MPa $\cdot$ m, corresponding to the critical coefficient of stress intensity $K_{1C} = 1$ MPa $\cdot$ m$^{0.5}$, $\tau_0 = 0.01$ MPa.

4. Analysis of the experimental results

IOP Conf. Series: Earth and Environmental Science **134** (2018) 012028 doi:10.1088/1755-1315/134/1/012028

In Figures 3a and 4 the fractures are produced experimentally and calculated numerically for different distances H and h. Solid lines correspond to tracks, calculated numerically; lines with triangle marks are experimental fractures; dashed lines are for calculated paths with the use of a perfect fluid as a working fluid, that is $p(s,t) = p(t)$. Points denote free surface. It is obvious that shapes of fractures calculated with the use of a plastic material exhibit less curvature and better compliance with experimental curves. As well as its curvature tends to lower with increase in distance from a developing fracture to a print press and from free surface.

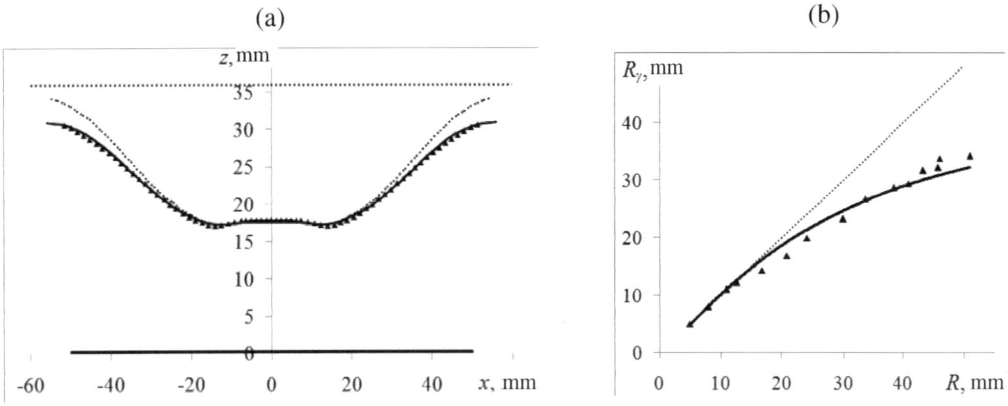

Figure 3. Experimental results on the specimen of $H = 36$ mm: (a) paths of axisymmetric fracture, developing from the initial spot in distance $h = 18$ from the available fracture with opening in the center $2U_0 = 0.2$ mm; (b) relationship of a length of fracture section stuffed with a plastic material versus its length.

As a rule, a delay of the working fluid front from fracture edge is observed at small values of normal stresses on the fracture contour from the side of the external compression field. In Figure 3b relationship of a length of a section filled with a plastic material R_γ versus fracture length R is shown. Dashed line is for the case when the working fluid fills completely fracture ($R_\gamma(t) = R(t)$), solid line is for numerical calculations, marks are for experimental data. Divergence between numerical and experimental data does not exceed 15%.

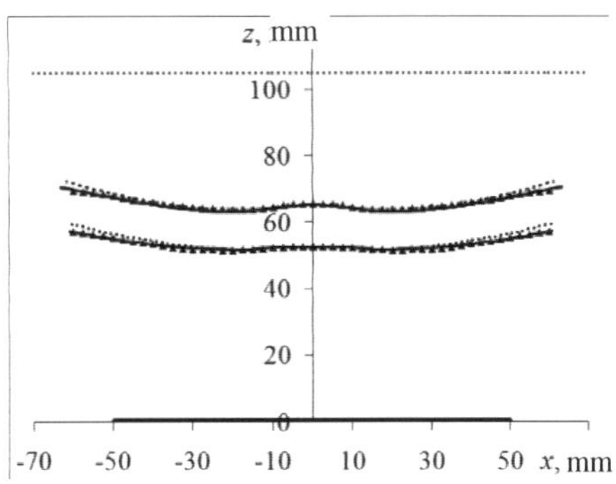

Figure 4. Paths of axisymmetric fractures induced in distance h = 53 mm and h = 65 mm from the available one in a specimen of H = 106 mm

Let consider development of the fracture in the vicinity of only one free surface. Figure 5a demonstrates the growth paths of axisymmetric fractures obtained experimentally (triangle marks) and by numerical computation (solid lines) nearby free surface. Free surface coincides with abscissa. In Figure 5b solid and dashed lines present relations of calculated volumes of the fracture and a plastic material versus fracture length. The experimental volume of plasticine pumped into a fracture is denoted by markers.

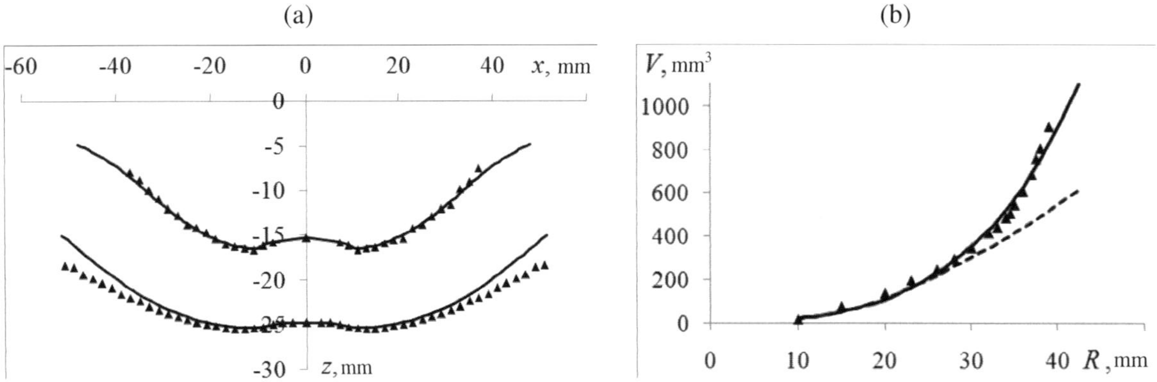

Figure 5. (a) Paths of axisymmetric fractures initiated in distance of 15 and 25 mm from free surface in a test specimen of $H = 106$ mm; (b) relationship of a volume of pumped plastic material (dashed line) and a volume of a fracture versus fracture length in the growth stage in distance 15 mm from free surface.

Comparison of theoretical and experimental results, reported in Figures 3–5 reveals that computation scheme described in [5, 6] enables to describe clearly parameters of fractures initiated in hydrofracturing transverse to a borehole.

5. Conclusion

The schemes are developed for experimental and numerical modeling of growth of axisymmetric fractures transverse to hydrofracturing borehole with the use of a plastic material in an active zone containing an earlier available disc-type fracture and free surface. Series of experiments were undertaken in different distances between fractures and distances to free surface. The comparative analysis of parameters of fractures produced experimentally with theoretical calculations showed their good compliance.

References

[1] Kyu NG and Tsygankov DA 2003 Method for directional failure of rocks by plastic substances, *Journal of Mining Science* Vol 39 No 6 pp 573–578
[2] Chernov OI, Freidin AM and Kyu NG 2001 Dimension stone production using hydraulic fracturing *Gorny Zh.* No 3 pp 71–75
[3] Grigoriev GA and Afanasieva TA 2012 Prospects for commercial-scale development of nonconventional gas resources in Russia *Neftegaz Geolog. Teor.Prakt.* Vol 7 No 2
[4] Gilaev GG, Afanasiev IS, Timonov AV, et al. 2012 The use of horizontal wells with multiple hydraulic fractures to develop low-permeability reservoirs in terms of pilot site at Priobsky deposit *Nauch-Tekh. Vestn. NKRosneft* No 2 pp 22–26
[5] Sher EN and Kolykhalov IV 2015 Determination of hydrofracture Geometry in a production

reservoir, *Journal of Mining Science* Vol 51 No 1 pp 81–87]

[6] Kolykhalov IV and Martynyuk PA 2016 Modeling growth of axisymmetrical fractures in fracking at multiple intervals, considering plastic properties of breakdown fluid, *J Fundament Appl Min Sci* Vol 3 No 1 pp 88–92 (in Russian)

[7] Crouch SL and Starfield AM 1984 Boundary element methods in solid mechanics, *J.Engineering Geology and Hydrogeology* Vol 17 pp 399–400

[8] Linkov AM 2008 Numerical modeling of fluid flow and a hydraulically induced fracture propagation *Journal of Mining Science* Vol 44 No 1 pp 24–42

IOP Conf. Series: Earth and Environmental Science **134** (2018) 012029 doi:10.1088/1755-1315/134/1/012029

The perspective of batch-wise removal of soil plug from pipes during trenchless installation

AS Kondratenko*, AS Smolentsev, VV Timonin and AYu Primychkin

Chinakal Institute of Mining, Siberian Branch, Russian Academy of Sciences, Novosibirsk, Russia

E-mail: *kondratenko@misd.ru

Abstract. Modern approaches to construction of utility lines are reviewed. The hole-making technology of pipe driving can be improved by introduction of concurrent removal of soil plug by batches, which essentially enhances performance efficiency. The minimum air pressure required to separate batches from soil plug and to move it along a pipe is determined. The feasibility of cyclic destruction and removal of soil plug during pipe ramming is experimentally proved.

It is commonly known that the trenchless method of laying underground utility lines is the most advanced and ecology-friendly. This method enjoys wide application as development density is increased and different-purpose communications expand. Many trenchless passages are made under roads and railways. This lays higher standards on the hole-making technology such as elimination of subsidence or buckling of roadway surface both during construction and operation of underground passage [1].

In pipe installation under roads at shallow depth, one of the most widely applied approaches is one-ended pipe driving as it the best way of eliminating subsidence or buckling of roadways surface [2]. This method can be implemented either as pipe ramming by air-driven, hydraulic or electric hammer, or using static force of hydraulic jacks.

Percussion driving features high penetration rate and eliminates construction of a retaining wall in a working pit. In this method of pipe hammering, the disadvantages are increased noise, which is not permissible in urban conditions, and vibration that can result in over-compaction of engineering sand and gravel cushions made under roads and railways, and in failure of foundations of close-spaced buildings [3].

During both dynamic and static process of pipe ramming, soil enters the pipe, gradually compacts and makes a soil plug. This prevents continuous enter of soil in the pipe. Soil is pressed in the hole walls, which induces radial deformation in enclosing soil body, and the pipe penetrates considerably much slower as in case of closed-ended pipe. So, well-timed removal of soil plug is an important engineering factor of efficiency of the pipe driving technology [4]. The currently available technologies of intermediate soil plug removal (warms, barrels) require using auxiliary equipment and, moreover, need extra forces and time to assemble and disassemble cumbersome and heavy facilities [5].

It seems to be most reasoned to remove soil plug from pipes by batches without introduction of auxiliary equipment. The implementation layout of such approach in pneumatic hammering of pipe is depicted in the Figure 1 [6].

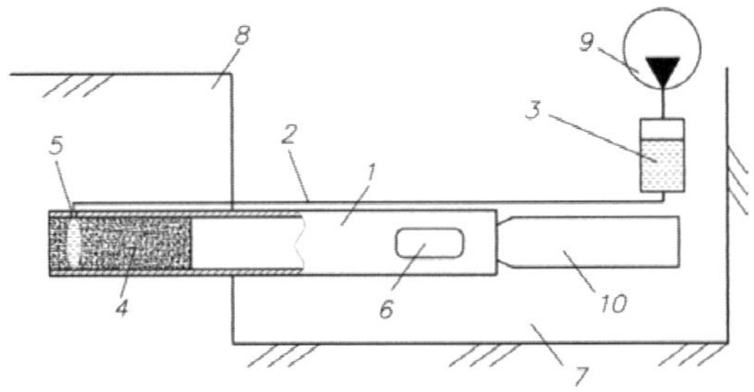

Figure 1. Layout of pipe driving with batch-wise soil plug removal: 1—driven pipe; 2—pipe line; 3—reservoir of pushing medium; 4—soil plug; 5—inlet for pushing medium; 6—outlet window; 7—working pit; 8—soil body; 9—compressed air source; 10—air-driven hammer.

In the proposed approach, a pushing fluid medium 3 is fed to the front end of the driven pipe 1 along the separate pipe line 2. When the soil plug 4 is formed, the pushing fluid medium is fed under pressure inside the pipe via the inlet 5. The pushing medium cuts off a batch of the soil plug and moves it towards the tail end of the pipe and pushes the soil batch through the outlet window to the working pit 7. After that, the pipe driving process continues until the next soil plug is formed. The driving and soil batch removal cycles rare repeated until the pipe enters the receiving pit. It is also possible to remove soil batches without full break in the air hammer operation by providing lower pressure percussion.

The process of batch-wise soil removal can be divided into 3 stages: formation of the soil plug, separation of a soil batch and removal of the soil batch to the tail end of the pipe. At the stage of the soil plug formation, it is required that the soil batch in the zone of its cutting-off is dense and fills the whole cross section of the pipe. This is required to create and maintain the pushing medium pressure sufficient to displace and remove the cut-off soil batch. At the same time, it is clear that in case of the over-compacted soil plug, it may be impossible to cut-off a batch, while in case of a low-coherent plug, a blue hole can appear, which results in the drop in the pressure of the pushing medium, and soil batch separation and removal becomes impossible. Thus, creation of suitable conditions for cutting-off and removal of soil batches is not a simple process and needs investigation.

For instance, according to the experimental data [7], soil plug is formed when pipe penetrates soil body by the value (6–10)D. If content of clay particles is high in soil, the soil plug is formed at the shorter length of the pipe penetration.

From the condition of equilibrium of elementary soil layer, the length of soil plug inside the pipe, L_p, is given by [8]:

$$L_p = \frac{D}{4f\varepsilon} \ln\left[1 + \frac{4f\varepsilon\sigma}{4c + \rho D}\right], \tag{1}$$

where D is the driven pipe diameter; f is the internal friction coefficient of soil; ε is the lateral earth-pressure coefficient; c is the cohesion of soil; ρ is the density of soil; σ is the stress in the soil body.

Soil breakage by liquid or gas flow is used in practice of injection of soil foundations with different chemical compounds [9], in creation of anti-seepage screens in soil by hydraulic or pneumatic fracturing [10] and in pulsed gas ripping. The batch separation pressure can be found as [11]:

$$p = \rho g h + \frac{\upsilon}{1-\upsilon}(\sigma - \rho g h) + \sigma_{\max}, \tag{2}$$

where h is the distance from the ground surface to the axis of the driven pipe; υ is Poisson's ratio; σ_{max} is the tension strength of soil.

In all mentioned practices, the rupture pressure for natural soil is not higher than 0.6 MPa.

The process of soil batch removal is a version of a tube transfer system. Such processes are implemented in drilling with coring [12], in transport of agricultural and construction materials [13].

When determining required pressure and flow rate of compressed air, it is necessary to take into account air leakage in the contact zone of soil and internal surface of the pipe, air loss connected with air seepage in soil and the limited cross section of the air feed pipe line. A sufficient condition for separated soil batch transfer is the relation [14]:

$$p = \frac{1}{K} \cdot \frac{\gamma D}{4\varepsilon f} (\exp \frac{4\pi f}{D} l - 1) + p_0, \qquad (3)$$

where K is the soil porosity; γ is the bulk weight of the soil plug; l is the soil batch length; p_0 is the atmospheric air pressure.

At this stage of batch-wise soil removal, it is important to preserve integrity of a soil batch, governed by the uniformity of the batch movement, density and moisture content of soil, roughness of the internal surface of the pipe and the change in the pipe cross-section. In case of vibro-hammering of a casing pipe, the other factors governing separation and removal of soil batches can be vibrations of the pipe and adjacent soil and elastic deformation of the pipe walls under impact. It is known that both internal and external friction coefficient can decrease under action of alternating load [15, 16]. On the other hand, this process remains yet to be studied better, and its positive effect on batch-wise removal of soil plug needs laboratory-scale investigation using physical models. Furthermore, vibration can affect integrity of a soil batch while removed.

In this manner, the proposed approach to trenchless pipe-casing driving with the batch-wise soil plug removal seems feasible but requires practical testing and multifactor laboratory experimentation and mathematical modeling.

The practicability of batch-wise soil removal by compressed air or air-and-liquid mixture was experimentally proved in conditions maximally close to real. The in situ tests were carried out on the testing ground of the Institute of Mining. It was required to make a hole with a diameter of 110 mm and 5 m long at a depth of 1.5 m from the ground surface, at an incline from 2 to 5 degrees. The static driving of pipe section used a hydraulic cylinder. The increment in length of the soil pipe was recorded while the pipe was driven in soil. The pressure of soil batch separation and removal was not higher than 0.6 MPa. Soil batch removal was carried out after the soil plug stopped growing in length by feeding compressed air or an air-and-liquid mixture along a separate pipe line. After the first batch was removed, the pipe driving process was re-started until the new plug was formed. The cycles were repeated down to the complete driving of the pipe in soil.

The tests proved efficiency of the proposed approach to pipe driving. The determined optimal length of a soil batch to be removed equals 3 diameters of a driven pipe.

Conclusion

The practical results obtained allow developing an efficient method of soil plug removal from pipes in the course of long trenchless pipe driving of high accuracy.

References

[1] Kondratenko AS 2008 Peculiarities of soil plug removal from casing pipes *GIAB* No 7 pp 326–331

[2] Kondratenko AS and Petreev AM 2008 Featurs of earth core removal from a pipe under combined vibro-impact and static action *J. Min. Sci.* Vol 44 No 6 pp 559–568

[3] Meskele T and Stuedlein AW 2013 Hammer–pipe energy transfer ratio for pipe ramming *Proc. No-Dig 2013* Liverpool NY

[4] Kondratenko AS, Timonin VV, Abirov AA, Gosmanov MK, Esenov BU and Zharkenov EB 2014 Safe trenchless horizontal and inclined hole-making technology *Vestn. KuzGTU* No 1 pp 40–45

[5] Gileta VP, Baris AV and Vanag YuV 2016 Evaluating layout and variables of air-powered drive of small-size pneumatic winch *J. Fundament. Appl. Min. Sci.* Vol 3 No 2 pp 25–30

[6] Kondratenko AS and Petreev AM 2010 RF Patent No 2399725 *Byull. Izobret.* No 26

[7] Gudavalli SR, Safaqah O and Seo H 2013 Effect of soil plugging on axial capacity of open-ended pipe piles in sands *Soil Mechanics and Geotechnical Engineering* pp 1487–1490
[8] Mitkina GV 1971 Infleunce of soil plug height on cload-carrying capacity of soil piles *Candidate of Engineering Sciences Disseration* Kuibyshev–Ufa (in Russian)
[9] Smagulova LK 2017 Chemical reifnorcement of soil *Simvol Nauki* No 6 pp 28–31
[10] Fateev NT, Sergeev SV, Karyakin VF, Gapon SV and Shchetinin OV 2007 Directional hydraulic fractuirng application to construction of enclosing structures in rock mass *Proc. Miner's Week-2007* pp 260–264 (in Russian)
[11] Winda AA 2011 Trenchless pipeline laying by the directional inclined drilling method. Hydraulic fracturing of sands and elimination of springs *Sfera Neftegaz* No 1 pp 104–105
[12] Ignatov AA 2012 *Analytical Research of Core Velocity in Viscous Fluid* Dnepropetrovsk: DonNTU (in Russian)
[13] Kril SI and Chaltsev MN 2010 Calculating procedures of basic parameters for pneumatic transport of granular materials in horizontal pipes *Prikl. Gidromekh.* Vol 12 No 4 pp 36–41
[14] Kershenbaum NYa 1968 *Vibro-Method of Horizontal Drilling* Moscow: nedra (in Russian)
[15] Aleksandrova NI and Sher EN 2010 Wave propagation in the ecodn pediodical model of a block-structured mediu. Part I: Characterisctics of waves under impulsive impact *J. Min. Sci.* Vol 46 No 6 pp 639–649
[16] Nikolaev SK and Korolev VA 1981 Role of temperature in thixotropic structurization of water-saturated clayey soil *Inzhenern. Geolog.* No 5 pp 37–47

IOP Conf. Series: Earth and Environmental Science **134** (2018) 012030 doi:10.1088/1755-1315/134/1/012030

Geo-navigation system for rotary percussion drilling in rocks of high and low electrical conductivity

AI Konurin, AP Khmelinin* and EV Denisova

Chinakal Institute of Mining, Siberian Branch, Russian Academy of Sciences, Novosibirsk, Russia

E-mail: *hmelinin@misd.ru

Abstract. The currently available drill navigation systems, with their benefits and shortcomings are reviewed. A mathematical model is built to describe the inertial navigation system movement in horizontal and inclined drilling. A prototype model of the inertial navigation system for rotary percussion drills has been designed.

1. Introduction

Drilling is one of the most important processes in mineral mining and in construction. The issue of deviation of drilling from the preset direction remains topical though the market offers plenty of modern geo-navigation systems [1–3].

This problem of navigation is also topical in rotary–percussion drilling in hard rocks using, for example, air hammer drills [4]. Borehole devices to measure deflection of hole during drilling use acceleration transducers, or magnetic intrusion detectors and gyroscopic pickups which are insufficiently strong and reliable in operation under impact load [5].

Geotechnology of mineral mining with blasting needs information on spatial position of holes and their deviation from drilling-and-blasting design since standard drilling and blasting designs and explosive analyses leave out a probability of drilling deviation. In the meanwhile, deflection of a blast hole from the drilling path results either in oversizes or in overgrinding.

The horizontal direction drilling currently widely uses such locating and tracking systems as Digi Track F2, Digi Track F5, SNS-100, 200, etc., which allow sufficiently high precision (to 0.1°) of determining hole deflection in vertical plane but are low-accurate in estimating deviation angle in plan [1–3]. Moreover, their effective distance is limited, and the horizontal directional drilling is impossible in hard rocks.

The systems of measurement and logging (Measurement While Drilling, MWD; Logging While Drilling, LWD) estimate vertical deviation of holes using magnetic intrusion devices and acceleration meters as a rule [6].

More accurate measurement of hole deflection is possible with the systems using downhole gyroscopic inclinometers (Gyro Tracer Directional, UGI-42, IES-54, etc. system GWD90 manufactured by Gyro Data Inc., Gyro Pulse by Shlumberger, gyro MWD by Scientific Drilling and others) [6]. Such systems are insusceptible to magnetic anomalies and have smaller error as compared with the magnetic sensors.

The gyro inclinometers ensure measurement of hole incline angle and azimuth as well as the turn angle of drill bit at an accuracy from 0.1 to 1 deg (at the incline not less than 3 deg). The systems MWD/LWD perform the same measurement at an accuracy from 0.25 to 3 deg.

The devices based on the gyroscopic inclinometers are as a rule built in the data transmission and control of MWD, which makes them universal but inapplicable in drilling assemblies unequipped with MWD/LWD.

2. Problems of geo-navigation system application

Under conditions of development of the Far North and Arctic areas, new problems arise in the sphere of geo-navigation.

The difficulties for the operation of geo-navigation systems in the high-arctic zones are presented by the extremely severe climate and the depression of useful signal strength: horizontal component of the magnetic field of the Earth and horizontal component of the angular velocity of the Earth, which are the basis for operation of magnetic intrusion detectors and gyroscopic inclinometers of compass type [5].

Drilling in high-conductance rocks cannot use magnetic intrusion detectors which are widely employed as one of additional channels of useful signals for the correction of readings of gyro detectors. This limits the use of the current geo-navigation systems in locating of blat holes and in control over drilling.

In all available geo-navigation systems as well in the devices described in [7, 8], the sensor is either arranged at the drilling tool or built in it, which subjects the detector to the increased mechanical loading and, consequently, shortens its life.

High price and operation complexity of the current systems limits their wide application in industry. Furthermore, they usually contain no feedback devices for the operational adjustment of the hole path while drilling.

3. Development of inertial navigation systems

Inertial navigation systems (INS) enjoy now wider application owing to their small size, high positioning accuracy self-sufficiency and relatively simple manufacture.

It is of the current concern to develop mathematical models of INS to improve their accuracy and interference immunity. The model of dynamics of an augmented state vector of an integrated system within a strapdown inertial navigation system and an on-line long-range navigation system has been proposed in [9, 10]. The estimation algorithm for nonobservable errors of acceleration meters and gyros for the correction of their readings in the further independent operation of the inertial navigation system is presented. The approaches to improving accuracy of acceleration meters and gyros within a strapdown inertial navigation system by hardware and software rare described in [11, 12]. The authors suggest periodical adjustment and calibration of the inertial measurement unit before a measurement cycle. It is also proposed to use the procedure of optimal estimation and correction of basic errors of the system by means of special modes of motion of a source in the framework of complexing measurements obtained from the inertial and other measurement systems. It has been validated that the combined application of such procedures ensures sufficient precision of a navigation system.

4. Development of inertial navigation system for rotary–percussion drilling

Majority of the development mathematical models of INS are meant for the measurement devices used in the aerospace navigation and, accordingly, suitable for operation at high velocities and for arrangement inside an encasement of an objects. Such systems are inapplicable to modeling navigation devices for rotary–percussion or horizontal directional drilling for cardinally different movement velocities and arrangement at a navigation object.

For this reason, a new mathematical model is required to describe INS movement in rotary–percussion drilling of inclined–horizontal holes in rocks of high and low conductance.

At the present time, there is no a scientific framework for the underground collection of navigation information and for the analysis of geo-navigation system movement parameters with regard to their operating differences from the aerospace navigation systems. In this case, the most significant are the electromagnetic parameters of a hosting medium, external impact in the form of blows and vibration, velocity and travel length of the system, hole incline angle and the ambient temperature.

In this connection, the geo-navigation system for the inclined–horizontal rotary–percussion drilling based on positioning principle and the mathematical model of the system movement in real conditions will ensure more reliable results refined using the readings of sensors based on different physical principle of operation. Integration of a number of measurement channels (based on gyro. Acceleration meters, tilt sensors, encoders, etc.) within a single system and the simultaneous processing of information from all measurement channels will enable using statistical processing of the results, which will improve the geo-navigation system accuracy and positioning reliability.

The inertial navigation systems based on accelerometer and gyro feature high accuracy within shorter time spans, which reduces with longer periods of operation. It is required to introduce additional measurement channels for the refinements of the positioning data, e.g. measurement channels operating on the ground surface, based on the use of GPS, magnetic intrusion and barometric detectors.

Such measurement channels are inapplicable in underground conditions. A geo-navigation system needs additional measurement channels serviceable in mines. Such as channels, rotation sensors, encoders, etc.

The authors of this paper have developed a structural layout of an inertial navigation system for rotary–percussion drilling (Figure 1). The layout includes a measurement probe composed of a unit of measurement sensors, a processing unit, a reception–transmission module as well as an operator's block comprising a stationary navigation unit, processing unit, displaying facility and a data reception–transmission module.

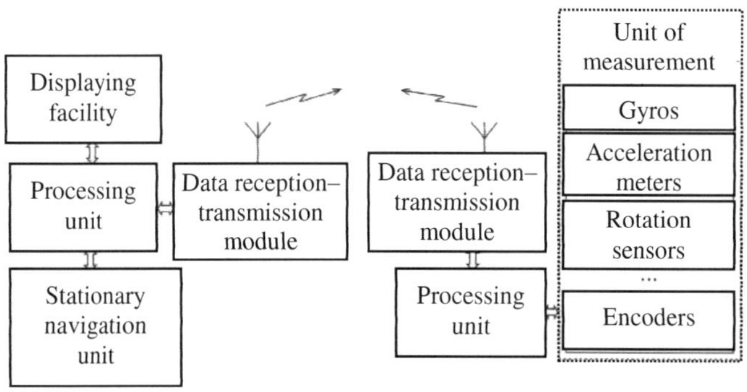

Figure 1. Structural layout of inertial geo-navigation system.

In the proposed system of geo-navigation, the measurement probe is advanced in an uncased hole or inside drill rods while drilling. Due to this, the error accumulated by the gyro is decreased owing to the reduction in the measurement time and in the period of stay of the measurement probe in the hole, and due to the elimination of impact loads exerted by rotary–percussion drilling on INS.

The operator's block is stationary mounted on a drill rig, or is arranged at the hole mouth; it informs on drilling parameters (rotation frequency, current hole length, blow frequency, etc.) and sends then, via the reception–transmission module, to the data processing unit. Due to this, INS becomes serviceable in underground conditions.

The geo-navigation data include angles of hole axis deviation from the accepted zero values of zenith (vertical) and azimuth (horizontal). These data are necessary for the hole positioning. After the processing, the data are displayed as a 3D trajectory of the hole by the displaying facility.

For the circuit optimization of a new INS, it is required to model its near-real operation mathematically. The mathematical model of INS operation should account for: ambient temperature, velocity and travel length, hole incline angle, shocks and vibrations on the hole and electromagnetic parameters of the hosting medium.

The mathematical model describes a measurement probe as a dynamic object, using differential equations to reflect continuous temporal variation of its state. The equations of motion are written in terms of components of the axes of the selected coordinate systems (CS).

1. The normal earthbound coordinate system. The origin is at any convenient point of the ground surface. The axis *X* (northward) and axis *Z* (eastward) are arranged in the horizontal plane, and the axis *Y* is directed upward (along the local vertical line).

2. The normal coordinate system. The origin is in the center of masses of the probe. The axes *X* (northward) and *Z* (eastward) are arranged in the horizontal plane, and the axis *Y* is upward. Further on, it is supposed that the axes of the normal earthbound CS and the normal CS are parallel. Their relative position of these systems is governed by the vector *r* between their origins.

3. The coupled coordinate system. The origin is in the center of masses of the probe. The axis *X* is oriented along the hole trajectory and is named the longitudinal axis. The axis *Z* is directed rightward relative to the probe advance is named the transversal axis. The axis *Y* lies in the plane of symmetry, is oriented upward (in horizontal drilling) and is named the normal axis.

The relative position of the coupled and normal CS is generally defined by 9 directional cosines and is determined mathematically based on the angles of Euler–Krylov. In this case, for the transition from the normal to the coupled CS, the sequence of turns is: turn by an angle of yaw (round the axis *Y*), turn by an angle pith (round the axisи *Z*) and turn by and angle of swing (round the axis *X*).

Below the authors present the calculation of the current positioning of a measurement probe in a hole while drilling.

Let a_x, a_y, a_z—accelerations in along the corresponding axes in the system related with the measurement probe; Ω_x, Ω_y, Ω_z—angular velocities of rotation around the corresponded axes in the system connected with the probe.

The turn angle α_i around an *i*-th axis is given by:

$$a_i = \int_{t_1}^{t_2} \Omega_i dt,$$

where t_1 and t_2—start and end of measurement, respectively.

The accelerometers compensate drift of gyros and determine the initial angles. The readings of accelerometers include free fall acceleration, noise acceleration of the probe. In case that *R*—vector of apparent acceleration measured by an accelerometer, let R_x, R_y, R_z—components of this vectors in *X*, *Y*, *Z*. Then, the turn angle α_i around an *i*-th axis axis measured by the accelerometer is:

$$a_i = \arccos\frac{R_i}{R}, \text{ where } R = \sqrt{R_x^2 + R_y^2 + R_z^2}.$$

The noise of gyro is due to the eigen noise of the detector (distributed uniformly) and due to the influence of the Earth rotation. Additional sensors are used to balance the drift of angles measured by gyros since this value cannot be balanced by accelerometers in strictly vertical and horizontal motion. The data on the current angles from accelerometer, gyro and other sensors, after fusion and filtration, inform on the inclination angles of the measurement probe (Euler–Krylov angles) during advance inside a hole and, accordingly, on the path trajectory of the hole.

The optimal way to calculate coordinates seems to be algorithm based on the mathematical models with the Rodrigues–Hamilton parameters for the kinematic equation constructed using the Rodrigues–Hamilton parameters and linear and integrable at any angles by Euler–Krylov.

(a) (b)

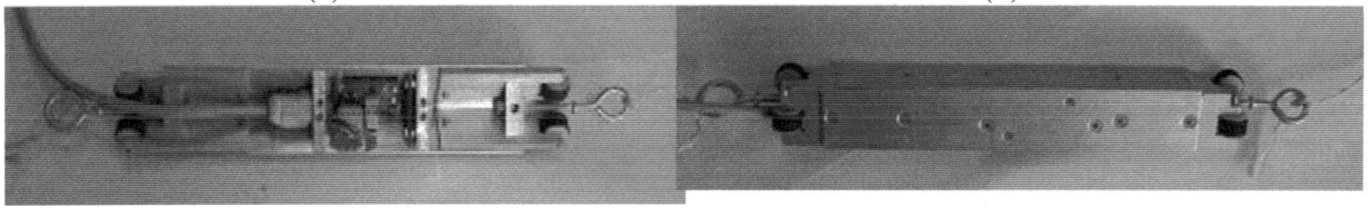

Figure 2. Pictures of the prototype INS (a) without the housing and (b) inside it.

The researchers of the Mine Geophysics Laboratory, Institute of Mining, have designed and manufactured a simplified prototype of INS (Figure 2) with the measurement channels using gyroscopic detectors.

Currently, the manufacture of the test bench for the INS prototype is in process.

5. Conclusion

The issues connected with the development methods and equipment towards resource saving and improved safety and efficiency in hard mineral mining, including drilling quality enhancement, become of increasingly higher concern.

This study aimed development of new approaches to locating and tracking in rotary–percussion drilling in rocks of low and high conductance is based on the integrated numerical and experimental research with regard to many parameters of INS operation.

The presented mathematical model of INS operation allows selecting optimal layouts of the equipment with a view to eliminating intersection and deviation of various-purpose horizontal drill holes, which, in its turn, will enable enhancing efficiency of hard mineral mining.

Acknowledgements

The study has partly been supported by the Novosibirsk City Executive Board, Grant for Young Scientists and Specialists for Research and Innovation Activities under Contract No. 16 dated July 4, 2017.

References

[1] Rybakov AP 2005 *Foundations of Trenchless Technologies (Theory and Practice)* Moscow: PressByuro (in Russian)

[2] *SNS-200 Location System of Horizontal Directed Drilling* (in Russian): http://sense-inc.ru/assets/templates/sense/doc/ SNS200-PRO.pdf

[3] *Eclipsecomplex*: http://russian.digital-control.com/products/discontinued/eclipse.html].

[4] Kondratenko AS, Timonin VV and Patutin AV 2016 Prospects for directional drilling in hard rocks *J. Min. Sci.* Vol 52 No 1 pp 129–134

[5] Scoblo VZ, Ropyanoy AYu and Lukht AI 2010 Gyroscopic orientation device KURS—tool for gyroscopic orientation of whipstock while drilling *Vestn. Assots. Bur. Podryad.* Vol 3 pp 31–34

[6] GyroPulsereal-Timegyro Orientation and Measurement while Drilling: http://www.slb.com/~/media/Files/drilling/brochures/mwd/gyropulse.pdf

[7] Paderina TV and Sokolov DA 2012 Boreholes survey by inclinometer under high-latitude conditions *Proc. XIV Conf. Young Scientists: Navigation and Movement Control* pp 95–101 (in Russian)

[8] Sokolov DA 2013 Borehole device of universal gyroinclinometer *Abstract of Candidate of Engineering Sciences Dissertation* Saint-Petersburg (in Russian)

[9] Paderina TV and Sokolov DA 2013 Variants of constructive realizing diametric scheme of gyroinclinometer for high-latitude systems of underground navigation *Izv. Vuzov. Priborostoenie* Vol 56 No 7 pp 21–26

[10] Shalimov LN, Fokin LA and Shipitsin AG 2005 Strapdown INS accuracy analysis and synthesis software *Aviakosm. Priborostr.* Vol 12 pp 15–21

[11] Vasilev PV, Meleshko AV and Pyatkov VV 2014 Accuracy improvement of correctable inertial navigation system *J. Izv. Vuzov. Priborostoenie* Vol 57 No 12 pp 15–21

[12] Afonin AA, Tyuvin AV and Sulakov AS 2014 Hardware and algorithmic methods of reducing errors of graviinertial complex systems inertial sensors *Mekhatron., Avtomatiz., Upravl.* Vol 12 pp 42–52

IOP Conf. Series: Earth and Environmental Science **134** (2018) 012031 doi:10.1088/1755-1315/134/1/012031

Stress state of rock mass under open pit mining in the influence zone of tectonic disturbances (in terms of the Oktorkoi Fault, North Tien Shan)

KCh Kozhogulov[1], OV Nikolskaya[1], AK Rybin[1,2] and SI Kuzikov[2]

[1]Institute of Geomechanics and Development of Subsoil,
National Academy of Sciences of the Kyrgyz Republic, Kyrgyzstan
[2]Research Station, Russian Academy of Sciences, Bishkek-49, Kyrgyzstan

E-mail: krsu@krsu.edu.kg

Abstract. The qualitative connection between the crack growth direction and the orientation of the main axes of horizontal deformations in rocks mass in the area of the Boordin gold ore province is revealed. The effect of the rock mass quality (RQD) and contact conditions of crack surfaces on the stability index of pit wall rock mass is evaluated, and the influence of the rock mass quality index on the pit wall stability is determined.

One of the key tasks of open pit mining is stability of pit walls. Open pit mining in mountainous regions is faced with complicated geology, tectonics and hydrology. Enclosing rock mass is a rule composed of intrusive and effusive rocks susceptible to hydrothermal and metasomatic alteration. Rock mass is heavily fractured, and density of fractures grows at the contacts of rocks and faults.

It is highly important to ensure stability of pit walls in upland regions in the influence zones of faulting. Prior to starting mining, it is required to determine directions of the ground surface deformation inside the mine field and in adjacent areas.

In Kazakhstan gold deposits mostly occur at different regional faults, dip-slips and thrust faults.

At the present time, underground mining has been started at Taldy-Bulak Levoberezhny deposit, and surveying is implemented with a view to open pit mining at Kurandjailyau deposit. These deposits and mined-out fields such as Aktyuz and Kutesay belong in the Boordin gold ore province (BGOP). Spatially and structurally, they adjoin the Oktorkoi Fault (Figure 1).

The main faults in the ground in this region strength north-eastward and rupture comparatively young Palaeogene–Quaternary formations. The Oktorkoi Fault together with lower-scale faults of Chuyskaya Depression (Issyk-Atinsky, Shamsinsky) are active faults. Out of Central Asia GPS network points within the distance of 10–20 km from BGOP, only 3 discrete survey points are situated to the west and east of the gold ore province. In Figure 1 velocity vectors are plotted in the framework of EURA-2008 system for 1995–2015. The northern and southern GPS points are more than 30 km away from the province and are beyond the geographic limits in Figure 1. The absolute values of the velocity vectors are 2.3–3.8 mm/yr horizontally and –0.5–1.1 mm/yr vertically. The velocity vectors have narrow orientation window of 34–36°. It is noticeable that the northern and eastern components of the vectors gradually lower in the line from the south to the north, and on a regional scale.

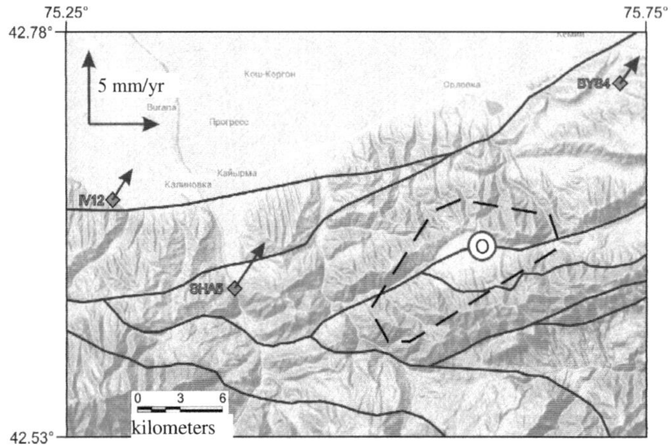

Figure 1. Boordin gold ore province—dashed polygon. Lines show active faults. O—Oktorkoi Fault; rhombs and signs—GPS points and their coded names; arrows—velocity vectors (mm/yr, EURA-2008, 1995–2015).

Moduli of difference (north to south) of the shown velocity vectors in three-dimensional space are IV12-BYS4 = 0.7, SHA5-BYS4 = 1.6, SHA5-IV12 = 1.2 mm/yr. The greater difference falls at the horizontal components (– 0.2, 1.2, 1.5 mm/yr, respectively) oriented at 20 to 36° (average ~30°). The vector difference moduli of the mentioned GPS points relative to the farther points are not higher than 3–4 mm/yr.

The averaged deformation picture in the area of BGOP based on the GPS data is shown in Figure 2. Here, based on the information from [1], the orientations and values of horizontal deformation velocity are given in the major axes with the averaging of data from GPS points within a radius of 25 km. The axes of contraction velocity are within a narrow range of azimuths 7–10° at the values of –31÷–66 (average –53)$\times10^{-9}\times$yr^{-1}, respectively; in the axes of elongation velocity, the values vary as 9–23 (averageв 17)$\times10^{-9}\times$yr^{-1}. The maximum velocities of deformation in the region of the Kirgiz Ridge are observed westwards this territory (~75°East longitude) and slightly exceed the presented values. The horizontal deformation velocity decreases from the west to the east and from the north to the south.

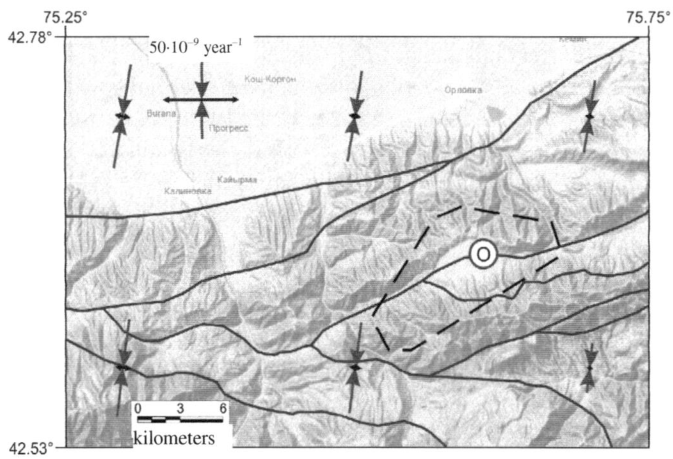

Figure 2. Arrangement of major axes of horizontal deformation in BGOP area (dashed polygon). Counter arrows—contraction; differently directed areas—elongation; the length of the arrows is proportional to the value of deformation relative to the reference of $50\times10^{-9}\times$yr^{-1}.

IOP Conf. Series: Earth and Environmental Science **134** (2018) 012031 doi:10.1088/1755-1315/134/1/012031

Regarding the data on seismology in the given area, the orientation of major deformation axes is similar, the azimuths of the contraction axes are 350÷0°, and deformation is -0.2×10^{-9}. The ratio of deformations in the axes of contraction and elongation is somewhat different. For instance, in the depth interval from 0 to 10 km, they are approximately equal; with depth the values grow along the contraction axes and reduce along the elongation axis to a ratio of 3 : 1, respectively [2].

With respect to the stress state in the discussed area, using the seismic data, it is possible to perform zoning of faults [3] by the value of normalized Coulomb's stresses τ_C / τ_f . It appears that BGOP occurs in the zone of faults subjected to the increased stresses.

Considering the direction of the earth crust movement, values of deformation and the increased stresses of active faults, the detailed analysis of jointing of geological medium within Kurandjaylyau was carried out.

With this end in view, it was chosen to examine a section of the future pit wall the exposed surface of which clearly shows orientations of joints, and it was possible to count the systems of joints for the statistical data collection. Three basic systems of joints were distinguished within the test section. The average spacing of joints, estimated based on there measurements in different points of the pit wall, ranged between 0.02 and 0.1 cm. the main direction of growth of joints coincided with the revealed direction of tensile deformations of the earth crust (Figure 3).

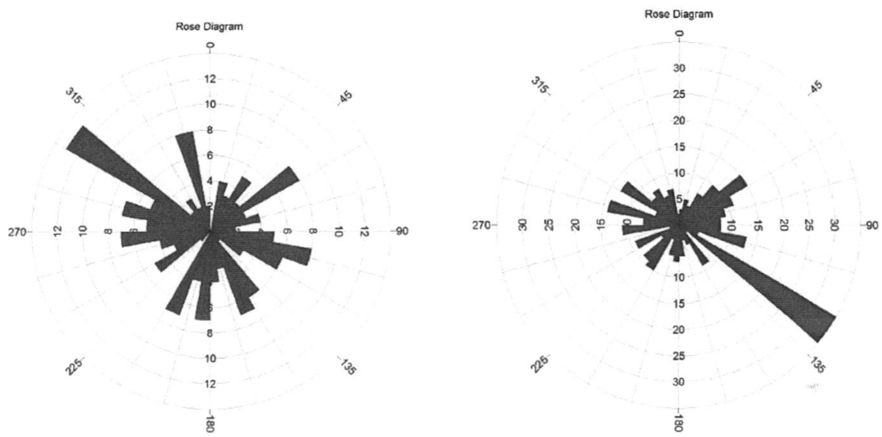

Figure 3. Illustrations of the principal orientation of jointing in the area under study.

The basic characteristics of rock mass jointing are the degree of jointing, average spacing of joints and a structural parameter.

In order to design safe pit wall in jointy rock mass, it is necessary to define the rock mass quality index *RQD*, rock mass rating and geological strength index *GSI* [4].

The international practice of rock mass quality estimation widely uses MRMR classification by Laubscher (Mining Rock Mass Rating) [5–7].

According to this classification, the rock mass rating is based not only on stress characteristics of rocks and their geomechanical behavior but also on the description of its jointing.

The index *MRMR* is given by:

$$MRMR = RMR \times K;$$

$$RMR = RRBS + JS + JC,$$

where *RRBS*—rock block strength; *JS*–index of joint number; *JC*—index of joint condition; *K*–persistence to account for weathering, joint orientation, rock mass stresses, blasting, groundwater.

Using the data on location of the major axes of horizontal deformation in the area of BGOP and directions of joints within the open pit mine under planning at Kurandjaylyau, it was assessed how tectonic faulting and jointing affected stability of enclosing rock mass around an open pit pit mine.

All values of strength characteristics determined from testing of drill core samples were recalculated for rock mass and corrected with regard to the degree of jointing in order to determine stability and slope of open pit wall.

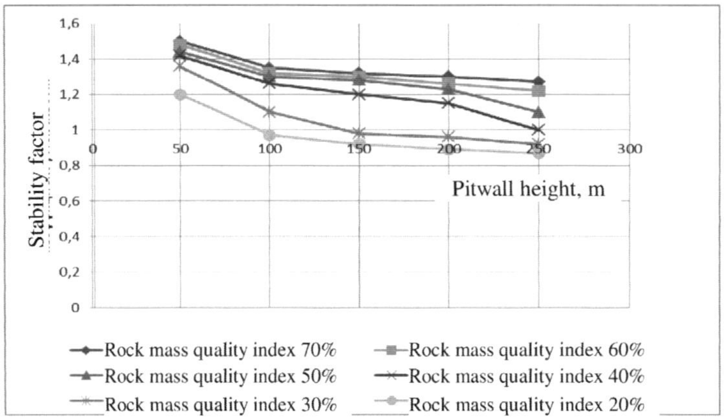

Figure 4. Stability factor versus depth of pit wall and rock mass quality index (σ_{com} = 210 MPa, pit wall slope of 45 deg).

(a) (b) (c)

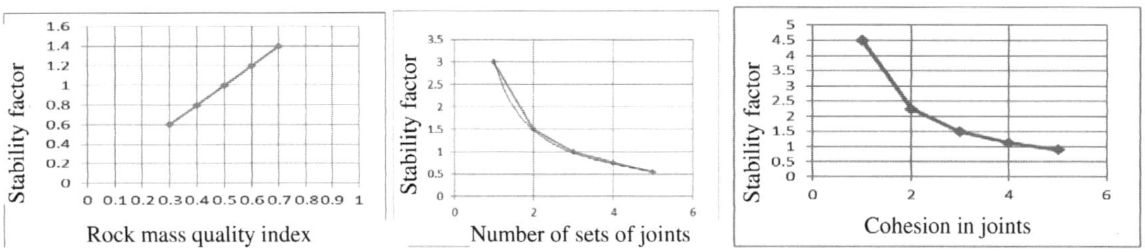

Figure 5. Stability factor of pit wall rock mass versus conditions of jointing: (a) rock mass quality index; (b) number of sets of joints; (c) cohesion in joints.

The average spacing of joints from the three measurements at different points of the slope is from 0.02 to 0.1 cm. the density of jointing is w = 2.5. The structural weakening coefficient $\lambda = 0.03$. By the rating *RQD*, the quality index of this rock mass is not higher than 0.2 [7].

The resultant characterization of stability of open pit wall rock mass is depicted in Figs. 4 and 5.

Conclusion

The direction of the ground surface deformation is the governing factor of growth and orientation of joints in rock mass.

Jointing and the conditions of jointing exert a determining influence on rock mass quality and pit wall stability by the geological index of rock mass quality.

References

[1] Mansurov AN 2016 The velocity field of the modern deformation of the Earth's crust of the Northern and Central Tien Shan according to the observations of the Central Asian GPS network *Vestnik KRSU* Vol 16 No 1 pp 157–160

[2] Rebetzkiy YuL 2016 Tectonic-physical zoning of active faults of the Northern Tien Shan *Geolog. Geofiz.* Vol 57 No 6 pp 1225–1250
[3] Sycheva NA 2016 Comparison of the estimates of the crustal deformation of the Northern and Central Tien Shan obtained on the basis of seismic and GPS data *Vestnik KRSU* Vol 16 No 1 pp 178–182
[4] Hoek E 2002 *Practical Rock Engineering* London: Institution of Mining and Metallurgy
[5] Laubscher DH 1990 A geomechanics classification system for rating of rock mass in mine design *J. South African Inst. Mining and Metallurgy* Vol 90 No 10 pp 257–273
[6] Laubscher DH and Jacubec J 2000 *The MRMR Rock Mass Classification for Jointed Rock Masses. Foundations for Design Brisbane* pp. 475–481.
[7] Bieniawski ZT 1989 *Engineering Rock Mass Classifications* New York: John Wiley & Sons

IOP Conf. Series: Earth and Environmental Science **134** (2018) 012032 doi:10.1088/1755-1315/134/1/012032

Prediction of slope stability based on numerical modeling of stress–strain state of rocks

KCh Kozhogulov[1*], VI Nifadyev[2] and SF Usmanov[2]**

[1]Institute of Geomechanics and Subsoil Development, National Academy of Sciences of the Kyrgyz Republic, Bishkek, Kyrgyz Republic
[2]Kyrgyz-Russian Slavic University, Bishkek, Kyrgyz Republic

E-mail: *ifmgp@yandex.ru,** krsu@krsu.edu.kg

Abstract. The paper presents the developed technique for the estimation of rock mass stability based on the finite element modeling of stress–strain state of rocks. The modeling results on the pit wall landslide as a flow of particles along a sloped surface are described.

The National Security Concept approved by the President of the Kyrgyz Republic on June 12, 2012 points at the risk of increase in number and scale of natural and induced emergencies. Also, it is emphasized that the methods and means currently available for responding and preventing such emergencies display insufficient efficiency.

Today, there are more than 5000 landslide-hazardous sites revealed in the Republic. Landsliding initiates both on natural mountainsides and on slopes of technical structures. Out of a few open pit mines in the Republic, Kumtor gold mine is the largest. It is highly important to ensure stability of pit walls, dumps and embankments. Moreover, it is necessary to maintain steadiness of canyons and hydroelectric dams. Naturally, it is required to predict stability of artificial slopes along mountainous roads in the country.

The landslide on the slope of Bishkek–Balykchy road on August 12, 2015 inflicted considerable damage to economy. By a lucky chance, no people were injured in that accident on the road with the heaviest traffic between two regions and the capital of the country.

Figure 1. Landslide, Bishkek–Balykchy road.

The Kyrgyz-Russian Slavik University develops a rock mass stability estimation procedure based on numerical modeling of stress state of rocks. Modeling involves the finite element method. The calculation uses a multi-processor computation cluster (super computer), which enables solving complex spatial and dynamic problems.

Landslide simulation is performed with the designed software system *LandslideModeller*. This system is mean for numerical modeling of slides and falls using structural dynamic equations discretized in space. To check the landslide prediction reliability, *LandslideModeller* algorithms were tested with the known numerical solutions from the available mathematical models of landslide processes.

The mathematical models available for calculating velocities and amplitudes of displacements [1–3] are grouped with regard to approaches to the mechanism of slide initiation and growth: я оползней:

—viscoplastic medium models;

—hydraulic models;

—granular medium models.

Small mud slides are described by the first group models. The second group models use complicated equations including dry friction between slide layers and friction between the slide body and floor. The common feature of the models in this group is the representation of a slide flow by a liquid with non-Newtonian properties. Such models sufficiently accurately describe flows saturated with clay or mud, with well coherent particles which cannon collide. The models based on the theory of granular media [4–6] use an assumption that a slide flow is composed of separate particles (grains) of different size. A granular medium is s set of discrete particles subjected to satisfy conditions that:

—the particles are sufficiently many to generate a flow;

—the high concentration of particles ensures their frequent collisions.

Within the period of interaction between particles, their collision is represented by a damping system of mass and spring with friction oriented along a line tangent to the contact point of particles. Motion of each particle in the landslide flow is described by the second-order differential equation with regard to the contact interaction of particles and taking into account external bulk forces. The effective friction coefficient is an increasing function of shear velocity, and material, which lays atop landslide at the beginning of the process, remains atop in the end of landsliding. In full scale conditions, this statement is valid for large particles on the surface of landslide at the beginning of its displacement. More complex models of granular media take into account interaction of particles depending on which forces—viscous or colliding—have higher influence on the landslide flow dynamics.

In the given study, landslide was modeled as a flow of particles over an inclined surface. Initially, it is assumed that the displacing portion splits into flow of particles downward the slope. Interaction with air on side edges is negligible. The flow motion is subjected to the force f gravity. The external mass inflow is assumed absent. The landslide motion is described using the system of Navier–Stokes equations and the law of mass conservation [7–8]:

$$\begin{cases} \dfrac{Du}{Dt} + \dfrac{1}{\rho}\dfrac{\partial P}{\partial x} = \mu \nabla^2 \vec{u}, \\ \dfrac{Dv}{Dt} + \dfrac{1}{\rho}\dfrac{\partial P}{\partial y} = \mu \nabla^2 \vec{u}, \\ \dfrac{Dw}{Dt} + \dfrac{1}{\rho}\dfrac{\partial P}{\partial z} = \mu \nabla^2 \vec{u} + g, \\ \dfrac{D\rho}{Dt} = 0, \end{cases} \tag{1}$$

where *x, y, z*—Cartesian coordinates; *u, v, w*—components of the velocity vector $\vec{u}$; *t*—time; *ρ*—flow density; *P*—pressure; *μ*—viscosity; *g*—acceleration of gravity.

The particle-in-cell method transforms basic motion equation to the equations for interacting particles. It is assumed that all interactions between particles are limited to a finite volume r_e, and no interaction takes place beyond this radius. In this case, computational complexity of recalculating an unknown function at each time step equals O(*NM*), where *N*—total number of particles; *M*—number of interacting particles. The unknown function is presented as a finite sum of Dirac's functions *δ*:

$$\phi(\vec{x})=\sum_{i}^{M} m_i \frac{\varphi_i}{\rho_i}\delta(\vec{x}-\vec{x}_i), \tag{2}$$

where m_i, ρ_i, $\vec{x}_i$—mass, density and position of a particle *i*, respectively.

In accordance with Eq. (2), the slide flow density is given by:

$$\rho(\vec{x})=\sum_{i}^{M} m_i \delta(\vec{x}-\vec{x}_i). \tag{3}$$

The pressure of the landslide flow is calculated using the equation of state:

$$P=P_0+k(\rho-\rho_0), \tag{4}$$

where P_0, ρ_0—pressure and density of landslide in rest.

For the momentum conservation equation, it is necessary to express a gradient's operator and a Laplace operator used to calculate forces of pressure and viscosity applied to particles. The pressure and viscosity forces are given by:

$$F_i^{press}=\sum_{i}^{M} m_i \frac{P_i+P_j}{2\rho_i}\nabla\delta_{press}(\vec{r}_i-\vec{r}_j), \tag{5}$$

$$F_i^{vis}=\mu\sum_{i}^{M} m_i \frac{u_i-u_j}{\rho_i}\nabla\delta_{vis}(\vec{r}_i-\vec{r}_j), \tag{6}$$

where $\vec{r}_i$, $\vec{r}_j$—positions of interacting particles *i* and *j*, respectively.

The weight functions for the pressure, viscosity and other members are calculated as:

$$\nabla\delta_{press}(\vec{r})=\frac{45}{\pi r_e^6}(r_e-|\vec{r}|)^3\frac{\vec{r}}{|\vec{r}|}, \tag{7}$$

$$\nabla\delta_{press}(\vec{r})=\frac{45}{\pi r_e^6}(r_e-|\vec{r}|), \tag{8}$$

$$\delta(\vec{r})=\frac{345}{64\pi r_e^9}(r_e^2-|\vec{r}|^2)^3. \tag{9}$$

Beyond the radius of interaction between particles, r_e, all functions equal zero. The boundary conditions are materialized with the help of immobile boundary particles. It is assumed that the flow particles locate at a distance *d* from the boundary particles. A particle *i* approaching the boundary closer than *d* is assumed to undergo the force of pressure in the line $\vec{n}(\vec{r}_i)$ from the side of the boundary particle. In this case, the force of pressure is:

$$F_i^{press}=m_i\frac{\Delta\vec{x}_i}{dt^2}=m_i\frac{(d-|\vec{r}_{iw}|)\vec{n}(\vec{r}_i)}{dt^2}, \tag{10}$$

where $|\vec{r}_{iw}|$—distance between particle *i* and the boundary particle.

This method needs no smooth solution and uses a nonuniform 3D grid for the modeling. Each time iteration includes the sequence of steps:

—determination of interaction radius r_e for each particle;
—calculation of flow density and particle velocities;
—recalculation of particle positions by the Euler scheme.

The algorithm has been implemented and then used for the 3D discrete modeling of landslide flow motion (Figure 2). The profile of the landslide is presented be a set of elements. The frontal plane of the landslide is inclined to horizon at an angle of 45°, the grains are 0.5 m in size. The maximum horizontal displacements are generated at the right bottom of the frontal plane of the landslide in all stages of the flow development. The maximum vertical displacements are observed in the upper portion of the model.

The displacements change their relative values as the landslide develops. The horizontal displacements exceed the vertical movements in the initial stages of the landslide activation, which the vertical displacements dominate over the horizontal displacements later on. Figure 3 demonstrates the results of modeling pitwall landslide.

Figure 2. Modeling landslide motion in the PIC method.

Figure 3. Modeling pitwall landslide.

Conclusion

The authors have developed the procedure to model landslide motion without the requirement of the solution smoothness and applicable with the nonuniform spatial grid. The modeling procedure makes it possible to estimate stability of a landslide, its volume and flow distance. Using the procedure, 3D discrete modeling of landslide and fall of pitwall has been carried out.

References

[1] Grigoriyan SS, Nilov NN et al 1983 Mathematical modeling of large rock falls and landslides *Inzh. Geolog.* No 6
[2] Emelyanova ED 1972 *Basic Regularities of Landslide Process* Moscow: Nedra (in Russian)
[3] Drannikov AM 1956 *Landslide. Types, Causes, Control* Kiev (in Russian)
[4] Campbell CS 1990 Rapid granular flows *Annu. Rev. Flu. Mech.* Vol 22 pp 57–92
[5] Jaeger HM, Nagel SR, and Behringer RP 1996 The physics of granular materials *Physics Today* Vol 1
[6] Savage SB, Hutter K 1999 The motion of a finite mass of granular material down a rough incline *J. Fluid Mech.* pp 177–215
[7] Bogomolov SV, Zaharov EV and Zerkaly SV 2001 *Mathematical Modeling of Landslide by the Method of Particle Flow. MDOZMF-001* Kherson (in Russian)
[8] Harada T, Koshizuka S and Kawaguchi Y 2007 Smoothed particle hydrodynamics on GPUs *Proc. Conf. on Computer Graphics International* Petropolis Brazil

IOP Conf. Series: Earth and Environmental Science **134** (2018) 012033 doi:10.1088/1755-1315/134/1/012033

Deformation of rock blocks with a cavity

AA Krasnovsky

Chinakal Institute of Mining, Siberian Branch, Russian Academy of Sciences, Novosibirsk, Russia

E-mail: visanta@ngs.ru

Abstract. On the basis of a system of singular integral equations connecting the boundary values of all stress and displacement components for an arbitrary simply connected region, relations are obtained for the stress and displacement components at the boundary of a rock block with a rectangular cavity. Examples of numerical implementation are given.

Numerical calculations in mechanics are mostly based on the theory of elasticity. The mathematical modeling permitting the theoretical analysis of the solution allows a more detailed estimation of the behavior of a specimen under loading.

Knowing stress state is required to calculate strength, rigidity and stability, including all components of stresses and displacements at the boundary and contacts in order to detect initiation of failure. In calculation of stresses and strains in the finite domains with an arbitrary hole, no analytical solution exists.

This paper addresses the issue of stress state determination at the boundary of a rectangular domain with a hole for the main boundary-value problems without factual finding of complex potentials $\varphi(z)$ and $\psi(z)$ in a computation domain. The author attempts to analyze simultaneously and uniformly all three basic problems of elasticity and to propose their implementation methods in the context of rock blocks with a weakening rectangular hole. The model is based on the system of singular integral equations connecting all boundary values of stresses and displacement [3].

Consider a rock block with a rectangular hole as schematically shown in Figure 1a. Given the geometrical and force symmetry, the problem is greatly simplified. In this case, thanks to the symmetry, it is sufficient to analyze a half or a quarter block which is a simply connected region [4, 5].

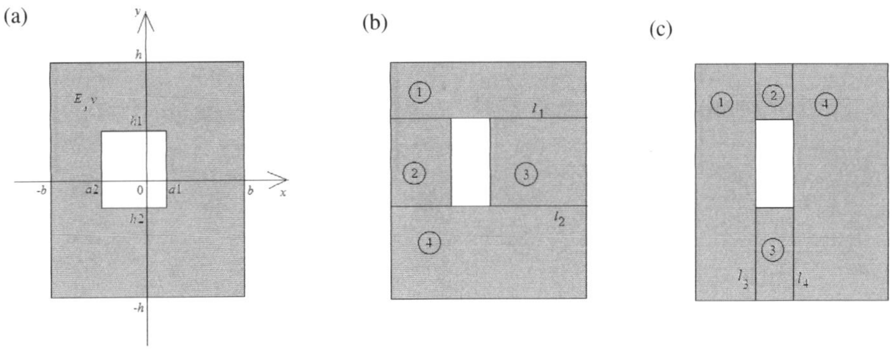

Figure 1. Schematic layouts of rock blocks with the rectangular holes.

Let us discuss the case of the asymmetrical weakening. This is a biconnected region, which induces certain complexity in the problem solving. For this reason, it is suggested to divide the block into four parts. Figs. 1b and 1c illustrate such division: l_1, l_2, l_3 and l_4 are the dividing lines and simultaneously the boundaries of the domains which compose the block and are the simply connected region each. With such division of the computation domain, it is possible to find the values of the normal and shear stresses and displacements at l_1, l_2, l_3 and l_4, depending on which boundary information is required, without determining stress state inside the whole region under analysis.

The boundary conditions can be set in different combinations in the framework of the basic problems of elasticity.

As an example, we suggest the variant of the boundary conditions below:

$$\sigma_n = \sigma_0 = -1,\ u = 0 \quad \text{at the boundaries } -b \le x \le b,\ y = -h \text{ and } y = h,$$

$$\sigma_n = 0,\ \tau_n = 0 \quad \text{at the boundaries } -h \le y \le h,\ x = -b \text{ and } x = b, \tag{1}$$

$$\sigma_n = 0,\ \tau_n = 0 \ \text{at the weakening hole boundary: } h2 \le y \le h1,\ x = a2 \text{ and } x = a1,$$

$$a2 \le x \le a1,\ y = h2 \text{ и } y = h1,$$

where σ_n, τ_n—normal and shear stresses; u—horizontal displacement.

In the general for for an arbitrary simply connected region, the system of singular integral equations connecting boundary values of all stresses and displacements has the form [3] and contains the function $f(t)$:

$$f(t) = i\int_0^t (X_n + iY_n)ds, \tag{2}$$

where X_n, Y_n—forces at the boundary in the line of the axes x and y, respectively; t—point at the boundary; i—unit imaginary number.

It is assumed that cohesion exists at the dividing lines–boundaries of the block parts:

$$\sigma_n^+ = \sigma_n^-,\quad \tau_n^+ = \tau_n^-,\quad u^+ = u^-,\quad \upsilon^+ = \upsilon^-, \tag{3}$$

which means continuity of the normal and shear stresses (the upper symbol identifies the corresponding part).

Based on [3], using the earlier obtained equations for all displacements and the function $f(t)$ at the boundary of a rectangular domain [4] and the condition (3), the relations connecting stresses and displacements at the boundaries and interfaces of the block parts have been written.

For the numerical implementation of the resultant system of equation, we use the dimensionless values and assume the dimensions of length as the half length of the weakening and the dimensions of stresses as the characteristic values of the boundary stresses. The calculations involved $h = 4$, $b = 3$, $\nu = 0.25$, $E = 10^4$ in the combinations below:

$$1)\ h1 = 3,\ h2 = 2,\ a2 = -2,\ a1 = 0;\ 2)\ h1 = -1,\ h2 = -3,\ a2 = 1,\ a1 = 2. \tag{4}$$

Figure 2 demonstrates the calculated deformation of the boundary of the analyzed block in different variants of geometry and location of the weakening (4).

IOP Conf. Series: Earth and Environmental Science **134** (2018) 012033 doi:10.1088/1755-1315/134/1/012033

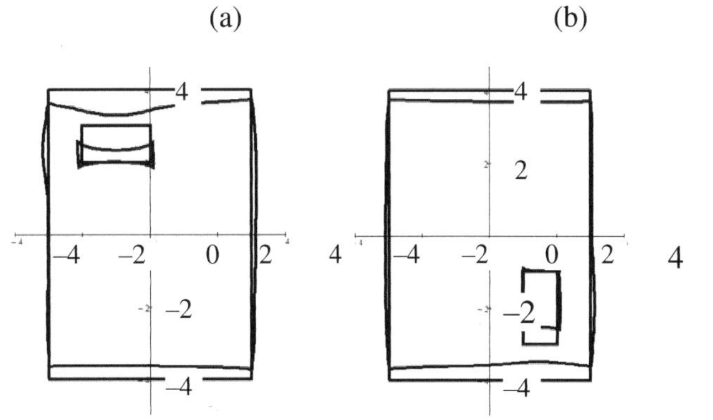

Figure 2. Deformation of the rock block boundary: a—$h1 = 3$, $h2 = 2$, $a2 = -2$, $a1 = 0$; b—$h1 = -1$, $h2 = -3$, $a2 = 1$, $a1 = 2$.

Figure 3 illustrates the calculated results for the horizontal and vertical displacements $u(x)$, $\upsilon(x)$ and the real and imaginary parts of the function f at the contact line l_1: $y = h1$, $-b \leq x \leq b$ (see Figure 1b) fitting the first combination in (4). The analysis of the function (2) allows estimating the respective stresses both at the contact lines and at other boundaries of the rock blocks. To this effect, it is sufficient to differentiate the obtained results for the real and imaginary parts of the function f .

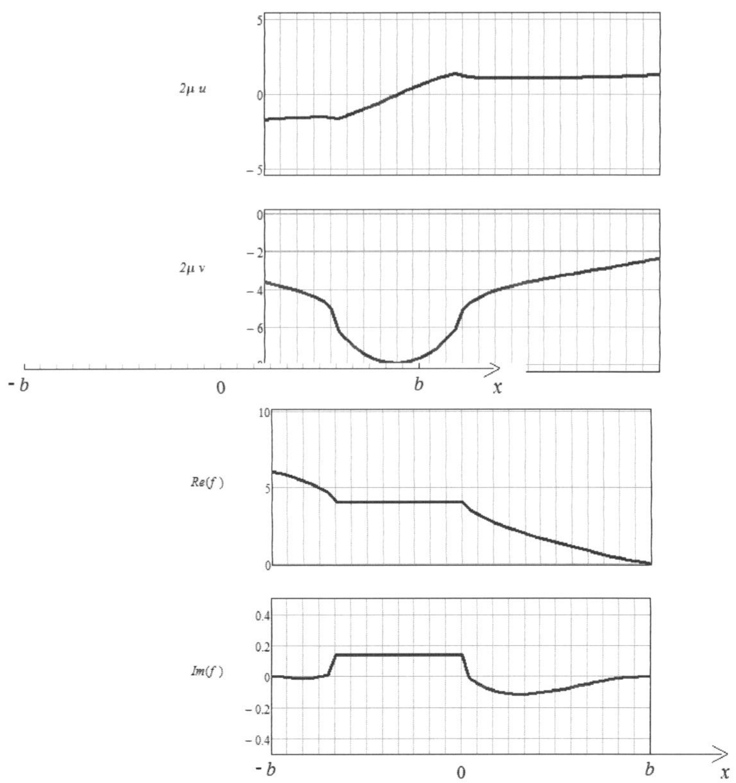

Figure 3. Displacements and the function f at the line l_1.

IOP Conf. Series: Earth and Environmental Science **134** (2018) 012033 doi:10.1088/1755-1315/134/1/012033

In this manner, the obtained system of singular integral equations and its implementation offers all necessary information on normal and shear stresses and displacement at the whole boundary of a rock block with a rectangular hole. The solution in the integral form makes it possible to analyze the solution itself and to vary the input parameters of the problem with a view to reaching the wanted results.

Conclusion

The author has obtained the relations connecting boundary stresses and displacements in the domains composing a rock block with a rectangular weakening. When the computational domain is divided into parts, these relations allow additional information on the stresses and displacements at the dividing lines without actual determination of stress state inside the whole domain. The research findings enable calculation of stress state in composite blocks with rectangular weakening.

References

[1] Muskhelishvili NI 1968 *Some Basic Problems of the Mathematical Theory of Elasticity* Moscow: Nauka (in Russian)
[2] Muskhelishvili NI 1968 *Singular Integral Equations* Moscow: Nauka (in Russian)
[3] Mirenkov VE 1978 Relation between stresses and shifts at the periphery of a working *J. Min. Sci.* Vol 14 No 3 pp 251–254
[4] Mirenkov VE 2007 Contact problems in rock mechanics *J. Min. Sci.* Vol 43 No 4 pp 370–381
[5] Krasnovsky AA 2017 Deformation of Rock Specimens with a Rectangular Opening *J. Fundament. Appl. Min. Sci.* Vol 4 No 1 pp 131–135 (in Russian)

IOP Conf. Series: Earth and Environmental Science **134** (2018) 012034 doi:10.1088/1755-1315/134/1/012034

Design model for bending vibrations of single-stage tunnel fan rotor

AM Krasyuk[1,2] and PV Kosykh[1]

[1]Chinakal Institute of Mining Siberian Branch Russian Academy of Science, Novosibirsk, Russia
[2]Novosibirsk State Technical University, Novosibirsk, Russia

Abstract. Using of one-mass model of tunnel fan rotor is justified for estimation calculation of the natural bending vibrations frequency during the design stage. It's shown that the evaluative computation of the main axial tunnel fan at the early design stage yields the acceptable accuracy. It is shown that after completion of the design, the mass of the stepped-type shaft differs from the mass of the calculated uniform-diameter shaft no more than by 40%. Inclusion of this additional mass in the estimation calculation makes it possible to improve the calculation accuracy. The region of the dimensionless rotor design parameters at which the relative difference of frequency in the evaluative and verification calculations is not higher than 5 % is determined.

A variety of tunnel fans employed in seven Russian subways tends to increase with the launch of new subway stations into service. Service life of some tunnel fans run down and their replacement or modernization with rotor module replacement is required [1]. The design activity on perfection of tunnel fans is continuous in character. Scientific approach and computation methods consider the effect of ever variable dynamics of aerodynamic ventilation parameters in development and updating of fans to develop new more efficient facilities.

Advance in design of new aerodynamic systems permitted to gain the wanted performance parameters of single-stage fan facilities. The most popular is the driving actuator – transmission shaft – fan shaft structural rotor layout. VO-21 fan developed at Institute of Mining SB RAS is typical of this design structure. RotorofVO-21 fan is shown in Figure 1.

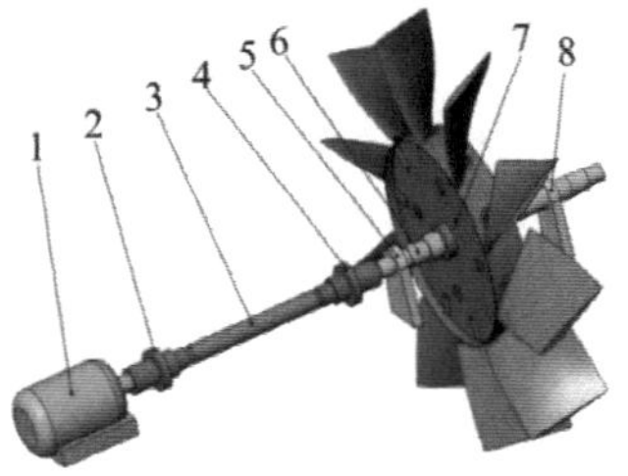

Figure 1. Rotor of VO-21 fan: 1 – electromotor; 2, 4 – compensating coupling; 3 – transmission shaft; 5 – front support; 6 – fan shaft; 7 – impeller; 8 – rear support.

Of actual significance are the design techniques to develop tunnel fan rotors for ventilation networks subjected to essential disturbances in consumption and pressure of an air flow under piston–like influence of train traffic.

In design work an important issue is evaluation of the first critical frequency of fan rotor rotation, as resonance bending oscillations used to arise in close approximation to it. This operation mode leads to breakdown of a fan facility or accelerated wear of its parts. That is the reason why it is preferable to prevent operation mode in vicinity of resonance frequencies.

The accurate calculation of the critical frequency implies consideration of a variety of factors and is possible exclusively when all the design parameters of the shaft are known, so after the design of the fan is completed, the checking calculations are performed for verification purposes. It requires a complicated mathematical model including all the required parameters. It is a common practice to employ different numerical methods: finite element method [2], matrix methods (Miclestadt-Prol method) or integral equation methods [3].

Checking calculation of the critical frequency can be also performed by means of discrete single-, dual- or multimass models. This approach enables to obtain an analytical expression for frequency, but it is less precise as compared to numerical calculations. It makes difficult to take into account a number of factors, such as fillets, spline ways, grooves, etc. affecting a critical frequency value. Nevertheless, this calculation method can be sufficient for rigid rotors, viz., for rotors rotating at frequency less than critical, provided that frequency margin is enough large, in other words, the critical frequency is by more than 15% higher than fan rotation frequency [3].

Thecheckingcalculation–nomatterwhethernumericaloranalyticalmethodisused–requiresknowledgeofallthedesign rotor parameters and can be realized only after the shaft design is completed. It can happen that the geometric shaft parameters approved in the design do not give satisfactory results, and the critical frequency appears too close to the operational frequency of the fan facility. In this case the re-design of the fan facility is imperative.

In order to prevent this situation it is preferable to perform an evaluative calculation of the critical frequency at the design stage when the final design parameters of the shaft are not known.

Beforeadesignergetsdowntoworkheisinformedoninertia-massspecificationsofanimpeller(m_2 is impeller mass, I is axial moment of impeller inertia; I_d is inertia moment of an impeller along axis normal to rotation axis), total length of fan shaft, a location of impeller hub web mounting, impeller supports, namely, lengths of fan shaft sections l_2, $l_3$are known a priori (Figure 2).

The designer determines the minimal diameter of the impeller based on actuator motor capacity and general recommendations on shaft design. Usually the minimum shaft diameter matches bearing journal diameter or is close to it. Then bearings are selected. Next it is feasible to make evaluative calculation of the critical frequency.

In the evaluative calculations for a single–staged tunnel fan it is handy to use the simplest single-mass discrete model (Figure 2). In some cases when length of transmission shaft is large as compared to rotor shaft a two-mass model is applicable [4]. Let determine a range of rotor parameters at which evaluative calculation of the lowest frequency of eigen shaft vibrations on a single – mass model gives satisfactory results. The single – mass model considers impeller mass m_2, and a shaft is assumed as a shaft with a constant cross-section of the minimum diameter equal to bearing journal diameter. Besides it is possible to consider rigidity of the front c_1 and rearc_2 supports and gyroscopic moment on the part of the impeller.

IOP Conf. Series: Earth and Environmental Science **134** (2018) 012034 doi:10.1088/1755-1315/134/1/012034

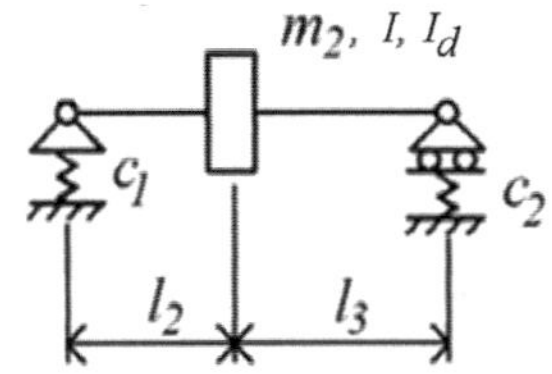

Figure 2. Calculation scheme of the single-mass model of fan rotor considering moments of impeller inertia and rigidity of supports.

The eigen vibration frequency of the system under consideration disregarding support rigidity and gyroscopic moment is determined from expression [5]:

$$p_1 = \sqrt{\frac{3EJ(l_2 + l_3)}{m_2 l_2^2 l_3^2}}. \quad (1)$$

To meet the static strength conditions the final design shaft is step–shaped with diameter increasing towards location of impeller hub mounting. As in the evaluative calculations the fan shaft diameter is assumed equal to the minimum diameter, viz., bore diameter of bearing, then rigidity of the ready-made shaft is actually much higher as compared to that of the shaft with the minimum constant diameter.

Analysis of available versions of axial tunnel fans with impeller diameter of 2.1 – 2.4 m reveals that in the final design version the stepwise shaft mass $m_в$ usually differs by no more than 40% from the mass of constant – diameter shaft assumed in the evaluative calculations. For example, since VO-21 fan shaft acquired a multi-staged shape, its mass increased by 23% as compared to the mass of the shaft of the uniform minimum diameter, assumed in the evaluative calculations.

Let assess the effect of shaft mass on a critical frequency value. Available publications report the well–known solution to account for a mass of uniform-cross-section beam under load vibrations after Rayleigh's method [5]. The shaft mass multiplied by certain factors is added to a localized mass. Eigen vibration frequency in such system is determined from equation:

$$p_в = \sqrt{\frac{3(l_2 + l_3)EJ}{(m_2 + a\gamma l_2 + b\gamma l_3) l_2^2 l_3^2}}, \quad (2)$$

where $a = \frac{1}{3}\frac{(l_2 + l_3)^2}{l_3^2} + \frac{23}{105}\frac{l_3^2}{l_3^2} - \frac{8}{15}\frac{l_2(l_2 + l_3)}{l_3^2}, b = \frac{1}{12}\frac{(2l_2 + l_3)^2}{l_2^2} + \frac{1}{28}\frac{l_3^2}{l_2^2} - \frac{1}{10}\frac{l_3(2l_2 + l_3)}{l_2^2}$ – factors determined from equation for kinetic energy of a vibrating beam; γ is linear mass of a shaft, EJ is bending rigidity of the shaft.

Introduce dimensionless parameters:

$$n = l_3 / l_2, \; k = \frac{m_s}{m_2} = \frac{\gamma}{m_2} l_2(1 + n),$$

where m_s is a shaft of uniform cross-section.

Then expressions a and b are:

$$a = \frac{2}{105}\frac{(1+n)}{n^2}(1 + 29n), b = \frac{2}{105}n^2 + n\frac{2}{15} + \frac{1}{3}.$$

Introducing them into expression (2) gives:

$$p_s = \sqrt{\frac{3l_2(1+n)EJ}{\left[m_2 + \frac{2}{105}\frac{(1+29n)}{n^2}km_2 + (\frac{2}{105}n^2 + n\frac{2}{15} + \frac{1}{3})\frac{km_2 n}{(1+n)}\right]n^2 l_2^4}} \,. \quad (3)$$

Transfer expression (1) with account for dimensionless values:

$$p_1 = \sqrt{\frac{3EJl_2(1+n)}{m_2 n^2 l_2^4}} \quad (4)$$

Relative difference of solutions with/without account for shaft mass is:

$$\delta_s = \frac{p_1 - p_в}{p_1} = 1 - \sqrt{\left(1 + \frac{2}{105}\frac{(1+29n)}{n^2}k + (\frac{2}{105}n^2 + n\frac{2}{15} + \frac{1}{3})\frac{kn}{(1+n)}\right)^{-1}} \quad (5)$$

This expression is valid at $n > 1$.

Assume if difference δ_s is less than 5 %, shaft mass insufficiently influences a value of the critical rotor frequency. In Figure 3 there is a shaded section in coordinates n–k where $\delta_s < 5\%$.

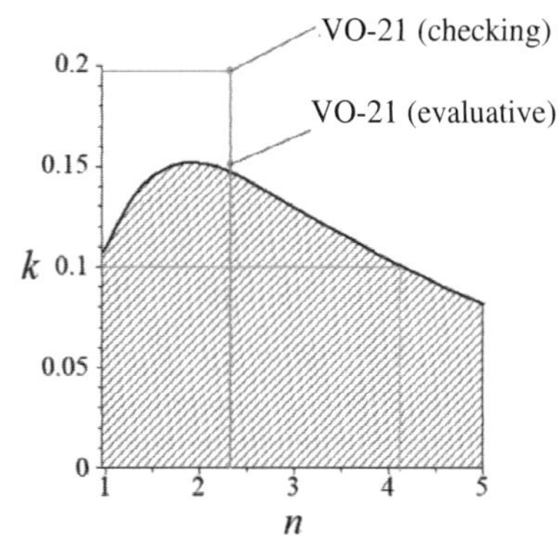

Figure 3. Rotor parameters domain when a relative difference of frequencies for evaluative and checking calculations does not exceed 5%.

The analysis of equation(3) revealed that to account for an increase in shaft mass it was enough to increase the uniform-cross-section shaft mass m_s by 40% and the resultant frequency would be really lower than an actual one.

Thus, the evaluative calculations for VO-21 fan gave frequency of 36 Hz (2160 rpm). This value exceeds more than twice its operative rotation frequency (1000 rpm). The final design frequency of fan shaft determined in checking calculations was 64 Hz (3840 rpm).

As bearings used to yield to a certain extent, the account for them in calculations leads to reduction in the critical frequency. Flexibility of a discrete bearing can be calculated by the procedure described in [6]. The yielding values established empirically for bearings of certain diameters can be helpful in this case [3]. This process is especially convenient as it is possible to determine the rigidity based on the known bearing diameters, so this process offers undisputable advantages for evaluative calculations.

In earlier publication [4] it is established that yielding of bearings negligibly affects a calculated value of the critical frequency of an axial tunnel fan provided that inertia of impeller turning is disregarded. However in the case when the impeller turning inertia must be taken into account by all means, it is obligatory to account for yielding of supports.

Conclusions

It is demonstrated that at the initial design stage of development of a main axial fan it is feasible to make evaluative calculation of frequency of eigen rotor shaft vibrations based on a single-mass model with acceptable precision. The researchers established the domain of design rotor parameters where relative difference of frequencies of evaluative and checking calculations does not exceed 5%.

Given that in evaluative calculations for a rigid rotor the bending vibration frequency of the shaft differs by more than 15% from the driving force frequency, then the checking calculation can be omitted as the actual difference of frequencies should be higher than 15%.

References

[1] Krasyuk AM 2006 *Tunnel Ventilation of Metro* Novosibirsk: Nauka (in Russian)
[2] Genta G 2005 *Dynamics of Rotating Systems* New-York: Springer
[3] Maslov GS 1968 *Calculating of Shafts Vibrations* Reference book Moscow: Mashinostroyeniye (in Russian)
[4] Kosykh PV and Krasyuk AM 2016 To the method of calculating the critical rotational speed of the main ventilation fan rotor *Mining Equipment and Electromechanics* No 5 pp 36–42 (in Russian)
[5] Timoshenko SP,Young DH and Weaver W 1974 *Vibration problems in engineering* NY: Wiley
[6] Beyzelman RD, Tsypkin BV and Perel LY 1975 *Rolling Bearings* Handbook Moscow: Mashinostroyeniye (in Russian)

Influence of different geological structures on stress–strain state of hard rock mass

NN Kuznetzov* and YuV Fedotova**

Mining Institute, Kola Science Center, Russian Academy of Science, Apatity, Russia

E-mail: *nikavalon@mail.ru, **fjulia@mail.ru

Abstract. The results of numerical simulation of stress–strain state in a hard rock mass area with the complex geological structures are presented. The variants of the stress value change are considered depending on the boundary conditions and physical properties of the model blocks. Furthermore, the possibility of in-situ stress formation under the influence of energy coming from the deeper Earth's layers is demonstrated in terms of the Khibiny Massif.

1. Introduction

Currently there exists a hypothesis that rock mass take energy from large faults, and horizontal components of the resultant stress fields are much higher that the vertical components [1–4]. The major horizontal stresses are oriented in perpendicular to the fault edges in this case. Rock mass structurally arranged as hierarchies of blocks are mostly composed of hardest rocks capable to accumulate huge energy until the critical threshold and failure as a consequence. The maximum stresses are observed in intrusions as the most illustrious representatives of hard rock masses [3]. It is typical for intrusions to contain predominantly vertical or subvertical structural discontinuities of various rank and genesis. These discontinuities, having different physical properties than enclosing rock mass, can change stress concentrations at the boundaries. The change in the stress concentration can be governed by the shape of a discontinuity, or by the properties of constituent blocks, or by the value of energy coming from the deepest levels of the Earth. This paper discusses modeling of two discontinuities in enclosing rock mass (both monolithic and composed of blocks having different physical properties) at the preset intake energy value with a view to checking the above mentioned hypothesis in terms of the Khibiny Massif.

2. Object of the research

The Khibiny Massif in the north–east of the Baltic Shield is one of the world's largest central-type, complex multi-phase intrusions with an area of 1327 km^2 (36×45 km). The Massif elevates to 900–1000 m over the surrounding valley. The Hercynian age is round 290 million years. The Massif features a variety of tectonic structures at different hierarchical scales, connected in time with the conical and radial faulting. There are two stages in the tectonics of the Khibiny Massif. The first stage coincides with the time of the conical faulting. This is the period of formation of apatite–nepheline deposits, and it predetermines basic tectonic elements of individual deposits. The second stage is the period of the regional and radial fault that cut the Khibiny Massif into an ore body and separate blocks, respectively. As a consequence, the initially single ore body was split and dislocated relative to its original position. Structurally, the zones of radial faults are very thick (from 100–150 m to 2–3 km), have steep dip angels, and are characterized by considerable deformation of rock mass and by multiple-stage formation [5].

Iolite–urtite rock mass makes a thick conical intrusion arc. The internal structure of the arc is nonuniform. Jointing of apatite–nepheline rock masses is represented by four sets of joints: gently dipping and diagonal joints of set 1 (more than 60%) and steeply dipping joints of sets 2–4 (totally less than 40%). General jointing decreases with depth. Steepness of set 1 joints grows. There area large-block and small-block joints, the large-block joints stretch from hundreds meters to a few kilometers, and their width ranges from 1 to 200 mm. The small-block joints are not longer than 10 m and have the width of fractions of millimeter. Some joints (both large- and small-block) have the fill of hydrothermal minerals.

3. Results and discussion

The analysis of influence of geological structures and physical properties on stress state of hard rock mass was carried out using the finite element method. For the conditions of the Khibiny Massif, three models of two geological structures were constructed: fault F1 and dyke D1. The other blocks were composed of enclosing rocks. The physical properties of rock blocks were chosen based on the test data [6] and from the handbook [7]; the distributed force from below was determined by modeling natural stress field with selection of values to match the in situ measurement results.

The first model is shown in Figure 1 (profile). The boundary conditions are: Figure 1b (variant one)—bottom and side boundaries are fixed, rock mass experiences only gravity, lateral earth pressure is created owing to fixation of the side boundaries; Figure 1b (variant two)—lateral sides are fixed, rock mass is subjected to gravity and distributed load of 750 MPa from below. The enclosing rock blocks have: bulk weight $w = 3.3$ g/cm^3, elasticity modulus $E = 80.0$ GPa, Poisson's ratio $\nu = 0.20$. The fault blocks (F1): $w = 1.5$ g/cm^3, $E = 5.0$ GPa, $\nu = 0.46$; the dyke blocks (DД1) composed of monchikite: $w = 5.0$ g/cm^3, $E = 120.0$ GPa, $\nu = 0.10$. The modeling yields the distribution of the major principal stress σ_1. In variant one (Figure 1b), in the enclosing rocks, σ_1 gradually grows with depth from 10 to 80 MPa. For the dyke, it is found that the stress gradually grows with depth from 30 to 110 MPa, and the increased compressive stresses to 150 MPa are observed at the dyke and fault intersection (elevation 3800 m). At the lower portion of the dyke (elevation 500 m and below), a zone of increased stresses forms. For the fault, σ_1 vary from 1 to 45 MPa and no relation with the depth is observed. On the whole, at the depth of 1000 m (elevation 4000 m), where mining operations are carried out, the principal stress σ_1 have the average value of 30 MPa. In variant two (Figure 1c), under the load applied from below, the principal stress also grows with depth though its value is much higher, from 30 to 770 MPa.

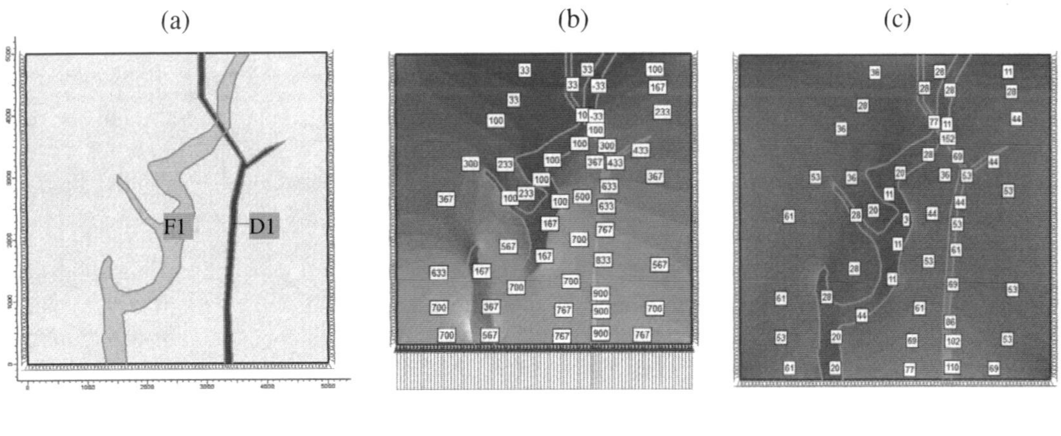

Figure 1. The first model (a) schematic fault F1 and dyke D1; (b) distribution of σ_1 (MPa) in the domain with the fixed boundaries; (c) distribution of σ_1 (MPa) under the load applied from below.

The stress state in the block in-between the fault and dyke is higher than in the external blocks. In the dyke, the stress σ_1 varies from 30 MPa at the top to 900 MPa at the bottom below elevation of 1000 m. At the dyke and fault intersection, σ_1 is 100 MPa, i.e. 50 MPa smaller then in variant one. In the fault at the top of the rock mass (above the elevation 4000 m), there are tensile stresses of – 30 MPa; at the bottom (below the elevation of 1000 m), there are compressive stresses of 700 MPa. On the whole, σ_1 increases with depth in the fault zone, thought there are areas (folds and pockets), where σ_1 can either grow or decrease considerably.

Summing up, in the variants within the first model, it is typical that the major principal stress σ_1 increases with depth in the blocks of enclosing rock mass and in monchikite dyke. The stress state in the fault depends on the fault geometry and on the physical properties of enclosing rock mass. The highest values of the stress σ_1 are observed in the dyke which is a strong stress raisor. The least stresses are found in the fault zones. At the top of the model (down to the elevation of 4000 m), the average stress value is 30 MPa.

The second model is depicted in Figure 2. Similarly to the first model, there are two geological structures, i.e. the fault and the dyke. These structures are divided into blocks with different physical properties. For the fault blocks F1: $w = 1.5$ g/cm^3, $E = 5.0$ GPa, $\nu = 0.46$; F2: $w = 1.8$ g/cm^3, $E = 8.0$ GPa, $\nu = 0.44$; F3: $w = 2.0$ g/cm^3, $E = 10.0$ GPa, $\nu = 0.41$; F4: $w = 2.2$ g/cm^3, $E = 12.0$ GPa, $\nu = 0.39$; F5: $w = 2.4$ g/cm^3, $E = 16.0$ GPa, $\nu = 0.35$; F6: $w = 2.6$ g/cm^3, $E = 20.0$ GPa, $\nu = 0.32$. For the dyke blocks D1: $w = 5.0$ g/cm^3, $E = 120.0$ GPa, $\nu = 0.10$; D2: $w = 4.7$ g/cm^3, $E = 115.0$ GPa, $\nu = 0.12$; D3: $w = 4.4$ g/cm^3, $E = 110.0$ GPa, $\nu = 0.15$. The other blocks are composed of enclosing rocks having $w = 3.3$ g/cm^3, $E = 80.0$ GPa, $\nu = 0.20$. The boundary conditions are set as in the first model: variant one—the side and bottom boundaries are fixed, and the model is only subjected to gravity (Figure 2b); variant two—the side boundaries are fixed, the model undergoes gravity and the load of 750 MPa applied from below (Figure 2c).

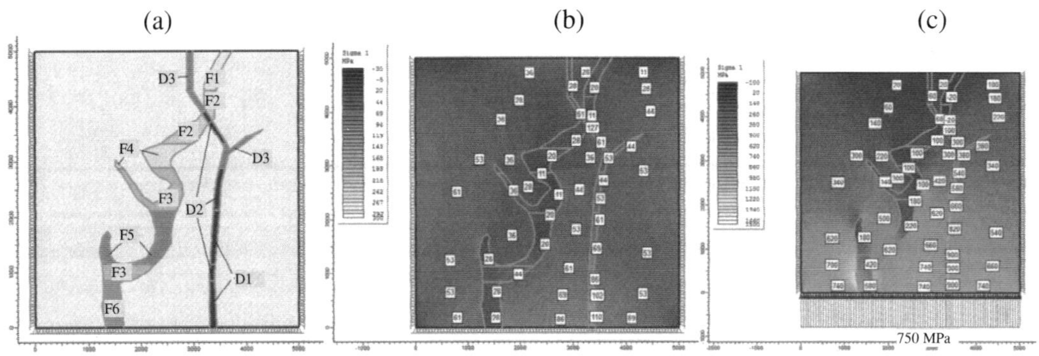

Figure 2. The second model: (a) F1–F6 are the blocks with different properties in the fault; D1–D3 are the blocks with different properties in the dyke; (b) distribution of σ_1 (MPa) in the domain with the fixed boundaries; (c) distribution of σ_1 (MPa) under the load applied from below.

In the second model, in variant one in Figure 2b, the distribution of the major principal stress σ_1 conforms with the distribution in variant one of the first model. Accordingly, the difference in the physical properties of rocks composing the fault and the dyke has caused no significant variation in the stress state of rocks. The comparison of variants two in the first and second model shows that in the latter case (Figure 2c), the stresses σ_1 in the enclosing rock blocks decrease by 20 MPa on average (the stresses change from 20 to 740 MPa with depth). The enclosing rock mass between the fault and the dyke undergoes lower stresses than in the first model. In the dyke, the values of σ_1 double (from 30 to 60 MPa) at the top (elevation 4000 m) and decrease by 50 MPa in the interval from 2000 to 4000 m. In the blocks composing the fault, at the top (above the elevation of 4000 m), tensile stresses are observed as in the first model. On the whole, the stress σ_1 in the fault zone increases by 20 MPa on average. Thus, in the considered variants of the second model, the behavior of the principal stresses σ_1 in many ways conforms with the stress behavior in the first model. At the same time, in the second model variant one, the difference in the properties of the dyke and fault rocks results in no change in the model rock mass state while in variant two with the load applied from below, the influence of the properties grows.

The third model is shown in Figure 3. There are two geological structures split into blocks with different physical properties totally coincident with the block properties in the second models. The enclosing rock blocks have the other properties: B1—w = 3.3 g/cm^3, E = 80.0 GPa, v = 0.20; B2—w = 2.9 g/cm^3, E = 60.0 GPa, v = 0.23; B3—w = 2.6 g/cm^3, E = 40.0 GPa, v = 0.26. The boundary conditions are the same as in the first and second models: variant one—the side and bottom boundaries are fixed, and the model is only subjected to gravity (Figure 3b); variant two—the side boundaries are fixed, the model undergoes gravity and the load of 750 MPa applied from below (Figure 3c).

In variant one of the third model, the values of the principal stress σ_1 in the enclosing rock mass gradually increase from 10 to 80 MPa with depth as in the first and second models. On the other hand, in the third model, at the elevations from 1000 to 4000 m, in all blocks composed of the enclosing rocks, it is observed as the stresses lower by 10 MPa on average. In the dyke, to the elevation of 2000 m, the stresses conform with the stresses in the first and second models on the whole but are higher by 30–40 MPa in the interval between the elevations of 500 and 2000 m. The stresses in the fault are the same as in the previous models. Regarding variant two of the third model the principal stress σ_1 in the outside enclosing rock mass blocks remains the same as in the second model and lowers by 80 MPa in the blocks in-between the fault and the dyke. In the dyke blocks, in the interval from 1000 t0 3500 m, the stresses jump by 200 Moa as compared with σ_1 in the second model. In the fault the stresses are higher by 40 MPa than in the second model.

(a) (b) (c)

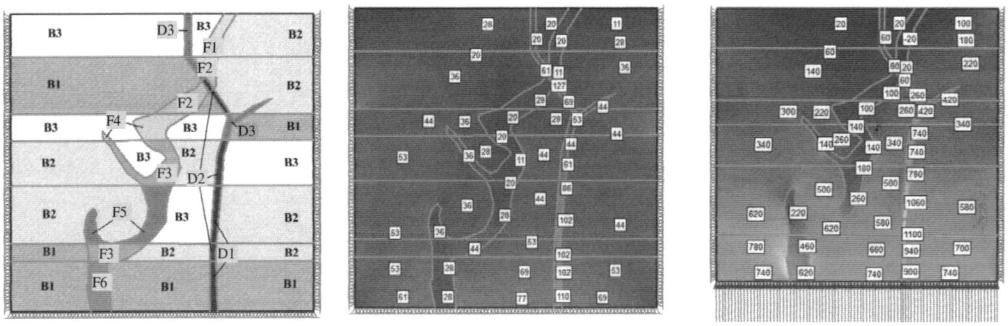

Figure 3. The third model: (a) F1–F6 are the blocks with different properties in the fault; D1–D3 are the blocks with different properties in the monchikite dyke; B1–B3 are the blocks of enclosing rock mass with different properties; (b) distribution of σ_1 (MPa) in the domain with the fixed boundaries; (c) distribution of σ_1 (MPa) under the load applied from below.

In this manner, in variant one of the third model, the behavior of the principal stress σ_1 to a higher degree corresponds to the second model. In the enclosing rock blocks with the decreased properties, the stresses are lower; in the dyke blocks adjoining the enclosing rock blocks, the stresses are higher. When the load is applied from below, the difference in the physical properties results in essentially different stress state in the model rock mass.

4. Conclusion

The numerical modeling of the rock mass stress state in terms of the Khibiny Massif conditions shows that the presence of such geological structures as faults and dykes can influence the stress behavior.

The dykes act as the stress raisors, and the faults generate stress field considerably governed by the geometry of the faults and by the properties of rocks the faults are composed of.

Furthermore, it has been found that in the geological medium structured as a hierarchy of blocks subjected to the action of gravity only, the difference in the physical properties of rocks has no essential influence on the stress state. In case when the energy enters the rock mass from the deeper levels of the crust, the difference in the properties of rock blocks can result in the excessive horizontal stresses generated in the dykes and in the fold/pockets of the faults, as well as at their boundaries. Mining operations in such zones are exposed to the highest hazard of dynamic events due to rock pressure.

References

[1] Kuzmin you and Zhukov VS 2004 *Modern Geodynamics and Variations of Physical Properties of Rocks* Moscow: MGGU (in Russian)

[2] Letnikov FA 2008 Synergetics of processes in the deep fault zones *Tectonophysics and Pressing Issues of the Earth Sciences* Moscow: IFZ RAS Vol 2 pp. 63–70 (in Russian)

[3] Markov GA 1982 Model of excessive horizontal stress formation in rocks under the influence of ascendant movements in the Earth's crust *Nature and Methodology of Tectonic Stress Determination in the Upper Part of the Earth's Crust* Apatity: KF AN SSSR pp 59–81 (in Russian)

[4] Rebetsky YuL 2008 Fault—A special geophysical body in the Earth's crust *Tectonophysics and Pressing Issues of the Earth Sciences* Moscow: IFZ RAS Vol 2 pp 123–131 (in Russian)

[5] Onokhin FM 1975 *The Specificity of the Khibiny Massif Structures* Leningrad: Nauka 1975 (in Russian)

[6] Kuznetzov NN and Fedotova YuV 2015 Study of the energy saturation of the Khibiny Massif deposit models *Proc. 19th Int. Conf. on Theory and Practice in Modern Science* Moscow: Pero pp 129–136 (in Russian)

[7] Melnikov NV, Rzhevsky VV and Protodyakonov MM (Eds) 1975 *Handbook (Cadaster) of Physical Properties of Rocks* Moscow: Nedra (in Russian)

Roofbolters with compressed-air rotators

MA Lantsevich*, AA Repin, VI Klishin and DI Kokoulin

Chinakal Institute of Mining, Siberian Branch, Russian Academy of Sciences, Novosibirsk, Russia

E-mail: *lama0350@mail.ru

Abstract. The specifications of the most popular roofbolters of domestic and foreign manufacture currently in operation in coal mines are discussed. Compressed-air roofbolters SAP and SAP2 designed at the Institute of Mining are capable of drilling in hard rocks. The authors describe the compressed-air rotator of SAP2 roofbolter with alternate motion rotors. From the comparative analysis of characteristics of SAP and SAP 2 roofbolters, the combination of high-frequency axial and rotary impacts on a drilling tool in SAP2 ensure efficient drilling in rocks with the strength up to 160 MPa.

The cost of underground coal mining is influenced by drilling for rock roofbolters. Low-pressure pneumatic machines (roofbolters) used in Russian coal mines perform rotary drilling with three- (or two-) stage telescopic feed (feeder). Table 1 describes the most popular roofbolters of the domestic and foreign manufacture that are currently in operation in underground coal mines.

Table 1. Compressed-air rotary drilling roofbolters.

Model	SBR	RAMBOR	WOMBAT	TYRBO	GOPHER	MQT130
Manufacturing country	Russia	Australia	Australia	Great Britain	Germany	China
Specifications						
Rock hardness, MPa	to 80	to 80	to 80	to 80	to 80	to 80
Hole diameter, mm	28–35	28-30	28–30	28–30	28–30	27–42
Hole length, m	to 3.5	to 3.5	to 3.5	to 3.5	to 3.5	to 3.5
Rotation speed, min^{-1}	700	900	–	600	850	300
Torque, NM	160	245	190	200	170	220
Feed force, kN	7.0	9.0	7.0	6.0	6.8	8.4
Pressure, MPa	0.63	0.63	0.63	0.63	0.62	0.63
Weight, kg	38	44	50	45	49	50

All of the listed roofbolters possess weak sides:

- Rotary method of drilling limits the application area of the machines to rocks with the hardness from 60 to 80 MPa;

- Pneumatic rotators exert static torque on the tool, which can shutdown the machine under the increased drag torque connected with the seizure of the drill rod or tool.

Figure 1a shows roofbolter model SAP designed at the Institute of Mining [1, 2]. The machine has been trialed at laboratory and commercial scale, and is recommended for manufacturing. The feature of the machine is the combination of pneumatic rotation and percussion devices which can operate independently and concurrently. The machine can perform rotation, percussion and rotary–percussion drilling if required, which allows application of SAP roofbolter in rocks of different hardness, up to 160 MPa and above. In rock mass with the hardness to 60 MPa, drilling is carried out without an air hammer, which makes the machine lighter and handy in operation.

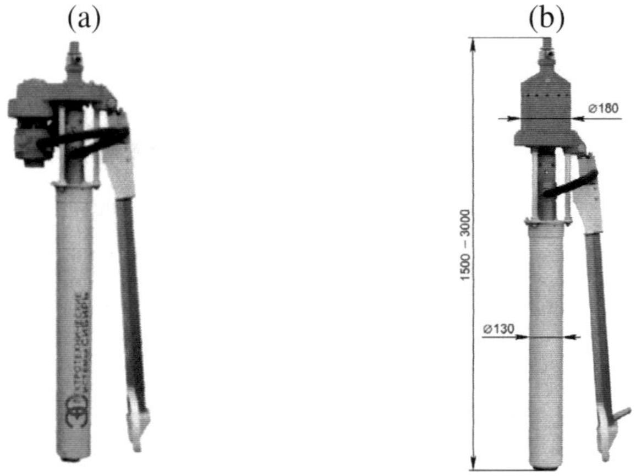

Figure 1. Roofbolters designed at the Institute of Mining: (a) model SAP; (b) model SAP2.

Tests of SAP show that rotary–percussion drilling rate is 2–3 times higher than rotary drilling rate in rocks with the hardness to 80–160 MPa [1]. Furthermore, in rotary–percussion drilling in rocks with the hardness to 80 MPa, wear of cutter bits is reduced by 3–4 times as against rotary drilling.

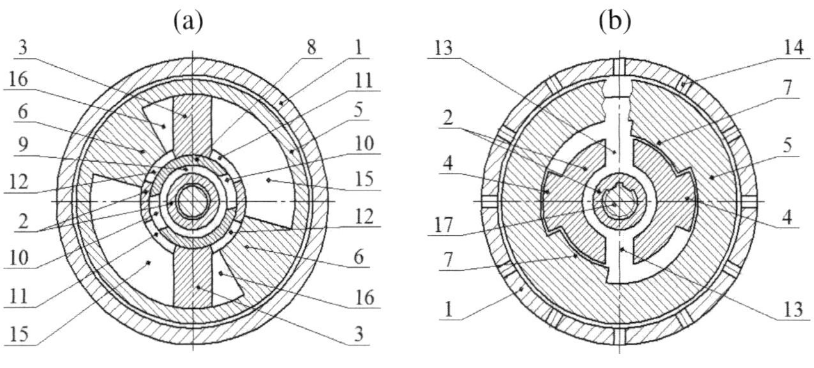

Figure 2. Cross sections of SAP2 rotor: (a) air feed lines; (b) lobes of hammer rotor and shaft rotor.

A fundamentally new design of the Institute of Mining—SAP2 roofbolter shown in Figure 1b—combines the pneumatic rotator and percussion device as SAP model. The new feature of the design is the swinging moving rotors of the rotator [3] that generates high-frequency percussion–rotation moments. Figure 2 depicts cross sections of air channels and rotor lobes of SAP2 roofbolter.

The cross-sections show the body 1, shaft rotor 2 and blades 3 and lobes 4, hammer rotor 5 with blades 6 and lobes 7, distributor 8 interacting with the hammer rotor. The shaft rotor and distributor have air feed lines 9–12 (Figure 2a, the outlet channels are provided but not shown in the figure) and the exhaust holes 13 and 14 (Figure 2b). The blades of the rotors generate pairs of working chambers 15 and 16. The shaft 17 transmits rotary percussion moments to the drilling rod and tool. The shaft also communicates with the impact piston.

SAP2 roof bolter is placed in a mine, the drill tool is pressed to the face by the feeder and the drill rod, and compressed air is then supplied to the impact piston and rotator. The impact piston, via the rod and the shaft, generates axial dynamic impacts on the drill tool.

Regarding the rotator, compressed air, along the air feed lines 9–12 (Figure 2a) in turns enters the pairs of working chamber 15 or 16 while the neighbor pairs 16 or 15 are connected to the atmosphere via the outlet channels and exhaust holes 13 and 14 (Figure 2b). Compressed air influence the blades 6 of the hammer rotor 5 and makes it carry out alternating motions relative to the shaft rotor 2 that is kinematically connected to the body by means of a free wheel unit (not shown in Figure 2). The alternating feed of compressed air in the pairs of the working chambers 15 and 16 is performed by the distributor 8.

At the end of the power stroke (clockwise) of the hammer rotor 5 (Figure 2), its lobes 7 collide with the lobes 4 of the shaft rotor 2, and the rotary percussion moment is transmitted, via the shaft 7 and the drill rod, to the rock-breaking tool. At the end of the idle motion (counterclockwise) of the hammer rotor, collision of the lobes is eliminated. The hammer rotor is stopped by compressed air and this soft impulse (reactive torque) is taken by the body 1 via the free wheel unit. Owing to comparatively high inertia moment of the body, as well as weight and length of the attachment with the control levers, vibration is not transferred to a miner's hands.

The pneumatic rotor using alternating motion of rotors is self-adjustable. In case of a low resistant moment on the rock-breaking tool and rod, they rotate at the maximum speed under the action of moments due to the pressure exerted by compressed air on the blades 3 of the shaft rotor 2 from the side of the working chambers 16 while the collision between the lobes 7 and 4 of the hammer rotor and shaft rotor is absent. In case of the high resistance moment, up to the seizure of the rock-breaking tool and rod, the air-driven rotator performs percussion mode, which liberates the tool and rod and allows the drilling process to be continued.

Table 2 compares SAP and SAP2 models; the analysis of the specifications yields some advantages of SAP2 roofbolter over SAP model:

—axial impacts and rotary–percussion torques intensify drilling;

—self-adjustable rotator in case of low loading allows no-impact moderate mode of operation and increment in torque on the tool under increasing resistance moment;

—capability of the rotator to operate at the complete shutdown of the drill rod unit it is recovered;

—smaller air consumption;

—no reducer is required;

—smaller weight.

Conclusion

The specifications of the most popular domestic and foreign roofbolters currently in operation in coal mines have bee discussed.

The authors describe SAP and SAP2 roofbolters designed at the Institute of Mining and present the pneumatic rotator using alternating motion of rotors.

The comparison of the specifications of SAP and SAP2 roofbolter shows that the axial rotary–percussion mode of drilling in SAP2 roofbolter combines high-frequency axial impacts and high-frequency rotary–percussion attacks on the rock-breaking tool, which ensures efficient operation of the machine in rocks with the hardness up to 160 MPa.

Table 2. Specifications of roofbolter models.

Roofbolter model	**SAP**	**SAP2**
Description		
Drilling mode	Rotary–percussion	Axial rotary–percussion
Rock hardness, MPa	to 160	to 160
Hole diameter, mm	28–30	30–32
Hole length, m	to 4.0	1.5–3.0
Feed force, kN	5.7	5.7
Rotator actuator		
Pneumatic	Rotation	Alternating motion rotors
Rotation speed, m^{-1}	2500	0–210
Air pressure, MPa	0.4–0.63	0.4–0.63
Air flow rate, m^3/min	3.0–4.0	1.5–3.0
Torque on rotor, Nm	13.6	100–300
Rotary percussion frequency, Hz	-	16–24
Power, kW	3.6	-
Weight, kg	11	18
Reducer	Planetary–cylinder	Absent
Reduction ratio	10	-
Weight, kg	11.8	-
Rod rotation speed, min^{-1}	0–250	0–210
Torque on rod, Nm	136	100–300
Air hammer:		
Blow energy, J	35	35
Axial blow frequency, Hz	30	30
Air flow rate, m^3/min	1.58	1.58
Weight, kg	6.9	6.9
Weight without drilling rods, kg	60	Not more than 500

References

[1] Klishin VI, Kokoulin DI and Fokin YuS 2007 Development of drilling equipment for coal mines *Ugol* No 4 pp 25–27

[2] Klishin VI, Kokoulin DI, Kubanychbek B and Gurtenko AP 2010 Drilling rigs for coal mines *Rudnik Budushch.* No 4 pp 32–38

[3] Tishkov AYa, Labuzov NN, Gendlina LI, Lantsevich MA, Levenson SYa and Eremenko YuI 2004 *Compressed-Air Drives and Vibration Exciters Using Alternating Motion of Rotors* Novosibirsk: Nauka (in Russian)

IOP Conf. Series: Earth and Environmental Science **134** (2018) 012037 doi:10.1088/1755-1315/134/1/012037

Influence of structural parameter included in nonlocal rock mass model on stress concentration around circular tunnel

SV Lavrikov, OA Mikenina and AF Revuzhenko*

Chinakal Institute of Mining, Siberian Branch, Russian Academy of Sciences, Novosibirsk, Russia

E-mail: *revuzhenko@yandex.ru

Abstract. A model of elastic body, including local curvature of elementary volume, is matched with a nonlocal model with a linear structural parameter in the differential approximation. The problem on deformation of rock mass around a circular cross section tunnel is solved numerically. The contours of the calculated stresses are plotted. It is shown that inclusion of local bends in the model results in expansion of influence zone of the tunnel and reduces stress concentration factor at the tunnel boundary.

1. Introduction

The problem on deformation of rock mass surrounding a tunnel is one of the basic problems in geomechanics. Many formulations of the problem have been analyzed [1–5]. This problem is very attractive for the use in the theoretical research as there are known exact solutions in some limiting cases. Such solutions can be test solutions for more complex models, for instance, taking into account internal structure of rock mass [6–8]. Internal structure is one of the fundamental properties of rocks and is a source of numerous equilibrium states and capacity of rock mass to accumulate some energy of external impact in the form of internal self-balancing stresses [9–11].

There many ways of taking into account internal structure of rock mass. The mathematical models with the internal variables in [12–14] describe behavior of the medium in terms of its structural elements. The non-Archimedean analysis methods are used in [15, 16] to describe multi-scale rock mass and its deformation and failure.

It is noteworthy that models of media having internal structure are used in the other research areas, such as mechanic of composites and nanomaterials, dynamics of crystal lattices [17, 18].

In this paper, in the framework of the linear elasticity model with a structural parameter [19, 20] and taking into account local curvature of a medium, the authors solve the problem on deformation of rock mass surrounding a mined-out tunnel.

2. Mathematical model

Let us use the approach from [19]. Let a discrete number of elastic particles seat at the points of a square lattice (plain-strain deformation). Let L_1, L_2—linear dimensions of the deformation domain; m, n—particles that fall on L_1, L_2, i.e. $L_1 / m = L_2 / n$—linear dimension of particle (diameter); $N = m \cdot n$—total number of particles. In a general case, at the particle contacts, it is possible to introduce sliding by plasticity laws, and dry or viscous friction. In case under analysis, we discuss an elastic model and assume that sliding is absent at the particle contacts. It is also assumed that the point moments are not transferred through the contacts. Then, for each particle (with a number i), four force vectors $\overline{T}$ and four displacement vectors $\overline{u}$ will be determined at the contact points A_i, B_i, C_i, D_i with the neighbor articles (Figure 1). One vector conforms with two scalar components. Temperature and other parameters can be included additionally. The total number of the scalar unknowns is: $4N-2m-2n$—displacements at the contacts; $4N-2m-2n$—forces at the contacts; $4m+4n$—displacements at the boundaries; $4m+4n$—forces at the boundaries. All in all, there are $8N+4m+4n$ unknowns.

For each out of N particles, there are two equations of equilibrium and one equation of moment of forces, which makes total $3N$ equations. Then, at each of $2m+2n$ boundary contacts, two conditions should be set (either in terms of forces or displacements, or their combination); i.e. there are $4m+4n$ conditions. Regarding constitutive equations, the four points A_i, B_i, C_i, D_i agree with the four vectors of displacements—8 degrees of freedom. Constitutive equations can only include such combinations that are independent of translation and rotation of particles, which means that three degrees of freedom should be withdrawn. Accordingly, we have five invariant combinations of displacement. There are also 8 degrees of freedom for the forces. The vector of sum of forces and the moment should equal zero. So, we have five degrees of freedom as a result. Consequently, the constitutive equations should relate five invariant combinations of displacements with five force characteristics. There should be five equations for a particle and $5N$ equations for all particles. We have $8N+4m+4n$ equations all in all. The system is closed and reduced to $8N+4m+4n$ algebraic equations for $8N+4m+4n$ unknowns.

In this manner, we need five constitutive equations for elementary volume. On the other hand, there are three constitutive equations for an elastic body. This means that the classical theory contains assumptions that have equal strength of two equations. Moreover, these equations are of equal significance as in the Hook law. Let us formulate such equations in explicit form.

We select five invariant combinations of displacements capable to be included in the constitute equations. The number of choices is unlimited. In out case, we take the closest variant to the linear elasticity:

$$
\begin{aligned}
&\frac{E}{2l}\left[u_1(A_i)-u_1(C_i)\right]=\frac{t_{11}(A_i)+t_{11}(C_i)}{2}-\nu\frac{t_{22}(B_i)+t_{22}(D_i)}{2};\\
&\frac{E}{2l}\left[u_2(B_i)-u_2(D_i)\right]=\frac{t_{22}(B_i)+t_{22}(D_i)}{2}-\nu\frac{t_{11}(A_i)+t_{11}(C_i)}{2};\\
&\frac{E}{2(1+\nu)}\frac{1}{2l}\left[u_2(A_i)-u_2(C_i)+u_1(B_i)-u_1(D_i)\right]=\frac{1}{4}\left[t_{12}(A_i)+t_{12}(C_i)+t_{21}(B_i)+t_{21}(D_i)\right];\\
&\frac{1}{2l}\left[\frac{u_1(A_i)+u_1(C_i)}{2}-\frac{u_1(B_i)+u_1(D_i)}{2}\right]=\xi\frac{1}{2}\left[t_{11}(A_i)-t_{11}(C_i)-t_{21}(B_i)+t_{21}(D_i)\right];\\
&\frac{1}{2l}\left[\frac{u_2(B_i)+u_2(D_i)}{2}-\frac{u_2(A_i)+u_2(C_i)}{2}\right]=\xi\frac{1}{2}\left[t_{22}(B_i)-t_{22}(D_i)-t_{12}(A_i)+t_{12}(C_i)\right];
\end{aligned}
\qquad i=\overline{1,N}, \qquad (1)
$$

where E, ν –Young's modulus and Poisson's ratio; $2l$ –linear dimension (diameter) of particle; ξ —structural parameter dimension of which is inversely proportional to dimension of stresses. The first three equations are the discrete analog of the Hook law. The last two equations are not formulated explicitly in the classical theory. Though, there is information about them. The classical theory involves (as a matter of course) the postulate on diffeomorphism [21]. More simply, it is assumed that all functions are sufficiently smooth. This means that any functions can locally be presented as a linear function, for instance: $u_1=a_{11}x_1+a_{12}x_2$, $u_2=a_{12}x_1+a_{22}x_2$, where $a_{11},...,a_{22}$—const. Placing this representation in (1) yields $\xi=0$. The contrary is valid, too. Consequently, the two equations from

(1) are "hidden" in the postulate of diffeomorphism. When $\xi \neq 0$ we obtain a linear elasticity theory with a structural parameters, and Eq. (1) describe local bends.

The system of equations of equilibrium is given by:

$$\begin{aligned} &t_{11}(A_i)-t_{11}(C_i)+t_{21}(B_i)-t_{21}(D_i)+X_1(O_i)=0, \\ &t_{12}(A_i)-t_{12}(C_i)+t_{22}(B_i)-t_{22}(D_i)+X_2(O_i)=0, \quad i=\overline{1,N}. \\ &t_{12}(A_i)+t_{12}(C_i)-t_{21}(B_i)-t_{21}(D_i)=0, \end{aligned} \tag{2}$$

Here, X_1, X_2—mass forces in the center of particle. The first two equations describe the zero main vector of forces and the third equation—zero main moment of forces.

The conditions of interaction of grains (total adherence) take on form (Figure 1):

$$\begin{aligned} &t_{11}(C_j)=t_{11}(A_i); \quad t_{12}(C_j)=t_{12}(A_i); \\ &u_1(C_j)=u_1(A_i); \quad u_2(C_j)=u_2(A_i); \end{aligned} \quad i, j=\overline{1,n(m-1)}, \tag{3}$$

$$\begin{aligned} &t_{22}(D_k)=t_{22}(B_i); \quad t_{21}(D_k)=t_{21}(B_i); \\ &u_1(D_k)=u_1(B_i); \quad u_2(D_k)=u_2(B_i); \end{aligned} \quad i,k=\overline{1,m(n-1)}. \tag{4}$$

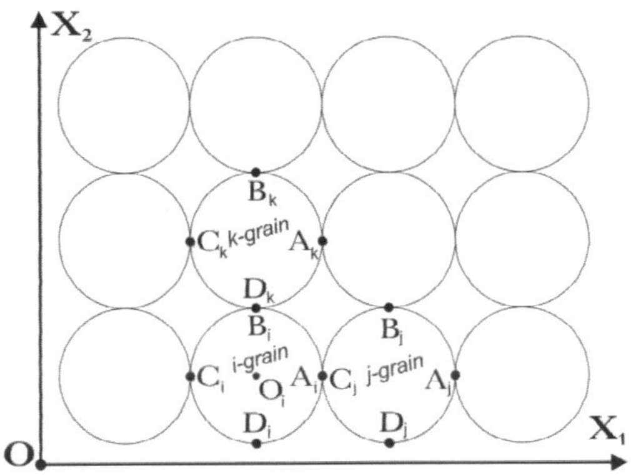

Figure 1. Internal structure of the medium: arrangement of contacts of particles.

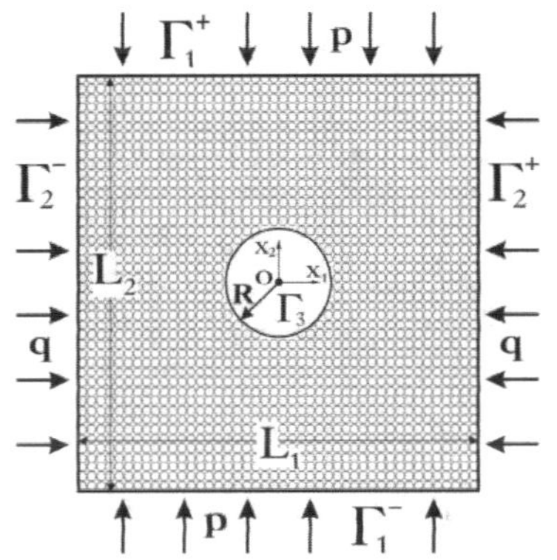

Figure 2. Problem on deformation of rock mass around a tunnel.

3. Problem formulation and calculation results

Regarding the formulation of the boundary value problem, let the calculation domain with the dimensions L_1, L_2 surround a tunnel with a radius R (Figure 2). We present the calculation domain as a set of discrete particles with a radius l. The calculation domain is subjected to gravitational and tectonic compression, and the tunnel boundary is free from stress. The boundary conditions at the boundary of the calculation domain are given by:

$$\begin{aligned} &t_{22}(B)\big|_{\Gamma_1^+}=t_{22}(D)\big|_{\Gamma_1^-}=-p; \quad t_{21}(B)\big|_{\Gamma_1^+}=t_{21}(D)\big|_{\Gamma_1^-}=0; \\ &t_{11}(C)\big|_{\Gamma_2^-}=t_{11}(A)\big|_{\Gamma_1^+}=-q; \quad t_{12}(A)\big|_{\Gamma_1^+}=t_{12}(C)\big|_{\Gamma_1^-}=0; \end{aligned} \tag{5}$$

Similarly, at the boundary of the tunnel:

$$t_{nn}\big|_{\Gamma_3}=t_{nt}\big|_{\Gamma_3}=0; \tag{6}$$

where t_{nn}, t_{nt} denote, respectively, the normal and shear forces at one of the points A, B, C, D.

The system (1)–(6) is a closed system of algebraic equations to be solved numerically using the Gauss method.

We select the dimensionless parameters of the problem (the values having dimension of length are related to L_1, the values having dimension of stress—to p):

$$p = 1, \quad q = 0{,}4, \quad E = 2000, \quad \nu = 0{,}25, \quad L_1 = 1, \quad L_2 = 1, \quad R = 0{,}26. \tag{7}$$

In the analysis of the influence exerted by the structural parameter ξ, it is first assumed that $\xi = 0$, which is case of the classical linear elasticity theory. The resultant contours of the vertical t_{22} and maximal shear $\tau_{\max} = 0.5\sqrt{(t_{11} - t_{22})^2 + 4t_{12}^2}$ stresses calculated from (1)–(6) are shown in Figures 3a and 4a, respectively. The structure has no influence in this case, the local bends are absent. The highest concentration of the stresses, both t_{22} and $\tau_{\max}$) is observed in comparatively small zones at the side boundaries of the tunnel.

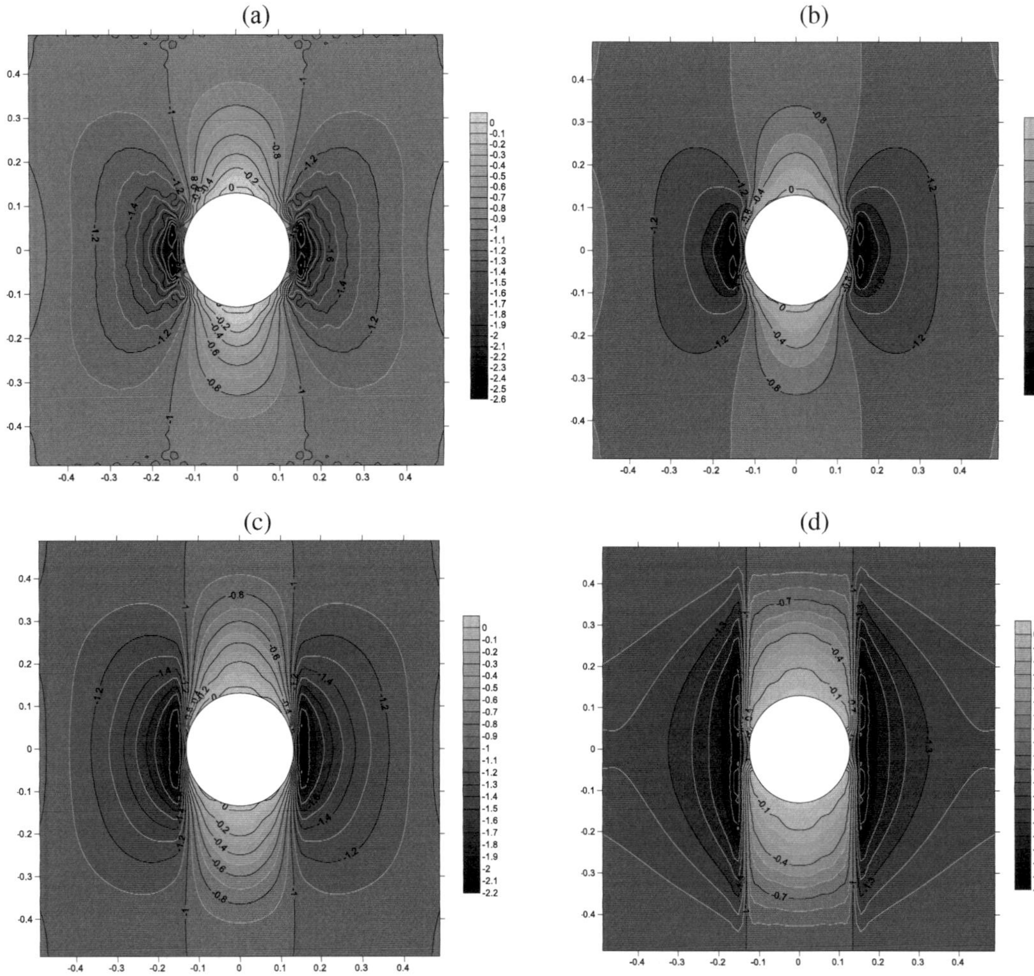

Figure 3. Contour lines of the stress t_{22} depending on the structural parameter: (a) $\xi = 0$; (b) $\xi = 5 \cdot 10^{-4}$; (c) $\xi = 5 \cdot 10^{-3}$; (d) $\xi = 5 \cdot 10^{-2}$.

Gradual increase in the value of the structural parameter due to local curvature results in considerable redistribution of stresses in the adjacent rock mass. It is seen that the stresses t_{22}, as ξ is increasing, concentrate in the consistently expanding rock mass zones in the horizontal direction (Figures 3b–3d). The maximal concentration of t_{22} changes from $t_{22} = -2.8$ (Figure 3a) to

$t_{22} = -1.9$ (Figure 3d). The maximum shear stress τ_{max} behaves similarly: it concentrates in the successively expanding zones leftward and rightward the tunnel in the vertical direction (Figures 4b–4d). The maximum concentration of τ_{max} changes from $\tau_{max} = 1.5$ (Figure 4a) to $\tau_{max} = 1.0$ (Figure 4d). In this manner, with the local curvature taken into account in modeling, the maximum concentration of stresses is "smeared", literally, in rock mass surrounding a tunnel.

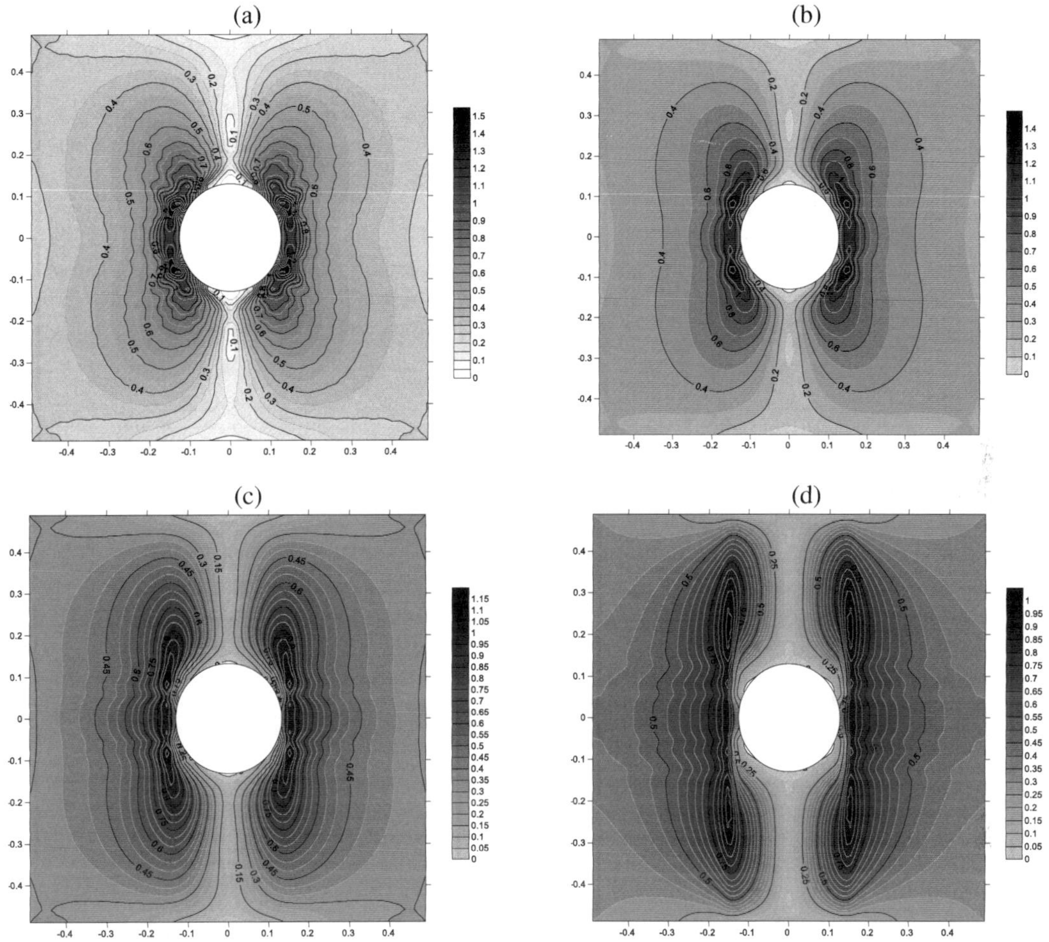

Figure 4. Contour lines of the stress τ_{max} depending on the structural parameter: (a) $\xi = 0$; (b) $\xi = 5 \cdot 10^{-4}$; (c) $\xi = 5 \cdot 10^{-3}$; (d) $\xi = 5 \cdot 10^{-2}$.

4. Conclusion

In the models of elastic deformation of rock mass surrounding a tunnel with taking into account local bends, the influence zones of the tunnel is expanded and the stress concentration factor at the tunnel boundary is lower.

5. Acknowledgements

The studies have been supported by the Russian Science Foundation, Project No. 16-17-10121.

References

[1] Shemyakin EI 1975 Two problems in rock mechanics arising out of the working of deep ore or coal deposits *Journal of Mining Science* Vol 11 No 6 pp 632–646

[2] Vinogradov VV 1989 *Geomechnaincs of Rock Mass Control near Tunnels* Kiev: Naukova Dumka (in Russian)
[3] Kolymbas D 2005 *Tunnelling and Tunnel Mechanics. A Rational Approach to Tunnelling* Springer-Verl;ag, Berlin, Heidelberg
[4] Lavrikov SV and Revuzhenko APh 1990 Deformation of a blocky medium around a working *Journal of Mining Science* Vol 26 No 6 pp 485–492
[5] Klishin SV and Revuzhenko AF 2017 3D discerete element approach to the problem on abutment pressure in a gently dipping coal seam *IOP Conf. Series: Earth and Environmental Science* 84 DOI: 10.1088/1755-1315/84/1/012003
[6] Kocharyan GG 2016 *Geomechanics of Faults* Moscow: Geos (in Russian)
[7] Ponomarev VS 2011 Problems connected with the analysis of energy-active geological medium *Geotektonika* No 2 pp 66–75
[8] Kurlenya MV, Adushkin VV, Garnov VV, Oparin VNm Revuzhenko AF and Spivak AA 1992 Alteranting response of rocks to dynamic impact *Doklady RAN* Vol 323 No 2 pp 263–265
[9] Kurlenya MV, Oparin VN, Revuzhenko AF and Shemyakin EI 1987 Some features of rcok mass response to near-range blasting *Doklady AN SSSR* Vol 293 No 4 pp 22–28
[10] Lavrikov SV and Revuzhenko AF 2016 Numerical modeling of elastic energy accumulation and release in structurally heterogeneous geomaterials *Journal of Mining Science* Vol 52 No 4 pp 632–637
[11] Lavrikov SV and Revuzhenko AF 2015 DEM code based modeling of energy accumulation and release in structurally heterogeneous rock masses *AIP Conference Proceedings* 1683 DOI: 10.1063/1.4932811
[12] Novozhilov VV and Kadashevich Yui 1990 *Microstresses in Construction Materials* Leningrad: Mashinostroenie (in Russian)
[13] Muhlhaus HB, Moresi L and Sakaguchi H 2000 Discrete and continuum modeling of granular materials *Constitutive Modelling of Granular Materials* D. Kolymbas (Ed) Springer-Verlag Berlin Heidelberg pp 209–224
[14] Lavrikov SV and Revuzhenko AF 2016 Deformation of geo-medium with considering for internal self-balancing stresses *AIP Conference Proceedings* 1783 DOI: 10.1063/1.4966423
[15] Revuzhenko AF 2012 *Mathematical Analysis of Functions of Non-Archimedean Variable: Application-Specific Mathematical Tool for Description of Structural Levels in a Geomedium* Novosibirsk: Nauka (in Russian)
[16] Lavrikov SV, Mikenina OA and Revuzhenko AF 2008 Rock mass deformation modeling using the non-Archimedean analysis *Journal of Mining Science* Vol 44 No 1 pp 1–14
[17] Kunin IA 1975 *Theory of Elastic Media Having Microstructure* Moscoe: Nauka (in Russian)
[18] Eringen AC 1967 *Theory of Micropolar Elasticity* Princeton University, Princeton–New Jersey
[19] Revuzhenko AF 2016 Version of the linear elasticity theory with a structural parameter *J. Applied Mechanics and technical Physics* Vol 57 No 5 pp 801–807
[20] Lavrikov SV and Revuzhenko AF 2017 Effect of structural parameter on stress concentration in a linearly elastic body *AIP Conference Proceedings* 1893 DOI: 10.1063/1.5007580
[21] Trusov PV 2009 Some isses of nonlinear mechanics of deformation body (A discussion) *Mat. Modelir. Sistem Protessov* No 17 pp 85–95

IOP Conf. Series: Earth and Environmental Science **134** (2018) 012038 doi:10.1088/1755-1315/134/1/012038

Conditions of efficient vibrodischarge of rock materials in modern mining and processing technologies

SYa Levenson*, LI Gendlina and EG Kulikova

Chinakal Institute of Mining Siberian Branch Russian Academy of Sciences
Novosibirsk, Russia

*E-mail:shevchyk@ngs.ru

Abstract. The paper reviews vibration feeders used to discharge storage reservoirs in mineral mining. In spotlight are vibrofeeders equipped with an active member of low flexural rigidity developed at Chinakal Institute of Mining. The authors present the results of the physical and numerical studies on vibratory discharge of cohesive rocks from a bunker.

The development of mining industry indispensably relates to introduction of advanced mining technologies, new brand machinery to mine and handle different mineral materials.

One of present-day mining problems is the arrangement of favorable conditions for uninterrupted discharge of bulk material from storage bins, bunkers responsible for continuous operation of the mechanized mining plant. Efficiency of granular material haulage in most depends on physic-mechanical properties of the material. Thereto, cohesive dispersed media are characterized with essential cohesion forces acting between particles, which hamper the material motion along with internal friction. Discharge of such materials from bunkers encounters the pipe-formation and frequent hanging challenges and remains a burning problem so far.

Vibration feeders are capable to operate at high productivity under heavy-duty mining conditions, herewith hygiene, sanitary conditions and operation and labor safety as well as automated discharge of bulky material from bunkers are provided [1]. At present more than 50 types of vibrohaulage machines are available, most of them are designed and manufactured for specific industrial conditions.

The typical design of a vibrofeeder consists of a rigid active member, elastic element system, connecting it to a base, and vibroexciter in a single-mass scheme as a rule. Versions of such machines are electrovibration feeder of PE-type, Mekhanobrtekhnika, Russia, (Figure 1a) [2]. The active member of the feeder is fixed to bunker walls through the elastic element system and is actuated by a single electric vibroexciter. This and similar machines operate efficiently at mines and concentrators on dosed feeding of granular materials not prone to compaction, with low moisture content from storage bins. However they are not designed to operate with cohesive materials specified with appreciable cohesion.

Figure 1. Vibration feeders of a single-mass scheme: (a) PE electrovibration feeder; (b) GZD vibration feeder.

GZD feeder, Chinese Break-Day Co. (Figure 1b) [3], is designed as a heavy-duty vibration machine of a single-mass type to operate in an orepass. It is convenient in maintenance, efficient to haul bulk materials, but not usable for cohesive, sticky materials.

High-power feeder of double-mass scheme is proposed by US company, General Kinematics (USA) (Figure 2) [4].

Its design allows adjustment of eigen frequency of its active member, automated fitting to variations in pressure of a material to be discharged without reduction in delivery rate and a rise in power consumption.

This machine is capable to cope with heavy-duty loads on transporting surface and operates efficiently to discharge bulk rock materials of low moisture content from storage bunkers to haulage facilities.

Figure 2. Double-mass feeder, General Kinematics Co.

The general imperfections of vibrofeeders with a rigid active member are high labor intensity of preparation and mounting operations, need in heavy under structure, as the carrying members of these machines suffer appreciable loads under dynamic unbalance conditions. Their application scope is restricted as they are not applicable to handle cohesive and sticky materials.

At Chinakal Institute of Mining SB RAS the researchers developed a series of vibrofeeders with the elastic active member [5, 6].

The active member made of steel sheet of low bending rigidity is readily placed on the foundation and performs bending vibrations in the operation mode. The vibration source is an electric inertial or pneumatic vibroexciter. The operational principle of these machines enables to realize greater variety of haulage modes as compared to feeders of conventional scheme. They are distinguished for simple construction compact dimensions, small mass, and low mounting and demounting costs.

First vibration machines were VLR-1, BLR-2, and VLR-3 feeders designed to discharge ore from ore passes and to load ore into haulage facilities. Imperfection of these feeders is a small zone of vibro effect on a bulk material because of essential attenuation of vibration in its transmission from a vibration source along an active member. This drawback does not permit to discharge efficiently cohesive heavy-duty materials. Herewith, the additional preparation of foundation or a frame is required to mount the active member.

Alternatively to VLR-type feeders "Volna" vibrofacilities are manufactured with a case-type support frame, and this notably facilitates mounting operation. As an unloading facility the machine is fixed on bunker walls with elastic elements. Volna–1.5P, Volna–2P, Volna–3P, Volna–4Pvibrofeedershave the same structure scheme, differ in overall dimensions and production capacity, and designed to handle different granular materials with low moisture content. Elastic active member of these machines is curved in a shape to improve material discharge at the expense of artificially created surface close in shape to natural slip surfaces which can be realized in a material under discharge with the aim to boost sliding of granular material. The pneumatic vibroexciters, designed at IM SB RAS, simple in design and high-reliable are employed as a vibration source. Figure3 presents the scheme of Volna–4P vibrofeeder used to draw broken ore from a block.

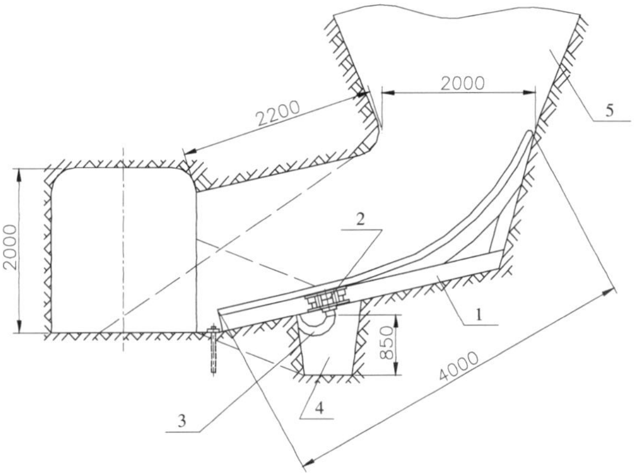

Figure 3. Scheme of Volna – 4P vibrofeeder to draw broken ore from a block: 1 – vibrofeeder; 2 – vibroexciter; 3 – air-supply hose; 4 – a hose pit; 5 – mine working.

The analysis of vibromachines revealed that the operating efficiency of available vibrofeeders sharply falls with increase in viscosity of granular media. These vibromachines are designed to handle bulk materials not prone to adhesion.

At IM SB RAS the research works are conducted on investigation into vibration handling of cohesive geomaterials and search for technical solutions to this problem.

To be particular, the use of extra vibroactuator appreciably intensifies vibration of a loading section of the elastic active member and reduces the effect of fading of bending vibrations on discharge of granular materials including cohesive ones. However, in this case the requirement to provide synchronous operation of vibroexciters is difficult to realize.

The principally new structure scheme is proposed as an alternative solution, where an active member of small bending rigidity is fixed in a suspended state on elastic support elements connecting it to a support frame [7]. The researchers propose a vibromachine where there is no contact between the active member and the foundation, unlike a vibromachine with elastic active member where the longitudinal component of vibrations is negligibly small because of friction between the active member and the foundation, and the output is realized due to bending vibrations. In the new version attenuation of vibrations is reduced along transporting surface and amplitude of vibrations on the loading section of the feeder is intensified.

The physical and numerical modeling techniques were applied to study dynamics of vibrofeeders. Physical experiments were carried out at the stand consisting of a vibrofeeder and a storage bin. The test cohesive material was clayey sand of 0.005…0.1 mm in particle size and 10% clay content. Cohesion was varied by altering moisture within 10–18%. Vibration parameters were measured by means of measurement–computing complex. The time of output of the measured volume of cohesive material was recorded.

The numerical modeling was performed by the finite element method in ANSYS software. Solution of the complete system of dynamics equations (Full Transient Analysis) was carried out with the use of implicit scheme of direct integration in time based on Newmark method [8].

It was experimentally demonstrated that the unattenuating longitudinal component appreciably intensifies vibrations of the loading section of the active member (Figure 4); in the outlet their amplitude remains practically the same.

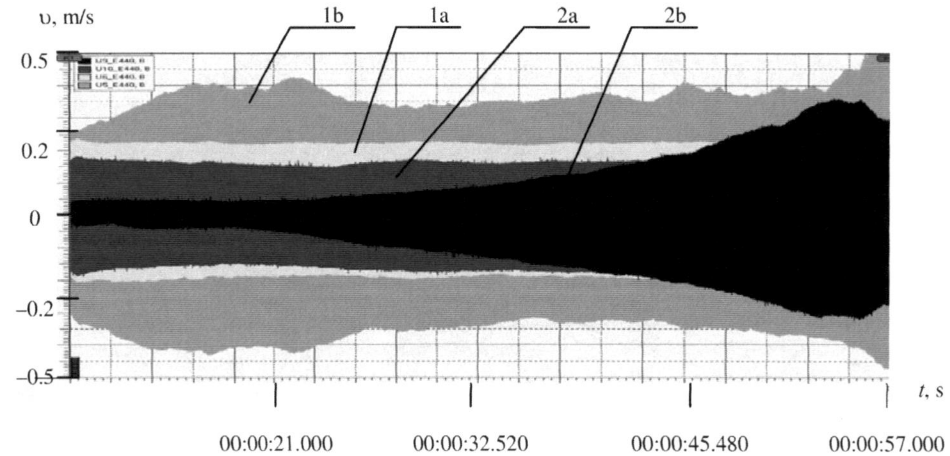

Figure 4. Oscillograms of longitudinal (a) and lateral (b) components of vibrovelocity, υ: 1 – a spot of driving force application, 2 – loading end.

It is established that to realize efficient cohesive material discharge and to meet admissible sinking-down of the active member under a load implies that ratio $\frac{l_{on}^3}{EI}$ should not exceed 0.025–0.030 (l_{on}, m, EI, Nm^2 – length and rigidity of a support, respectively).

A vibrofeeder prototype to discharge granular and cohesive materials from a bunker was developed on the base of experimental results (Figure 5). A curved active member 1 of small bending rigidity is inside a case-type frame 8. Its loading and discharging end are rigidly connected to elastic support elements fixed on the frame. One of design support version is shown in Figure 5b. It consists of elastic metallic plates 2 and 3, rubber–fabric plates 4 and 5, and mounting blocks 6 and 7 to fix plates to the active member and the frame, respectively.

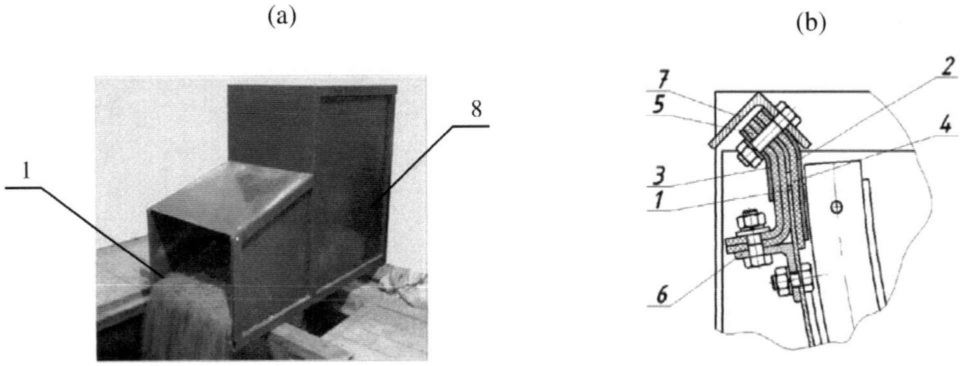

Figure 5. Pilot vibrofeeder to discharge granular and cohesive materials: (a) general view; (b) arrangement of support component in the loading end: 1–active member; 2, 3–elastic metallic plates; 4, 5–elastic rubber-fabric plates; 6, 7–mounting blocks; 8–frame.

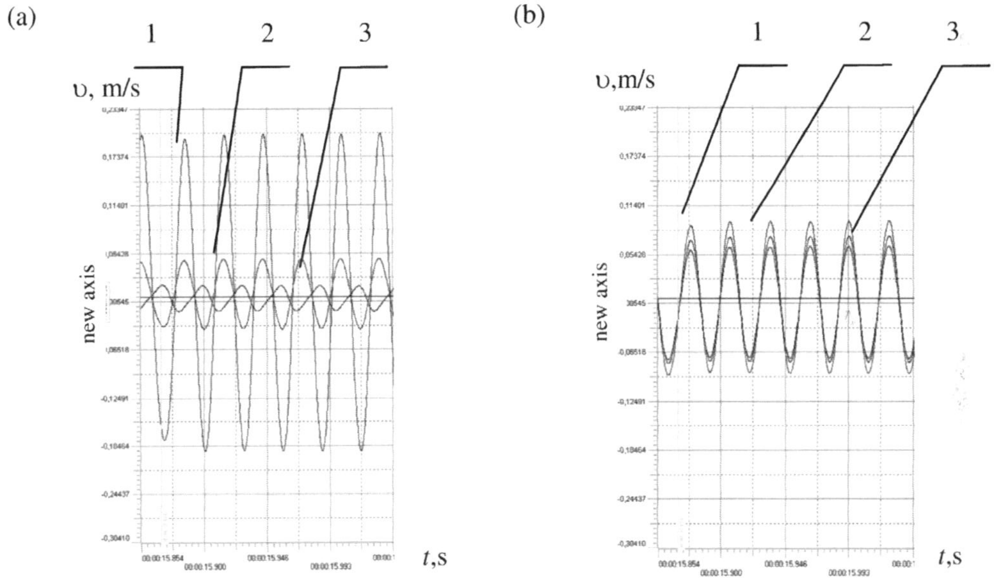

Figure 6. Vibrovelocity v of lateral (a) and longitudinal (b) vibrations of the active member: 1–in a spot of the driving force application; 2–in the middle of the active member; 3–in the loading end.

The laboratory feeder tests showed its high performance. The discharge of clayey sand of 10% clay content and 13% moisture proceeded uniformly with none of sticking. Sticking or compaction of the feed in loading end of the active member was not observed, vibration in this section was confirmed by oscillograms of the active member vibrations recorded in the course of output (Figure 6).

The output capacity amounted to 55–60 t/h at driving force parameters cited in the Table 1.

IOP Conf. Series: Earth and Environmental Science **134** (2018) 012038 doi:10.1088/1755-1315/134/1/012038

Table 1. Feeder specifications

Overall dimensions, mm:	
–length	1500
–width	664
–height	1050
Area of output window, m^2	0.27
Mass , kg	160
Vibration frequency, Hz	35
Driving force, N	2400
Capacity of vibroexciter, W	950

Conclusion

The elastic supporting elements in the feeder design enable to increase intensity of vibrations in the loading end of the active member of a small bending rigidity, thus providing conditions for efficient discharge of not only bulk materials, but cohesive materials as well.

References

[1] Molotilov SG, Vasil`yev YeI, Kortyelyev OB, et al 2000 *Intensification of Loading and Haulage Operations at a Quarry* Novosibirsk: SB RAS (in Russian)
[2] Electric vibrating feeders (vibrating feeders)–Site access mode: http://www.mtspb.com/production_current.php?id=4&id_group=22
[3] Vibrating feeder–Site access mode: http://www.break-day.com/ru/vibrating_feeder.htm
[4] Vibratory Rock Screens / Feeders – Site access mode: http://www.generalkinematics.com/mining/proddesc.cfm/productid/94
[5] Tishkov AYa, Gendlina LI and Levenson SYa 1992 Vibratory equipment with flexible active member to mining operations *News of the Higher Institutions Mining Journal* No 10 (in Russian)
[6] Levenson SYa, Gendlina LI, Eremenko YuI, Kulikova EG and Morozov AV 2008 Vibratory equipment for underground and open cast mining *Proceedings of the VI International Scientific and Technical Conference "Technical Equipment for the Mining and Oil and Gas Industry"* Ekaterinburg (in Russian)
[7] Gendlina LI, Levenson SYa, Alesik MYu and Kulikova EG 2013 About vibratory equipment parameter effects on the process of coherent rock material discharge from the bunker *Mining Equipment and Electromechanics* No 1 (in Russian)
[8] Lukashevich AA 2003 *Modern Numerical Methods of Structural Mechanics* Khabarovsk: KhGTU (in Russian)

IOP Conf. Series: Earth and Environmental Science **134** (2018) 012039 doi:10.1088/1755-1315/134/1/012039

Difference in rockburst hazard in ore and coal mines

AV Lovchikov

Mining Institute, Kola Science Center, Russian Academy of Sciences, Apatity, Russia

E-mail: vocson@goi.kolasc.net.ru

Abstract. In the Russian mining and engineering literature, in most cases, there is no difference in the assessment of the rockburst hazards in metal and coal mines. Nevertheless, it exists, in view of the difference in geological and geotechnical conditions of coal and ore deposits. Since ore deposits occur in the solid magmatic or metamorphic rock masses, the strongest induced earthquakes are much more powerful in ore mines than in coal mines. The main difference of rockbursting lies in the difference of natural stress state: gravity stress state in the coal fields and gravity-and-tectonic stress state in ore mines. The actual stresses are mostly vertical in the first case and horizontal in the second case, which conditions the difference in rockburst hazard in coal and ore mines.

The Russian mining and engineering literature and regulation standards are somewhat conservative in rockburst-hazard estimation in ore mines, which is connected with the history of the phenomenon. The first rockbursts in Russia were recorded in the 1940s in coal mines. Scientific men, mostly from the VNIMI Institute (S.G. Avershin, I.M. Petukhov and others) solved the problem to a certain degree and developed mine safety rules under conditions of rockburst hazard, and risk of rockbursts in coal mines was abated.

Rock bursts in ore mines in Russia started later, in the 1960s. The theoretical framework of rockburst prediction and prevention, earlier developed for coal mines, were extended to ore mines, which is wrong as geological conditions of coal and ore mines differ considerably. Some of those rules have still been in use unreasonably in the regulatory documents provided by the Federal Environmental, Industrial and Nuclear Supervision Service of Russia [1].

Ore deposits occur in magmatic and metamorphic host rocks, and both the ore and rocks possess high strength properties and deformation characteristics. Coal deposits occur in sedimentary rocks, and their characteristics are much lower. Table 1 describes strength of rock mass of some most rockburst-hazardous coal and ore deposits in Russia.

It follows from Table 1 that enclosing rock mass of ore deposits has strength by an order of magnitude higher than host rock mass of coal deposits. The stronger rock mass feature increased tectonic stresses, as shown in [3], i.e. strong rock masses are saturated with much energy. This means that the energy of the strongest seismic events in ore mines should be higher than in coal mines, and this fact is proved by practice. Table 2 compiles data on the strongest induced bursts (induced earthquakes) in ore and coal mines in Russia [4–6].

Evidently, power (energy) expressed in terms of magnitude of the strongest induced earthquakes is one–two orders of magnitude (by tens–hundreds times) higher in ore mines than in coal mines, which proves the above statement on higher energy level at ore deposits. At the same, after having analyzed the study [6], this inference is not evident. The matter is that the author of [6] made some geomechanical faults. In many cases, the author omits the scale on which the magnitude of an event was determined. Moreover, in the reference seismic catalogs, the magnitude m_b of earthquakes is determined by the teleseismic records of P wave [7] hundreds and thousands kilometers away from the earthquake hypocenter. In geomechanics the local magnitude scale M_L (by Richter) is used, and this scale is only applicable at the distances to 600 km from the earthquake hypocenter [7]. The most erroneous values by the P wave records are given by the earthquake depth. Thus, the study [6] presents mainly incorrect, higher values of magnitudes of seismic events than, for instance, in [4] where the values of the magnitudes M_L of induced earthquakes in mines are checked with respect to a number of sources. Anyway, the presented examples show that the energy of seismic events in ore mines (i.e. at the depth to 1 km below surface) is for a number of reasons higher than in coal mines in Russia.

Table 1. Strength of rock masses hosting ore and coal deposits in Russia.

Ore deposits				Coal deposits [2]		
Deposit	Rocks	Density, t/m^3	Compression strength, MPa	Deposit	Rocks	Density, t/m^3
Lovozero, rare metals	Foyaite	2.65	160–243	Argillite	2.50–2.90	20–70
	Urtite	2.75	247			
	Lujavrite	2.6–2.96	187–332	Siltstone		
Khibiny, apatite	Apatite–nephelene ore	2.8–3.1	60–150	Hardest coal (Kizelov basin)	~1.3	30–50
	iolite–urtite and rischorrite	2.8–2.9	100–250			
Tashtagol, iron ore	Magnetite (ore)	3.3–4.9	120–140			
	Syenite, skarn, prophyrite	2.6–3.2	50–180			
North Ural, bauxite	Bauxite (ore)	2.7–3.2	32–220			
	Limestone (enclosing rocks)	2.6–2.7	45–155			

Table 2. Some of the strongest rockbursts and induced earthquakes in ore and coal mines in Russia.

Ore deposit [4]			Coal deposit [5, 6]		
No.	Time and place of event	Magnitude M_L	No.	Time and place of event	Magnitude M_L
1	17 August 1999, Umbozer Mine, Lovozero rare metal deposit, Kola Peninsula	5.1	1	28 August 1973, Lenin Mine, Kizelov Basin, Ural	2.8±0.5
2	16 April 1989, Kirov Mine, Khibiny apatite deposit, Kola Peninsula	4.3	2	16 July 20111, Pechora Coal Basin	3.4

The most significant difference in the ground conditions of coal and ore deposits is the natural stress state. For coal deposits occurring in sedimentary rock mass, the characteristic stress state is of gravity nature, when maximum stresses are generated by the weight of overlying strata. For ore deposits enclosed in magmatic or metamorphic rock masses, it is typical to have mixed gravitational-and-tectonic stress state when the maximum stresses are induced by horizontal tectonic forces conditioned by the modern orogeny. The stresses in rock mass surrounding an underground excavation at a certain depth H below surface in case of gravitational and gravitational-and-tectonic stress state are illustrated in Figure 1.

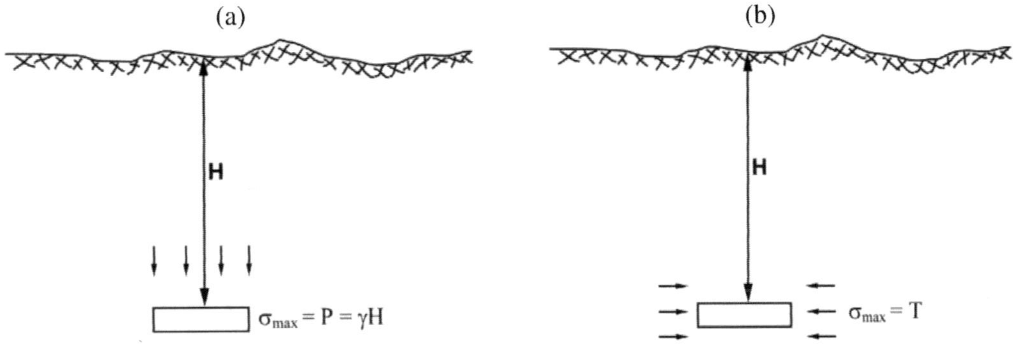

Figure 1. (a) Gravitational stress state; (b) gravity-and-tectonics stress state.

The maximum stresses in the roof of the excavation in case of the gravitational stress state (Figure 1a) are governed by the weight of overlying strata:

$$P = \gamma H, \tag{1}$$

where γ—density of overburden rocks, t/m^3; H—depth below surface, m.

In the gravity-and-tectonics tress state, the maximum stresses are generated by the tectonic component T (Figure 1b) which is oriented horizontally and, as practice shows, considerably exceeds P at the respective depth below surface. The value of the tectonic stresses T can only be determined by experimental measurements in rock mass. As a rule, the maximum horizontal stress in rocks under the gravity-and-tectonic stresses is 5–10 higher than the vertical stress conditioned by the overlying rock weight. As a consequence, rockburst hazard at ore deposits is generally governed by the tectonic stresses. Table 3 gives the characteristic of the gravity-and-tectonic stress state assessed at the most rockburst-hazardous ore deposits in Russia [8].

Table 3. Natural stress state of rock mass at the most rockburst-hazardous ore deposits in Russia [8].

Ore deposit	Maximum horizontal compressive stress, MPa	Ratio of vertical stress to the maximum horizontal stress in rock mass
Khibiny (Kirov Mine	20–80	1 : 3–20
Lovozero	40–70	1 : 5–10
North Ural (bauxite mines)	25–50	1 : 2–2.5
Tashtagol	60–90	1 : 2–5

It is evident from Table 3 that the most rockburst-hazardous ore deposits in Russia experience sharply nonuniform stress state generated by gravity and tectonics, where the vertical stress is the lowest value. Taking into account that the compression strength at the deposits listed in Table 3 is usually σ_c = (50–200) MPa and in many cases is close to 100 MPa and the minimum coefficient of tectonic stress concentration in the roof and floor in mines is k = 2–3, it becomes clear that in the adjacent rock mass, the rockbursting condition is fulfilled [8]:

$$\sigma_a \geq 0.8\sigma_c, \tag{2}$$

where σ_a—maximum actual stresses at the mine boundary; σ_c—compression strength of rock mass.

In this manner, it is natural that the tectonic stresses are the main cause of rockbursts at the listed deposits. The tectonic forces generate such stress level in rocks surrounding a mine that could be induced by the gravity forces at a depth of 2–3 km below surface.

So, the main different between natural stress states at coal and ore deposits is the vertical direction of the maximum stress at the coal deposits (Figure 1a) and the horizontal direction of the maximum stress at the ore deposits (Figure 1b). This factor has many implications. First of all, there is no notion of critical depth in terms of rockburst hazard at ore deposits. The critical depth is any mining depth at which:

$$\sigma_{cr}>0.5, \tag{3}$$

where σ_{cr}—stresses causing dynamic events of rock pressure.

Under gravitation stress state, the critical stresses are readily calculated from:

$$P=\sigma_{cr}=0.5. \tag{4}$$

There is no critical depth under the gravitational-and-tectonic stress state as the level of the critical stress is not defined by the depth but by the value of the tectonic forces. The tectonic forces in the horizontal plane have a definite orientation and to a certain degree ate independent of depth. The critical stresses in rocks around an underground excavation are not conditioned by depth but depend on the direction of the excavation relative to the line of the tectonic forces. Figure 2 shows the location of epicenters of strong seismic events (M_L > 1) in plan view of mining operations (with the horizontal lines of the daylight surface) in the west wing of Karnasurt Mine at Lovozero rare metal deposit. As seen in Figure 2, the earthquakes take places at the depths from 50 to 600 m below surface, i.e. they are independent of depth. Some of the earthquakes in Karnasurt Mine are described in [9].

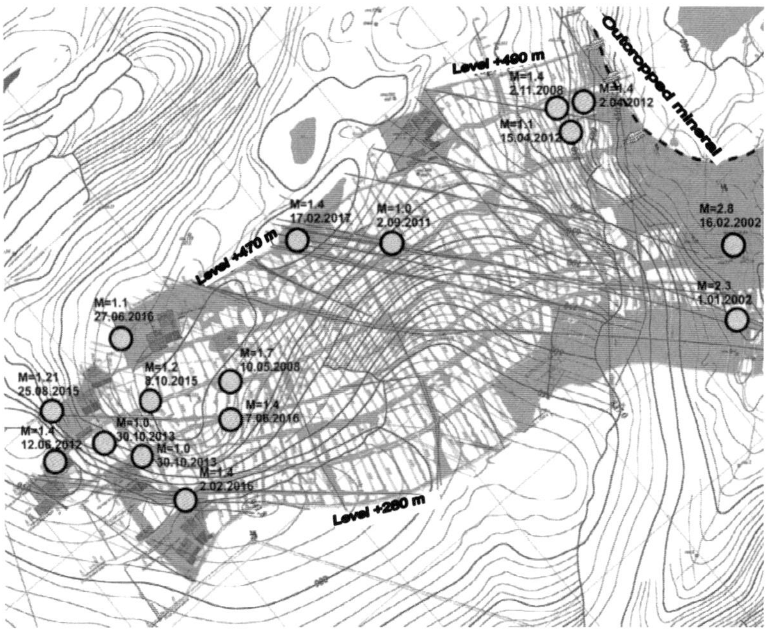

Figure 2. Mining plan for the west wing of Karnasurt Mine with the surface horizontals and locations of epicenters of induced earthquakes (M—magnitude; 27.06.2016—date as dd.mm.yr).

Under the gravitational-and tectonic stress state, there are no such notions as the zone of abutment pressure, the zone of stress relief, the protecting ore body (or layer) and some other provisions borrowed from the coal mining practice. These notions have been formulated for the conditions of the vertical gravitational forces that govern the rockburst hazard in coal mines and mechanically adopted in the ore mining regulations [1].

The stress state of rocks mass during mining in protective ore bodies under the gravitational-and-tectonic stress state is analyzed in [10]. It is found that under dominant horizontal stresses, mining with 'protecting bodies' does not reduce rockburst hazard in rocks but generates, vice versa, conditions for the initiation of the strongest geodynamic events. At the same, the discussed notions remain actual in coal mining under conditions of the natural gravitation stress state. The orientation of the maximum stresses in rock mass is the basic difference in rockburst hazard events in ore and coal mines.

Conclusions

1. The induced earthquakes are much stronger in ore mines than in coal mines because of the higher strength of ore-enclosing rock mass of magmatic or metamorphic nature than the strength of coal-hosting sedimentary rocks.

2. Coal deposits occur in rock masses mostly subjected to gravitational natural stress state when the maximum stresses governed by the weight of the overlying rocks act in the vertical plane. Ore deposits generally experience gravitational-and-tectonic stress state when major principal stresses are generated by the tectonic forces and are oriented horizontally. This is the key difference in the rockburst hazard events in coal and ore mines.

3. Some notions mechanically taken from the rockburst-hazardous coal mining practice and adopted in the guidelines for the rockburst-hazardous or mining should be withdrawn.

Acknowledgements

The study has been supported by the Russian Foundation for Basic Research, Project No. 15-05-04323a.

References

[1] *Federal Rules and Regulations on Industrial Safety: Safe Mining at Rockburst-Hazardous Deposits* Approved by Order of the Federal Service for Environmental, Technological and Nuclear Supervision as of December 2 2013 No 576
[2] *Handbook (Cadastre) on Physical Properties of Rocks* Moscow: Nedra 1975 (in Russian)
[3] Markov GA and Savchenko SN 1984 *Stress State and Rock Pressure in Structures of Mountainous Relief* Leningrad: Nauka (in Russian)
[4] Lovchikov AV 2016 *Strongest Earthquakes and Tectonic Rock Bursts in Mines in Russia* Palmarium Academic Publishing Saarbrucken Germany
[5] Lomakin VS and Khalevin NI 2002 Tectonic rock bursts in the actual seismicity of the Urals *Geodynamic Safety in Development of Subsoil and the Earth's Surface* Apatity: KNTs RAN (in Russian)
[6] Adushkin VV 2016 Tectonic earthquakes of anthropogenic origin *Phys. Solid Earth* No 2 p 173
[7] Methods for estimating seismic impacts (gudielines) *Vopr. Inzh. Seismol.* 1993 issue 34 pp 5–94
[8] Lovchikov AV 1997 Mine tectonic shocks in Russia: Statistics and analysis *J. Min. Sci.* Vol 33 No 2 pp 120–128
[9] Lovchikov AV and Rakhimov RG 2004 Near-surface earthquakes at Karnasurt Mine *GIAB* No 6 pp 88–93
[10] Lovchikov AV and Savchenko SN 2015 Protective deposits in the field of tectonic forces *J. Fundament. Appl. Min. Sci.* Vol 2 pp 128–132 (in Russian)

IOP Conf. Series: Earth and Environmental Science **134** (2018) 012040 doi:10.1088/1755-1315/134/1/012040

Constructing a generalized network design model to study air distribution in ventilation networks in subway with a single-track tunnel

IV Lugin

Chinakal Institute of Mining, Siberian Branch, Russian Academy of Sciences,

Novosibirsk, Russia

E-mail: ivlugin@misd.nsc.ru

Abstract. In focus are the features of construction of the generalized design model for the network method to study air distribution in ventilation system in subway with the single-track tunnel. The generalizations, assumptions and simplifications included in the model are specified. The air distribution is calculated with regard to the influence of topology and air resistances of the ventilation network sections. The author studies two variants of the subway line: half-open and closed with dead end on the both sides. It is found that the total air exchange at a subway station depends on the station location within the line. The operating mode of fans remains unaltered in this case. The article shows that elimination of air leakage in the station ventilation room allows an increase in the air flow rate by 7–8% at the same energy consumption by fans. The influence of the stop of a train in the tunnel on the air distribution is illustrated.

Tunnel ventilation as the basic component of the life-support system in the subway should meet the requirements and standards of SP Subways [1] in routine and emergency situations. Mathematical network model of the ventilation system was elaborated on the base of the reviewed design and construction documents on air flow motion in acting subways with the objective to investigate air distribution in subway infrastructure units in terms of Novosibirsk Metro as a case study. To verify compliance of design data and parameters of operating units the researcher undertook field survey of ventilation facilities in stations and track sections. The specified data served the base to calculate aerodynamic resistance of network branches [2, 3]. The subway ventilation network represents the integrated system where operating fans exert substantial influence on airflow parameters in different points of the network. The objective of the present research is to identify components constitutive for air distribution and fan operation modes in the network.

The generalized scheme of the ventilation network is worked out. It includes repeating modules: station passenger underpasses, station ventilation rooms, ventilation chambers in track sections, and connecting track tunnels to establish general air-distribution regularities in terms of the ventilation network in Novosibirsk subway (Figure 1). Aerodynamic resistance in sections of these modules is computed as averaged values for respective sections of the ventilation network consisting of Novosibirsk subway stations and tunnels.

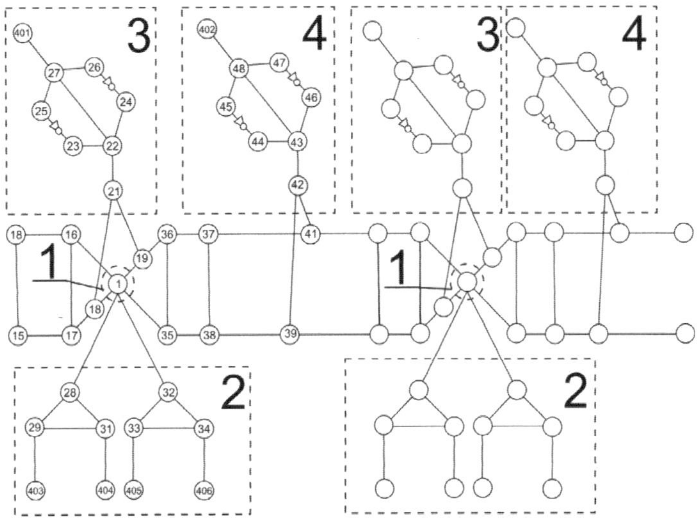

Figure 1. Generalized computed ventilation system scheme of a subway line: 1 – station platform; 2 – station pedestrian passways; 3 – station ventilation chamber; 4 – ventilation chamber in a track section; circled numbers are numbers of modules of the computed ventilation network

The application of the generalized model permits to eliminate influence of specific peculiarities of different stations on air distribution.

The standard station has two entrance halls with two exits per a hall. In exits there is one row of 4-valve doors with preset opening angle of 20°, in the entrance to the booking hall there is also a row of 4-valve doors with preset opening angle of 20°. The principal resistance source is doors, it amounts to 0.03 кμ (sections 403-29, 29-28, etc.).

The standard station ventilation chamber is equipped with two fans without diffusers, there are built-in splitter silencers made of slag blocks before and after fan facilities, there is a single kiosk, air passes are in both platform ends with an additional air pass through under-platform channel. The principal resistance is outlet of the ventilation kiosk and amounts to about 0.003 кμ (section 401-2); exits to tracks: 0.01 for the nearest exit and 0.013 for the far exit.

The standard ventilation chamber at a track section is equipped with two fans without diffusers; there are built-in splitter silencers made of slag blocks before and after fan facilities, there is a single kiosk, air passes to tracks are through both tunnels separately through a channel between tracks. The principal resistance is outlet of the ventilation kiosk and amounts to about 0.003 кμ; exits to tracks are 0.0025 кμ per each exit.

The standard one-track tunnels are lined with concrete tubing of 5.1 m in diameter, the ventilation connexion is 20 m in length, half-track section length is 500 m.

There are 8 stations per a line it is the minimal quantity for the proper control of mutual influence of air distribution in the line [4].

The subway line exists in two versions: semi-open with one dead end and another end coming out to open air and dead-type confined with both dead ends.

The ventilation network operating at summer mode approved for Novosibirsk subway: VOMD-24 station fans operate in a draft ventilation mode, one per a ventilation chamber; the second fan is switched off, its gate is closed; fans in track sections are out of operation, their gates are open; the gates of civil defense significance are open.

Operation of tunnel fans was modeled with the use of averaged aerodynamic characteristic of fan facilities calculated based on full-scale measurements. Thereto, resistance of inlet and outlet elements was taken into account in aerodynamic characteristic of fan facilities with respective improvement of accuracy of calculations and better compliance with the actual object. This characteristic is described by relationship:

$$P = -0.0488 \cdot Q^2 + 1.31 \cdot Q + 55,$$

where P is static pressure, Pa; Q is air flow rate, m^3/s.

The use of the static characteristic is justified by the fact that the dynamic component of the fine pressure developed by VOMD-24 tunnel fan is practically completely lost at fan outlet to the discharge section of the ventilation chamber because of imperfection of outlet components and lack of a diffuser.

The air exchange at station platform and its components: air in-takes from a tunnel and passenger passways of the station were selected as basic parameters of air distribution.

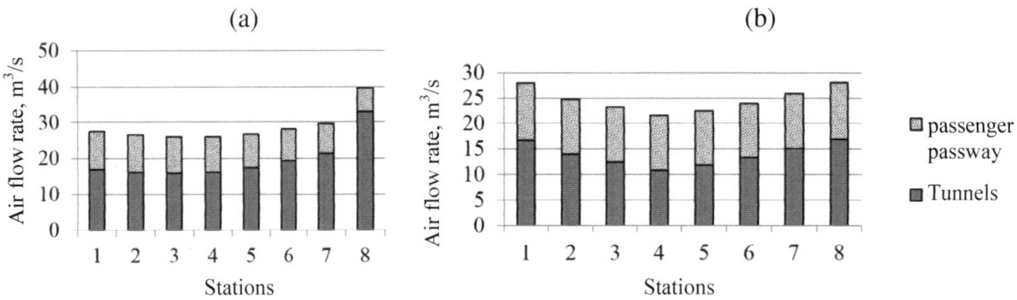

Figure 2. Air flow rate through the station platform room: half-open line, (a) an outlet to open air at the 8-th station, (b) dead line.

In Figure 2a the cumulative air exchange diagram is plotted for a station of a half-open line. It is explicit that air flow rate tends to grow in approach to the open air exit which locates after the 8-th station. Growth of air flow rate is realized thanks to air supply from a tunnel, the air supply volume is twofold greater than volumes in the first and blind-end stations. Herewith, air flow rates of the fans in the entire line do not actually change and range within 45.7–45.9 m^3/s.

In Figure 2b the cumulative air exchange diagram is plotted for stations of the blind (butting) line. It is obvious that air flow rate in opposite stations is approximately the same (small difference is due to unsymmetrical air outlets on the way from station ventilation chambers) The air flow rate in blind sections is higher thanks to air supply from the tunnel. The air flow rates for fan facilities in the entire line are the same and amount to 45.7 m^3/s.

Influence of variations in resistance of block sections:

–under variations in resistance of the second idle fan in the station ventilation chamber from 0.666 to 150 кμ in the half-open line the air flow rate at stations increased by 7–8%, thereto the air flow rate in the fans remained within 45.1–45.9 m^3/s, in other words, it altered inappreciably. Thus, reduction in leakage through the idle fan provides an essential air exchange boost at the station without increase in energy consumption for ventilation;

– mounting of an additional row of doors increases resistance of sections with doors from 0.03 to 0.06 кμ, herewith in the half-open section of the dead-end station the air flowrate lowered by 7.3%, and at the eighth closest-to-open-air station it increased by 9.1%. Production capacity of fan facilities remains the same within 45.6–45.9 m^3/s;

– enclosure of entrance hall at the forth station (Figure 3) leads to notable redistribution of air exchange at neighboring stations: air exchange is lower by 13.5% at the third station and is higher by 16.1% at the fifth station. The fan performance did not change and remained within 45.7–45.9 m^3/s;

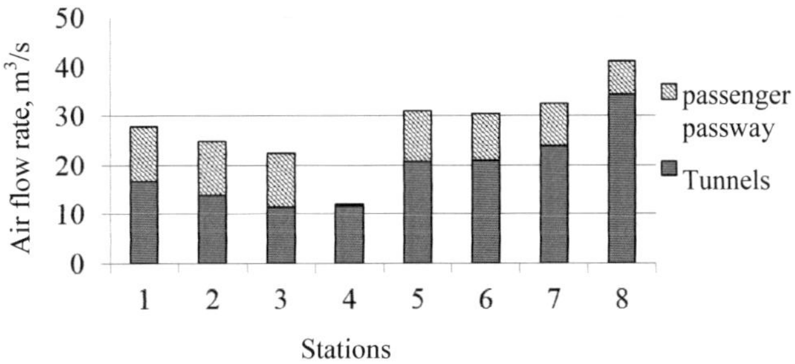

Figure 3. Air flow rate through station platform: half-open line, closed entrance hall at the forth station

– variation in resistance of tunnel sections (in terms of a train made a stop in a tunnel) was examined in sections between stations 1 and 2, 3 and 4, 7 and 8 in half-open line (Table 1) and stations 1 and 2, 3 and 4 for a dead-end line (Table 2) considering its symmetry (Figure 4). Air distribution was modeled provided that the train stopped on both sides from the ventilation chamber; the air flow rates were considered at all four tunnel sections. It was concluded, air exchange at stations did not actually change, or changed by no more than 2–3%. The fan performance remained the same within 45.7 for the dead-end line and 45.7–46 m^3/s for the half-open line. Air flow rates in the test tunnel section altered appreciably, the resultant difference was about 1200% with an air flow reversed in discrete cases.

Table 1. Half-open line

Train location	Branches	Basic flow rate	Version			
			1)		2)	
			Q, m^3/s	Δ, %	Q, m^3/s	Δ, %
Track section 1–2*	a	–9.9	–4.7	52.5	–10	–1
	b	0.6	5.5	–816.7	0.5	16.7
	c	–9.9	–14.6	–47.5	–9.8	1
	d	0.5	–3.9	–680	0.7	–40
Track section 3–4	a	–9.6	–4.5	53.1	–9.6	0.0
	b	0.4	5.2	–1200	0.5	–25.0
	c	–9.6	–13.9	–44.8	–9.6	0.0
	d	0.4	–3.7	–825	0.3	25.0
Track section 7–8	a	–17.5	–8.1	53.7	–14.3	18.3
	b	–10.2	–2.6	74.5	–5.5	46.1
	c	–17.6	–23.5	–33.5	–19.6	–11.4
	d	–10.2	–15.0	–47.1	–13.3	–30.4

*track section between stations 1 and 2, hereinafter - respectively; Q – air flowrate, m^3/s; Δ – difference between the basic and actual flow rates, %

Table 2. Blind line

Train location	Branches	Basic flow rate	Version 1)		Version 2)	
			Q, m^3/s	Δ, %	Q, m^3/s	Δ, %
Track section 1–2*	a	–9.2	–4.5	51.1	–9.7	–5.4
	b	1.8	6.3	–250.0	1.3	27.8
	c	–9.2	–13.9	–51.1	–8.9	3.3
	d	1.8	–2.7	–50.0	2.2	–22.2
Track section 3–4	a	–6.6	–3.5	47.0	–8.1	–22.7
	b	4.3	7.1	–65.1	2.5	41.9
	c	–6.4	–9.2	–43.8	–5.2	18.8
	d	4.4	1.8	59.1	5.8	–31.8

*track section between stations 1 and 2, hereinafter – respectively; Q – air flowrate, m^3/s; Δ – difference between basic and actual flow rates %

Conclusions

The principal aerodynamic resistance objects of subway with two-track tunnel are station doors, kiosks of ventilation chambers, exits to the tunnel in the station ventilation chambers.

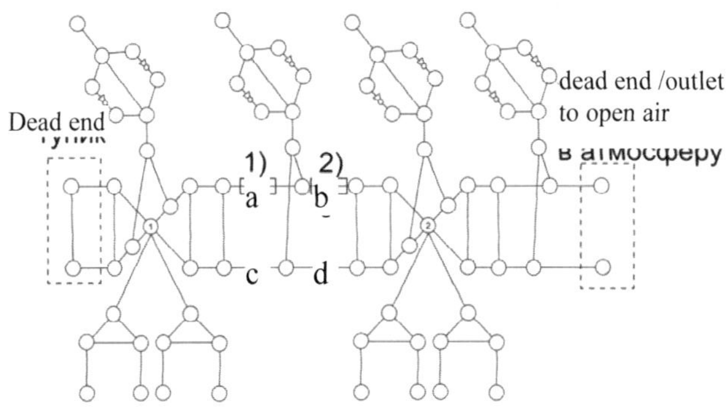

Figure 4. A train in tunnel section: 1), 2) are variants of train location in the tunnel; a, b, c, d are branches of the ventilation network where variations in the air flow rates were monitored.

Reduction in leakage through an idle fan facility allows higher air exchange at stations by 7–8% without extra energy consumption.

Variation in aerodynamic resistance of tunnel sections when trains stop in them affects insufficiently the air exchange at stations and the fan operation mode, but it is capable to alter an air flow rate in neighboring and parallel tunnel sections up to 1200% depending largely on a train location in the line.

The line topology, variations in branch resistance beyond ventilation chambers do not exert any appreciable effect on the fan operation mode.

References

[1] *RF Construction Regulations SP 120.13330.2.12 Subways* 2013 Approved by the Ministry of Regional Development of the Russian Federation Effective January 1 2013[
[2] Tsodikov VYa 1975 *Ventilation and Heating in Subways* Moscow: Nedra (in Russian)
[3] Idelchik IE 1992 *Handbook on Hydraulic Resistances* MO Steinberg (Ed.) Moscow: Mashinostroenie (in Russian)
[4] Krasyuk AM and Lugin IV 2003 Interconnection of ventilation modes at subway stations *GIAB* No 4 pp 199–203

IOP Conf. Series: Earth and Environmental Science **134** (2018) 012041 doi:10.1088/1755-1315/134/1/012041

Infiniteness of stresses for regions with angular points—A class of ill-conditioned problems

VE Mirenkov

Chinakal Institute of Mining, Siberian Branch, Russian Academy of Sciences, Novosibirsk, Russia

E-mail: mirenkov@misd.nsc.ru

Abstract. The current theory of failure is connected with the stress intensity factors at angular points. All known solutions with the infinite stresses at some points are the elasticity problem solutions according to their authors. The analysis of such solutions shows that though these are the fundamental solutions of the elasticity theory, the use of them to model subproblems of the elastic theory is unpromising.

1. Introduction

In view of the dozens of monographs devoted to the theory of elasticity, it might be expected that the logic and style of the presentation are set. Nonetheless, the elasticity research is yet faced with the known difficulties defined by the diversity of the problems and the solution method, sophisticated mathematical apparatus and the method of presentation. As a rule, having derived basic relations, the researchers turn to examples of solving specific problems while the issues of the qualitative theory are left aside. The given research breaks the rule and addresses the qualitative principles in the first place. These principles rest upon the confession that deviation from from the theoretical framework of elasticity, no matter how small, is an evidence of improper formulation of the problem.

The author focuses on a plane elastic problem in polar coordinates for regions with angular points of the boundary. Here belong finite and infinite domains with fractures, piecewise-homogenous wedge structures, half-planes under point force or subjected to indentation with a rigid press-tool etc.

2. Point force application at the apex of a wedge

To start with, let us discuss a partial solution of a plane problem with the function of stresses given by:

$$\varphi = r\theta(A_1\sin\theta + B_1\cos\theta), \tag{1}$$

which conforms with the simple radial stress state (see the figure 1). According to [1], the stresses are normal and shear:

$$\sigma_r = \frac{2}{r}\left(A_1\cos\theta - B_1\sin\theta\right), \quad \sigma_\theta = 0, \quad \tau_{r\theta} = 0, \tag{2}$$

The stress distribution (2) is valid for an infinite wedge-shaped domain with the boundary conditions set as the zero normal stresses σ_θ and shear stresses $\tau_{r\theta}$ at the wedge boundary $\theta = \pm\alpha$ (figure 1):

$$\sigma_\theta = 0, \quad \tau_{r\theta} = 0 \quad \text{at } \theta = \pm\alpha. \tag{3}$$

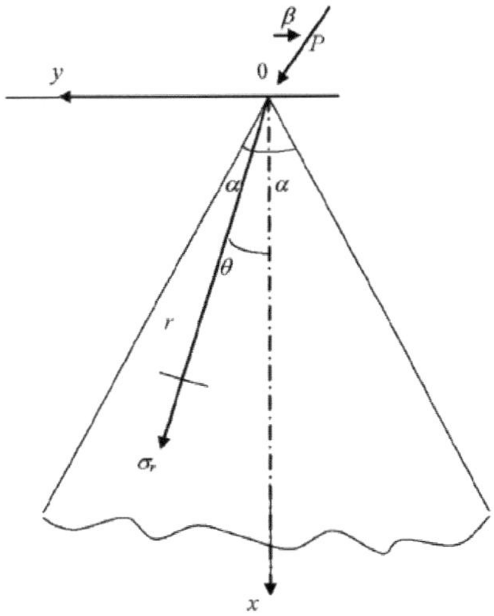

Figure 1. Pattern of calculation to fit the solution (1)

It is readily checked that the function (1) is the solution of a boundary value problem of elasticity. At the same time, this solution determines the stresses (2) that to zero as $r \to \infty$ and infinitely increase when $r \to 0$, i.e. indefinite stresses. On the other hand, the boundary conditions (3) assume that a wedge with an apex angle 2α is free from external forces, i.e. unloaded. Then where the stresses (2) come from? The source of the stress state (2) in the body under analysis can be either in the coordinate origin or at infinity.

In order to answer the question, let use analyze the behavior of the solution (2) at infinity. We take a part of the wedge as a sector limited with the section $r = R$, and the action of the withdrawn part is replaced by the response $\sigma_r(R)$. Calculate the components of the resultant vector of forces which are external relative to the wedge sector under analysis and are applied to its cambered boundary:

—x-component

$$I_x = \int_{-\alpha}^{\alpha} \sigma_r \cos\theta ds, \tag{4}$$

–y-component

$$I_y = \int_{-\alpha}^{\alpha} \sigma_r \sin\theta ds. \tag{5}$$

With regard to (2) and $ds = Rd\theta$, we find:

$$I_x = A_1(2\alpha + \sin 2\alpha), \quad I_y = -B_1(2\alpha - \sin 2\alpha). \tag{6}$$

Thus, when A_1 and B_1 are not concurrently zero:

—the analyzed wedge sector is not in the equilibrium and the resultant solution is impossible to interpret as the solution of a boundary value elastic problem;

—the values of I_x, I_y are independent of r, accordingly, at infinity, despite $\lim_{r\to\infty} \sigma_r = 0$, the resultant vector of external forces is non-zero.

In other words, the analyzed wedge is subjected at infinity to the action of the resultant vector of forces (6). This resultant force has no counter-balance and the solution is, therefore, senseless. Given any equilibration of the vector of the external forces applied at infinity to the wedge $0 \le r \le \infty$, $-\alpha \le \theta \le \alpha$, the solution (2) has sense and the constants A_1 and B_1 are determined unambiguously only in this case. The only possible place for an equilibrating vector is the coordinate origin ($r = 0$), i.e. the uncertainty point of the solution (2). The present research author believes this chain of reasoning is more natural and correct than in [1–5]. In all of the known textbooks on the theory of elasticity, this chain of speculations is inverted. The primary element there is the point force applied at the wedge apex and the solution in form of (1) is selected for this force. The direction of the point force P is characterized by the angle β (refer to the figure 1). From the condition of equilibrium of the wedge, the constants A_1 and B_1 are found in the form of:

$$A_1 = -\frac{P\cos\beta}{2\alpha + \sin 2\alpha}, \quad B_1 = \frac{P\sin\beta}{2\alpha - \sin 2\alpha}. \tag{7}$$

The most interesting seems to be the partial solution of (4)–(7) when $\alpha = \pi/2$ and $\beta = 0$. This so-called solution of the problem on a point force directed in perpendicular to the boundary of an elastic half-plane (Flamant solution):

$$\sigma_r = -\frac{2P}{\pi}\frac{\cos\theta}{r}, \quad \sigma_\theta = 0, \quad \tau_{r\theta} = 0. \tag{8}$$

With the known stress distribution (8), it is possible to find the related displacements of points at the half-plane boundary:

$$\begin{aligned} u &= -\frac{2P}{\pi E}\cos\theta \ell n r - \frac{(1-\nu)P}{\pi E}\theta\sin\theta, \\ \mathrm{v} &= \frac{2P}{\pi E}\sin\theta \ell n r - \frac{(1-\nu)P}{\pi E}\theta\cos\theta + \frac{(1+\nu)P}{\pi E}\sin\theta. \end{aligned} \tag{9}$$

The elasticity problem solutions should ensure continuous extension to the boundary irrespective of our way from the body to a point of this boundary [5].

Regarding stresses, it can emphasized that:

$$\lim_{r\to 0} \sigma_r = 0 \quad \text{at} \quad \theta = \pm\frac{\pi}{2},$$
$$\lim_{r\to 0} \sigma_r = -\infty \quad \text{at} \quad \theta = 0,$$

i.e. the limit of σ_r is different in different directions from $\theta = \pm\frac{\pi}{2}$ to $\theta = 0$ to $r = 0$.

The same is valid for the displacements:

$$\lim_{r\to 0} u = -\frac{(1-\nu)P}{2E} \quad \text{at} \quad \theta = \pm\frac{\pi}{2}, \quad \lim_{r\to 0} u = \infty \quad \text{at} \quad \theta = 0;$$

$$\lim_{r\to 0} \mathrm{v} = \pm\infty \quad \text{at} \quad \theta = \pm\frac{\pi}{2}, \quad \lim_{r\to 0} \mathrm{v} = 0 \quad \text{at} \quad \theta = 0.$$

Despite the solutions (8) and (9) are of no significant interest in applied problems such as the problem discussed above, theoretically, the solution for the point force is the fundamental solution in the theory of elasticity. Its fundamentality is connected with its ability to be integrated over an arbitrary boundary of a body and with obtainability of a finite solution everywhere at the boundary subject the resultant vector and resultant moment are zero, i.e. the problem is properly posed. In this sense, an example of an ill-posed problem, included in [1–5] too, is distributed load applied to s finite section of the boundary of a half-plane with the non-zero resultant vector, or indentation of a press-tool in the half-plane.

3. Solutions for V-shaped domains

Let in a finite V-shaped domain with an apex angle $2\theta > \pi$ similarly to (1):

$$\varphi(r, \theta) = r^{1+\varphi} F(\theta), \tag{10}$$

where

$$F(\theta) = A\cos(\varphi+1)\theta + B\cos(\varphi-1)\theta + C\sin(\varphi+1)\theta + Д\sin(\varphi-1)\theta,$$

A, B, C, $Д$—arbitrary constants; φ—constant less than unit. For the stress state symmetrical relative to the axis $\theta = 0$ of the wedge, it can be set that $C = Д = 0$. Given asymmetry, $A = B = 0$. Finally, after analysis of the stress state, we have:

$$\sigma_r = -A\varphi r^{\varphi-1}\left[(1+\varphi)\cos(1+\varphi)\theta + (3-\varphi)\frac{\cos(\varphi+1)\alpha}{\cos(\varphi-1)\alpha}\cos(\varphi-1)\theta\right],$$

$$\sigma_\theta = -A\varphi(\varphi+1)r^{\varphi-1}\left[\cos(\varphi+1)\theta - \frac{\cos(\varphi+1)\alpha}{\cos(\varphi-1)\alpha}\cos(\varphi-1)\theta\right],$$

$$\sigma_{r\theta} = -A\varphi r^{\varphi-1}\left[(\varphi+1)\sin(\varphi+1)\theta + (1-\varphi)\frac{\cos(\varphi+1)\alpha}{\cos(\varphi-1)\alpha}\sin(\varphi-1)\theta\right], \tag{11}$$

$$u = \frac{A}{2G}r^{\varphi}\left[-(\varphi+1)\cos(\varphi+1)\theta + (\varphi+1-\tfrac{\varphi}{1+\nu})\frac{\cos(\varphi+1)\alpha}{\cos(\varphi-1)\alpha}\cos(\varphi-1)\theta\right],$$

$$v = \frac{A}{2G}r^{\varphi}\left[(\varphi+1)\sin(\varphi+1)\theta - (\varphi-1+\tfrac{\varphi}{1+\nu})\frac{\cos(\varphi+1)\alpha}{\cos(\varphi-1)\alpha}\sin(\varphi-1)\theta\right].$$

By way of example, consider a wedge $2\alpha = 270°$. For such angle, $\varphi = 0.545$, i.e. the singularity of the stresses, as per [10], is $1-\varphi = 0.455$. The classical solution (11) determines the pre-deformation angle $2\alpha = 270°$ as the post-deformation angle of 180°, which means the assumption of smallness of deformation is no more valid. On the other hand, the opposite sign stress state, which induces displacement of the boundaries $\theta = \pm 135°$ and results in material–material penetration, contradicts the physical sense of the problem.

In this manner, the solutions (11) in the interpretation by the authors [4–10] are senseless. Indeed, on the one part, the V-shaped domain is free from any force:

$$\sigma_r = 0, \quad \tau_{r\theta} = 0 \quad \text{at} \quad \theta = \pm\alpha \tag{12}$$

but is deformed and one of the backbone hypothesis of the linear elasticity theory is violated; on the other part, the solution is found accurate to a constant.

Aiming to find a real ground of the solution (11), we take a sector with a radius $r = R$ out of a V-shaped domain and replace the action of the withdrawn potion by the responses in accordance with (11). In this case, it is assumed that $2\alpha = 270°$, $\varphi = 0.545$. At the radial boundaries of the selected sector, the forces are absent as per [5–10], i.e. the condition (12) is fulfilled. It is readily checked that the sector is out of equilibrium when subjected to the forces σ_r and $\tau_{r\theta}$ applied at the cambered boundary, which means that the problem formulation is ill-posed in [5–10]. Actually, the resultant vector of external forces applied to the selected sector is:

$$Q_x = 2.888AR^{\varphi}.$$

In order to eliminate this deficiency, an additional condition should be set. Such condition may be a force applied at the apex of the V-shaped domain, or any other permissible influence. In the meanwhile, the statement in [5–10] suggests no additional forces applied at the boundary or inside the domain. Let us loop the setting of the boundary conditions with the assumption that the apex is rigidly fixed ('nailed') at a point at the coordinate origin. In this case, the selected sector is balanced by the unknown response of the nailing. Mathematically, this condition means that $u(0,0) = v(0,0) = 0$. Now, the constant A can be determined in terms of the nailing response – Q. For such kind of fixation at a point, the classical elasticity theory has no definition of a response as against the case when the singularity of stresses is equal to unit (so called point force).

IOP Conf. Series: Earth and Environmental Science **134** (2018) 012041 doi:10.1088/1755-1315/134/1/012041

For the V-shaped domains described by the stress function (1), the physical sense is supported by introduction of a hypothetical 'point force' (or determination of the constant A_1). In case of (1) the science has no even a hypothetic structure to explain the very existence of the 'solution'. Thus, the constant A in (10) is indefinite.

The presented analysis of the solutions for the domains with an infinitely remote point and infinite stresses displays weakness of complementary explanations aimed to justify the results which drop out the framework of the theory of elasticity in [5–10]. It seems to be worthy of adding that when points with infinite solutions are withdrawn, the withdrawn portion should be replaced by responses based on an infinite solution. This results in a partial stress distribution over rounded boundary, which never exists in the applied problems, and the size of the withdrawn portion is uncertain.

4. Connection between the solutions (1) and (10)

For the V-shaped domains, irrespective of the boundary conditions, the singularity of stresses (strains) is determined from the equations [10]:

$$\sin 2\varphi\alpha = \pm\varphi \sin 2\alpha,$$

where "±"»—respectively, in the case of symmetric ($\beta = 0$) and asymmetry ($\beta = 90°$) of the problem. In particular, it is simple to obtain the singularity for a plane with a half-infinite cut (V-shaped domain with an angle $\alpha = \pi$) using the mapping of the half-plane with the so-called 'point force' solution on a wedge with an apex angle 2π. The mapping of a half-plane on a wedge with an apex $2\alpha = 2\pi$ is given by the relations:

$$r = \sqrt{r_1},\ \theta = 0.5\theta_1, \tag{13}$$

where r_1, θ_1—polar coordinate system connected with the wedge $2\alpha = 2\pi$. Given the mapping (13), the boundary condition and the solution for the half-plane convert, respectively, to the boundary conditions and the solution for a plane with a half-infinite cut. The boundary condition in the half-plane is set as the force P, accordingly, the boundary condition in the V-shaped domain is the imaging force P.

Consequently, assuming that the force P has a physical sense, its imaging becomes physically real. As has been demonstrated above, this interpretation represents the response that balances forces at infinity for a wedge. The common interpretation of the solutions at a crack tip (a V-shaped domain with an apex angle $2\alpha = 360°$), stating that the tip is free from forces, is incorrect. In point of fact, the tip of a cut is subjected to a concentrated impulse—the imaging of a 'point force' applied to a half-plane.

This investigation offers a most complete explanation of the sense, inter-relation and place of indefinite stresses in the theory of elasticity.

5. Mathematical cuts

Let us fix on a problem on a plane with a weakening in the form of a finite straight-line cut. Irrespective of the approached to solving the problem, the results coincide; we focus on the solution using the conformal mapping, e.g. [5]. The solution in [5] is incorrect as the conformality of mapping is violated at the angular points, which automatically brings infinite stresses and is inadmissible. This could be the end but the authors [1–10] permit using this inelastic solution at the distance from the cut tips. In this solution, it is essential that the tips of a cut (for any cracks) are 'nailed' at the points where $u = v = 0$. It is a subconscious comprehension that in the elastic solution, when the cut widens, its tips displace toward the cut center. On the other hand, when the tips are 'nailed', the responses arise at the nailing points (and such notions is absent in the theory of elasticity), and such responses can be infinite. Deformation of a cut with the 'nailed' tips is a complex structure and cannot be modeled by a plane with a mathematical cut that opens under normal forces. The theory of failure based on the stress intensity factor is no more satisfactory. Stress intensity factor is an absolute number divided by zero, which is wrong. The discussed classical solutions are mathematically ill-conditioned and physically inadmissible

6. Conclusion

The known solutions with the infinite stresses are impossible to assume elastic as they violate the provisions of the elastic model of a medium.

The singularity of stresses in all known solutions is connected with the conformal mapping of the analyzed domains and is the result of the rigid fixation of the domains at the angular points.

In case of mathematical cuts, large deformation or penetration of a material into a material in the vicinity of the tips of a cut prove ill-posedness of the solutions which are mathematically and physically senseless.

It is incorrect to map domains with angular points on domains with smooth boundaries inasmuch as the conformality of the mapping is violated.

References

[1] Samul VI 1982 *Elements of Theory of Elasticity and Plasticity* Moscow: Vyssh. Shkola (in Russian)
[2] Terebushko ON 1984 *Basic Principles of Theory of Elasticity and Plasticity* Moscow: Nauka (in Russian)
[3] Timoshenko SP, Goodier JN 1970 *Theory of Elasticity* McGraw Hill Book Company
[4] Filonenko-Borochich MM 1947 *Theory of Elasticity* Moscow: Gostekhizdat (in Russian)
[5] Miskhelishvili NI 1966 *Some Basic Problems of Mathematical Theory of Elasticity* Moscow: Nauka (in Russian)
[6] Uflyand YaS 1967 *Integral Transformations in Problems of the Elasticity Theory* Leningrad: Nauka (in Russian)
[7] Ivlev DD 1967 Theory of quasi-brittle failure *Prikl. Mekh. Tekh. Fiz.* No 6 pp 88–128
[8] Belonosov SM 1962 *Basic Plane Static Problems of Elasticity for Simply Connected and Biconnected Domains* Novosibirsk: SO AN SSSR (in Russian)
[9] Novozhilov VV 1950 *Theory of Elasticity* Leningrad: Sudprom (in Russian)
[10] Bogy DB 1971 Two edge-bonded elastic wedges of different materials and wedge angles under surface tractions *J. Appl. Mech. ASME* Vol 38 No 32 pp 377–386

System approach to automation and robotization of drivage

VV Zinov'ev[1, 2], AE Mayorov[1, 2], AN Starodubov[1, 2] and PI Nikolaev[1, 2*]

[1] Institute of Coal of the FRC CCC SB RAS
Kemerovo, Kemerovo Region, Russia
[2]T. F. Gorbachev Kuzbass State Technical University
Kemerovo, Kemerovo Region, Russia

E-mail:*mrnikolaev@mail.ru, rector@nstu.ru

Abstract. The authors consider the system approach to finding ways of no-man drilling and blasting in the face area by means of automation and robotization of operations with a view to reducing injuries in mines. The analysis is carried out in terms of the drilling and blasting technology applied in Makarevskoe Coal Field, Kuznetsk Coal Basin. Within the system-functional approach and using INDEFO procedure, the processes of drilling and blasthole charging are decomposed into related elementary operations. The automation and robotization methods to avoid the presence of miners in the face are found for each operation.

The present day coal industry is ranked as the industry most hazardous for life and health of miners though the fatal injury factor in industry is reduced from 1.2 man/thous in 1992 down to 0.06 man/thous in 2016. In Kuzbass this factor in the mining industry remained the highest as compared to other industries. In [1] imperfection of coal mining technology is named as a reason in25% of accidents in Russian collieries. The most coal and gas outbursts, fires, roof caving and other emergency events fall on face zone of a development working. To lower overall injury rates in drivage it is actually reasonable to develop geotechnologies enabling to eliminate presence of miners in a face area.

Further extensive development of the coal industry in Kuzbass is planned mainly in Tersinsky geological-economic region, the Kuznetsk Coal Basin [2].Kusheyakoskoe, Uvalnoe and Makarievsky deposits are explored in details and are closest to the developed infrastructure locations. The first two of them have been opened already, and the tender is invited for Makarievsky deposit. In Makarievsky deposit the perspective productive seams of more than 1 meter in thickness occur in Verkhnebolokhonsky coal seam subseries. Most of them are insufficiently aged, of simple structure, thin or average in thickness. Coals in these seams are of vitrinite composition with dominating lean coalsand anthracites [3], applicable in metallurgy, electric-power industry and special energy-technological processes. The main restriction in development of these deposits is nearby occurrence of table water springs and natural park "Kuznetsky Alatau". To initiate Makarievsky deposit development requires proper geotechnologies eliminating presence of miners in a face area, provided that environment standardsare fulfilled.

In [4] the researchers propose the procedure for substantiation of geotechnologies with no human participation in a face area. The procedure based on the system approach and simulation modeling makes it possible to analyze all feasible variants of automation of technological operations and to select efficient solution, excluding human-factor errors.

First it is reasonable to describe a geotechnology conventional for the given mining-geological conditions. Verkhnebalakhonsky sub-series (Makarievsky deposit) are characterized with a formation sequence of sandstone (45.4% lithologic composition), aleurolite(48%) and subordinated coal seams of about 850 m in total thickness [3]. Coalseamsarepronetoshock bumps, coal and gas outbursts. The principal ground water zone in primary deposits is in the level of 100-150 m; in greater depths it attenuates. Therefore, undertheseminingandgeologicalconditionsthecutting, hydraulicorshieldprimary mining processes are not rational or even impossible because of hardness of host rock (6-8 under Protodiakov's scale). Theprimary boring-blastingmining isexpedientin this case.

Next, theprocedureimpliesanalysisoftraditionalgeotechnologiesbyIDEF0 methodology within the system approach with the aim to find variants of their executors.

The drivage sequence under conditions of Makarievsky deposit in terms of the system- functional approach is shown in Figure 1:

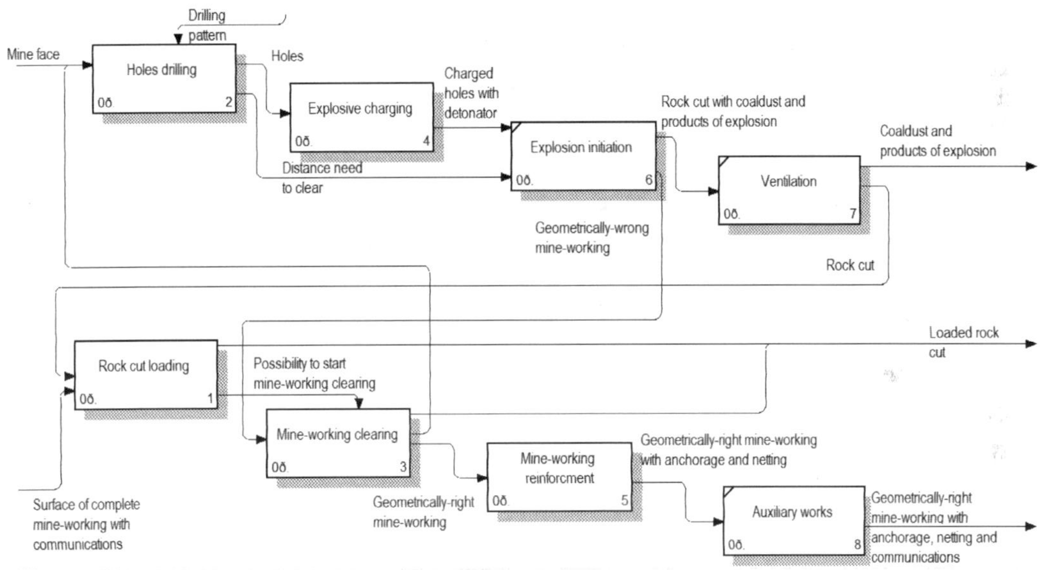

Figure 1.Drivage in terms of the system-functional approach.

Tofindpossibleexecutorsofeachoperation requires to decompose each functional block in the given diagram and to preserve interrelations between blocks at all the levels.

The firs functional block to decompose is "making blastholes". Its function is to transform the face front into two aspects: blasthole making and rock mass planned for extraction at a certain distance in compliance with the approved boring pattern. Decomposition of this operation is presented as a system of functional blocks representing blasthole-boring operation (Figure 2).

The first operation to be fulfilled is survey of local environment for proper orientation and dislocation of respective equipment in the fore-breast area. In the conventional boring-and-blasting process it is a miner who is to survey the environment by means of his sensory organs. First attempts to transmit partially this composite function to a mechanical executor are reported in [5], where a drill rig, capable to identify walls of a working by means of hydraulic jack, is described. In [6] the author sets forth a procedure for mounting a video camera on mining equipment. In blasthole making the camera mounted on a drill rig can exclude miner's participation in a face area, but it does not contribute to precision of positioning of a drill rig in fore-breast area. The state-of-the-art methods for scanning of a mine field imply application of laser scanners. They can be mounted on a drilling machine [11], on a separate platform with a mobile mine cartographer [7], or miners can employ them as a manual instrument [8].

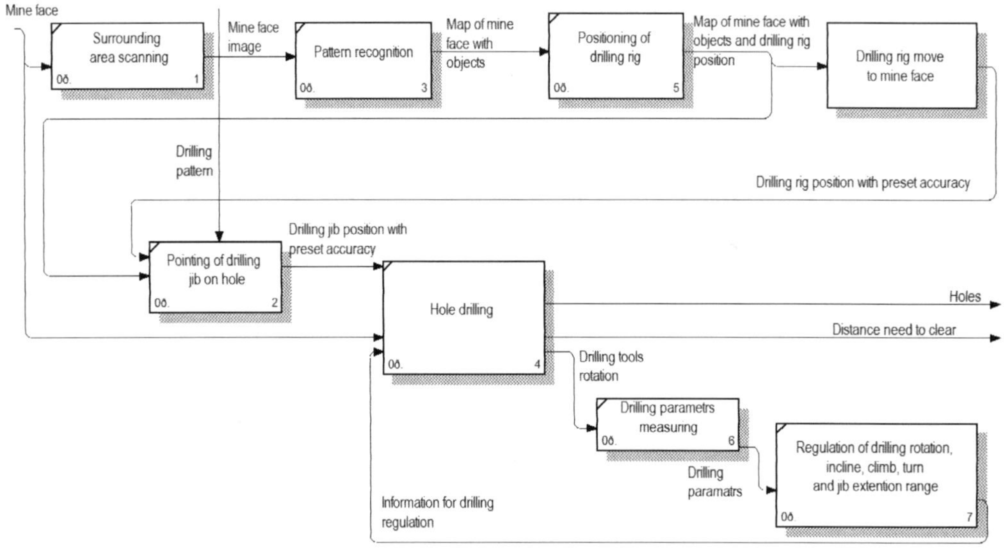

Figure 2. Blasthole-boring in terms of the system-functional approach.

In conventional geotechnology a miner is engaged in processinginformation (identificationofminingobjectsforfurtherwork: aface, roof, walls, other objects if required). Recentlyattemptsaremadetotransferthisfunctiontoacomputersystemarrangedas "aneuronnetwork" [9], but so far robotization of this operation is rather expensive and implies contribution of professionals in artificial intelligence.

Spatial non-operator orientation of a drill rig can be executed by means of a position-control system. There are actual cases in practice [10]: domestic"RealTrack" mine, RTLServiceCo. and "SBGPS", Granch Co., and foreign "ConnectedMining", CISCO Co., and "BeckerMiningSystem", BeckerGMbH, all the above systems are based on wireless translation inside mine workings. Access points of wireless network and individual position-control means mounted on concerned machines and in a mine working are interrelated to identify a position of a facility.

In conventional mining process a drill rig is operated by a miner in a cabin. The simplest modification of this operation is a remote control in a safe distance from a face. The main imperfection of this control process is the lack of complete visual information on an operation area [13]. That is the reason why this variant is proposed only in the cases of urgent need to remove an operator from a hazardous spot. A few companies propose automated control of mine machinery dislocation [11]: RCS 5 computer control system of AthlasCopco Co., Automine and Automate systems of Sandvik Co., Minegem control system of Caterpillar Co. Advantages of these computer control systems are less wear of discrete parts and motors thanks to optimal operation modes and an precise motion of a drill jig at a wanted distance. In this case the main function of an operator is to watch operation of a machine and he is in power to switch off an automated control at a remote mode.

In the conventional mining process a miner is to align a drill rig manipulator to a blasthole spot according to the boring pattern. RCS5 computerized blasting control system, Athlas Copco Co., provides the maximum automation of positioning of several drill rig manipulators in a computed optimal route according to the boring pattern. The automated drilling system "AutoMineDrilling", SandvikCo., is equipped with a control system "Sandvik Intelligent Control Architecture (SICA)" allows automated positioning with precision up to 5 cm.

Boring parameters are measured by sensors mounted on a manipulator with a hydraulic perforator to measure velocity of drill rotation, its removal, turning and extension length. ThecomputercontrolsystemsRCS 5 and AutoMine Drilling for the drilling process receive sensor information and adapt drilling parameters in view to save drill steel consumption, to prevent steel sticking, to improve drilling velocity. The computer control systems are capable to manage the drilling process and to detect variations in properties of rocks to be drilled, fractured zones. These functions provide prompt adjustment of blasthole length, their location in order to improve the coefficient of blasthole use [11].

The global producers, including Athlas Copco, Sandvik, Aramine, Robodrill,etc. propose a variety of drill machines, which can be classified by availability of operator's cabin, arm or boom (1-4 in number) a boom with a basket for miners.

Thesecondfunctional block to be decomposed is "blasthole charging". A hollow blasthole is filled with an emulsion explosive and a detonator. Decomposition of this process can be represented as a system of functional blocks in Figure 3.

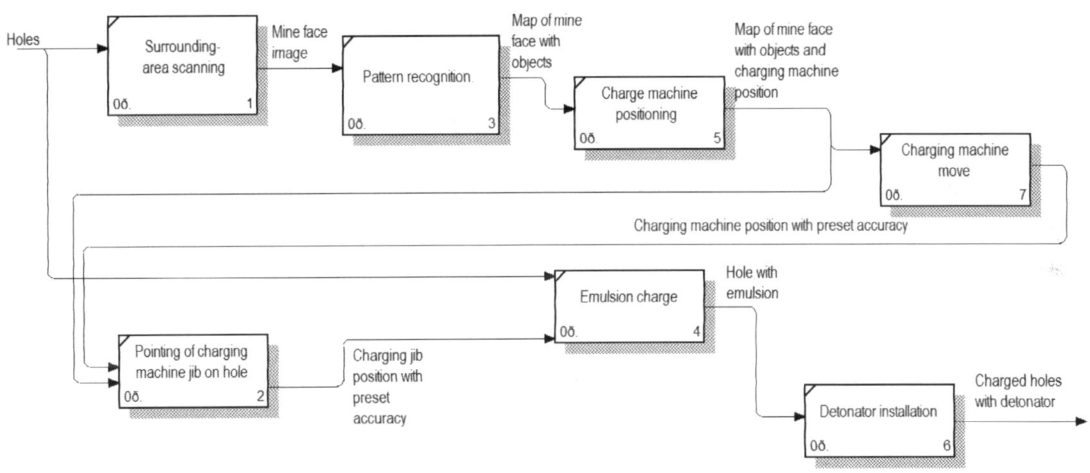

Figure 3. Blasthole charging in terms of the system –functional approach.

Proposals on unmanning of a face area in operations "surrounding area scanning", "pattern recognition", "charge machine positioning", and "charging machine move" are identical to similar operations described in section "blasthole drilling".

Operation "emulsion charge" involves a few stages: hose insertion, blasthole purging, oxidizer heating, explosive emulsification, explosive charging, explosive cooling. In the conventional process preparation of explosive emulsion is automated in mixing-charging systems, but introduction of a hose into a blasthole, control of explosive amount are performed by a miner from a basket of a charging machine. The mixing – charging systems can be connected to one or two hose baskets. Machines of

Normet Co. and Orica Co. are equipped with microprocessor control system based on CAN-bus technology, this allows automation of hose supply and dosing of explosive emulsion fed to a blasthole, but a miner remains in a face area to operate a charging machine.

Up to date the unmanned versions of "detonator installation" operation are not available because of complex manual manipulations executed by a miner. The single facility required is a charging – mixing machine with a basket to deliver a miner to blastholes filled with explosive emulsion.

Automation versions under consideration indicate that the present-day producers of underground drilling-blasting equipment pay more attention to automation and robotization of mechanical operations and remain cognitive functions: observation, recognition of objects in a space and orientation in a space to an operator. The complete unmanning of a face area implies transition of the above operator's functions to mining machines. The efficiency of proposed automation and robotization versions is verified by imitating modeling, and test results on imitating models are used to elaborate recommendations on development of novel geotechnologies.

The present work is supported by the Russian Foundation for Basic Research, project No. 16-47-420490 "Modeling and Substantiation of Underground and Open-Underground Coal Mining Processes with no Regular Presence of Miners in Face Areas" (2016-2017).

References

[1] Kochetkova EA 2001 Method of labor protection control efficiency evaluating based on dependence between traumatism risk and cash investments. *Candidate of Technical Sciences dissertation*(technical, 05.26.01). (inRussian)

[2] KlishinVI, Fedorin VA and Shakhmatov VYa 2015 Evaluation of coal field mastering in Tersinskiy geology-economical region of Kuzbass *Vestn. Kuzbass. Gos. Tekh. Univer.* No 5 pp 205-211 (in Russian)

[3] Cherepovskiy V 2003 *Russian coal base. Volume II. Coal basins and coalfields of Western Siberia (Kuzneckij, Gorlovskij, Zapadno-Sibirskij basins; coalfields of Altay Kray and Altay Republic)* Moscow Geoinformtsentr (in Russian)

[4] Nikolaev PI and Zinovyev VV 2016 Procedure to evaluate robotic underground geotechnologies without permament presence of miners in the face area *Bulletin of Kuzbass State Technical University* No 4 pp 26-33 (in Russian)

[5] USSR Author's certificate no 875019 (11) MPK E 21 S 11/02. Drill carriage 1981 *Byull. Izobr.* No 39

[6] Konyukh VL 2000 *Mine robotics* Kemerovo: Kuzbassvuzizdat (in Russian)

[7] Martin Magnusson 2006 *3D Scan Matching for Mobile Robots with Application to Mine Mapping* Department of Technology at Orebro University 17, Orebro

[8] Ovcharenko AV and Udoratin VV 2015 Operative underground cavities research with laser 3D scanning *Bulletin of Geology Institute, Komi Scientific Center, Ural branch, RAS* No 4 pp 20-25 (inRussian)

[9] Lavrov SS 2005 Wavelet transform of hidden neurons of a multilayer perspetron for geological pattern identification *Min. Inform. Analyt. Byull.*No 5 pp 170-172(inRussian)

[10] Vaganov VS and Urusov LV 2016 Review of methods to arrange data networks for multi-purpose safety systems in coal mines *Vest. Nauchn. Tsentra Bezop. Rabot Ugol. Prom.* No 3 pp 72-81 (inRussian)]

[11] Rusin EP, Freydin AM and Tapsiev AP 2007 Automation of underground mineral mining *GIAB* Vol 10 No 12 pp 235-244 (in Russian)

IOP Conf. Series: Earth and Environmental Science **134** (2018) 012043 doi:10.1088/1755-1315/134/1/012043

Empirical mode decomposition method in deformation measurement data processing

MD Parushkin[1,2], SV Panov[1]*, VM Semibalamut[3] and YuN Fomin[1,3]

[1]Institute of Laser Physics, Siberian Branch, Russian Academy of Sciences, Novosibirsk, Russia
[2]Institute of Mining, Siberian Branch, Russian Academy of Sciences, Novosibirsk, Russia
[3]Siberian Division of the Geophysical Service of the Russian Academy of Sciences, Novosibirsk, Russia

E-mail: *lss@laser.nsc.ru

Abstract. The paper describes the method of empirical mode decomposition and its application in the analysis of data obtained by laser measurement of deformation. The issues of the method adaptation to actual data are discussed. The described method allows tracing monotonous trend and slow term of deformation much more efficient as against the conventional techniques. Furthermore, the method is also good in removal of noise and high-frequency components from signals.

1. Introduction

During long-term continuous operation of laser deformograph installed in a gallery of Talaya seismic station in the Baikal Rift Zone, vast experimental information has been accumulated [1, 2]. The observation data are mostly processed using the popular and common methods of geophysical and deformation measurement data analysis. The main problem connected with the application of these classical approaches is their applicability chiefly to linear and stationary systems. On the other hand, the international practice has advanced in the recent decades in the analysis of nonlinear but stationary and deterministic systems, or linear but nonstationary systems (for instance, Wigner–Ville distribution, wavelet analysis, etc.). However, majority of actual physical processes and systems are nonlinear and nonstationary. In this connection, the analysis of recorded data has to be simplified, especially relative to the a priori set basis of analyzed signal.

One of the key conditions of the correct presentation of nonlinear and nonstationary data is forming an adaptable basis governed by the content of the data. Such approach is implemented in the Hilbert–Huang Transform procedure (HHT), which is the decomposition of signal into empirical modes and application of Hilbert Transform to the resultant components of the decomposition. By now there is no rigorous mathematical validation of the method; at the same time, the positive results of the method application to a wide range of practical tasks enables expecting improvement of the situation in the nearest future.

The empirical mode decomposition method was for the first time been proposed for the analysis of surface typhoon waves by Huang in 1995. The method was generalized for the analysis of arbitrary time series in 1998 [3, 4].

IOP Conf. Series: Earth and Environmental Science **134** (2018) 012043 doi:10.1088/1755-1315/134/1/012043

The method of decomposition is based on the assumption that all data under analysis are composed of various oscillation processes and at any time a signal can contain a set of overlapping oscillations. Each process, either linear or nonlinear, stationary or nonstationary, is an oscillation that is to a certain degree "symmetrical" relative to a local mean value and, consequently, has extremums and zero intersections. Such oscillation processes can by presented using intrinsic mode functions (IMF) given that:

—the numbers of extremums and zero intersections differ by 1 at the maximum;

—the average value between the mode function envelopes governed by the local maximums and local minimums is zero for all points.

The modes represent oscillation processes; at the same time, unlike harmonics resultant from Fourier transform of a discrete signal, each IMF can have variable amplitude and frequency as functions of time. Any arbitrary signal can be divided into a set of mutually orthogonal intrinsic mode functions which are not set analytically and are exclusively determined by the analyzed sequence.

2. EMD algorithm

The brief description of EMD below seems to be sufficient for understanding the algorithm of EMD. To get a more profound insight into the method and its peculiarities, it is possible to address the works of the method's author—Huang [3].

In a general case, the algorithm is composed of sequential operations aimed at identifying mode functions, starting with high frequencies, in a signal. Each IMF identification step includes:

1. Detection of all local extremums of the signal and formation of amplitude–coordinate vectors for maximums and minimums;
2. Identification of upper and lower envelopes of the signal using the found local maximums and minimums by means of "natural" cubic spline. Determination of average values function between the maximum and minimum envelopes—m_1;
3. Finding IMF function in the first approximation using the formula: $h_1(k) = y(k) - m_1(k)$;
4. Repeating steps 1–3 and replacement of the initial signal by h. allows the second approximation for IMF function: $h_2(k) = h_1(k) - m_2(k)$;
5. Similarly, the third and next approximations of IMF are found. As the number of iterations is being increased, the functions h_r and m_r tend to their invariable forms. The cessation criterion is chosen as a limit of normalized square difference between two subsequent iterations, or is set as maximum iteration number (as a rule, 7–109, which conforms with the limit of 10^{-4}—10^{-5};
6. The value of h obtained during the last iteration is assumed the first, highest frequency mode function (I_i) $c_1(k) = h$ contained in the initial signal y.

When the obtained mode is withdrawn from the signal, low-frequency components are left:

$$r_1(k) = y(k) - c_1(k).$$

The processing of the resultant function using the same procedure produces the second mode function (I_i) - c_2 and so on:

$$r_2(k) = r_1(k) - c_2(k).$$

Finally, the signal is decomposed in the n-dimensional mode empirical approximation with the resultant excess r_{n+1}:

$$y(k) = \sum_n c_n(k) + r_{n+1}(k)$$

The decomposition may be continued until the maximum "straightening" of the excess so that only the trend of the signal is left with no more than 3 extremums out of which it is impossible to derive another IMF.

As a result, the initial signal is decomposed with respect to an adaptive basis obtained from the analyzed data. Although undefined analytically, it satisfies all requirements of a basis. This means that it is complete, convergent, orthogonal and unique. Huang's statement on the decomposition uniqueness seems disputable as the empirical process of decomposition, due to its adaptivity, is uncontrolled in a general case. Even a monotone local component of a signal, under a certain influence of destabilizing factors (noise, impulse interference, etc.), can split into two or three IMF during decomposition. Naturally, the said local component can be identified in summing-up of these functions but this requires that user has some a priory knowledge on the signal composition.

3. Application of EMD method to deformograph data

The practice of introduction of EMD method in processing of deformation measurement data points at the high application potential of the method, including identification of slow deformation changes (plus a monotone trend), detection of nonstationary disturbances in signals, regularization of data and withdrawal of any nature high-frequency noise from signal.

The main difficulty in the application of EMD in processing of deformograph data is the breaks from minutes to days long in the signal. Such breaks worsen the method efficiency, especially when plotting envelopes using cubic splines. When finding first (high-frequency) mode functions, local extremums in the zone of breaks are far from each other as compared with the neighbor extremums; for this reason, under interpolation of envelopes by spline, considerable false bursts can appear in the mode functions, which are then transferred to the excess for the further decomposition. This effect accumulates during later steps, which completely impairs the result. In order to avoid the situation, it was decided to add the signal with the Gaussian noise at a level not higher than 0.01–0.05 of standard scatter of initial signal.

Addition of noise in the initial signal, aside from the problem of breaks in the signal, enables overcoming another rather frequent problem connected with the division of one or a few of closely spaced harmonics into different neighbor mode functions. This situation often takes place in processing of actual data and is uncontrollable in a general case owing to adaptivity of the decomposition basis. The problem can be solved by operating pairs of modes rather than separate modes. On the other hand, there already exists the method of EEMD (Ensemble Empirical Mode Decomposition) [5] which is the multiple variation of the initial signal by means of adding the Gaussian noise, finding of mode functions in each case and their averaging over an ensemble. Depending on the number of variations, the level of noise is varied, and as the ensemble of variations grows, the level of noise can be increased as in the averaging the value of the resultant noise is inversely proportional to the ensemble size root. Thus, when averaging the ensemble of 100 variations, the signal can be added with the Gaussian noise at the level of 25% of standard scatter of the initial signal. Varying the level of the added noise and the size of the ensemble makes it possible to solve the problem connected with the mix of the mode functions. The payment for this is the considerable computer time of the signal processing.

The result of detection of slow deformation component using the described method is depicted in Figure 1. The upper curve in the figure shows the unfiltered deformation signal 85 days long recorded by a measurement arm of the deformograph installed in the Talaya seismic station gallery (Baikal Region). The curve in the middle is the result of the low-frequency Fourier filter operation at a frequency of 0.6 1/day. The slow deformation component is detected sufficiently well only in the center of the curve. At the edges there are so-called "marginal" effects 10–12 days long. The empirical decomposition method allows the total trend and the slow component detection along the whole signal, which is illustrated by the lower curve. The curve is the result of summation of the three last mode functions and the decomposition excess shown in Figure 2.

After the analysis of the modes IMF_{11}–IMF_{13} and the excess R, it has been concluded that except for the monotone drift (compression of the deformograph arm), there is a gradual oscillation of low frequencies in the signal. The cause of such behavior requires additional analysis, which is beyond the

IOP Conf. Series: Earth and Environmental Science **134** (2018) 012043 doi:10.1088/1755-1315/134/1/012043

scope of this paper; on the other hand, detectability of such behavior of deformation deserves attention.

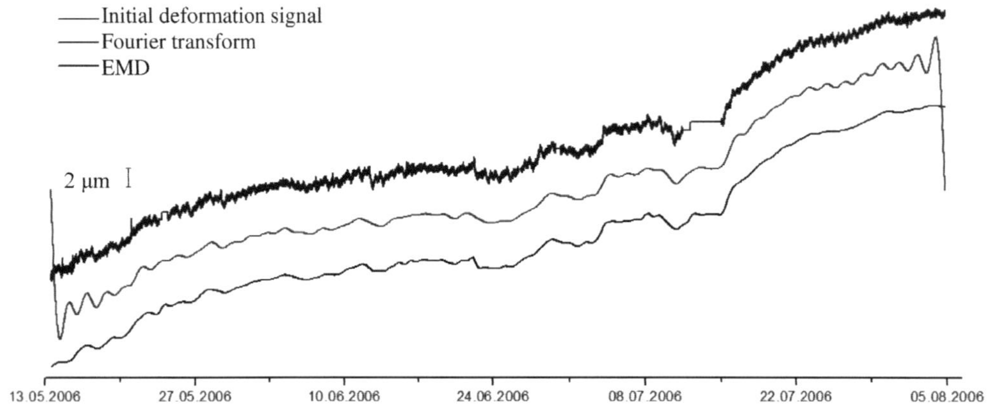

Figure 1. Comparison of the results of slow deformation component detection by the empirical mode decomposition method and digital filtration.

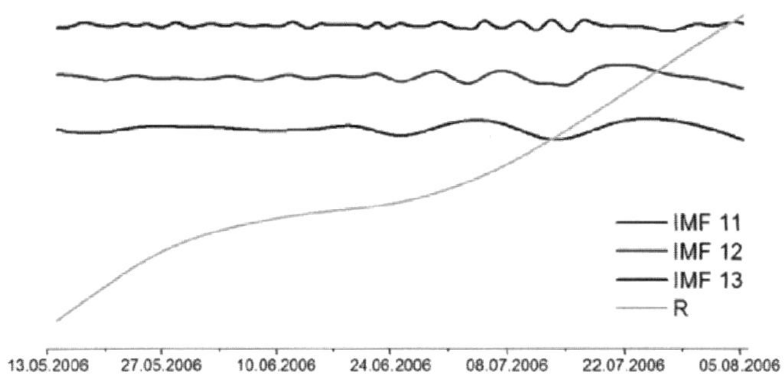

Figure 2. Monotone trend and low-frequency modes in deformation signal by EMD method.

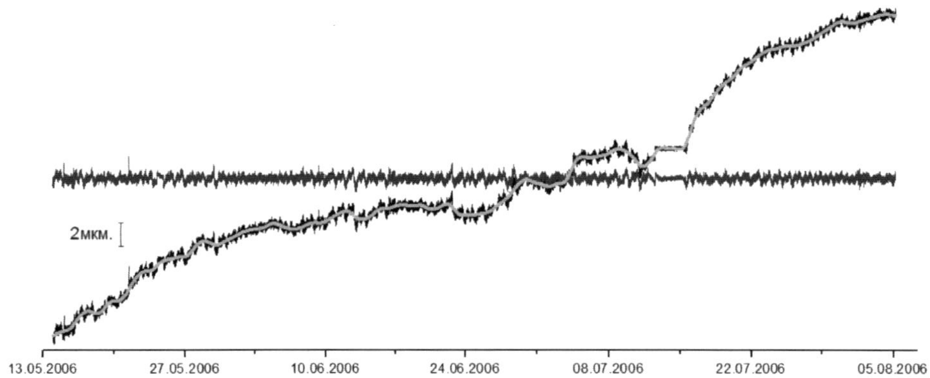

Figure 3. Results of erasing of monotone trend and low-frequency deformation component by EMD method.

The result of removal of monotone trend from the signal is shown in Figure 3. Aside from the straightened plot with the observable tidal oscillations, the figure demonstrates the initial signal. The absence of the "marginal" effect of EMD detection of the monotone part is seen.

IOP Conf. Series: Earth and Environmental Science **134** (2018) 012043 doi:10.1088/1755-1315/134/1/012043

Aimed to prove the efficiency of EMD method and the absence of "marginal" effects, Figure 4 shows the deformation signal and the identified slowly changing part and the similarly calculated slow part of signal from the expanded data for 7 days toward each side. Complete coincidence of the signals is evident in the figure.

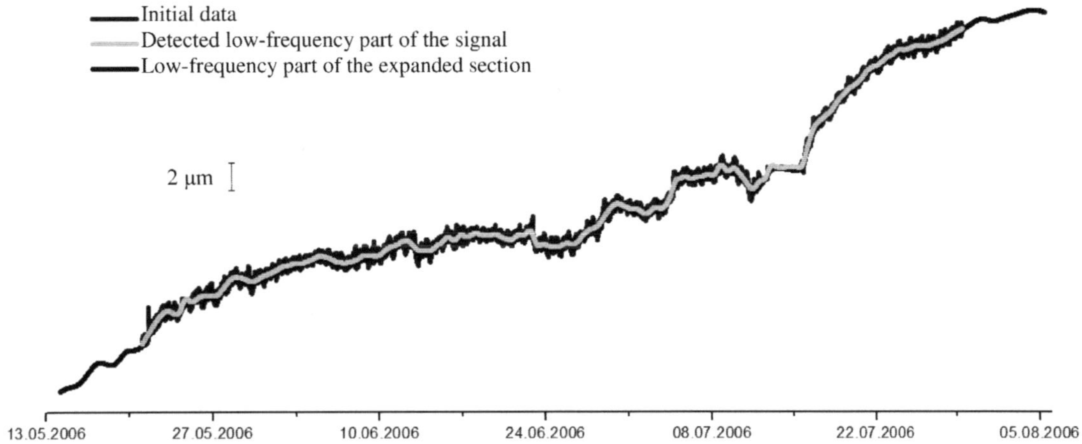

Figure 4. Absence of marginal effects in EMD method application.

The key potential of EMD in processing of deformograph data lies in the method applicability to studying nonstationary processes in the range of ultra long period oscillations of earth. Inside this range, in the course of continuous recording of deformation in the Earth's crust in order to trace precursors of earthquakes, a repeating effect was identified. In the deformation curves plotted for each recording interferometric arms, separately, an unusual disturbance was observed against the background of tidal oscillations 1.5–2 days before an earthquake. This effect is described in more detail in [6–8]. The lithospheric disturbances were clearly recorded as trains of deformation noise in the band of periods of oscillations 0.5–2 hours long; however, due to impermanence of the disturbances, it is very difficult to analyze their structure using classical methods.

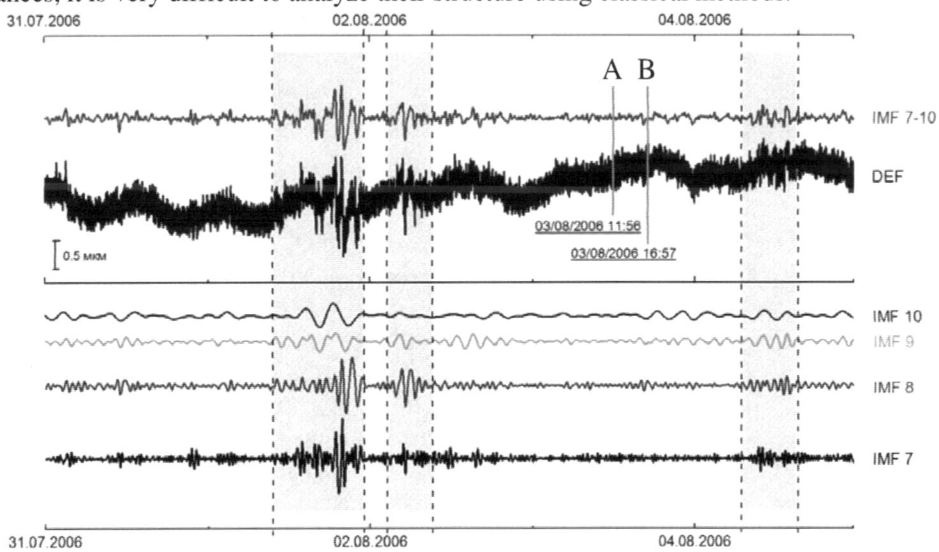

Figure 5. Excitation of ultra long period oscillations on the eve of regional earthquakes: A—earthquake on 3 Aug 2006, 11:58 a.m., φ = 55.40, λ = 110.41, K = 12.0; B—earthquake on 3 Aug 2006 16:57 p.m., φ = 55.41, λ = 110.40, K = 12.2.

An illustration of EMD application to a section of deformation plot 5 days long, containing the described above effect is given in Figure 5. At the top of the figure, there is the unfiltered deformation plot with clearly seen deformation excitations against the background of half-day tide, which have comparable amplitudes with the tides. The second curve in the upper part of Figure 5 is the superposition of four modes in which during decompositions the mentioned disturbances were detected; these modes separately are presented at the bottom of the figure. It is possible to characterize the onset and duration of oscillations (grey-colored rectangles in the figure), and the amplitude–frequency dynamics of the event; also, it is possible to identify the most representative range. It is seen that the effect appears at different time in different modes and has the largest amplitude in mode 8, which conforms with the range of periods of 1–2 hours for this decomposition. Inside this mode, there are traceable harmonics with the periods of 62, 73 and 85 min. The effect is present in the higher frequency modes (e.g. 5 and 6) but is less expressed and has much smaller amplitude. The application of EMD method has allowed identification of the wanted signal in the test section and the analysis of its behavior in different frequency ranges (modes), which later on can offer additional information required for the interpretation of the event and for the determination of its prediction potential.

Finally, the result of EMD application for the removal of high frequencies, including different nature noise, from the signal is depicted in Figure 6. To this effect, first three modes IMF_1–IMF_3 were withdrawn from the initial signal. The method is efficient and yields to none of the most widespread rivals, is free from the marginal effect and is good in processing of single bursts. In particular, EMD has proved the suggestion [9] that the daily variation of amplitude of the high-frequency component in the deformation signal is not of the noise nature as it is totally absent in the first identified mode.

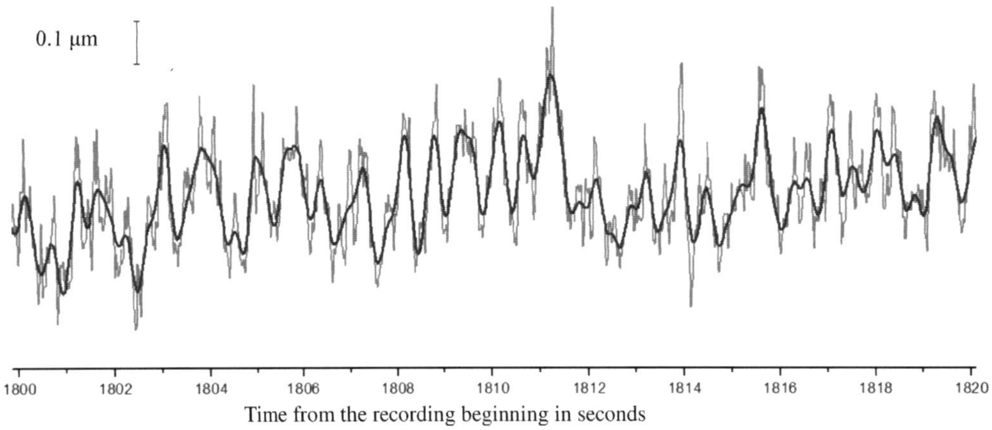

Figure 6. Filtration of noise from the signal by EMD.

4. Conclusion

Despite being comparatively new, the method of empirical mode decomposition has already some modifications and is widely applied in various research areas connected with nonlinear and nonstationary data. The authors believe that the method presents a promising tool for operation with geophysical data and deformographs, in particular, to detect local features of signals and their internal structure, to remove noise, to carry out regularization and to identify slow part in a signal, including a monotone trend. It is highly expected that the method can be efficient in studying excitation of ultra long period oscillations before earthquakes as this phenomenon is of short duration and is nonlinear both in terms of amplitude and frequency.

Acknowledgments

The studies under the Section on Application of EMD Method to Deformograph Data of this paper have been supported by the Russian Science Foundation, Project No. 16-17-10121.

References

[1] Panov SV, Parushkin MD, Semibalamut VM and Fomin YuN 2016 Heterodyne lase deformograph for geophysical research *Current Challenges of Modern Seismology* pp. 293–299 (in Russian)

[2] Orlov VA, Panov SV, Parushkin MD, Semibalamut VM and Fomin YuN 2013 Capabilities of highly sensitive laser measurements inn studying seismic and geodynamic events *Pros. Int. Conf. Geodynamics and Stress State of the Earth's Interior* Novosibirsk: IGD SO RAN pp 25–31 (in Russian)

[3] Huang NE and Shen Z *The Hilbert–Huang Transform and Its Applications* World Scientific Publisjing Co. Pte. Ltd. Singapore 596224

[4] Huang NE, Shen Z, Long SR, Wy MC, Shih HH, Zheng Q, Yan N-C, Tung CC and Liu HH 1998 The empirical mode decomposition and the Hilbert spectrum for nonlinear and non-stationary time series analysis *Proc. R. Soc. Londo Ser. A* 454 pp 903–995

[5] Wu Z and Huang NE 2009 Ensemble empirical mode decomposition: A noise-assisted data anaysis method *Advances in Adaptive Data Analysis* Vol 1 No 1 pp 1–41

[6] Orlov VA, Panov SV, Parushkin MD and Fomin YuN 2008 Connection of the Earth's seismicity and solar activity by the precision deformograph data *Pros. Int. Conf. Geodynamics and Stress State of the Earth's Interior* Novosibirsk: IGD SO RAN pp 31–40

[7] Orlov VA, Panov SV, Parushkin MD and Fomin YuN 2008 Solar activity, oscillations of the inner core of earth and seismicity of the planet *Proc. IV Int. Symp. Geodynamics of Intracontinental Orogenes and Geoecology* Moscow pp 321–326 (in Russian)

[8] Orlov VA, Panov SV, Parushkin MD, Fomin YuN, Tokmoldin SZh, Nevmershitsky IS, Uzbekov MB and Tereshkin DO 2011 Spaced highly sensitive deformation observations in studying ultra long period osicllations of earth *Pros. Int. Conf. Geodynamics and Stress State of the Earth's Interior* Novosibirsk: IGD SO RAN pp 145–150 (in Russian)

IOP Conf. Series: Earth and Environmental Science **134** (2018) 012044 doi:10.1088/1755-1315/134/1/012044

Evaluating influence of luminescence excitation and recording on the efficiency of apatite recovery from Khibiny low-grade ore

DN Pavlishina[1*], BA Vlasov[1,2] and SV Tereshchenko[1, 2]

[1]Mining Institute, Kola Science Center, Russian Academy of Sciences, Apatity, Russia
[2]Apatity Division, Murmansk Arctic State University, Apatity, Russia

E-mail: [*]shibaeva_goi@mail.ru,

Abstract. X-ray luminescence separation efficiency depends on a number of factors among which the most important is the algorithm for excitation and recording of luminescence in ore particles. The comparative evaluation of efficiency of one- and two-way algorithms for luminescence excitation and registration is carried out on apatite-bearing ore. It is found that the one-way algorithm performance yields no change in the X-ray luminescence separation efficiency in terms of the analyzed ore type. The choice of this algorithm and the sustained admissible separation level make it possible to reduce a number of parts and size of a separator, to increase its specific productivity and to decrease its cost.

An amount of the main valuable component (phosphorus pentoxide) in Khibin apatite ore tends to lower with time under a number of factors including the depletion of rich and easy-to-handle ore reserves, complication of mining and geological conditions, application of high-performance mining machines and ever-increasing obligations of miners for ecological safety of mining operations. Moreover, so far substantial mining waste dumps can be evaluated as technogenic reserves consisting of valuable component (VC)-bearing rocks of cut-off grade. Nowadays a dump material grade can be comparable to a poor ore grade, taking into account the evolution of technological parameters in development of mineral deposits with time. In terms of Khibin deposit it is explicit that the cut-off grade has been lowered from 12 to 2–4% P_2O_5 (Figure 1). All the above provides the grounds to foster mastering of ore processing techniques, to search ways of economically efficient, ecofriendly low-grade ore beneficiation.

One of promising trends in Khibin apatite processing is the X-ray luminescence separation (XLS)as a bulk material pre-treatment stage. Expedience of application of apatite luminescence as an attribute of separation is justified by apatite capacity to be luminescent intensively under certain conditions unlike the rest rock-forming minerals [1].

Considering this apatite property along with sufficient statistical research work on establishment of relation between VC content in a test ore lump and a recorded signal (Figure 2) made it possible to single out exclusively the material which VC content exceeds a preset cut-off grade level from the entire ore flow under processing. Gangue and poorly mineralized rocks are rejected from the concentrator feed, thus improving a grade of the feed and lowering energy costs per a ton of produced apatite concentrates along with less volume of resultant finely-ground tailings of apatite flotation [2, 3]. The last factor indirectly influences a service life of tailing dumps and diminishes the negative mining impact on environment [4].

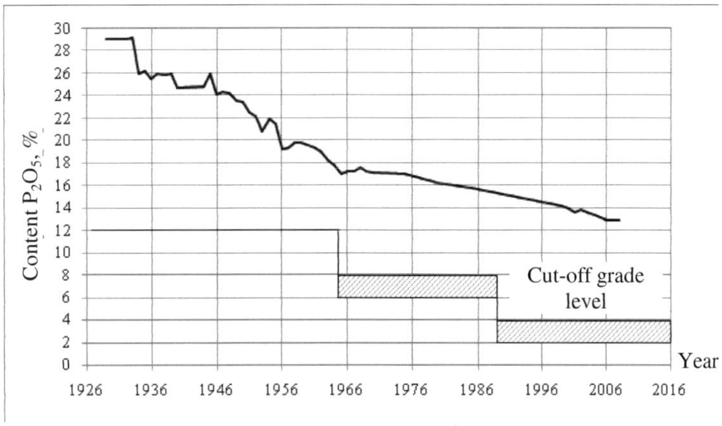

Figure 1. Variation in P_2O_5 content in Khibin mined ores.

Investigation into XL characteristics of Khibin apatite-bearing ores and feasibility to realize XLS in their processing have been the study objects at Mining Institute, Kola Science Center, RAS, since 1982 [5]. The research findings [1–3, 6–9] justify perspectiveness of XLS application and reasonability of full-scale tests of this process.

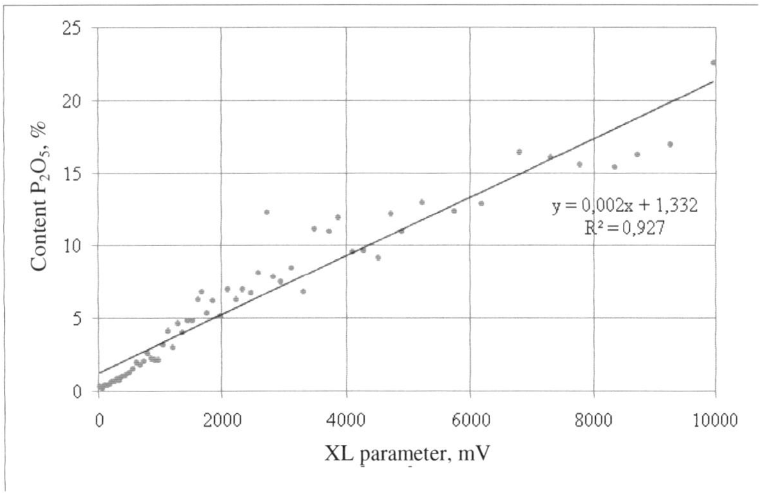

Figure 2. Correlation dependence of XL parameter versus P_2O_5 content in ore lumps.

Transition from laboratory tests to pilot and semi-commercial tests implies special attention to choice of XLS equipment.

In view to RF import-replacement government strategy the feasibility was considered to apply Russian machinery capable to handle bulk material of more than 20 mm in size to separate apatite-bearing ores:

–UAS-type radiometric separator, AO VNIIKhT, Moscow [10];

–X-ray luminescence separator, OOO EGONT, Saint Petersburg [11];

–SNR-10 radiometric separator, AO Soyuztsvetmetavtomatika, Moscow [12].

The above separators are equipped with one-way (for separators nos. 1 and 2) and two-way (for separator no. 3) algorithms for excitation and registration of X-ray luminescence of mined bulk ore.

In [1] the study object was the current run-of-mine ore; the researcher assessed the separation efficiency of scheelite, apatite, and fluorite ores and found that the separation was improving with increase in number of measured sides of ore lumps,

Separation efficiency was assessed under procedure described in [1]:

$$Э = \frac{П}{M}$$

where $П$ is parameter of the attribute for separation, a dimensionless unit.

$$\Pi = \frac{\sum_{i=1}^{n} |(z_{i\phi} - \alpha)\gamma_{i\phi}|}{\alpha \sum_{i=1}^{n} \gamma_{i\phi}}$$

n is number of fractions in the study object; z_{if} is VC content per a fraction; α is average VC content in the study object; γ_{if} is the fraction yield from all the elementary volumes of the study object; M is contrast parameter.

$$M = \frac{\sum_{i=1}^{n} |(y_i - \alpha)\gamma_i|}{\alpha \sum_{i=1}^{n} \gamma_i}$$

y_i is VC content in an elementary volume; γ_i is yield of an elementary volume of the study object.

Evaluation of separation efficiency by one- and two-way algorithms for excitation and registration in terms of the technical specifications of the test separators was studied on Partomchopp poor apatite-bearing ore specimens, be precise, two main apatite-nepheline and sphene-apatite-magnetite species of –50+20 mm in size.

Figure 3 demonstrates variations in registered luminescence signals of lumps by one- and two-ways algorithms for excitation and registration. Green color denotes the selected separation threshold, excess of its magnitude indicates a necessity to refer a single analyzed volume to an ore component of the ore flow under separation.

The comparative analysis of XL signals from apatite-nepheline ore by one- and two-way algorithms for excitation and registration reveals the actually identical behavior of graphs. Table 1 summarizes results of XLS realization by one- and two-way algorithms for excitation and registration.

Analysis of separation results in transition from two-way to one-way algorithms for excitation and registration of luminescence of apatite-bearing ore lumps indicated their negligible variation:

–yield of XLS tailings increases by 1.9–3.2 % and by 0.77–1.4% for apatite-nepheline and sphene-apatite-magnetite specimens, respectively;

– growth of VC loss in XLS tailings is 4.84–5.58% and 1.88–2.77% for apatite-nepheline and sphene-apatite-magnetite specimens, respectively;

– dilution of XLS concentrate with gangue and weak mineralized rocks is not higher than 3%.

IOP Conf. Series: Earth and Environmental Science **134** (2018) 012044 doi:10.1088/1755-1315/134/1/012044

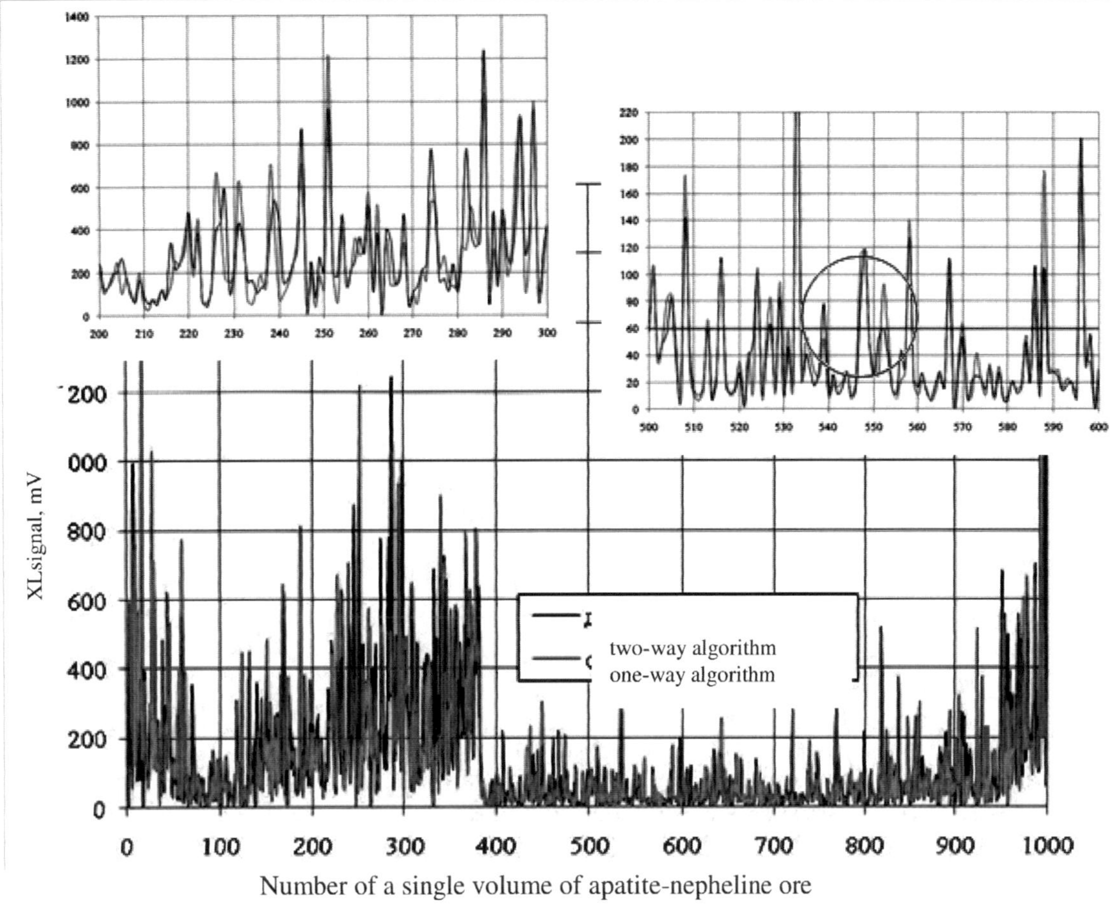

Figure 3. Record diagram of a X-ray-luminescence signal of apatite-nepheline ore lumps

Table 1. Results of XLS realization by one- and two-way algorithms for excitation and registration

Separation products	Two-side survey		One-side survey		One-side survey	
	Yield, %	P_2O_5content, %	Yield, %	P_2O_5content, %	Yield, %	P_2O_5content, %
Apatite-nephelineore						
XLS concentrate	45.9	9.4	49.1	9.1	47.8	8.9
XLS Tailings	54.1	1.3	50.9	1.3	52.2	1.2
Primaryore	100.0	5.7	100.0	5.3	100.0	5.3
Sphene-apatite-magnetiteore						
XLS concentrate	45.9	8.8	49.1	8.9	47.8	9.0
XLS Tailings	54.1	1.6	50.9	1.6	52.2	1.6
Primaryore	100.0	6.0	100.0	5.9	100.0	6.1

The efficiency of the selected algorithm can be presented graphically as contrast and concentration curves which can be plotted under the known procedure [5] (Figure 4). In the Figure it is obvious that tailing yield and VC content in the ore product vary under the fixed VC content in separation tailings.

If P_2O_5 content in XLS tailings is at 2% level, the efficiency of separation for Partomchorr apatite-nepheline ore is

– Э = 0.90 in two-way survey;

– Э= 0.84–0.89 in one-way survey.

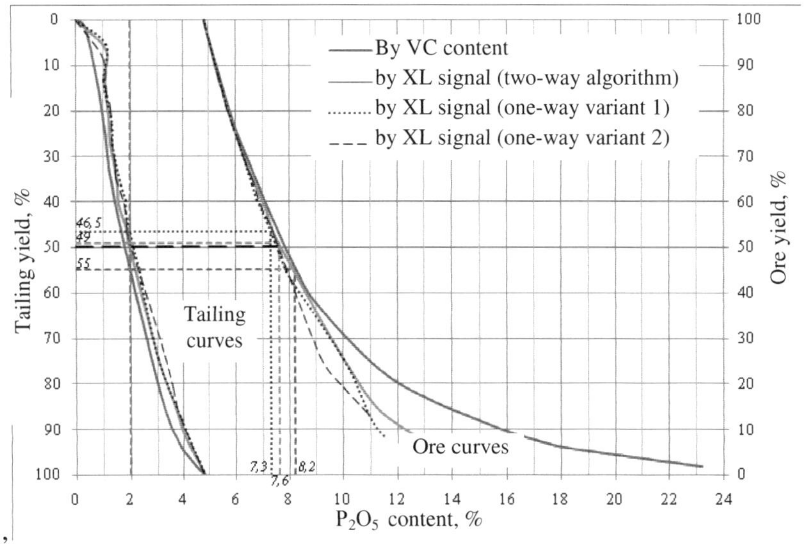

Figure 4. Curves of contrast and concentration for Partomchorr apatite-nefeline ore.

To conclude, negligible variations in technological parameters of X-ray luminescence separation prove feasibility of its realization with measurement of luminescence on only one surface of the test lump; possible application of XL separators with one-way algorithm for excitation and registration. Realization of one-way algorithm for excitation and registration in the separator can allow support of the admissible separation level, a reduced number of parts in the complete set, reduced overall dimensions of the separator, and feasibility to adjust specific separator performance and its cost.

References

[1] Tereshchenko S 2002 *The Main Provisions of X-Ray Luminescent Separation of Mineral Raw Materials*. Apatity: KFPetrGU (inRussian)

[2] Tereshchenko S, Marchevskaya V, Chernousenko E, Rukhlenko E, Pavlishina D and Smolnyakov A 2014 Ecological and technological aspects of ore quality management based on separation in: Ecological strategy for mining industry—shaping a new outlook for mineral mining: *Collected Papers of th Russian Academy of Sciences, Institute of Mining, Kola Science Center RAS* Saint-Petersburg: Renome (in Russian)

[3] Tereshchenko S, Marchevskaya V, Chernousenko E, Rukhlenko E, Pavlishina D and Smolnyakov A 2015 Integrated ore preparation in processing of low-grade apatite–nepheline ores *GIAB* No 1

[4] Tereshchenko S, Marchevskaya V and Pavlishina D 2016 Ways to reduce the negative environmental impact of mining industry *Vestn. Kol. Nauch. Tsentra RAN* No 4

[5] Lagov B, Tereshchenko S, Kaitmazova T, Yuriev V and Valshchikov A 1984 Studies into radiometric separation of Khibiny low-grade apatite ores in *Laboratory and Proof Tests and Dressing of Minerals. Domestic and International Experience* VIEMS Issue 5 (in Russian)

[6] Tereshchenko S, Marchevskaya V, Maslov A, Golovanov V and Pogrebnyak O 1998 High-

quality ore separation from diggings—one of the conditions of rational ore concentration *Vestn. Murmansk Gost.Tekh.Univer.*Vol 1 No 3

[7] TereshchenkoS, Denisov G and Marchevskaya V 2005 *Radiometric Methods of Sampling and Separation of Mineral Raw Materials* Saint-Petersburg: MANEB (inRussian)

[8] Tereshchenko S and Marchevskaya V 2002 Rational Subsoil Use Based on Radiometric Preconcentration of Minerals *Marksheider. Nedropolz* No 3 (inRussian)

[9] Tereshchenko S, Marchevskaya V, Chernousenko E, Pavlishina D and Rukhlenko E 2015 Investigation of priority trends in mineral processing *GIAB* No 6

[10] JSC VNIIHT [official website]. URL: http://www.vniiht.ru

[11] Innovative Ecological Mineral Dressing Technologies [official website]. URL: http://www.egont.ru

[12] SouzCvetMetalAvtomatica [official website]. URL: http://www.scma.ru

IOP Conf. Series: Earth and Environmental Science **134** (2018) 012045 doi:10.1088/1755-1315/134/1/012045

Effect of steel structure and defects on reliability of parts of impact mechanisms

AI Popelyukh[1], AA Repin[2]*and SE Alekseev[2]

[1]Novosibirsk State Technical University, Novosibirsk, Russia
[2]Chinakal Institute of Mining, Siberian Branch, Russian Academy of Sciences, Novosibirsk, Russia

E-mail: *repin@misd.ru

Abstract. The paper discusses selection of materials suitable for manufacturing critical parts of impact mechanisms. It is shown that in order to extend life of parts exposed to high dynamic loading, it is expedient to use medium- and high-carbon alloy-treated steels featuring low impurity with nonmetallic inclusions and high hardening characteristics. Application of thermally untreated parts is undesirable as steel having ferrite–pearlite structure possesses low fatigue strength. Aimed to ensure high reliability of parts with a hardness of 42–55 HRC, steel should be reinforced by thermal treatement with the formation of multicomponent martensite–bainite structure. High-quality production should include defectoscopy and incoming material control.

Extension of service life of impact machines is a complex task. Early failure of parts can be the result of engineering errors, or disregard of technology of manufacture and operation of the machines. One of the main causes of early failure of impact mechanism parts is the improper selection of material. Parts of impact machines should possess a set of properties, among which the critical characteristics are strength, hardness, impact toughness, crack and wear resistance [1]. Generally, an increase in hardness and strength of steel decreases its impact hardness and fatigue failure strength; for this reason, the main objective of equipment design is rational selection of materials and creation of a structure capable to ensure optimal combination of the above-listed properties (Figure 1).

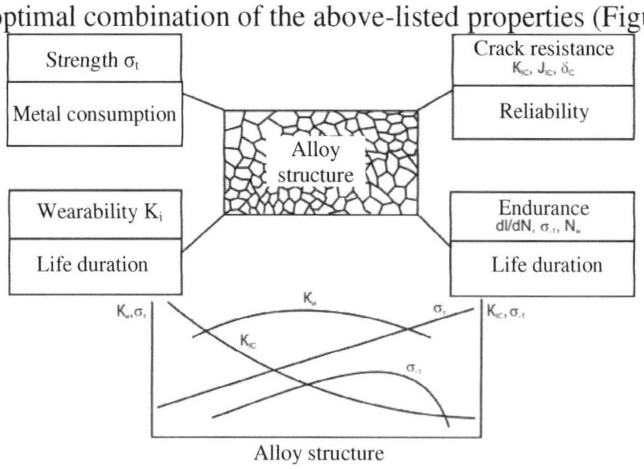

Figure 1. Correlation of the main properties and structure of steel.

Practically, the choice of steel for manufacturing different parts is governed by many factors: operating conditions and target life of mining machines, economic efficiency, availability of wanted roll stock, etc. Which steel should be used—cheap alloy-free or expensive alloyed—is a question of individual choice; however, it is desirable to manufacture parts of impact machines from medium- and high-carbon alloy-treated steels with high hardenability for thermal treatment. This condition is obligatory for critical parts subjected to cyclic or dynamic loading.

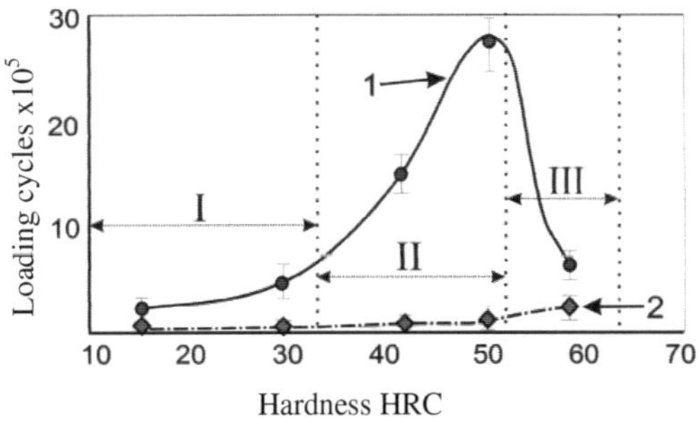

Figure 2. Fatigue life of steel 45: 1—general endurance; 2—initiation of fatigue cracks.

Figure 2 depicts tests data on fatigue life of steel 45 specimens subjected to multiple dynamic compression [2]. If steel after quenching and tempering has hardness less than 30 HRC, its endurance is low and weakly depend on the hardness level (zone I in Figure 2). In medium-carbon steel with the structural hardness from 35 to 55 HRC, endurance grows together with the strength characteristics (zone II). Generation of low-temperature tempered martensite structure with high level of internal stresses in steel decreases steel endurance due to drop in the length of stage of fatigue crack initiation and growth (zone III). An optimal combination of strength and fatigue resistance is a feature of steel having disperse structure of fine pearlite. Steel with ferrite–pearlite structure weakly resists fatigue failure. Fatigue cracks rapidly grow towards interfaces of ferrite and cementite plates, which result in quick breakdown of parts with the formation of brittle fracture surfaces (Figure 3). The use of steel possessing high hardenability reduces portion of ferrite–pearlite structure in the total volume of hardened parts and allows higher strength of steel at the simultaneous increase in the steel crack resistance.

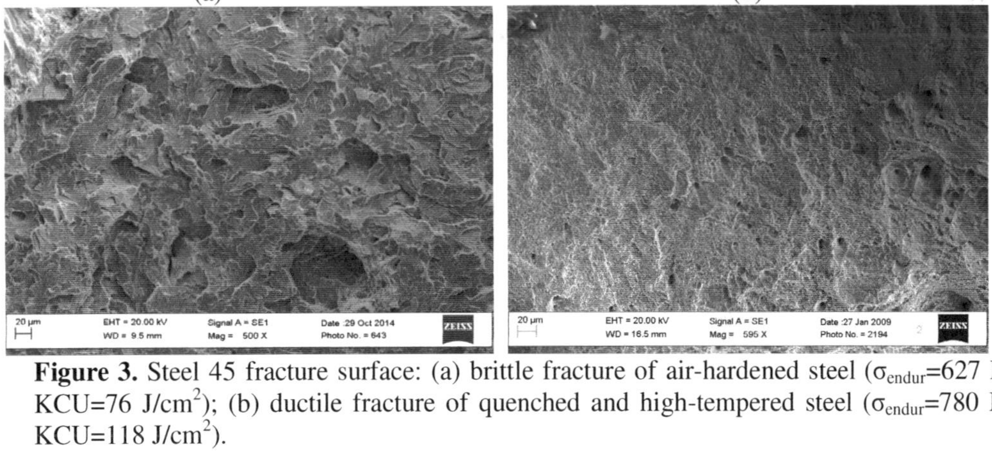

Figure 3. Steel 45 fracture surface: (a) brittle fracture of air-hardened steel (σ_{endur}=627 MPa, KCU=76 J/cm^2); (b) ductile fracture of quenched and high-tempered steel (σ_{endur}=780 MPa, KCU=118 J/cm^2).

It is noteworthy that optimization of steel structure at the stage of experimental development is not a guarantee of high performance of mining machines of mass production. Expert appraisal of steel quality and the analysis of failure causes in machine parts undertaken at the Destructive Testing Laboratory of the Novosibirsk State Technical University point at the tendency toward degradation of rolled steel of large thickness and diameters. Iron plates thicker than 20 mm, rolled and forged products having diameters larger than 200 mm may contain very many internal defects. Sometimes there are cracks up to a few millimeters long (Figure 4). For this reason, in order to ensure high reliability of products, machine-building plants should accomplish incoming inspection of rolled metals.

(a) (b)

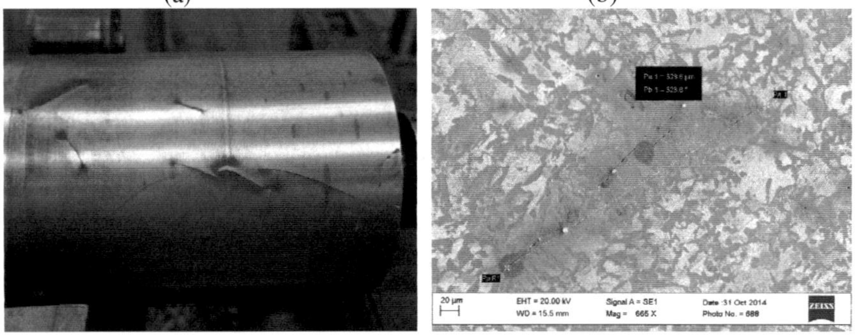

Figure 4. Cracks in (a) forged steel 20X24A and (b) Ø 250 rolled steel 45.

Another aspect of concern during manufacture of mining machine parts is high content of nonmetallic inclusions in steel (Figure 5). High impurity with nonmetallic inclusions in steel is explained by overwear of metallurgical equipment and general degradation in steel production culture since most of metallurgical plants produce construction steels for which impurity is not a critical factor. Low performance characteristics of steel with the high content of inclusions are conditioned by the fact that the external loading results in the localization of internal stresses at the inclusions, which initiates fast fatigue cracking. Furthermore, diffusion processes at the inclusions alter chemical composition and strength characteristics of steel, whereas different coefficients of thermal expansion of an inclusion and basic metal induce extra thermal stresses. Production of pure steel free from nonmetallic inclusions is an expensive process. Practically, it seem more important not to minimize the impurity due to nonmetallic inclusions but to produce steel containing inclusions of a certain type, shape, size and morphology so that to affect properties and performance of steel minimally.

(a) (b)

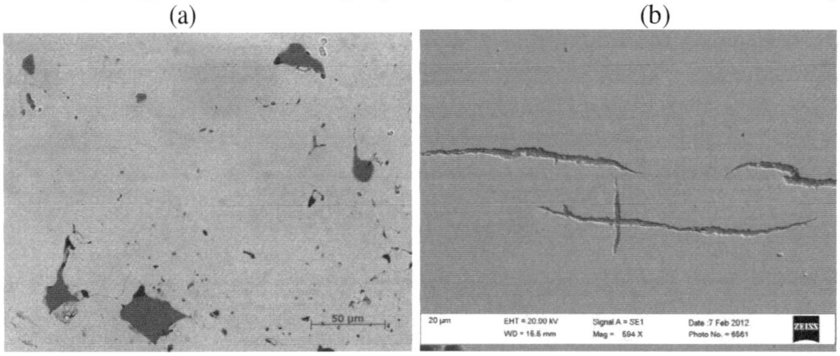

Figure 5. Nonmetallic inclusions in steel 45 in the form of (a) points and (b) strings.

The experiments carried out at the Novosibirsk State Technical University prove that the presence of large inclusions in steel results in the decrease in life of critical parts of impact machines by 2 times and more [2–4]. Statistics shows that the inclusions exert very high aggravating effect on fatigue endurance of highly strong steels. In this case, up to 70% of fatigue cracks causing failure are initiated by nonmetallic inclusions. Under low-cycle loading (target life of parts is less than 10^5 cycles), when specific loads are close to the yield point of steel, the main source of fatigue failure is not the inclusions but surface defects; for this reason, for parts intended for operation under static or low-cycle loading, impurity of steel due to nonmetallic inclusions is not critical. Thus, for manufacturing parts for long service life (more than 10^7) under loading, it is recommended to use steels with the inclusions of small size (less than 20 μm) and low content (not higher than the second number in the scale by State Standard GOST 1778-70). In case of cyclic tensile or alternating loading, the most critical parameters are the size and shape of inclusions (small size and spherical shape are preferred) while their morphology (or chemical composition) has no significant influence on fatigue endurance of steel. For parts meant for operation under dynamic compression, high durability can be ensured using steel having hard inclusions [4]. Inclusions highly affect performance of parts in aggressive media when corrosion processes intensify initiation and growth of fatigue cracks (Figure 6) [5]. Unfortunately, Russian standards have no regulations on content of inclusions in structural alloy-treated rolled steel. In this case, mining machine building plants would maintain high quality of manufactured parts by placing special requirements on impurity of steel due to nonmetallic inclusions (not higher than the second number in the scale), specifying these requirements in steel delivery contracts and by undertaking verification checks of rolled steel in accordance with the State Standards GOSR 1778-70: Steel. Metalographic analysis of nonmetallic inclusions.

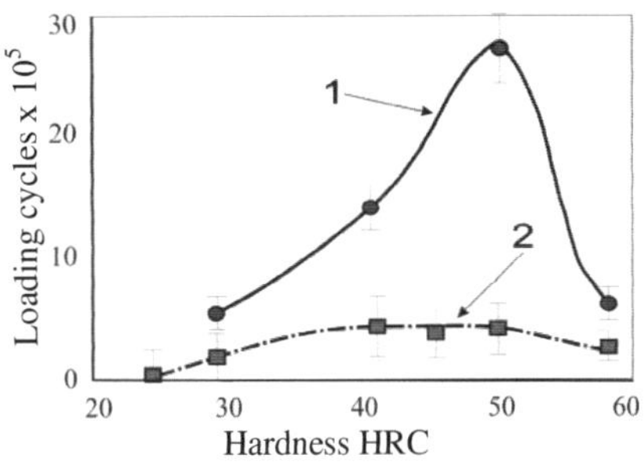

Figure 6. Fatigue endurance of steel 45: 1—air tests; 2—water tests.

High reliability and endurance of impact mechanism parts subjected to loading can be ensured using new advanced methods of thermal treatment and generation of multicomponent super-strong steel having high fracture toughness. As a rule, such methods are based on formation of mixed structure consisted of regulated amount of martensite, bainite and residual austenite [6–12]. Figure 7 depicts one of the treatment cycles. Thermal treatment generates gradient structure in steel (Figure 8). Martensite ensures high strength characteristics while bainite and austenite have high fracture toughness. Comparing to conventional tempering technology, the proposed thermal treatment approach provides a double increase in impact toughness and a triple increase in fatigue endurance of steel at the similar strength level.

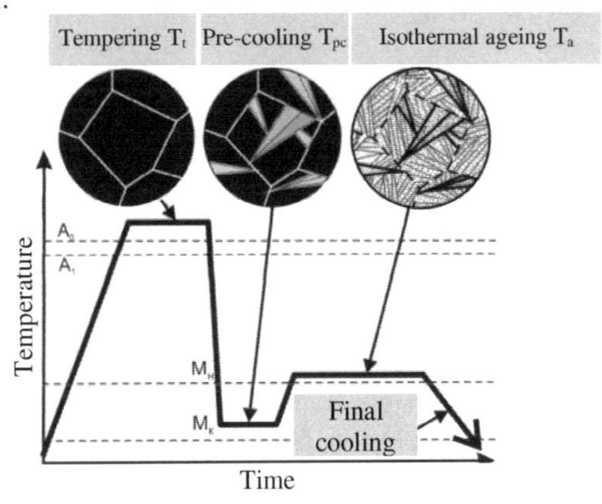

Figure 7. Thermal treatment and formation of martensite–bainite structure.

Figure 8. Mixed structure of steel 40X2H2MA: dark regions—martensite; lighter regions—bainite.

Another promising way of enhancing reliability and endurance of parts of impact mechanisms is application of new nano-structured bainitic steel possessing high strength and fracture toughness [13, 14]. New-generation nano-structured bainitic steel contains: carbon 0.8–1.2%; silicon 1.5–2%; manganese 1.5–2% and chromium 1.5–2%; they also can be additionally alloyed with cobalt, molybdenum and aluminium. Thermal treatment circuit of such steel includes heating up to austenite transformation temperature, subsequent accelerated cooling down to a temperature 5–10 degrees higher than martensite transformation temperature and, then, long isothermal ageing in a chamber furnace under the same temperature for a few hours. The resultant steel has a structure consisted of nano-bainite and residual austenite, possessing high strength (up to 2300 MPa), high fracture toughness (to 30 MPa·√m) and plasticity (to 20%). It should be mentioned that thermal treatment methods include accelerated cooling down to a temperature of 150–300°C and, thus, require purchasing low-temperature salt-bath furnaces.

Conclusions

1. Depending on basic criteria laid on a material, different types of thermal reinforcement should be used. Low tempering allows production of highly strong steel (hardness higher than 52 HRC). High-hardness steel (42–52 HRC) can be reinforced by thermal treatment with mixed austenite transformation. Isothermal quenching with the formation of lower bainite ensures optimal combination of fracture toughness and strength in 38–42 HRC steel

2. It is undesirable to use thermally untreated parts as ferrite–pearlite steel has low fatigue resistance. It is inexpedient to carry out thermal treatment of steel if brittle structure of low-tempered martensite with the hardness higher than HRC 55 is formed.

3. Nonmetallic inclusions in steel appreciably decrease its fatigue endurance. They are affect badly especially on high-strength steel parts exposed to multi-cycle loading. In low-strength parts under static or low-cycle loading, impurity of steel with nonmetallic inclusions is not a critical factor.

4. In order to provide high-quality production, machine building plants should carry out defectoscopy and incoming inspection of delivered rolled steel.

References

[1] Popelyukh AI, Teplykh AM, Terentiev DS and Ognev AYu 2009 Enhancing structural strength of percussion tools *Obrabot. Metal.* No 4 pp 15–16

[2] Repin AA, Alekseev SE and Popelyukh AI 2012 Enhancing reliability of parts of percussion machines Vol 48 No 4 pp 669–674

[3] Repin AA, Alekseev SE, Popelyukh AI and Teplykh AM Influence of nonmetallic inclusions on endurance of percussive machines *J. Min. Sci.*, Vol 47 No 6 pp 798–806

[4] Popelyukh AI, Veselov S, Munkueva DD, Timonin VV and Karpov VN 2017 Influence of nonmetallic inclusions on failure resistance of stee under multiple dynamic compression *Obrabot. Metal.* No 2(75) pp 67–78

[5] Popelyukh PA, Nikulina AA and Popelyukh AI 2013 Effect of external environment on

reliability of parts of mining machines under dynamic compression *Nauchn. Vestn. NGTU* No 4(53) pp 133–138

[6] Popelyukh PA, Popelyukh AI and Yurkevich MR 2013 Combined thermomechanical treatment of steel with martensite–bainite trasformation of austenite *Obrabot. Metal.* No 2 pp 62–68

[7] Popelyukh AI, Bataev AA, Teplykh AM, Ognev Ayu and Golovin ED 2011 Method of thermal treatment of instrumental steel with mixed martensite–bainite transformation of austenite *Stal* No 4 pp 69–73

[8] Popelyukh PA, Nikulina AA, Popelyukh AI and Yurkevich MR 2016 RF Patent No 2588936 *Byull. Izobret.* No 19

[9] Nie YH, Fu WT, Hui WJ, Dong H and Weng YQ 2009 Very high cycle fatigue behaviour of 2000-MPa ultra-high-strength spring steel with bainite–martensite duplex microstructure *Fatigue Fract. Engng. Mater. Struct.* Vol 32 pp 189–196

[10] Guhui Gao, Han Zhang, Zhunli Tan, Wenbo Liu and Bingzhe Bai 2013 A carbide-free bainite/martensite/austenite triplex steel with enhanced mechanical properties treated by a novel quenching–partitioning–tempering process *Materials Science & Engineering* Vol 559 pp 165–169

[11] Alanemea KK, Ranganathana S and Mojisola T 2010 Mechanical behaviour of duplex phase structures in a medium carbon low alloy steel *Journal of Minerals & Materials Characterization & Engineering* Vol 9 No 7 pp 621–633

[12] Wei DY, Gu JL, Fang HS, Bai BZ and Yang ZG 2004 Fatigue behavior of 1500 MPa bainite/martensite duplex-phase high strength steel *International Journal of Fatigue* Vol 26 pp 437–442

[13] Weng Y, Dong H and Gan Y 2011 *Advanced Steels: The Recent Scenario in Steel Science and Technology Metallurgical Industry* Press Beijing and Springer Verlag GmbH Berlin

[14] Bhadeshia HKDH 2010 Nanostructured bainite *Proc. Royal Societ. A* Vol 466 pp 3–18

IOP Conf. Series: Earth and Environmental Science **134** (2018) 012046 doi:10.1088/1755-1315/134/1/012046

Synthesis of engineering designs of drilling facilities

K Porozhsky

Ural State Mining University, Yekaterinburg, Russia

E-mail: porogski.k@ursmu.ru

Abstract. The article sets forth key principles of engineering of drilling equipment based on successive analysis of the goals of the production method, technologies of its implementation and conditions of mineral mining using a new approach to systematization of drilling methods. Potential advancement in the technologies and equipment of drilling is illustrated in terms of oil-well drilling.

The modern drilling facilities are the sophisticated equipment. The goal of their synthesis is generation of a structure such that to fulfill pre-set requirements [1]. The requirements are governed by the purpose, assumed mining technology and operational behavior of the system in the given application environment.

Engineering up-to-date drilling equipment is one of the most important objectives in the extractive industry for mining is nearly almost connected with drilling. The purpose and conditions of drilling, well design and making technology are diverse and continuously modified.

The drilling purposes determine process requirements imposed on the drilling results. There are three main purposes:

1. Geological exploration;
2. Surface borehole mining (oil, gas, groundwater, in-situ leaching of metallic and nonmetallic minerals or hydraulic mining);
3. Service drilling (observation, blasting, water drawdown, degassing, ventilation, water drainage, engineering geology, construction of roads, buildings and structures).

The functional efficiency of reaching the set goals is evaluated using essentially different criteria. In particular, for expendable wells, the main criterion is the quality of data on localization and properties of a mineral while for production wells, the main criteria are life duration and efficiency of reservoir–well connection; for service wells, the key criterion is the minimum cost of the whole package of drilling, blasting, excavation, transportation, crushing, ventilation and other processes. At the same time, the generalized indicators of drilling efficiency are energy cost and prime cost per unit product. In mineral mining, drilling equipment should be selected based on the feasibility study of minimum cost of all process flows: drilling, blasting, excavation, transportation, ore crushing before beneficiation with regard to ore breakage quality [2].

Based on the aforesaid, the first stage of drilling equipment selection is detailed analysis of the mineral mining process and estimation of role of drilling within this process. The second stage is the analysis of drilling efficiency required at each phase of mineral mining. As a rule, the common technical criterion of drilling efficiency is drilling path precision. At the third stage of selecting drilling equipment, drilling conditions are studied with an emphasis laid on the factors that considerably influence the efficiency of drilling and govern the choice of a drilling technology.

IOP Conf. Series: Earth and Environmental Science **134** (2018) 012046 doi:10.1088/1755-1315/134/1/012046

When selecting drilling technology, the method of rock destruction is usually mean (rotary, rotary–percussive and others). In the meanwhile, in hole-making, of no less importance than rock breakage at the hole bottom are the processes of drilling cuttings removal (drill mud), strengthening of hole walls, sampling, monitoring of holes, lowering and lifting of drilling tools. For instance, in deep oil and gas well drilling, the borehole cleaning quality appreciably affects drilling efficiency and requites much energy and material input than bottomhole breakage. It also important to analyze preparation of drilling tools for operation (assembling, disassembling, transportation) and auxiliary operations connected with the transportation and handling of drilling tools and materials. Thus, not only method of bottomhole breakage, but efficient combination of all rational operations listed can characterize a drilling technology in a specific hole.

In this connection, when synthesizing drilling facilities for a specific well in the specific drilling conditions, it is necessary to evaluate rational methods to implement each process. Selection of the implementation method should account for the role of the process in reaching the drilling purpose and the portion of energy input of the process in the total energy consumption of the hole drilling [3, 4].

Table 1 presents a variant of classification of drilling methods. The holistic description of a drilling method is the combination of methods of implementation of all involved operations (process functions). For example, Table 1 describes the most popular method of exploration hole drilling in oil production (shadowed areas). Even using this simplified classification, there are dozens possible scenarios of implementing all process functions. It is noteworthy that Table 1 gives a generalized description of drilling method. Characterization of each process functions needs more detailed description of the function implementation methods. Such classification is given in the description of rock destruction technique in Table 2. In the system-based approach to the synthesis of drilling facilities, such detailed tables should be compiled for every process function. The limited size of a paper is unsuitable for presentation of a detailed classification of drilling methods. There are much more variants of implementing the function. Literature informs on such methods of treating the bottomhole as explosive, heat, plasma, laser, electric arc, vibration etc., which are seldom practiced in drilling.

Table 1. Classification of drilling methods.

Operation (function)	Classification feature	Variants of implementation				
		1	2	3	4	5
Bottomhole breakage	Bottomhole impact method	Percussion	Rotation	Driving	Giant jet	Hybrid technique
Bottomhole cleaning	Drilling cuttings removal method	Mechanical	Hydraulic	Pneumatic	Compactio n into walls	Hybrid
Sampling	Type of sample	Bottomhole coring	Drill mud	Sample from walls	Reservoir fluid	Hybrid
Casing	Method of casing	Casing pipes		Flushing fluid	No casing	Compactio n of walls
		Per intervals	Concurrently			
Lowering and lifting	Method of lifting	Using drill string, per sections	Continuous flexible pipe column	Rope, cable, wire, hose	Fluid or gas flow	Hybrid
Borehole investigation	Type of investigation	Directional surveying	Measurement of drilling parameters at bottomhole	Properties of rocks	Reservoir pressure	Oil or gas rate

Table 2. Classification of bottomhole breakage methods.

Operation (function)	Classification feature	Variants of implementation				
		1	2	3	4	5
Bottomhole breakage	Bottomhole impact method	Percussion	Rotation	Driving	Giant jet	Hybrid technique
	Bottomhole configuration	Circle-shaped	Ring bottomhole	Stepped	Hybrid	
	Positioning of the impact tool	Ground surface	Bottomhole	Hybrid		
	Structural design	Roller cutter	Diamond	Carbide	Jet	Hybrid

The choice of a rational drilling technology should be based on the comprehensive analysis of influence of a drilling method on the drilling result. Efficiency of exploration drilling is evaluated by the core quality. For example, high-quality coring in hard fractured rocks is possible by rotary diamond-tool drilling with back-flow and by core sampler driving in soft rocks.

Efficiency criteria of a drilling method can be assumed [3]:

—quality of data on a mineral (core quality, drilling trajectory precision, quality of access to a mineral, quality of stratum investigations);

—generality (destruction of rocks of any drillability, efficiency of all drilling processes);

—energy input of data acquisition and hole making (drilling rate, energy cost);

—drilling cost (to be summed up with energy cost, drilling rate, wear of tools and materials consumption).

It is impossible to design a high-efficiency drilling tool without the comprehensive and holistic analysis of the modern trends in the sphere of the drilling equipment and technologies. There are many driving forces of the development process, and they can be grouped as [3, 4]:

—Production and ecology safety while drilling;

—Varied drilling conditions;

—Advanced technologies of mineral mining;

—Cost of drilling;

—Ergonomics.

Technological development is the second in terms of significance but the first in terms of influence exerted on equipment engineering. Drilling equipment (as any other equipment) and drilling technology can only evolve simultaneously. Equipment is only an implementer of a technology.

Regarding drilling technology in oil production, the development trends are:

1. Directional drilling:
   (a) Horizontal drilling, including far out-of-horizon drilling;
   (b) Branching
   (c) Multihole drilling;
   (d) Inclined drilling in high-viscosity oil production
2. Pressure control drilling (underbalanced drilling);
3. Casing drilling;
4. Casing while drilling;
5. Small-diameter drilling.

The methods suitable for these areas of technological development are compiled in Table 3 together with the efficiency criteria dependent on the process operation. Some modern technologies aim to improve efficiency of a number of process functions. For example, casing drilling allows simplifying the hole structure and lifting-lowering operations, reduces drilling time and energy input and can shorten assembling period of drilling rig [5, 6]. Underbalanced drilling essentially enhances quality of formation exposing [7].

Table 3. Hole-making processes, efficiency criteria and improvement methods.

Hole-making process	Efficiency criterion	Efficiency improvement methods		
		1	2	3
Bottomhole breakage	Energy input	Reduction in hole diameter and hole structure simplification	Increase in drilling rate (new models of rock-breaking tools, optimization of drilling modes)	Reduction in energy input (use of bottomhole motors, rock softeners, lubrications)
Drilling trajectory control for optimal drilling-in	Accurate drilling positioning	Adjustable bottomhole assemblies	Bottomhole telemetric systems	Improvement of communication between bottomhole and ground surface assembly
	Drilling adjustment by real conditions	Advanced acoustic sounding in drilling (radar)		
Hole cleaning, removal of drilling cuttings and formation exposing	Hole and reservoir contact area	Cluster drilling	Horizontal drilling	Multihole drilling, including horizontal and branched drilling
	Preservation of natural reservoir permeability	Balanced or underbalanced drilling	High-quality removal of solid particles from fluid	Polymeric fluids, oil-base or salt-base fluids. Foam systems
	Improvement of reservoir permeability	Impulse wave treatment of oil and gas reservoirs in the course of exposing	Hydraulic sources	Hydraulic fracturing after exposing
Hole strengthening	Efficiency of formation segregation and hydraulic resistances in the reservoir–hole system	Selective injection tools (selective shut-off tools)	Cementing process improvement (lip-type, selective, step-wise)	Extension pipes with cement-inflatable packers
	Decrease in material consumption (pipes, cement)	Mono-diameter drilling Expandable pipes	Cementing control and adjustment stations	Casing drilling

Table 3. Continued

Lowering-lifting operations	Reduction in energy input by: —decreasing the number of lowering/lifting of pipes	Increase of rock-breaking tool endurance	Casing while drilling—decrease in the number of casing strings	Bottomhole orienting devices
	—decreasing the labor content of lowering/lifting	Reduction of the number of operations in pipe extension by using overhead drive	Mechanization and automation	Continuous flexible casing string
Sampling б	Energy cost and quality of sampling	Detachable core barrels	Investigation of properties of host rocks and reservoir during drilling	
Equipment assembling/disassembling and transportation	Time cutting and energy cost reduction	Clustered and multihole drilling	Diminution of drilling equipment weight	Modular and self-propelled equipment

Progression in each of the listed spheres requires accelerated development of the related tools. The known engineering solutions of drilling facilities for the traditional drilling technologies are described in the corresponding literature [3]. In particular, wash rotary rotary drilling needs a rotary head, mud circulation system and a lifting/lowering device. The up-to-date requirements imposed on drilling tools to reach higher efficiency of the drilling process are listed in Table 4. Selecting the ways of synthesizing design solutions should be based on the production data of drilling [8]. According to [9], for the assessment of drilling rig efficiency in the specific geological, technological and climatic conditions, it is most convenient to use the method of quantitative estimation based on the expert's appraisal and determination of an integral efficiency criterion. On the other hand, the most significant criterion of drilling assembly efficiency is the quality of control and the controllability as the index of the equipment ergonomics [10].

It is recommended to synthesize design solutions following the steps below [11]:

1. Based on the functional and diagnostic analysis results, identify two functions, essential for the engineering system, featuring inadequate implementation;
2. Formulate a synthesis problem: How to implement the two identified functions using a single functional bearer?
3. Formulate a wanted bearer (image);
4. Function-oriented search for solutions;
5. If necessary, solve additional problem on adaptation of the obtained solution.

Let as discuss synthesis of a design solution for a gear of drilling facilities. We take two most essential functions: hole-making and lowering/lifting. A gear capable to participate in the both functions is the overhead drive. The key objective of a new design synthesis for an overhead drive is improvement of its performance quality based on the multipurpose usefulness criterion. In particular, an overhead drive can be quipped with tools for operation with casing pipes for the purpose of casing drilling. In inclined hole drilling in bitumen or for coal mine drainage, overhead drives are readily connected with the rack feed systems and allows drilling with the additional axial load. In principle, overhead drives in the modern drilling facilities allows efficient implementation of up to 20 drilling operations [3], and functional capabilities of this gear will enjoy increasing demand as new technologies are developed.

Table 4. New standards placed by drilling technologies on design and parameters of drilling equipment.

Drilling process	Task	Development trend (task solution)	Design requirements
Hole-making: bottomhole breakage and cleaning	Horizontal drilling;	Increase in drilling rate; Increase in rock-breaking tool life by means of optimization of drilling modes	Equipment of drilling facilities with a computerized station for geological exploration and production research of hole-making process; improvement of control over rotary feeding mechanism and mud system
	Cluster drilling	Breakdown rate reduction, drilling path precision control	Overhead drive application
	Multihole drilling, including horizontal branched drilling;	Downhole propeller engine application	Increase in capacity and limit pressure of mud pumps. Engineering of a standard-size line of triplex pumps having a wide process control range
		Bottomhole telemetry system application	Equipment of drilling facilities with gland seal swivels with lubrication pumps for cable communication with bottomhole telemetry system
	Increased drilling depth	Equipment assembling and transport reduction	Cluster drilling facilities; Mobile facilities
	Increased horizontal branch length	Increase in capacity of drilling rigs and rotating units (rotor or overhead power drive). Application of light-alloy drill pipes	Facilities with a capacity of 500–1000 t
		Quality removal of solid phase from mud fluid	Application of multistage mud systems
Hole strengthening	Hole design simplification	Casing drilling	Application of overhead drive and special pipe extension system. Tower height reduction. Equipment of drilling facilities with attachments for lowering/lifting integral drilling tools and with a pipe pushing system
	Reinforcement of horizontal hole portions with extension pipes		Equipment of drilling facilities with a computerized station for for geological exploration and production research of hole-making process

Table 4. Continued.

Downhole surveying	Analysis of properties of rocks and minerals during hole-making	Application of geophysical equipment (bottomhole telemetry system) with the downhole tool (Logging While Drilling, Measurements While Drilling)	Equipment of drilling facilities with a computerized station for for geological exploration and production research of hole-making process. Lowering/lifting hardware should ensure lowering/lifting rate from 0.05 m/s
	Sampling (coring)	Application of detachable core barrels	Equipment of drilling facilities with auxiliary winch for detachable core barrels
Formation exposing	Preservation of natural buttonhole zone permeability	Balanced or underbalanced drilling	Equipment of drilling facilities with rotary preventer, blowout detection system and overhead drive, and with a system of continuous mud circulation during pipe extension, if possible. Application of pipe pushing system
	Enhancement of bottomohole zone permeability	Impulse wave treatment of oil and gas formations while exposing	Application of bottomhole pulse emitters
Lowering/ lifting operations	Lowering/liftin g reduction	Increase in the travel length, hole design simplification, casing drilling	Equipment of drilling facilities with a computerized station for for geological exploration and production research of hole-making process, improvement of control over rotary feeding mechanism and mud system. Application of overhead drive
	Reduction in operations in lowering/lifting	Application of continuous flexible pipe string Cutting time of pipe extension while drilling, reduction in operations of pipe extension by means of using overhead drive	Equipment for lowering/lifting flexible pipe strung, sealing of hole mouth, formation treatment, lowering/lifting with casing pipes. Application of overhead drive and lowering/lifting mechanization system
	Enhancement of labor safety and efficiency	Full-scale mechanization and automation of lowering/lifting	Mechanization and lowering/lifting control system to ensure coherent and safe performance of machinery
Branching	Ratholing from operating wells to maintain production well stock	Application of flexible pipe string Application of overhead drive	Engineering mobile facilities for hole drilling and repair, including overhead drive

One of the objectives of synthesizing design solutions is overcoming discrepancies between different operations when implemented. It is known that the increase in drilling capacity by means of boosting capacity of drilling tools, expansion of tower, mechanization and automation of operations and improvement of working environment results in the growth of weight and size of the drilling facilities, which complicates transportability and extends time of assembling of the machinery. The integrated solution of the problem connected with the enhancement of the efficiency of drilling facilities is possible by way of increasing the life of the rock-breaking tool and using casing drilling technique, which will allow reduction in tower height and shortening time of drilling facilities transportation and assembling. The use of the overhead drives will enable applying high-efficient directional horizontal and inclined drilling methods in mineral mining, as well as make it possible to reduce accident rate and lowering/lifting time and labor content.

References

[1] Hubka V and Eder W Ernst 1988 *Theory of Technical Systems* Springer-Verlag Berlin Heidelberg
[2] Propsects for foreign mobile drilling units with hydroperforators in open-pit mining *Gornaya Promyshlennpost* 1998 No 1 pp 36–40
[3] Porozhsky KP (Ed) 2013 *Driling Facilities* Yekaterinburtg: UGGU
[4] Porozhsky KP and Epshtein VE 2012 Key objectives and ways of imporvement of oil and gas well drilling equipment *Proc. X Int. Conf.: Production Facilitites in Mining and Petroleum Industry. Kubachek's Memory Lectures* pp 233–235
[5] Fontenon K and Strickler Molina 2010 Casing while drilling "smear effect" imporves wellbore stability *World Oil* Vol 227 No 3
[6] Andreev NL 2010 Casing drilling by intervals in permafrost *Nauka Tekhnika Gas. Prom.* No 4 pp 15–18
[7] Tagirov KM and Nifantov VI 2003 *Underbalanced Oil Drilling and Exposing* Moscow: Nedra-Biznescentr
[8] Polovinkin AI 1988 *Engineering Art Elements: University Tutorial* Moscow: Mashinostroenie
[9] Krokhmal NI 1991 Efficient installation configuration for drilling facilities based on quantification of its performance *Cand. Tech. Sci. Dissertation* Moscow
[10] Porozhsky KP 2010 Ergonomics as an intergated criterion of drilling equipment performance *Upravl. Kach. Neftegaz. Komplekse* No 2 pp 12–15
[11] Feigenson NB 2008 Iporvement of tools for analysis and synthesis of the third-stage evolution engineering systems *TRIZ Master Dissertation* Saint Petersburg

IOP Conf. Series: Earth and Environmental Science **134** (2018) 012047 doi:10.1088/1755-1315/134/1/012047

Research into robotic automation of drilling equipment by the Institute of Mining, UB RAS

AS Regotunov* and RI Sukhov

Institute of Mining, Ural Branch, Russian Academy of Sciences, Yekaterinburg, Russia

E-mail: *pochta8400@inbox.ru

Abstract. The article discusses the issues connected with the development of instrumentation for the express-determination of strength characteristics of rocks during blasthole drilling in open pit mines. The trial results of the instrumentation are reported in terms of the drilling rate–energy content interrelation determined in the analyses of experimental drilling block data and by the digital model of rock distribution in depth versus drilling complexity index.

The main modern trends in improvement of open pit mining processes for the period to 2025–30 include partial robotic automation and remote control. Regarding drilling of blastholes, the improvements can be achieved by solving a key problem connected with automation of drilling control, based on express-analysis of physical properties of rocks.

The Institute of Mining, UB RAS, is engaged in R&D on design of the instrumentation for the express-information on strength characteristics of rocks [1].

The instrumentation is based on the theory that the torque generated by the drill rig rotator motor is proportional in value to the strength and resistance of rocks to be overcome by the rotator gear. On this basis, the rock strength criterion can be the value of the rotator motor torque and the power consumed by removal of drill cuttings from hole.

Thus, keeping the data on rock mass strength current is based on the knowledge of power spent by the drill rig motors in drilling.

Designed jointly by the Institute of Mining and Avtomatika Science and Production Association), a dedicated set of instrumentation is placed on an operator's board. During drilling, special sensors record electrical parameters of drill rig motors. Electrical signals of the sensors are converted into a digital model of rock mass areas having different strength characteristics.

The instrumentation set has been trialed in blasthole drilling by drill rig SBSH-250MNA-32 in open pit mines of Uralasbest. As a result, the instrumentation has offered a huge data base on energy content of drilling in rocks of different petrographic composition.

In particular, according to the analysis of data on drilling block 1, energy consumed per 1 m-long drilling is 3–4 kW·h, energy consumed by removal of drill cuttings from the hole is to 0.5 kW·h and auxiliary operations consume round 1 kW·h of energy. The fluctuations in the energy content are governed by the external influence, including faulting and facial nonuniformity. At the same time, the data interpretation shows that the block compose of rocks having the same average strength (category VII–XII with drilling energy content to 2 kW·h/m) has a pocket composed of stronger rocks (diorite, drilling energy content higher than 2 kW·h/m). The presence of this pocket has been proved after blasting when the pocket appears undamaged (Figure 1).

Figure 1. Block 1 after blasting with an unshattered diorite pocket.

Furthermore, the trial results allow finding comparatively tight correlation between the drilling energy content *e* and the drilling rate *v* for rock of different strength (Figure 2). The correlation can be approximated as:

$$v = 15.6e^{0.255}.$$

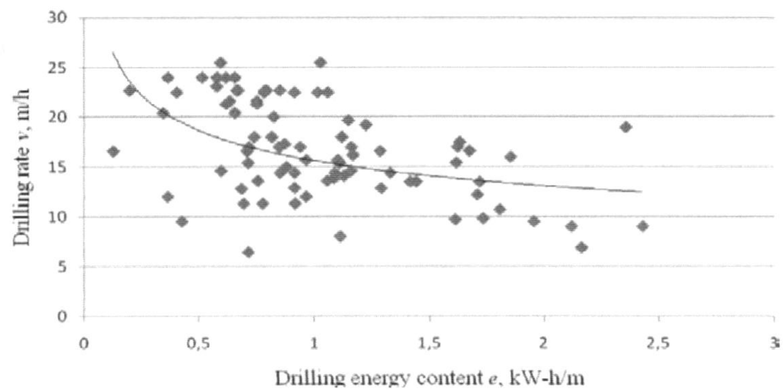

Figure 2. Drilling energy content–drilling rate curve.

It is seen in Figure 2 that the energy content of drilling lowers with the higher drilling rate. Thus, this is where the practical significance of the found correlation lies: in drilling different petrographic varieties using the same drill bit, it is possible to cut-down energy consumption by 1.5–2.0 times through the change of the drilling rate.

By expressing rock strength in terms of a drilling complexity index Π_6, based on the data on energy consumption, it is possible to visualize distribution of strength characteristics in rock mass by constructing a plane model (Figure 3).

With such model and based on the data on power consumed by the drill rig motors, a rock mass can be zoned by categories of drillability and blastability, which means that it is possible to adjust drilling mode and blasting design for each separate hole.

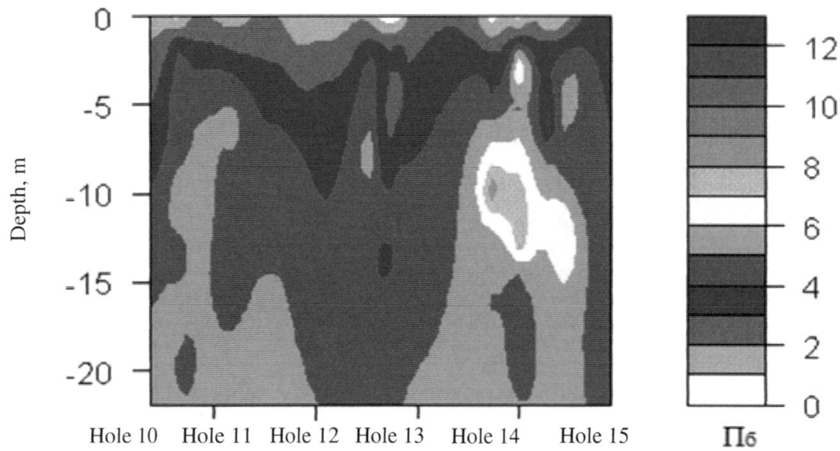

Figure 3. Model of rock strength distribution per depth by the drilling complexity index Π_6 by Rzhevsky.

Conclusions

1. Availability of promptly refined and adjusted data on strength characteristics of rocks allows applying energy saving modes of drilling, efficient size drilling equipment and optimized drilling pattern.

2. The designed software/hardware for the on-line determination of varying strength characteristics in rocks allows optimization of blast designs as early as the project stage, improvement of fragmentation quality and cutting down the drilling and blasting cost 1.2–1.3 times.

Acknowledgments

The paper has been prepared based on the fundamental research findings and has been supported by the Ural Branch of the Russian Academy of Sciences, Project No 15-11-5-7.

References

[1] Regotunov AS and Antonov VA 2015 Regressive modeling of experimental crystalline rocks break-down by rock burst *Problemy nedropolzovaniya* No 2 pp 3745 (in Russian)

Down-the-hole hammering machines for blasthole drilling

AA Repin* and SE Alekseev

Chinakal Institute of Mining, Siberian Branch, Russian Academy of Sciences, Novosibirsk, Russia

E-mail: *repin@misd.ru

Abstract.

Downhole pneumatic drill tools developed in Russia are widely used in blasthole drilling [1]. Percussion drilling consumes less energy in breaking medium-hard and very hard rocks, enables saving power of performance and allows using lighter equipment. Such drilling tool are highly productive reliable and simply maintained. Drilling rate is independent of the hole length, which allows long-hole drilling to be carried out in underground mines. It is also possible to drill larger diameter (to 350 mm) holes using lighter equipment. Al that combined with the simplicity and relatively low price make these machines popular both in open pit and underground mining.

A feature of down-the-hole drill hammers is that their length and diameter are strictly limited by the diameter of a hole or the size of an underground opening.

Efficiency of drilling in medium-hard and very hard rocks depends on the value of blow energy A given by:

$$A = p_{av} \cdot F \cdot s_p,$$

where p_{av} is the average pressure in the power stroke chamber before exhaust; F is the effective area of impact piston on the side of the power stroke chamber; Saint-Petersburg is the impact piston travel length determined as the distance between the total power stroke length and the length of the travel between the exhaust start and the blow on a tool.

The blow energy depends on the value of the effective area F.

The hole diameter strictly limits the diameter of the down-the-hole hammer body, it is not a simple task to increase the effective area of the impact piston. In this connection, it seems important to find how effectively cross-section area the hammer housing is utilized for the formation of the effective areas of impact piston. In most domestic and foreign designs, the utilization factor in this case is 80–90% [2].

The Institute of Mining has developed a layout to ensure 100% utilization of a cross-section area for the creation of the effective area F of an impact piston (Figure 1) [3–5].

The down-the-hole hammer structure includes two power stroke chambers and an intermediate idle stroke chamber. The annular head power stoke chamber 1 and the tail power strike chamber 2 are connected via the channel 3. The chambers are operated by the air feed in the annular head chamber and exhaust from the tail chamber. In-between there is the idle stroke chamber 5 connected to the compressed air line. return air from the power stroke chambers is exhausted along the channel 5 in the wall of the cylinder 6 to the atmospheric air chamber 7. The impact piston 8 has three impact stages

IOP Conf. Series: Earth and Environmental Science **134** (2018) 012048 doi:10.1088/1755-1315/134/1/012048

with the diameters D_1, D_2 and D_3. The largest-impact stage diameter D_1 is equal to the cylinder diameter D_c. Compressed air can flow to the power stroke chambers through the clearance fits A and B. when the impact piston moves, compressed air enters the power stroke chamber, which allows alternate motion of the impact piston and blow on the anvil block at the end of each operating cycle.

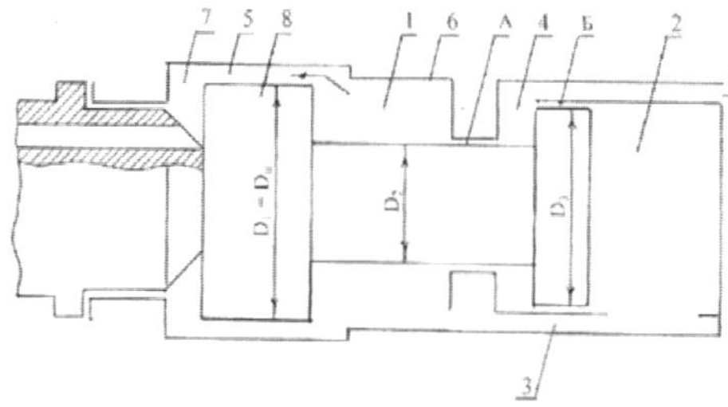

Figure 1. Layout of down-the-hole hammer.

In this layout, the power stroke area S_p is composed of effective areas from the sides of the head and tail power stroke chambers, while the idle stroke area S_i exists only on the side of the compressed air line:

$$S_p = \frac{\pi}{4}(D_1^2 - D_2^2 + D_3^2); \qquad S_i = \frac{\pi}{4}(D_3^2 - D_2^2);$$

It is important that irrespective of idle stroke area from which the force applied to the impact piston constantly resists the force from the side of the power stroke chambers, the condition below is always fulfilled:

$$S_p - S_i = S_c = \frac{\pi}{4}D_c^2,$$

where S_c is the cross section area of the cylinder.

Thus, in the described layout, the power stroke area grows with the idle stroke area.

The analysis of the 50 years-long development in percussion drilling shows that compressed air pressure is continuously increased. The leading foreign companies had long ago passed to the compressed air pressure of 0.9–1.8 MPa [6]. At the present times, the designs of the increased air hammers and bits are being constantly improved and the driving pressure is grown. For example, Atlas Copco COP 64 DTH hammer for holes with a diameter of 156–178 mm uses pressure of 1.2–2.5 MPa.

Technical and economic advantages of increased pressures are: higher drilling rate and extended life of DTH hammers (drilling rate grows from 90–150 mm/min to 800–1000 mm/min in hard granite, while life grows from 300–500 drilling meter to 5000 m and more); possibility to essentially increase drilling depth; reduction in energy content of rock destruction. On the whole, cost per drilling meter is decreased too. High-capacity DTH hammers successively compete with the other drilling equipment for instance, auger drill rig capacity connected with the axial pressure on the too and, consequently, with the weight of the rig, has nearly come to an end and fails to ensure productivity of DTH hammers.

The application of the high-pressure compressed air opens wide horizons for structural improvement of DTH hammers, engineering of modern automated drill rigs and using wider range of exchangeable drill tools for the complete compliance with the technology purposes.

Structural features of the machines sources by compressed air of increased pressure are: closed type, i.e. perforation-free housing; exhaust of total air to bottomhole; valve-less air distribution, as a rule, which allows utilizing energy of expansion of compressed air at much higher potential when the compressed air pressure is high. Efficiency of such machines is highly essential.

Foreign-manufacture increased pressure drill rigs enjoy increasingly higher application in Russia. Atlas Copco delivers drilling rigs ROC L6, ROC L7, ROC L8, ROC L9 and Mustang A-32 CNS with DTH hammers COP 34, COP 54, COP 54 and ODEX system for casing while drilling. However, it should be mentioned that the imported drill hammers and bits are expensive and the price tends to grow higher. It is of the current concern to design domestic import-substituting machinery.

To this effect and using the described layout, DTH hammer model PV170 was designed and tested on a laboratory and commercial scale. Test drilling was carried out using SWDB165 drill rig of Swedish design and Chinese manufacture, at the compressed air pressure of 1.38 MPa. The drill rig is a property of Novosibirskvzryvprom. The tests were conducted at Borok quarry near Novosibirsk, in drilling granite with the strength of 120–140 MPa (Figure 2). The DTH hammer demonstrated high efficiency.

Later on, the drill rig and DTH hammer were tested at Koen quarry in Toguchin district of the Novosibirsk Region, in drilling diabase with the strength of 140–160 MPa (Figure 3).

The next test stage was operation of PV170M DTH hammers with Swedish drill rigs CM760D at the compressed air pressure of 1.6 MPa and ROC L8 at the compressed air pressure of 2.4 MPa by Sibir Mining and Technology Company (Figure 4 and 5, respectively). The operation was performed at Iskitim open pit mine (Novosibirsk Region) in drilling marmorized limestone with the strength of 110–130 MPa.

Figure 2. DTH hammer PV170 and drill rig SWDB165 at Borok quarry, Novosibirsk.

Figure 3 DTH hammer PV170 and drill rig SWDB165 at Koen quarry.

Figure 4. Tests with drill rig CM760D at Iskitim open pit mine, Novosibirsk Region.

Figure 5. Tests with drill rig ROC L8 at Iskitim open pit mine, Novosibirsk Region.

All in all, 22 holes with the diameter of 170 mm and 16.5 m long were drilled. The drilling rate was 400 mm/min. At the Koen quarry, 13 holes 13 m in length were drilled at the compressed air pressure of 1.2 MPa and drilling rate of 266 mm/min. During the testes, total drilling length made 645 m at the sustained efficiency of DTH hammers.

The modern downhole drill hammers driven by the compressed air of increased pressures should possess essential life. Some DTH hammers of foreign manufacture gave life to 5000 meter and more.

Durable operation of a machine, alongside with strength of basic parts, is governed by wear of friction parts. In air-driven percussion machines, impact piston is a part of the air distribution system. This movable part has seats with clearances the sizes of which depend on the diameter. It is

IOP Conf. Series: Earth and Environmental Science **134** (2018) 012048 doi:10.1088/1755-1315/134/1/012048

unavoidable that compressed air leaks in the clearances. In the described layout, compressed air leaks in the clearances A and B (Figure 1). The influence of the enlarged clearances under wear of contact surface on the performance of DTH hammer was studied.

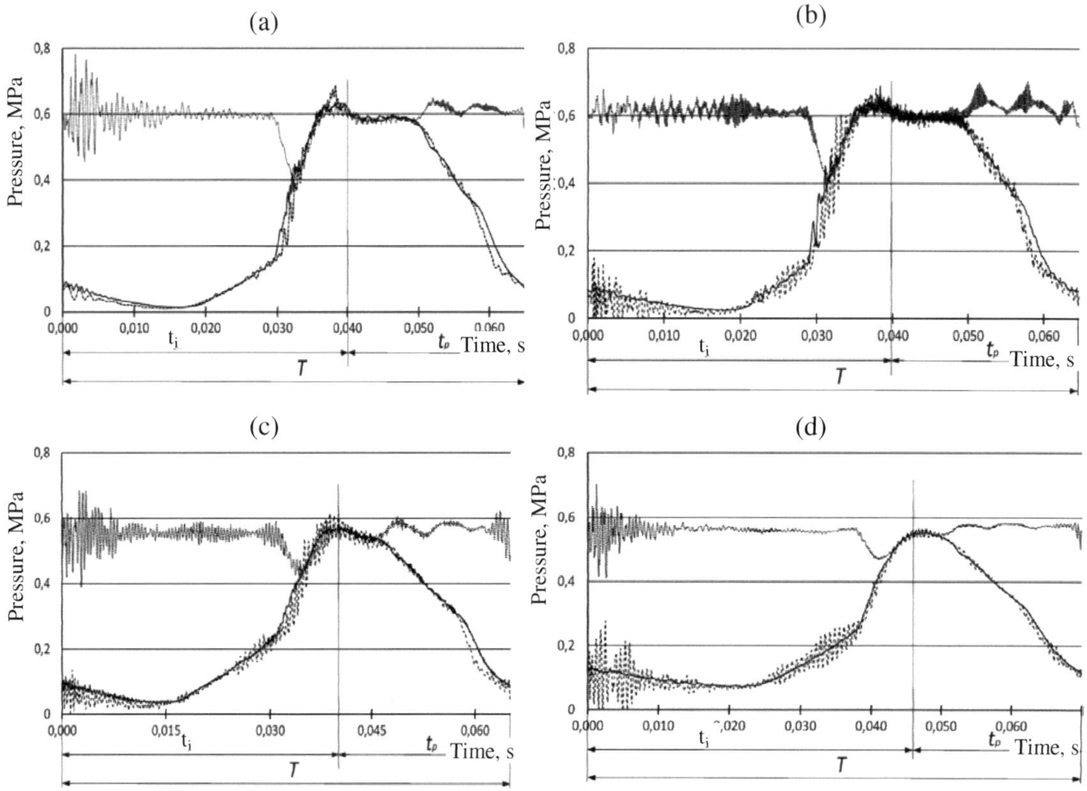

Figure 6. Pressure charts on power stroke and idle stroke chambers: (a) S_{AB} = 0.587 cm^2; (b) S_{AB} = 0.747 cm^2; (c) S_{AB} =1.207 cm^2; (d) S_{AB} = 1.527 cm^2; ············—pressure p_i in the idle stroke chamber, MPa; -----------—pressure p_{hp} in the head power stroke chamber, MPa; ————pressure p_{tp} in the tail power stroke chamber, MPa; t_i—idle stroke time, s; t_p—power stroke time, s; T—cycle duration, s.

The studies were carried out on vertical laboratory test bench GD-251 with the recording of pressure charts in the two power stroke chambers and idle stroke chamber (Figure 6). During the tests, the total cross-section area of the clearances was varied as S_{AB} = 0.587; 0.747; 1.207 and 1.527 cm^2. The compressed-air pressure was 0.6 MPa. The beginning and end of the cycle were determined by a blow marker.

Using the pressure charts, the basic characteristics of the hammer were found (Table1).

Table 1. Basic characteristics of DTH hammer model PV170 at varied total cross-section area of clearances.

Clearances cross-section area S_{AB}, cm^2	Blow energy A, J	Blow frequency n, min^{-1}	Recoil coefficient k_o	Blow velocity v, m/s	Impact piston travel L, mm	t_i/t_p
0.587	208.84	922.37	0.21	4.95	82	1.45
0.747	206.95	928.07	0.26	4.92	84	1.47
1.207	160.09	929.04	0.29	4.32	73	1.61
1.527	159.55	857.14	0.29	4.33	71	2.03

The change in the blow energy A and blow frequency n as function of S_{AB} is shown in Figure 7.

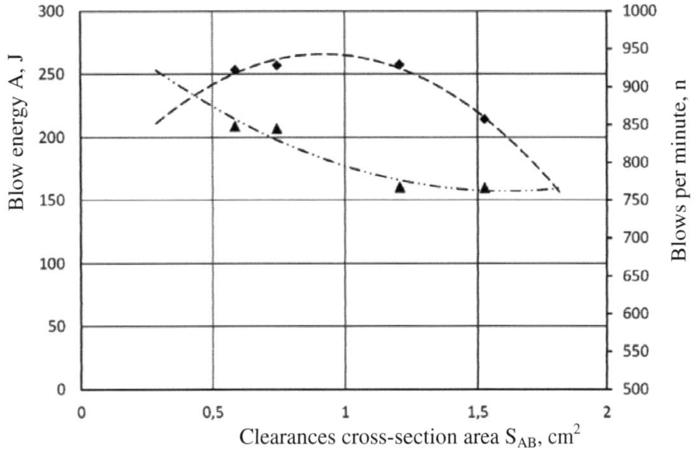

Figure 7. Curves of blow energy, blow frequency and clearances cross-section area in DTH hammer: ▲—unit blow energy, J; ♦—blow frequency per minute.

Evidently, the value of the total cross-section area of clearances has an essential influence both on the unit blow energy and frequency. These characteristics directly govern performance of an air-driven hammer. The cross-section area of clearances depends on the correct attachment of movable joints, quality of manufacture and, largely, on the wear of friction parts. In order to reduce the value of the total cross-section area of clearances, it is recommended to make grooves on adjacent surface, which decreases wear of the surfaces and air leakage. It is also possible to apply different coatings and sealers of immovable joints of the structure.

The proposed layout is suitable for the import-substituting manufacture of DTH hammers with wide pressure range of compressed air and high performance. The simple form of the impact piston without inside cavities, as well as the rational manufacture technology provide the required strength of the part under high impact loading. The influence of the growth of clearances under wear of movable parts is evaluated, which allows more accurate determination of service life of the structure.

References

[1] Ivanov KI, Glazunov VN and Nadion MF 1963 *Modern Hard Rock Drilling Techniques* Moscow: Gos. Izd. Lit. Gorn. Delu (in Russian)

[2] Abramenkov ED and Abramenkov DE 1993 *Pneumatic Mechanisms of Percussion Machines* Novosibirsk: NGU (in Russian)

[3] Alekseev SE RF Patent No 2090730 *Byull. Izobret.* 1997 No 26

[4] Repin AA, Alekseev SE ad Pyatnin GA RF Patent No 2343266 *Byull. Izobret.* 2009 No 1

[5] Repin AA, Alekseev SE and Karpov VN RF Useful Model No 121854 *Byull. Izobret.* 2012 No 31

[6] Gubenkov EK 1979 Drilling in permafrost *Metrostroi* No 7

IOP Conf. Series: Earth and Environmental Science **134** (2018) 012049 doi:10.1088/1755-1315/134/1/012049

Hydraulic fracturing stress measurement in underground salt rock mines at Upper Kama Deposit

EV Rubtsova* and AA Skulkin

Chinakal Institute of Mining, Siberian Branch, Russian Academy of Sciences, Novosibirsk, Russia

E-mail: *rubth@misd.ru

Abstract. The paper reports the experimental results on hydraulic fracturing in-situ stress measurements in potash mines of Uralkali. The selected HF procedure, as well as locations and designs of measuring points are substantiated. From the evidence of 78 HF stress measurement tests at eight measuring points, it has been found that the in-situ stress field is nonequicomponent, with the vertical stresses having value close to the estimates obtained with respect to the overlying rock weight while the horizontal stresses exceed the gravity stresses by 2–3 times.

Accident-free operation of Upper Kama potassium mine is impossible without prompt information support on stress–strain state of the salt rock mass. One of a few techniques applicable to the experimental determination of in site stresses deep in the salt rock mass is hydraulic fracturing. Capacities of the other known approaches are limited due to the structural features and nonlinearity of deformation of such geomedia under loading.

In 2015–2016 in underground mines of Uralkali, an experimentation series on determination of actual stress field parameters was undertaken using measurement-and-computation equipment Gidrorazryv (hydraulic fracturing in Russian, engineering design by the Institute of Mining, SB RAS) [1]. The first stage experiments showed that it was necessary to design a special drilling tool capable to ensure drilling at high geometric accuracy to satisfy hydrofracturing standards such as absence of deviation of drilling or misalignment of hole with jog formation, inadmissibility of jamming or ejection and solidification of clayey bands in holes, etc. to this effect, for the purpose of hydraulic fracturing stress measurement, the drilling tool, drilling assembly, splitter and hole gauge have been designed [2].

An important point of the hydraulic fracturing stress measurement is selection of locations and designs of measurement stations. When selecting the locations, the distance between the test site and the mining front was taken into account, and it was also required that the measurement area was not undermined or overmined and was suitable for placement of drilling equipment. Figure 1 shows the locations of the measurement stations in mines SKRU-1, SKRU-2 and SKRU-3 of Uralkali.

The measurement station design (number of holes and their directions) was first of all connected with the calculation procedure of the actual stresses by the data of hydraulic fracturing tests in a number of wells. Traditionally the hydraulic fracturing stress measurement presupposes testing in three wells: one vertical and two horizontal. Horizontal drilling in accordance with the the hydraulic fracturing stress measurement standards is more laborious that vertical. In Mines SKRU-1 and SKRU-2, horizontal drilling appeared impossible due to the work situation as compressed air was inavailable for cleaning wells from drill cuttings. In this connection, the procedure offered in [3] was used; it allows determining values of vectors of principal stresses in rocks mass in the plane of drilling by the test data obtained in three arbitrarily directed wells so long as the hydrofracture is orthogonal to the well axis.

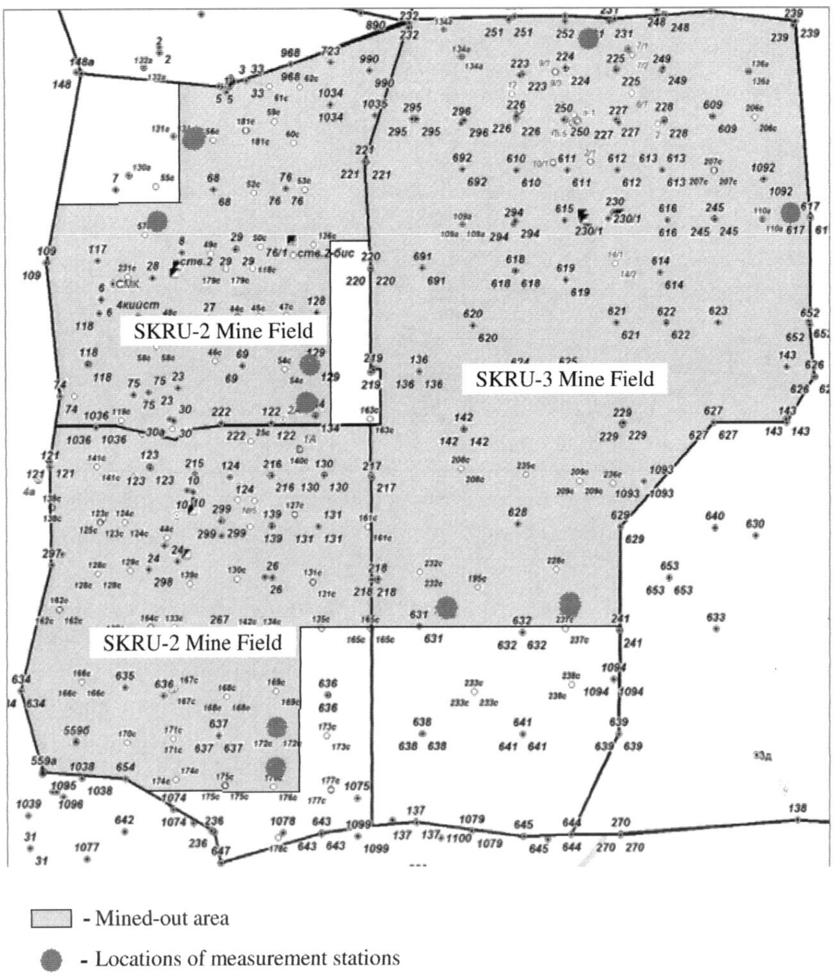

Figure 1. Arrangement of measurement stations in the fields of mines SKRU-1, SKRU-2 and SKRU-3.

In order that the calculation procedure of the actual stresses was applicable, in the proposed design of a measurement station (Figure 2), three differently directed wells were drilled in the orthogonal plane of the axis of an underground excavation: one well was drilled strictly vertical (in line of supposed orientation of one of the principal stresses), and the other two wells was inclined at the angles of 30 and 60 deg to horizon in order to ensure removal of drill cutting from the well. With a view to ensuring a preset direction of the hydrofracture, in the well wall a disk initiation split was made orthogonally to the well axis in the zone between packers using a special device—splitter [2].

During the hydraulic fracturing stress measurement, we recorded: P_c—breakdown pressure; P'_s—shut-in pressure; P_r—reopening pressure; P''_s—shut-in pressure under the second loading; P'_r—reopening pressure under the third loading; P'''_s—shut-in pressure in the third loading. The rupture strength of the rock mass was defined as the difference of P_c and P_r.

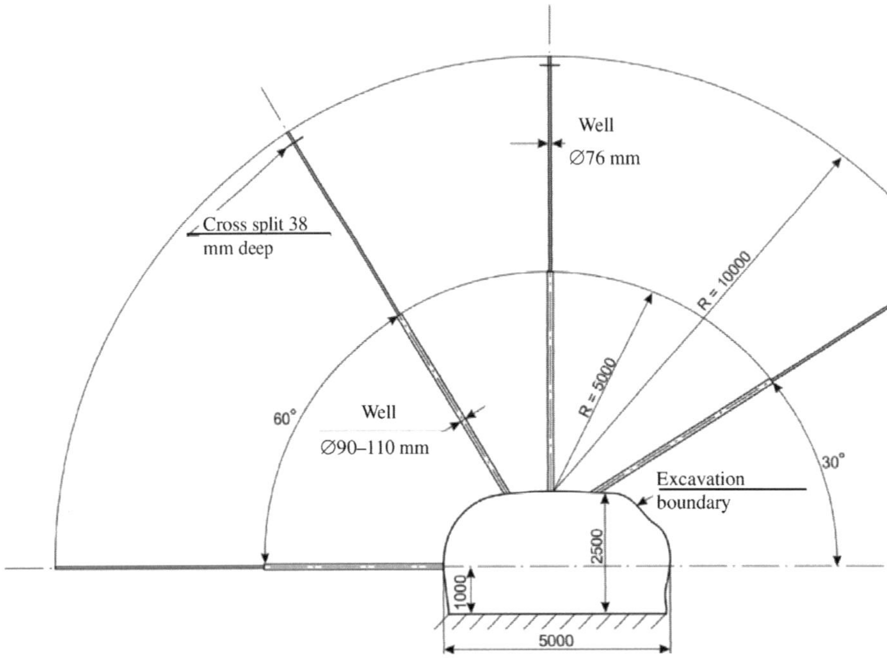

Figure 2. Design of hydraulic fracturing stress measurement station.

The hydraulic fracturing stress measurement is illustrated in this paper y the data from measurement station 8 in SKRU-2 mine. Measurement station 8 was installed in the main haulage way far from the mining front and other excavations, and was not undermined or overmined. The design was as in Figure 2 (one vertical well and two inclined wells). Figure 3 presents the pressure–time diagram plotted during hydraulic fracturing in the well drilled at the angle of 60 deg to horizon at a depth of 10 m from the haulage way boundary.

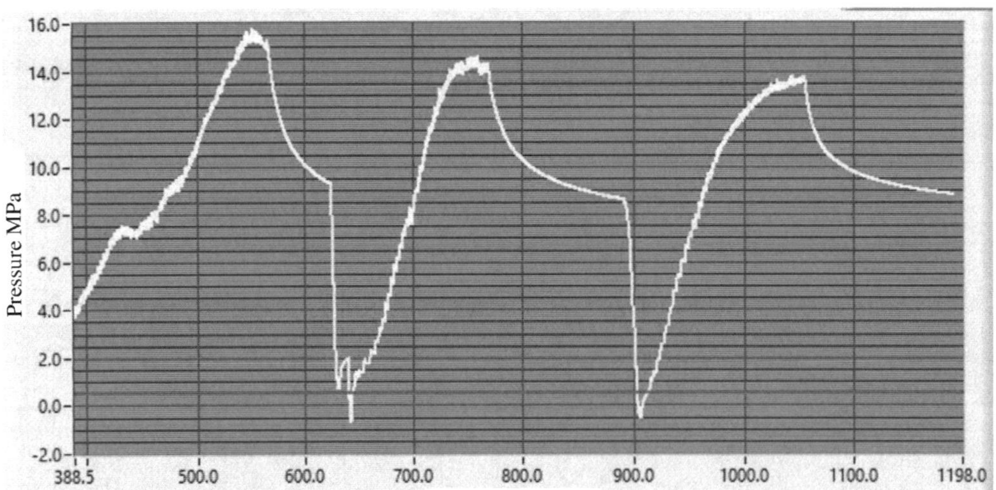

Figure 3. Experimental pressure–time diagram (mine SKRU-2, measurement station 8, well 60°, depth 10 m).

The table 1 below compiles the test data obtained in the three wells of measurement station 8.

Table 1. Hydraulic fracturing stress measurement test log at measurement station 8.

Drilling data		Loading stage 1		Loading stage 2		Loading stage 3		
Well orientation relative to horizon	Distance from the haulage way boundary to the fracture	Breakdown pressure P_c, MPa	Shut-in pressure P'_s, MPa	Reopening pressure P_r, MPa	Shut-in pressure P''_s, MPa	Reopening pressure P'_r, MPa	Shut-in pressure P'''_s, MPa	Rock rupture strength T_c, MPa
90°	10.0	10.8	6.5	12.4	7.4	11.9	7.1	1.6
90°	9.0	10.5	9.0	13.0	10.8	12.5	10.3	2.5
60°	10.0	15.3	11.1	14.3	9.8	13.5	9.7	1.8
30°	7.0	13.1	9.2	10.9	8.6	10.8	8.0	2.3

After the processing of the experimental data obtained at measurement stations 8 and 7 20 m apart in the SKRU-2 mine field, the actual principal stresses in the salt rock mass were evaluated as: the vertical stress $\sigma_h = 6.7–7.2$ MPa, the minor horizontal stress $\sigma_{\min} = 9.6–12.4$ MPa and the major horizontal stress $\sigma_{\max} = 16.4–19.0$ MPa.

All in all, in mines SKRU-1, SKRU-2 and SKRU-3 at the Upper Kama deposit 75 hydraulic fracturing stress measurement tests were performed at 8 measurement stations. It is found that the rock mass is subjected to the action of nonequicomponent stress field where the vertical stress is near the calculated value due to the weight of overlying strata and the horizontal stresses exceed the gravity stress by 2–3 times.

The implemented research series proves applicability of hydraulic fracturing in the experimental determination of stress field parameters at deep levels in salt rock masses.

References

[1] Leontiev AV, Rubtsova EV, Lekontsev YuM et al 2010 Measuring-computing complex Gidrorazryv *J. Min. Sci.* Vol 46 No 1 pp 89–94

[2] Lekontsev YuM, Rubtsova EV and Skulkin AA Drilling tool and a device to make measurement holes in salt rocks *Proc. XIII Int. Conf.: Subsoil Use. Mining. Trends and Technologies of Mineral Exploration and Extraction Geoecology: Interexpo GEO-Sibir-2017* Novosibirsk: SGUGiT Vol 3 pp 129–133 (in Russian)

[3] Rubtsova EV and Skulkin AA 2013 Theoretical framework of hydrofracturing stress measurement *GIAB* No 5 pp 188–191

IOP Conf. Series: Earth and Environmental Science **134** (2018) 012050 doi:10.1088/1755-1315/134/1/012050

Substantiation of artificial composite-structure roof construction in top-down cut-and-fill stoping

GD Rukavishnikov, SA Neverov* and AA Neverov

Chinakal Institute of Mining, Siberian Branch, Russian Academy of Sciences, Novosibirsk, Russia

E-mail: *nsa_nsk@mail.ru

Abstract. It has been found efficient and safe to use top-down cut-and-fill stoping with saving of the artificial backfill cost and material through the use of a solid nonuniform material composed of two components (layers). It is shown that the artificial roof can be advanced after a stope without essential change in the stope condition as compared with the classical top-down slice method, which provides higher safety of mining.

Majority of mines in Norilsk ore province use different versions of top-down cut-and-fill method. Such geotechnologies are generally applied in the conditions of difficult ground, complicated geomechanics and great depths. Alongside key advantages in terms of enhanced safety and minimized ore loss and dilution as well as fully mechanized preparation and stoping, these technologies have some shortcomings such as high material and labor cost of backfilling.

A trend in expanding the application field of the top-down cut-and-fill methods in medium quality ore is an attempt to reduce material and labor inputs in stoping and backfilling operations.

This research is focused on the efficiency of top-down cut-and-fill with the formation of an artificial reinforced layer of increased strength in the bottom of backfill.

The proposed engineering solution consists in caving of high-strong backfill at the above-lying level subsequent to the transition to the lower-lying level and formation of a new artificial composite-structure roof for the lower-lying level by means of mixing the caved backfill and concrete mix with the improved characteristics. This approach ensures reduction in inputs of basic materials used to prepare backfill (by 34%) and, accordingly, saves backfilling expenditures.

The idea of the proposed engineering solution lies in the safe horizontal cutting of ore under the protection of artificial composite roof made by blasting and caving of the above-lying backfilled layer and further impregnation of the caved heap with cement binder. The diagrams of the variants of top-down cut-and-fill are shown in Figure 1.

The geomechanical evaluation of the proposed method involved modeling and computation of stress state in ore body and backfill with the assessment of their stability in the finite element-based solution of 3D problem of elasticity 1–3].

According to the lithology of Talnakh–Oktyabrsky deposit, ore bodies and enclosing rock mass were simulated by hard high-modulus geological materials; the table 1 below describes the physical properties of the simulation materials. The problems assumed that backfill bears a certain load and has cohesion with ore and rocks. Based on that, backfill was modeled by a low-modulus elastic material (table 1) which ensured continuous fill of mined-out void. The mechanical behavior of the reinforced highly strong cemented backfill was determined in terms of the reduced modulus of elasticity.

IOP Conf. Series: Earth and Environmental Science **134** (2018) 012050 doi:10.1088/1755-1315/134/1/012050

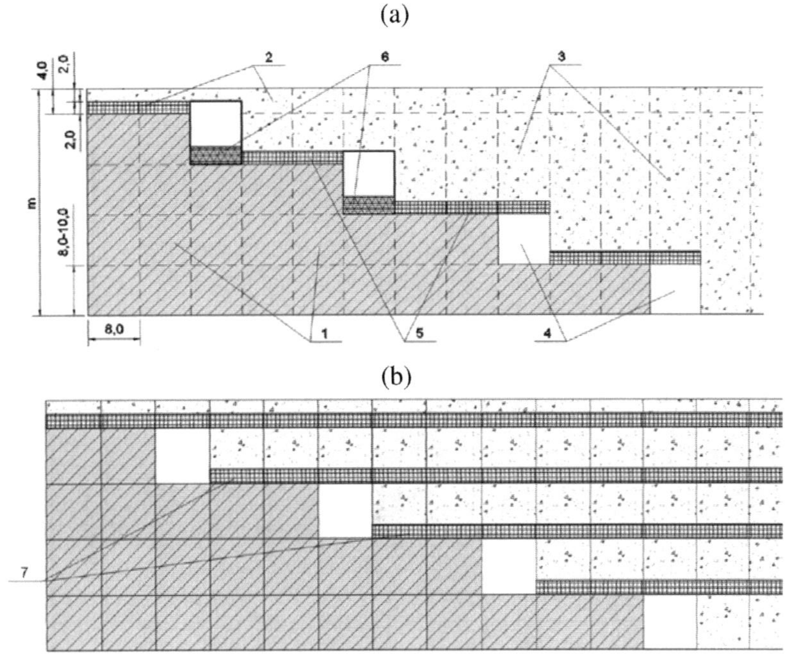

Figure 1. Top-down cut-and-fill (a) with the artificial composite roof and (b) with the formation of artificial backfill roof of different strength: 1—layers; 2—artificial roof under protective cover; 3—backfill; 4—mined-out void; 5—load-bearing composite structure layer; 6—above-lying layer caving for impregnation with the cemented binding mix; 7—current high-strength reinforced backfill layer.

Numerical estimation was carried out for standard conditions of a persistent ore body with a thickness of 36 m at a depth H of 1000 and 1600 m.

Regarding grades of strength, cemented backfill is presented by mix M30 intended for backfilling in the main layers and mix M100 for making the load-bearing reinforced and composite backfills.

Table 1. Physical properties of ore, host rocks and backfill

Description	Young's modulus, MPa	Poisson's ratio	Density, kg/m^3	Cohesion C, MPa	Internal friction angle φ, deg
Host rock mass	60000	0.25	4000	15	35
Ore body	50000	0.22	2700	18	35
Main layer backfill	1500	0.3	1800	1.2	25
Composite structure backfill	2800	0.3	2000	1.7	25
Load-bearing reinforced backfill	3500	0.3	2200	2.0	25

Figures 2–4 allow comparing the numerical modeling results on stress state of ore, enclosing rock mass and backfill and stability of structural elements of the top-down cut-and-fill variants.

IOP Conf. Series: Earth and Environmental Science **134** (2018) 012050 doi:10.1088/1755-1315/134/1/012050

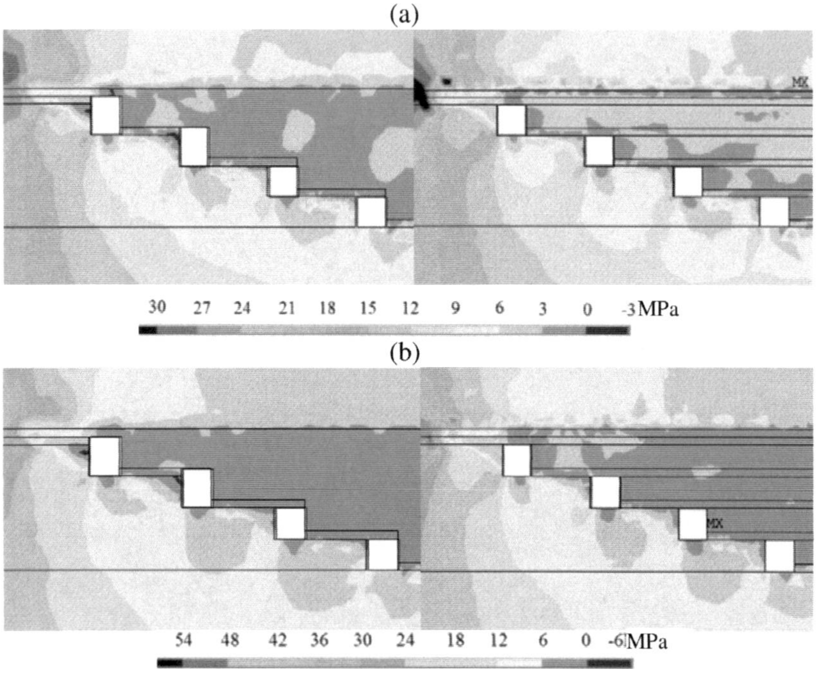

Figure 2. Distribution of minor principal stresses in the top-down cut-and-fill with the artificial composite roof and under different-strength reinforced backfill at the depths of (a) 1000 m and (b) 1600 m.

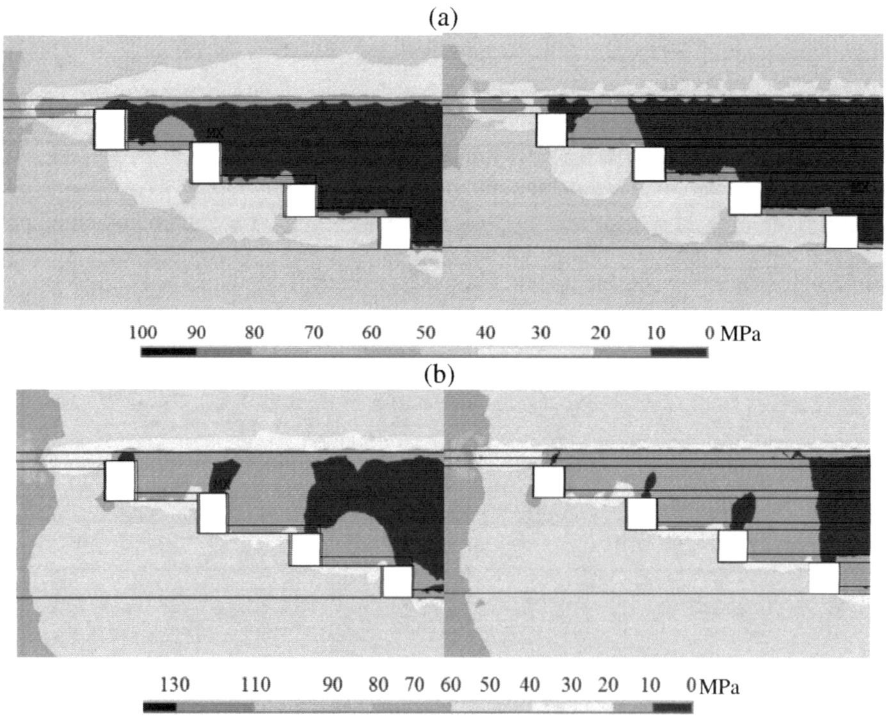

Figure 3. Distribution of major principal stresses in the top-down cut-and-fill with the artificial composite roof and under different-strength reinforced backfill at the depths of (a) 1000 m and (b) 1600 m.

IOP Conf. Series: Earth and Environmental Science **134** (2018) 012050 doi:10.1088/1755-1315/134/1/012050

(a)

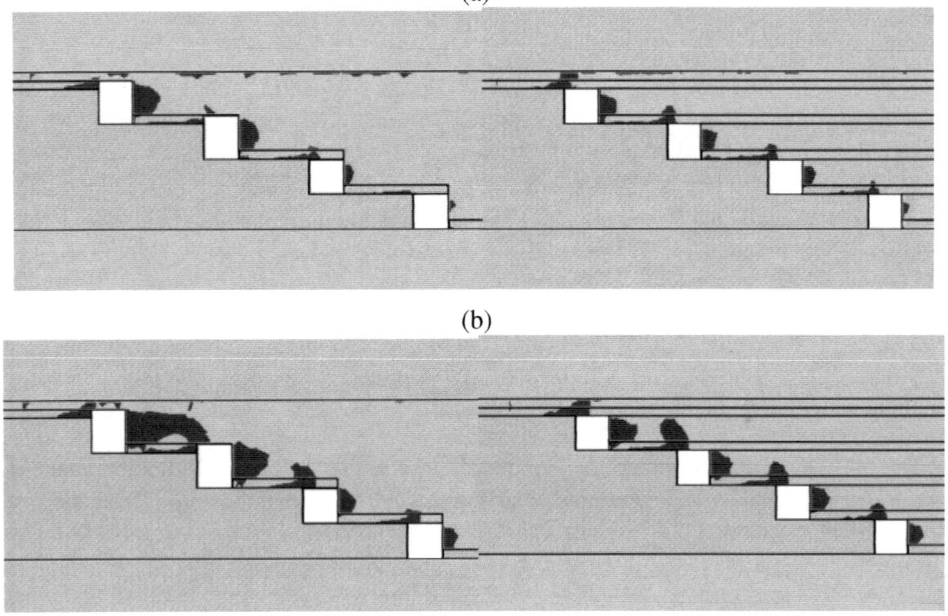

Figure 4. Prediction of probable collapse zones in the structural elements of the top-down cut-and-fill with the artificial composite roof and under different-strength reinforced backfill at the depths of (a) 1000 m and (b) 1600 m.

The results of the calculated stress state for the elements of the variant with the artificial composite structure roof in Figure 2 are indicative of tensile stresses in sidewalls and in roof irrespective of the depth of mining. In the top-down cut-and-fill with the formation of a different-strength backfill, no zones of tensile stresses are observed.

The change in the major compressive stresses in the discussed variants of the cut-and-fill method is reflective of the influence exerted by the backfill structure and the depth of mining on the stress state of the backfill, ore and rock mass (Figure 3).

At the depth of 1000 m, no hazardous concentration of the compressive stress is observed in ore body and enclosing rock mass. In the backfill, at the edges of cutting zone, maximum compressive stresses reach 14–16 MPa, which is much higher than the compression strength of backfill. On the other hand, the backfill will experience triaxial compression, which will favorably contribute to its strength.

The increased in the mining depth from 1000 to 1600 m entails higher major compression in ore body and rock mass by 1.5–1.7 times. The most adverse conditions are in the backfill at the edges of cutting zone, where the concentration zones of compression up to 20–25 MPa arise. In the backfill compressive stresses vary from 6.0–9.0 MPa in the center to 12–14 MPa at the edges of cutting zones. At the edges of the roof made of an artificial composite-structure layer as well as in the reinforced backfill, concentration of maximum compressive stresses higher than 25 MPa is observed.

The increased concentration of major tensile stresses τ_{max} in the ore body, rock mass and backfill is traced in the same zones as the major compression. Thus, stability of ore, rocks and backfill in the discussed process charts of the top-down cut-and-fill method will be governed by the increased compressive and tensile stresses.

Based on the estimates of the stress state in the structural elements of the top-down cut-and-fill system, their stability was assessed using the Mohr–Coulomb criterion [4–5]. The results are depicted as the critical zones of probable failure.

Figure 4 shows the pictures of instability zones (black-colored) in ore body, rock mass and backfill for the variants of the composite roof and the load-bearing reinforced backfill at the depths of 1000 and 1600 m.

It has bee found that given the assumed physical properties of the model media, the areas most susceptible to probable failure in the structural elements of the top-down cut-and-fill method are the walls of the cutting layer from the side of the backfill and the corner zones of the composite-structure roof and the load-bearing reinforced backfill from the side of the ore body. The roof in the cutting layer remains stable owing to the higher rigidity and strength of the composite structure and reinforce backfill. As mining transfers to deeper level of 1600 m, the zones of instability and probable failure expand.

Thus, by the condition of safety, the process chart of the top-down cut-and-fill variant with the artificial roof of composite structure backfill is equal to the classical scenario of different-strength backfilling and can efficiently be used in deep-level mining of gently dipping ore strata. As against the classical variant of artificial roof formation, the proposed engineering solution is advantageous for the ability to ensure the desired integrity and continuity of the load-bearing backfill layer owing to elimination of multi-stage operation.

The scaled-up feasibility study of the top-down cut-and-fill variant with the artificial composite-structure roof as against the classical scenario has shown:

—reduction in material and labor cost of making load-bearing highly strong backfill layer more than by 2 times;

—essential shortage of time of backfilling;

—increased productivity by 15–20%;

—possible saving of backfilling cost by 20–25%.

Conclusion

The authors have evaluated and compared the variant of the top-down cut-and-fill method with the formation of an artificial composite-structure roof and the classical variant with the formation of backfill made of different strength mixes. At the level of project solutions, the efficiency of the proposed approach has been proved in terms of copper ore mining at Oktyabrsky deposit.

References

[1] Zienkiewicz O 1971 *Finite Element Method in Engineering Science* McGraw Hill
[2] Fadeev AB 1987 *Finite Element Method in Geomechanics* Moscow: Nedra (in Russian)
[3] Reiter K and Heidbach O 2014 3-D geomechanical–numerical model of the contemporary crustal stress state in the Alberta Basin (Canada) *Solid Earth* No 5 pp 1123–1149
[4] Baklashov IV, Kartozia BA et al 2004 *Geomechanics: University Textbook* Moscow: MGGU (in Russian)
[5] Kazikayev DM 2005 *Geomechanics of Underground Mining of Ore: University Textbook* Moscow: MGGU (in Russian)

Analysis of dynamic parameters of mine fans

E Yu Russky

Chinakal Institute of Mining Siberian Branch Russian Academy of Science, Novosibirsk, Russia

E-mail:geomining@mail.ru

Abstract. The design of the rotor of an axial fan and its main units, namely double leaf blades impeller and the main shaft are discussed. The parameters of a disturbed mine air flow under sudden outbursts are determined and the influence of disturbances on frequencies of axial fan units is assessed. The scope of the assessment embraces the disturbance effect on the blades and on the torsional vibrations of the main shafts. The dependences of the stresses in the elements of the rotor versus the disturbed air flow parameters are derived.

Service reliability as the prime specification of the main mine fan facilities in most depends on the ultimate safety factor and vibration level of fan modules, which are under permanent influence of airflow disturbances, induced by unsteady aerodynamic forces arising as a consequence of irregular flows in interaction with ribs of the body, blades of guiders and straighteners [1], as well as disturbances induced by sudden outbursts and blasting.

The main module of the mine axial fan is rotor. Its reliability determines fan performance as a whole. Rotor parts are main shaft and an impeller consisting of a body and working blades [2]. Let consider the main mine fan facility of VO–series [2], which impeller has 8 double leaf welded–design blades. The main fan shaft is connected to transmission shaft through coupling; the transmission shaft is coupled kinematically with electromotor shaft. The fan schematic is shown in Figure 1.

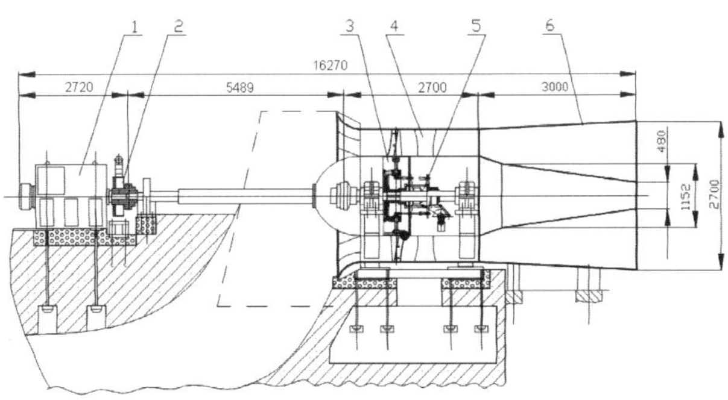

Figure 1.VO fan schematic: 1–synchronous electromotor of 630 kW, 750 rpm, 6.0 kV; 2–electromagnetic brake; 3–impeller with double leaf turning-in-motion blades; 4–unturning blades of a straightening unit; 5–mechanism for concurrent turning of impeller blades; 6–diffuser.

Under consideration is interaction of a disturbed air flow with a double blade of an axial fan.

Figure 2 presents the scheme of a double leaf blade of VO-36K axial fan [2]. Specifications of VO-36K fan are diameter within blade ends D = 3.6 m, static pressure P_{sv} = 2550 Pa, electromotor capacity N= 2000 kW.

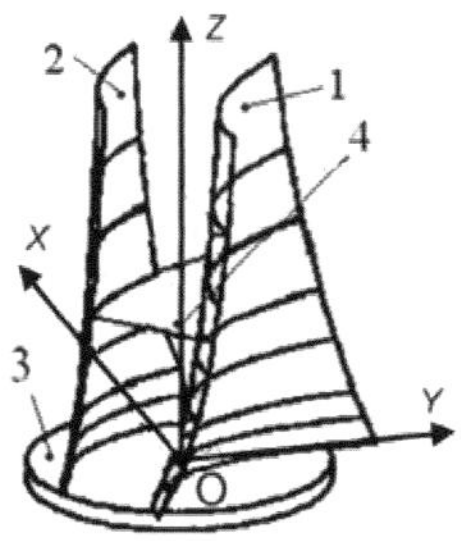

Figure 2. Double leaf blade (schematic): 1–large blade, 2–small blade, 3–rotable base, 4–partition.

Eigen frequencies of a double leaf blade for VO-36K are calculated in Ansys package (Figure 3; (a, b)–bending–type vibrations, (c)–torsion vibrations).

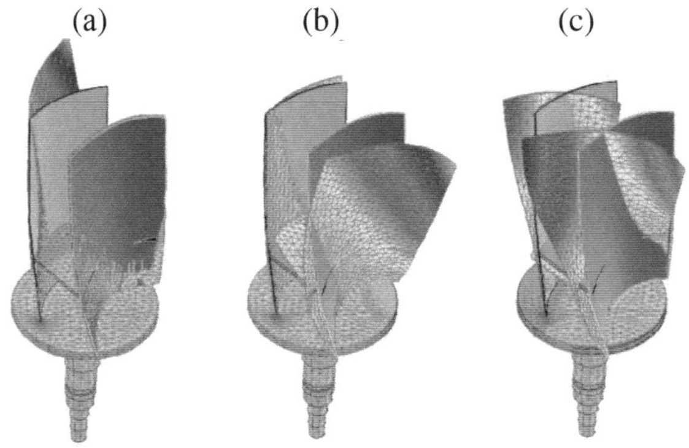

Figure 3. Forms of eigen vibrations of RK double leaf blade: (a) the first vibration form (66.5 Hz frequency), (b) the second vibration form (101.5 Hz frequency), (c) the third vibration form (165.1 Hz frequency).

Eigen frequencies of vibrations are 66.5 Hz in the first vibration form, 101.5 Hz in the second, and 165.1 Hz in the third (Figure 3). The principal exciting frequency is the rotor rotation frequency equal to 10 Hz. Appreciable detuning of eigen frequencies from the principal exciting ones is observed.

Along with the common loads: centrifugal forces, air flow resistance, thrust and inertia of an electromotor the fan blades experience cyclic-in-time disturbing loads which are capable to induce additional dynamic stresses in blades [3]. Coincidence of a disturbing force frequency with one of eigen blade frequencies generates resonance phenomenon, characterized with a substantial increase in vibration amplitudes. Resonance phenomena are also observed when the eigen frequency is not equal, but multiple to the exciting force frequency.

Resonance frequency values for air flow oscillations from ribs of a guider, straightener blades and loss of rotating self-oscillations can be written as [5]:

$$\omega_n^{(P)} = n\,N_P\omega,\quad \omega_n^{(CA)} = n\,N_{CA}\omega,\quad \omega_n^{(VO)} = n\,N_{VO}(1-\alpha)\omega,$$

where $n = 1, 2, 3, \ldots$ is number of exciting force harmonics; α is angle between incoming air flow and the blade profile chord, $0 < \alpha < 1$, rad; $\omega_n^{(P)}$, $\omega_n^{(CA)}$, $\omega_n^{(VO)}$ are resonance frequencies of airflow oscillations from ribs of guider, straightener blades, and loss of rotating self-oscillations, rad/s; ω is angular frequency of rotor rotation, rad/s.

We have to plot a vibration diagram in order to determine the influence of disturbing frequencies on blade vibrations (Figure 4).

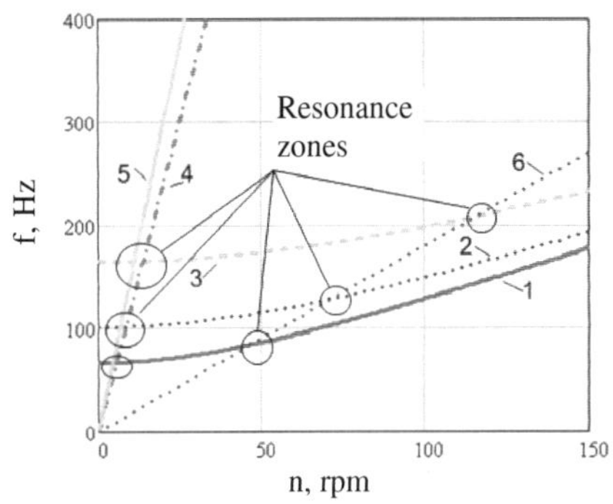

Figure 4. Relationship of eigen frequencies of double leaf blade vibrations and disturbing frequencies vs. the electromotor rotation number: 1–the first eigen frequency; 2–the second eigen frequency; 3–the third eigen frequency; 4–disturbing frequency of guider's blades; 5–disturbing frequency of straightener's blades; 6–disturbing frequency of a vortex separation.

It follows from the analysis of the curves, when a fan is actuated, the blades pass through several resonance zones (Figure 4). This situation does not lead to a rise of dangerous stresses and strains in the blade section structure because of a short time of being in these zones and a small-scale energy of disturbed air flow on the part of ribs of a guider, straightener blades, and vortex separation disturbances. In the case of run-down operation of the fan a brake for the fan electromotor should be provided to cut down time of being in resonance zones.

Calculation and analysis of torsion vibrations of transmission shafts of fan facilities is practiced to evaluate the maximum stresses and torsion angles (amplitudes) of the transmission shaft cross-sections versus angular velocity of electromotor rotor in the periods of run-up and run-down of the fan and also when aerodynamic disturbances, strong in rotation moment, be precise, an explosion or a sudden methane outburst affect a fan facility operation. Structural schematic of a fan for evaluation of torsion vibrations is shown in Figure 5.

The mathematical model of fan rotation is described by a differential equation set for the mechanical system with four degrees of freedom [5].

Calculation results on torsion deformation of VO-36K transmission shaft are reported in Figure 6.

The graph of relationship of tangential stresses in a material of the transmission shaft versus time in the launch of VO-36K fan facility is presented in Figure 7.

The excess of the maximum tangential stresses (τ_{max}) over stresses at nominal frequency (τ_n) is:

$$\frac{\tau_{max}}{\tau_H} = \frac{30{\cdot}10^6 \text{ MPa}}{6.05{\cdot}10^6 \text{ MPa}} \approx 5 .$$

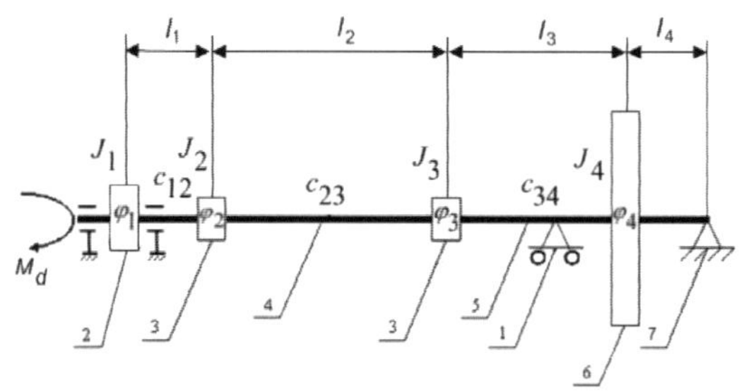

Figure 5. Structural scheme of a fan facility: 1–radial bearing; 2–actuating electromotor; 3–gear couplings; 4–transmission shaft; 5–main rotor shaft; 6–impeller; 7–angular-contract bearing; M_d–torsion moment of electromotor; φ_i–angular coordinate if the i-th cross-section; $c_{i,\ i+1}$–torsion rigidity of a shaft section between the i-th and (i+1)–th cross-section; J_1–inertia moment for electromotor rotor; J_4–inertia moment for fan rotor; J_2, J_3–inertia moments for gear couplings.

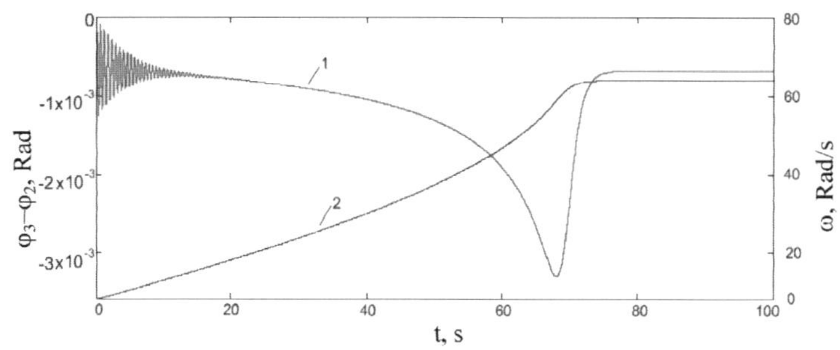

Figure 6. Relationship of torsion deformations of transmission shaft of the rotor (curve 1) and frequency of rotor rotation (curve 2) versus time in start of VO-36K fan facility.

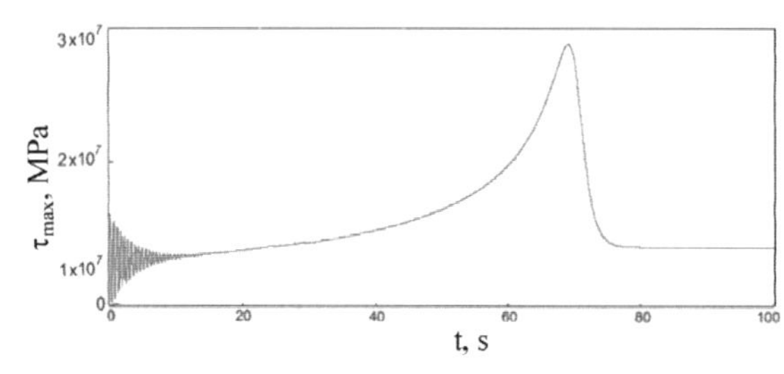

Figure 7. Relationship of tangential stresses in torsion of the transmission shaft of the rotor versus time in the start of VO-36K fan facility.

In a fan run down the resistance moment in an impeller against an air flow is much less than the electromotor moment in start, that is why vibration amplitudes are not significant $3.3 \cdot 10^{-13}$ rad.

Therefore, it is possible not to use a brake for cutting down the run-up time, viz., a time of being in a resonance zone. The brake will be helpful for technological needs, for example, to repair a fan facility.

The experimental evidence of the present research enables to infer:

– service resource of transmission shaft and rotor shaft mainly depends on a number of fan starts and a number of sudden outbursts;

– it is possible not to use the brake to cut down the time of running down, viz., to cut time of being in a resonance zone. The brake is helpful for technological needs, let, fan repair;

– to reduce probability of rotation loss it is preferable to exploit a fan with impeller blade mounting at angles not exceeding the rated parameters.

Considering a short-term start mode (less than15 s) and a small probability of high-power outbursts in the vicinity of a fan facility, the service life of transmission shaft should be determined in terms of a number of the fan starts and expectant sudden outbursts.

References

[1] Russky EYu 2015 Vibrational reliability of rotors of main mine fans *GIAB* No 7 pp 168–174

[2] Klepakov IV and Rudenko VA 1986 Design of a new line of axial main mine fans, in: *Theoretical and Operational Problems of Stationary Mine Installations* Donetsk: Fedorov VNIIGM pp 110–121 (in Russian)

[3] Manushin EA and Surovtsev IG 1990 *Design and Strength Analysis of Turbomachines of Gas-Turbine and Combined Installations* Moscow: Mashinostroenie (in Russian)

[4] Krasyuk AM, Lugin IV, Popov NA and Russky EYu 2015 Substantiation of parameters and estimation of strength for basic structural units of axial tunnel fan *Journal of Mining Science* Vol 51 No 6 pp 1139–1149

[5] Kovchin SA and Sabinin YuA 2000 *Theory of Electric Drive*: university textbook Saint-Petersburg: Energoatomizdat 2000 (in Russian)

[6] PanovkoYaG 1960 *Internal Friction under Vibrations in Elastic Systems* Moscow: Fizmatgiz (in Russian)

[7] Russky EYu 2015 *Strength Analysis of Axial Mine Fans Rotors* Vestn. KuzGTU No 2 pp 31–34

[8] Krasyuk AM and Russky EYu 2013 Analysis of stress state and frequency properties of impellers of axial main mine fans *GIAB* No 8 pp 152–156

IOP Conf. Series: Earth and Environmental Science **134** (2018) 012052 doi:10.1088/1755-1315/134/1/012052

Development of a robotic system of nonstripping pipeline repair by reinforced polymeric compositions

LA Rybalkin*

Chinakal Institute of Mining, Siberian Branch, Russian Academy of Sciences, Novosibirsk, Russia

E-mail: *leonid.rybalkin@gmail.com

Abstract. The article considers the possibility of creating a robotic system for pipeline repair. The pipeline repair is performed due to inner layer formation by special polyurethane compositions reinforced by short glass fiber strands. This approach provides the opportunity to repair pipelines without excavation works and pipe replacement.

A well-developed pipeline network is necessary for the stable and safe extraction of solid minerals by a mining enterprise. In accordance with the purpose mine pipelines are divided into the following categories: fire-irrigation, mine drainage, filling material, compressed air (up to 16 bar), degassing, air conditioning systems.

In mines the pipeline is subjected to a constant negative impact - the mine atmosphere and slightly acidic formation water, oxygen dissolved in the water, and wandering electrical currents, resulting in steel pipes subjected to constant wear, aging and destruction. To replace unsuitable parts of the pipeline and repair the equipment expensive operations are carried out. They are performed to deliver the necessary materials and new sections of steel pipes to the mine field with the subsequent replacement of an unusable pipeline section.

This article presents the idea of a promising technology of pipeline repair by creating an inner layer with special polyurethane compounds reinforced with short fiberglass strands. Unlike traditional means of repair, the proposed approach provides a significant reduction in the cost of work by eliminating the preparation and delivery of pipeline sections. In contrast to relining, cost reduction is achieved through continuous filling of the pipeline interval. There is no need for preliminary blowing of the hose. Inexpensive chemical reagents of Russian and Chinese production are used. Repair can be done directly from the mine field without the necessity to stop the process of mining.

Recently, solutions from polyurethanes are most widely used in comparison with all polymer compositions. These compositions are actively used in underground construction. Also, they are employed in mining in case of high formation pressures and temperatures to strengthen and isolate the wells' walls, to create anti-filtration barriers from formation water and to isolate the walls of production [1–4].

The employed synthetic resin compositions can be one-component or two-component. In the first case, the polymerization of the composition usually occurs, when it interacts with water. In the second case two components are mixed immediately before entering the working interval, where they polymerize, entering into chemical reactions between themselves and water. The advantage of the first approach is the possibility of using relatively simple pumping equipment. The disadvantages are the limited storage time of the compositions, the complexity of the polymerization control, and high

IOP Conf. Series: Earth and Environmental Science **134** (2018) 012052 doi:10.1088/1755-1315/134/1/012052

sensitivity to the water saturation of the medium. The advantages of two-component compositions are their lower cost and dependence on water saturation, the ability to control the polymerization time of the composition by simply changing the concentration of the activator in one of its components.

Implementation of this approach is possible through the development of special technical means (Figure 1).

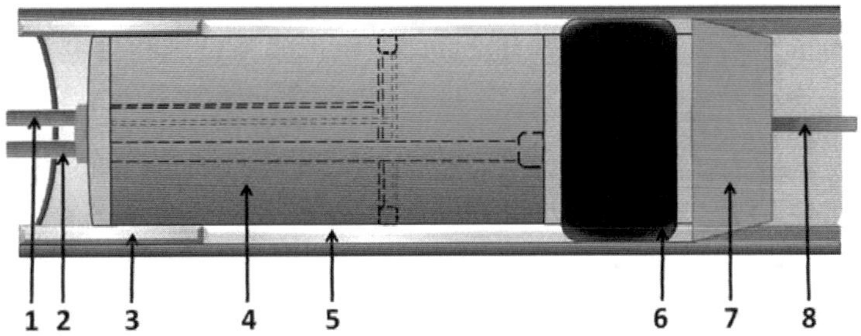

Figure 1. General view of the pipeline's repair device: 1 – two-channel line of the fluid supply, 2 – line of the fluid supply, 3 – polymeric shell, 4 – body, 5 – polymer layer, 6 – hydraulic packer, 7 – preliminary cleaning of the walls, 8 – pneumatic breakdown mechanism

The body 4 must be made of a smooth solid material, for example, a steel pipe in which the holes for formulations' supply through the 2-channel line 1 are made. For a reliable fixation of the body, a rubber coupling or hydraulic packer 6 can be used. The pressure will be controlled by line 2.The transportation of the device along the pipeline can be performed by a pneumatic breakdown mechanism 8, with preliminary cleaning of the walls 7 [5–6]. This device will make it possible to form a polymer layer 5 that solidifies inside pipeline and creates a polymeric shell 3.

The operation of the pipeline interval's repair can be performed according to the following scheme (Figure 2).

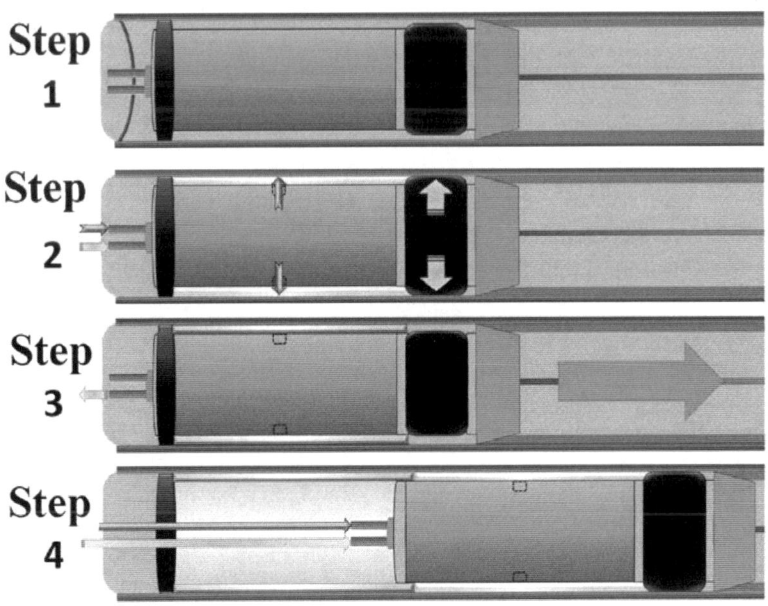

Figure 2. Pipeline repair operation.

Step 1. The beginning of the pipe is sealed with a rubber clutch, preventing the polymer from leakage outwards.
Step 2. The packer is being inflated, thus, sealing the work interval and centering the device in the pipe. The polymer composition is placed in the space between the device and the pipe, the pressure is monitored by a manometer.
Step 3. After the polymer is supplied, the time after which the composition will lose its plastic properties is detected. At this point, the pressure in the packer is released and the device is moved further along the pipeline, so that the body is adjacent to the newly created layer.
Step 4. After the device is moved to the next section of the pipeline, steps 2 and 3 are repeated.

The device suggests separate injection of the components of the composition and their mixing in the interval between the inner wall of the pipeline and the repair device. This feature will reduce the requirements for the used pumping equipment, as well as help to avoid problems with the contamination of the operating hydraulic equipment associated with the high reactivity of polyurethanes with water.

The resulting inner layer will be smooth, which will reduce the possibility of pipeline contamination. Reinforcing the composition with fiberglass threads will prevent the destruction of the layer under mechanical loads on the pipe.

Conclusion

The development of the proposed method and the creation of technical means for repairing the pipeline through the creation of an inner layer with special polyurethane compounds reinforced with short fiberglass strands will significantly reduce costs in case of pipeline section's repair and replacement and will extend its service life.

References

[1] Kurlenya MV, Serdyukov SV, Patutin AV and Shilova TV 2017 Development of chemical compositions for impervious screens in rocks *IOP Conf. Ser.: Earth Environ. Sci.* Vol 53 012021
[2] Shilova TV and Serdyukov SV 2015 Protection of operating degassing holes from air inflow from underground excavations *J. Min. Sci.* Vol 51 No 5 pp 1049–1055
[3] Serdyukov SV, Shilova TV and Drobchik AN 2016 Polymeric insulating compositions for impervious screening in rock masses *J. Min. Sci.* Vol 52 No 4 pp 826–833
[4] Kurlenya MV, Serdyukov SV, Shilova TV and Patutin AV 2014 Procedure and equipment for sealing coal bed methane drainage holes by barrier shielding *J. Min. Sci.* Vol 50 No 5 pp 994–1000
[5] Timonin VV and Kondratenko AS 2015 Process and measuring equipment transport in uncased boreholes *J. Min. Sci.* Vol 51 No 5 pp 1056–1061
[6] Kondratenko AS, Petreev AM 2008 Features of the earth core removal from a pipe under combined vibro-impact and static action *J. Min. Sci.* Vol 44 No 6 pp 559–568

IOP Conf. Series: Earth and Environmental Science **134** (2018) 012053 doi:10.1088/1755-1315/134/1/012053

Development of technical means for directional hydraulic fracturing with shearing loading of borehole walls

LA Rybalkin*, AV Patutin and DV Patutin

Chinakal Institute of Mining, Siberian Branch, Russian Academy of Sciences, Novosibirsk, Russia

E-mail: *leonid.rybalkin@gmail.com

Abstract. During the process of mineral deposits' mining one of the most important conditions for safe and economically profitable work of a mining enterprise is obtaining timely information on the stress state of the developed massif. One of the most common methods of remote study of the geomechanical state of the rock massif is hydraulic fracturing of the formation. Directional hydraulic fracturing is a type of the method employed to form cracks across production wells. This technology was most widely used in the gas industry to extract gas from shale formations. In mining, this technology is used to set up filtration screens, to integrate degassing, to soften the hard roof of coal seams. Possible practical appliance is the expansion of the application field of this technology to intensify the production of viscous oil, to leach non-ferrous metals, to create in the rock massif anti-filtration screens for various purposes, as well as to measure stresses acting along the wells.

Known methods of lateral cracks' formation are based on the use of gap initiators created by jet or mechanical slitters. This approach has no alternatives for cased gas and oil wells. However, in uncased wells it is possible to perform hydraulic fracturing without slot initiators.

Chinakal Institute of Mining SB RAS has carried out a number of works [1–6] showing the operability of the approach in creating a crack across the borehole due to additional tangential loading of its walls. Figure 1 shows the schemes of hydraulic fracturing with tangential loading of the well walls, with an anchorage without an indentor.

The fluid pressure (P) pushes the device out of the well with a force equal to $\pi R^2 \cdot P$. This force is transferred to the anchor. The force generates tangential loading on the anchor's contact with the well, which, together with the fluid pressure on the face, makes the model's material stretch along the well axis. In the variant with an indentor this loading (S_i) is concentrated along the exit line formed by the indenter of the edge crack and its value is reaching $PR/2$.

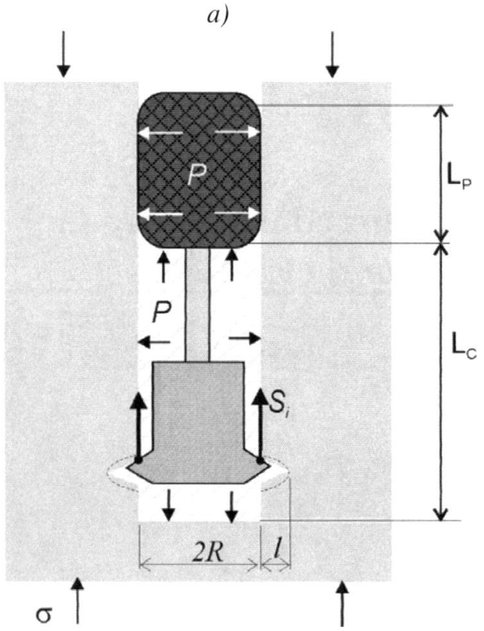

Figure 1. Scheme of the loading on the borehole surface during hydraulic fracturing: (a) with tangential loading on the borehole walls; (b) anchorage without indentor.

A device for directional fracturing was developed based on the performed works. The device is a combination of a mechanical anchor (anchor) and pressure sealer (packer). A general view of the downhole equipment is shown in Figure 2.

The anchor element is made of a split disc spring 2 with a pointed outer edge. Unlike the more common wedge anchors, the system is small along the borehole axis, which makes it possible to reduce the length of the fracture interval to half the diameter of the borehole. The small contact area of the anchor with the rock provides high local tangential loading of the well walls along its axis. All the conditions reduce the probability of longitudinal fracturing and contribute to the formation of a transverse crack. Sealing devices 3 and 4 are made in the form of a section mounted into each other, which, due to the different rigidity of the used rubber, firmly covers the gap interval without the possibility of displacing the device relatively to the borehole axis.

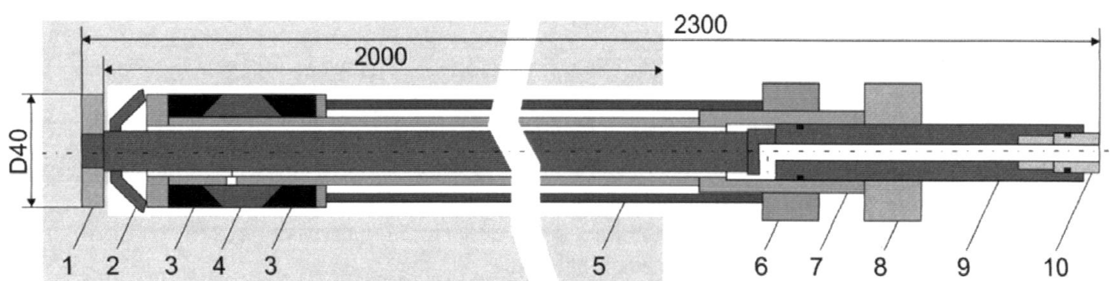

Figure 2. General view of a downhole device used for setting screens at a depth of up to 200 cm from the side of the mine: 1 – detent; 2 – anchor; 3 – rigid rubber; 4 – soft rubber; 5 – packer's pipe; 6 – packer's nut; 7 – anchor's pipe; 8 – anchor's nut; 9 – heavy-duty pipe; 10 – inlet manifold

A heavy-duty pipe 9, a nut 8, a pipe 7, a split disc spring 2 and a detent 1 form the anchor part of the device. The spring 2 is made of 4–5 mm thick steel. The outer diameter of the spring in the transport position is 44–46 mm. The anchor is actuated by tightening the nut 8 on the thread of the pipe 9. This reduces the distance between the detent 1 mounted on the pipe and the end flange of the pipe 7. As a result, the spring 2 is compressed; its outer edge is pressed into the walls of the well.

A nut 6, a pipe 5 and sealing elements of different stiffness 3, 4 form the packer part of the tool. The external sealers 3 are made of rigid rubber with a conical ward at one of the ends. The inner sealer 4 is made of soft rubber with conical protrusions at both ends. The packer is actuated by tightening the nut 6 on the external thread of the pipe 7. As a result, the sealers 3 and 4 are compressed at the axis, increasing in diameter and overlapping the annular clearance between the pipe 5 and the well walls.

After installation of the tool a hydraulic collector 10 is screwed into the pipe 9. High-pressure hoses are connected to the hydraulic manifold to supply two-component working fluids. The working fluid enters the bottomhole fracturing interval through the annular clearance between pipes 9 and 7. The pressure of the working fluid generates additional compression of the anchor spring 2, as well as the tangential loading at the contact of the anchor element with the rock.

After the hydraulic fracturing the pipe 9 is unscrewed (left thread) from the detent 1 and extracted from the borehole. The remaining part of the device is used for long-term sealing of the fracture interval, measurement of pressure in the screen and pumping in it a filler to compensate for leaks. For this purpose, a special adapter is installed in the mouth of pipe 7. Upon completion of the work, the device is dismantled from the well, in which a non-retractable detent 1 and a disk spring 2 remain.

Conclusion

The developed construction will make it possible to create a transverse fracture of uncased wells without carrying out an additional creation of an initial crack by a slotter. The experimental verification of this method shows that when fracturing is performed with indenter a transverse crack was formed, which indicates the efficiency of the method of directional transverse fracturing.

References

[1] Patutin AV, Timonin VV, Kondratenko AS and Rybalkin LA 2014 Integrated deep-borehole study of coal beds *J. Fund. Appl. Min. Sci.* Vol 1 pp 23–26 (in Russian)
[2] Lekontsev YuM and Sazhin PV 2014 Directional hydraulic fracturing in difficult caving roof control and coal degassing *J. Min. Sci.*Vol 50 No 5 pp 914–917
[3] Lekontsev YuM and Sazhin PV 2008 Application of the directional hydraulic fracturing at Berezovskaya Mine *J. Min. Sci.*Vol 44 No 3 pp 253–258
[4] Kurlenya MV, Altunina LK, Kuvshinov VA, Patutin AV and Serdyukov SV 2012 Froth gel for gas-bearing coal bed hydrofracturing in mine conditions *J. Min. Sci.* Vol 48 No 6 pp 947–953
[5] Kurlenya MV, Patutin AV, Rybalkin LA, Serdyukov SV and Shilova TV 2016 Laboratory studies into directional hydraulic fracturing with shearing loading of hole walls *J. Fund. App. Min. Sci.*Vol 3 pp 103–107 (in Russian)
[6] Serdyukov SV, Kurlenya MV, Patutin AV, Rybalkin LA and Shilova TV 2015 Experimental test of directional hydraulic fracturing technique *J. Min. Sci.* Vol 52 No 4 pp 615–622

IOP Conf. Series: Earth and Environmental Science **134** (2018) 012054 doi:10.1088/1755-1315/134/1/012054

Influence of mutual position of a noncircular tunnel and a dynamic impact source on stress distribution in lining

AS Sammal* and SV Antsiferov

Tula State University, Tula, Russia

E-mail: *sammal@mm.tsu.tula.ru

Abstract. Based on the analytical solution of the corresponding dynamic plane elasticity problem, general regularities of the formation of maximum stresses in lining of a noncircular cross-section tunnel are found and studied at different locations of a dynamic impact source. The paper shows that the closely spaced position of the source has a significant effect on the stress state of the underground structure, which in turn depends on the frequencies of the waves radiated by this source. It is revealed that when the dynamic impact source is removed deep in rock mass for a distance of the order of seven average radii of the tunnel cross section, the estimated stresses in the tunnel lining coincide with the stresses found from the solution of the problem on the wave propagation in the corresponding direction from infinity.

Application of blasting in mining and in construction is, as rule, connected with the additional safety measures to protect nearby structures since, for example, mine support (lining) is exposed in this case to static loading and essential dynamic effect. Currently, the common approach to predicting stress state in underground structures under dynamic impact is based on the analysis of rock mass and support interaction. Rock mass is simulated by a linearly deformable isotropic medium, and dynamic effects are considered as diffraction of stationary and nonstationary P- or S-waves at unsupported holes.

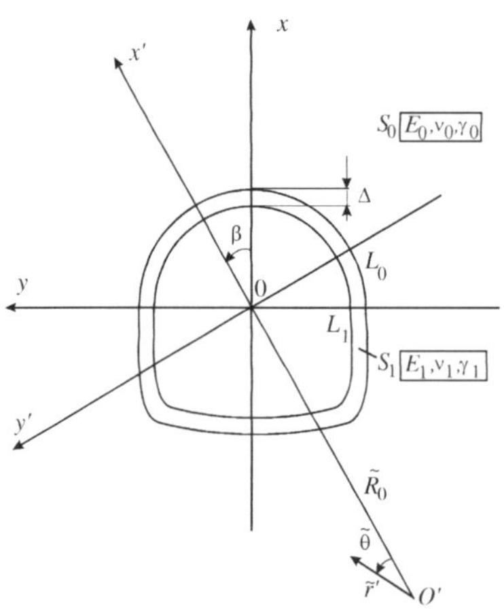

Figure 1. Analytical scheme.

In the framework of this approach, the original method of calculation of underground structures (mine support and lining of tunnels and buried pipelines) of arbitrary cross-section under dynamic loading has been developed at the Tula State University. The method is based on the mathematical modeling of the stress state formation in arbitrary cross-section lining and enclosing rock mass as a uniform deformable system under harmonic compression–tension or shear waves. To this effect, the solution of a dynamic plane elasticity problem of diffraction of harmonic waves propagating (including nearby the wave source) in a linearly deformable medium S_0, modeling rock mass, on a ring S_1 of the other material supporting a hole of arbitrary shape (with a single symmetry axis), modeling the lining, is used. The analytical model is shown in Figure 1.

Here, the ring S_1, modeling lining, has an arch thickness Δ and is made of material possessing a specific weight γ_1 and deformation characteristics represented by the deformation modulus E_1 and Poisson's ratio ν_1. The infinite uniform medium S_0, modeling rock mass, has different γ_0, E_0, ν_1. The ring and the medium deform jointly, i.e. the vectors of stresses and displacements at their interface L_0 are continuous. The inward interface boundary of the ring, L_1, is free from external forces. The source at the point $0'$ at the distance $\tilde{R}_0$ from the origin of the coordinate system $x0y$ emits harmonic compression wave with an angular frequency $\tilde{\omega}$ in the direction of the axis $0x'$ making an arbitrary angle β with the vertical axis $0x$ (the angle is counted anticlockwise).

The formulated problem is solved given the stresses and potentials are independent of the angle and the shear stresses are absent, in the descending wave, in the polar coordinate system, using the theory of functions of a complex variable, apparatus of conformal mapping of special cylindrical (Bessel) functions and complex series, the boundary disturbance method modified for the iterative process of finding complex potentials governing stresses and displacements at the points of the medium and the ring based on recurrent relations, which allows arbitrary number of approximations and enables higher accuracy of the calculation.

The described solution is implemented as a software system for efficient calculation of maximum dynamic stresses in lining under wave propagation (i.e. construction of envelopes of stress epures).

First, the obtained solution and its computer implementation were tested by means of comparing the calculated data and the available results reached by the other researchers in solving conformable dynamic problems, which can be presented as special cases in the framework of the developed method. to begin with, the results obtained by Savin [1] when determining dimensionless (related to the basic stress state in descending wave) maximum normal tangential stresses within the whole tame of the wave travel at the boundaries of circular, elliptical and square support-free holes in a infinite medium during propagation of harmonic compression wave were reproduced. Then, the results of Fotieva and Garaychuk [2] in the determination of stress at the boundary of an arbitrary cross-section tunnel in rock mass under propagation of P- and S-harmonic waves, as well as the calculated data on circular tunnel lining obtained by Erzhanov et al [3] were simulated. The actual and calculated stresses exhibited total coincidence in all occasions, unexceptionally.

Below in this paper, the calculations of tunnel lining (see the geometry in Figure 2) subjected to the dynamic impact of P-waves emitted by the source arranged in the horizontal straight matched with the axis $0y$ at relative distances $\tilde{r}_0 = \tilde{R}_0 / R_0 = 2,\ 7$ (here, $R_0 \approx 4$ m—average radius of the tunnel).

The calculation input data are: rock mass is composed of siltstone with $\gamma_0 = 18$ kN/m^3, deformation modulus $E_0 = 12000$ MPa and Poisson's ratio $\nu_0 = 0.3$; lining is made of concrete V20 with $\gamma_1 = 24$ kN/m^3, $E_1 = 27000$ MPa, $\nu_1 = 0.2$. The angular rate $\tilde{\omega} = 1265$ s^{-1} $\approx$ 200 Hz. The calculations involve 4 approximation, the series of the zero iteration has $N = 8$ members.

The calculated stresses were the maximum stresses at the inward and outward boundaries of the lining for the whole time of the wave travel.

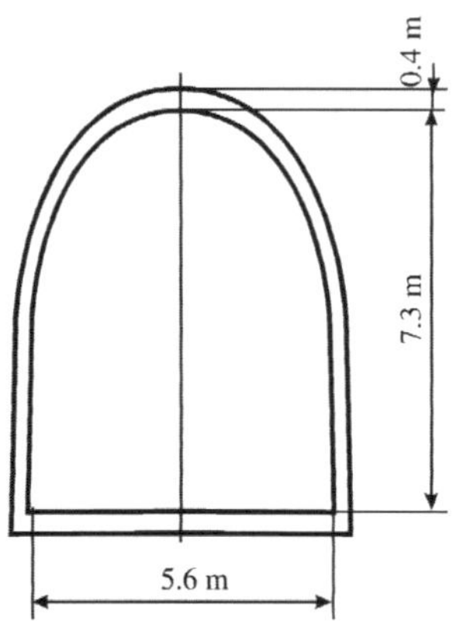

Figure 2. Geometry of the cross-section of the tunnel lining.

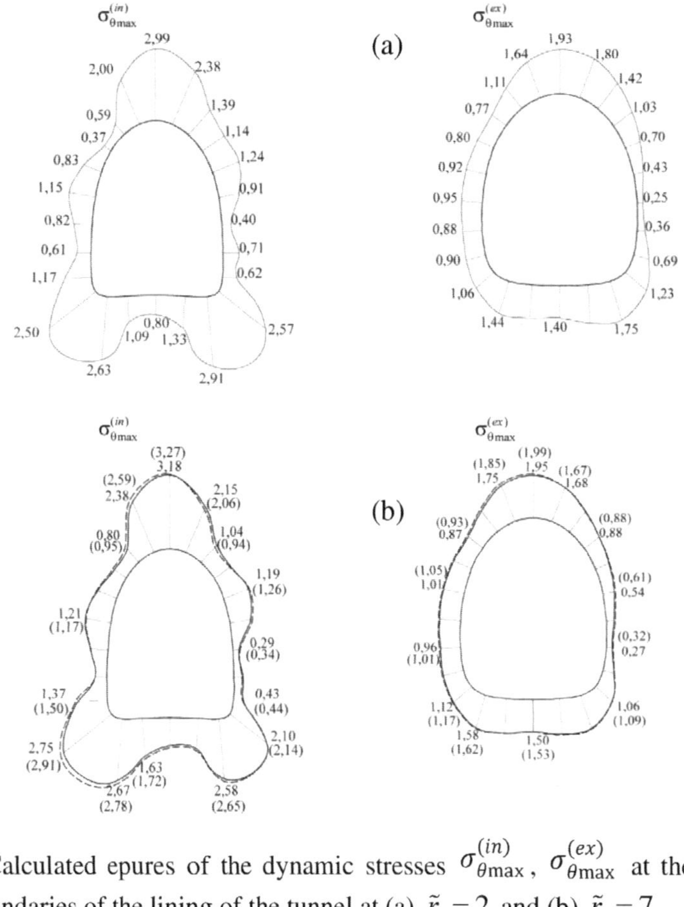

Figure 3. Calculated epures of the dynamic stresses $\sigma_{\theta\max}^{(in)}$, $\sigma_{\theta\max}^{(ex)}$ at the inward and outward boundaries of the lining of the tunnel at (a) $\tilde{r}_0 = 2$ and (b) $\tilde{r}_0 = 7$.

Figures 3a and 3b show the calculated epures of the normal tangential stresses $\sigma_{\theta\max}^{(in)}$, $\sigma_{\theta\max}^{(ex)}$ at the inward and outward boundaries of the lining, respectively. For the comparison, for the case when the source is at the distance $\tilde{r}_0 = 7$ (Figure 3b), the dashed lines (stress values are in brackets) show the data calculated for the compression wave propagation at infinity.

IOP Conf. Series: Earth and Environmental Science **134** (2018) 012054 doi:10.1088/1755-1315/134/1/012054

The epures of stresses shown by the solid and dashed lines in Figure 3b are almost identical. This allows drawing a conclusion that in this example with the source at the distance $\tilde{r}_0 \geq 7$, the dynamic effect of the emitted waves is sufficiently accurately determined as the effect of plane harmonic waves propagating at infinity.

At the second stage of the research, the multi-variant calculations revealed basic regular patterns in the formation of stress state in underground structures of noncircular cross-section under propagation of harmonic P-waves emitted by the closely located source. Special attention was given to the practically essential issue connected with the estimation of effect of the source location on the maximum stresses in the underground structure in different ground conditions.

In the second-stage calculations, the data of the support were the same as in the above example, the other data were varied: ω_0 = 0, 1; 2; 5; $\lambda^2 = \dfrac{\mu_1}{\gamma_1}\dfrac{\gamma_0}{\mu_0}$ = 4; 50. The source location was set by the polar coordinates $\tilde{r}_0$, β in the system with the origin at the center of the cross-section of the lining. All in all, 5 alternative locations of the wave source were analyzed: in the vertical axis under the tunnel (β = 0); above the tunnel (β = 180°), on the left in the horizontal axis (β = 90°), and in the intermediate directions characterized by the angles β = 45, 135°.

The scope of the analysis embraced the dependence of the normal tangential stresses (stress raisors) σ_θ at the inward boundary of the lining at the points of the arch and top of the side walls (curves 1) and in the cradle portion, including the point of the wall and cradle juncture (curves 2) on the relative source distance $\tilde{r}_0$.

The influence of the harmonic wave source location on the stress state in the tunnel in hard rock mass (λ^2 = 4) is illustrated by the curves in Figure 4 plotted by the processing data of multi-variant calculations.

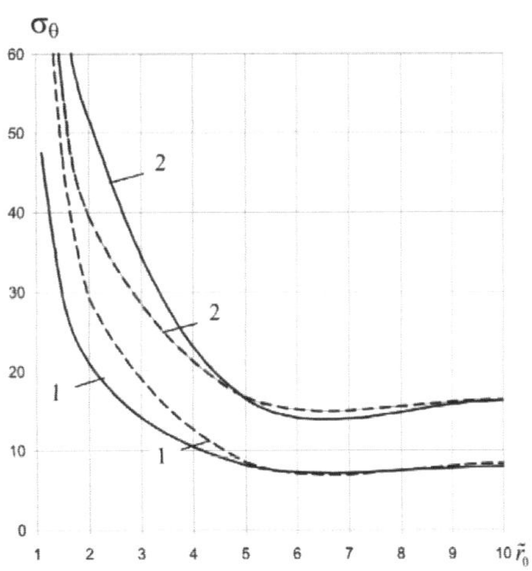

Figure 4. Dependence of the maximum tangential stresses σ_θ in the lining arch and at the top of the side walls of the tunnel (curves 1) and in the craddle of the lining (curve 1) in hard rock mass on the relative distance $\tilde{r}_0$ to the spource at β = 45° (solid curves) and 135° (dashed curves).

Conclusion

The implemented analysis has shown that close location of a wave source has an essential influence on stress state of an underground structure, conditioned also by the emitted wave frequency. For example, at a low frequency $\omega_0 = 0.1$, the dependence of the maximum stresses on the dimensionless distance $\tilde{r}_0$ is maximum. With an increase in the frequency ω_0, the curves become flatter and when $\omega_0 > 5$ the change in the source location from $\tilde{r}_0 = 2$ to infinity results in the change in the stress state no more than by 10%.

On the whole, the research findings imply that that increase in the distance between the source and the tunnel not always results in the decrease in the maximum stresses in the lining. At the same time, when $\tilde{r}_0 > 6$, the analyzed curves flatten out, as a rule, and the calculated stresses of the lining coincide with the stress values obtained in solving analogous problems on propagation of waves at infinity. The sources of equal capacity, placed in the symmetry axis $0x$ above and below the tunnel generate the same maximum stresses in the structure. With the increase in the frequency ω_0, the influence of the source distance $\tilde{r}_0$ on the maximum stresses decreases, while the effect of the wave direction (angle β) increases. Furthermore, the influence of the distance to the source is weaker in soft rocks as compared to hard rock mass.

References

[1] Savin GN 1951 Stress Concentration near Openings Moscow: Tekhizdat (in Russian)
[2] Fotieva NN and Garaychuk VG 1972 *Algorithm and Program for Determining Stresses in the Vicinity of Mining Workings under Stationary Dynamic Impacts* Moscow: NIIOSP (in Russian)
[3] Erzghanov GS, Aytaliev HM and Aekseeva LA 1989 *Dynamics of Tunnels and Underground Pipelines* Alma-Ata: Nauka (in Russian)

Formulations and algorithms for problems on rock mass and support deformation during mining

VM Seryakov

Chinakal Institute of Mining, Siberian Branch, Russian Academy of Sciences, Novosibirsk, Russia

E-mail: vser@misd.nsc.ru

Abstract. The analysis of problem formulations to calculate stress-strain state of mine support and surrounding rocks mass in rock mechanics shows that such formulations incompletely describe the mechanical features of joint deformation in the rock mass–support system. The present paper proposes an algorithm to take into account the actual conditions of rock mass and support interaction and the algorithm implementation method to ensure efficient calculation of stresses in rocks and support.

Two approaches to problems on stress state of mine support are currently widely applied. In the first approach, loads on support outback are determined based conditions of support installation and deformation, ore using various hypotheses of rock mass and support interaction. Then, the problem of stress state of mine support under the action of the determined loads is solved. In the second approach, the problem on joint deformation of mine and support interface is formulated. In this case, there are two stages of deformation. The first stage is free deformation of a tunnel boundary. The second stage is engagement of the tunnel boundary and support and their joint deformation [1–4].

So, in the first stage of deformation, free displacements of a tunnel boundary are found. These displacements are governed by external forces and connected with the partial release of initial stresses existing before the tunnel drivage. Thus, this partial stress release is to be determined at the first stage of the problem solution. The rest forces corresponding to the complete liberation of the initial stresses from the tunnel boundary affects both enclosing rock mass and support, and governs the stress state of the support and adjacent rock mass. The second stage is the problem on deformation of a nonuniform medium composed of rock mass and support having different physical properties. And here we have that determination method of the forces exerted on this nonuniform medium as well as their mechanisms remains yet uncertain.

Applied problem solving involves usually simpler formulations [1–4]. First, the problem on the adjacent rock deformation is solved as the problem on unsupported tunnel. Then, support subjected to normal compressive stresses applied to its outback surface is considered. By solving these two problems given the additional condition on the equality of the tunnel boundary and support outback displacements, it is assumed that formulated problem has been solved. The degree of approximation due to separate consideration of deformation of enclosing rocks and support is difficult to estimate.

The values of tunnel boundary displacement before support installation depend on the process variables of driving and supporting [5]. These values can be obtained from the analysis of is situ data of rock deformation in case of different designs of support systems. In [5], it is suggested to estimate tunnel boundary displacement before engagement with support by numerical solution of three-dimensional elasticity problems on stress state of rocks surrounding a tunnel. With the known value of lag between the support and face, it is possible to evaluate free displacements of the tunnel boundary, u_0, v_0, which are the oX- and oY-components of the displacement vector of the tunnel boundary before the support installation (in plane problem). Some formulations, which take into account that support is installed some time after partial relaxation of enclosing rocks from initial stresses, determine the support outback displacements u_p, v_p using the relations: $u_p = u–u_0$; $v_p = v–v_0$, where u, v—projections of the displacement vector of unsupported tunnel [6, 7]. With such formulation, displacements of tunnel boundary with regard to its joint deformation with support have the same values as in the case of a tunnel without support. In other words, support has not influence on overall displacement of tunnel boundary. To estimate the effect exerted by the degree of such approximation on calculated data of stress state of support and adjacent rock mass, the problem formulation should more exactly condition their interaction.

Below we discuss a problem formulation assuming the deformation of rocks and support is described by an elastic body model. In this case, the first stage problem on deformation of adjacent rock mass can be solved by subjecting the external boundaries of the computational domain to the action of stresses conforming with the occurrence depth of the tunnel and by setting condition of the tunnel boundary to be free from external load. With the X-axis oriented horizontally and the Y-axis vertically, at the tunnel occurrence depth H we have:

$$\sigma_y = -\gamma H; \ \sigma_x = -\lambda\gamma H; \ \tau_{xy} = 0. \tag{1}$$

Here, σ_y, σ_x, τ_{xy}—normal and shear stress components in the plane oXY; γ—bulk weigh of overlying strata; λ—lateral earth pressure coefficient. Before the moment when support is installed, the tunnel boundary displaces by some a priori known values determined from the solution of a boundary value problem involving gradual increment in the stresses σ_y, σ_x. at the boundary of the computation domain. The first stage is assumed completed when the a priori known and calculated values of the unsupported tunnel displacement reach satisfactory agreement. In this manner, after the first stage of solution, we know stress state of adjacent rock mass and external forces that displace the tunnel boundary by the determined value.

By determining the dormant portion of external forces as the difference of the initial force and the the force applied at the first stage of deformation, we determine the second stage stresses at the computational domain boundary. At this stage, the external loading is applied to deform a nonuniform medium composed of rock mass and support subjected to the condition of their rigid engagement. This problems solution yields stress state of the support and additional stresses and strains corresponding to the second stage deformation. The external forces on the adjacent rock mass–support system at the second deformation stage can be given by $\sigma_y = -\alpha\gamma H$; $\sigma_x = -\alpha\lambda\gamma H$ [1, 2]. The coefficient α is less than unit. We have said earlier that it is difficult to implement this algorithm in the problem on deformation of enclosing rock mass and tunnel support and, therefore, simplifications are involved. These simplifications are connected with various assumptions on the value of the coefficient α.

Let us apply the proposed algorithm to estimate the degree of approximation in a traditional problem formulation on deformation of enclosing rock mass and support in a single tunnel o the assumption that the support is installed 'instantaneously'. In case of the plane strain deformation, the computational domain is a half-plane with tunnel the boundary of which is in contact with a material attributed with mechanical properties of support. The inner surface of the supported tunnel is free from the external load, and the external boundary of the computational domain is subjected to forces conforming with the level of initial stresses in the center of the tunnel when being driven, i.e. Eq. (1).

This problems solution is a sump of solutions of two problems [8]. The first problem is on initial stresses in a nonuniform medium of enclosing rock mass and support. The external boundary of this nonuniform medium is set to experience external forces corresponding to the conditions (1). The second problem is the determination of additional stresses induced by the tunnel drivage. In this problem, the external boundary of the computational domain is free from forces, and the internal boundary of the supported tunnel is subjected to the normal and shear forces which vanish when summed with the forces obtained from the solution of the first problem.

The traditional formulation of problems in case of 'instantaneous' installation of support in a tunnel shows that the tunnel is driven in the initial stress state different from the stress state generated by the rock mass weight and tectonics. This disagrees with the actual conditions of tunnel drivage and installation of support system in it. Support is installed in rock mass which is in the initial stress state governed by the overlying rock weight and the related tectonic external forces. After 'instantaneous' installation of support, its inner boundary is free from the external forces and normal and shear stresses.

The finite element calculation [9] of stress state of rock mass–support system in the traditional formulation of 'instantaneous' installation of the support is illustrated in Figure 1. The calculations involved the input data on the geometry of the tunnel, mechanical properties of rocks and support and the initial stress field. For rocks: Young's modulus E = 50000 MPa; ν = 0.25. For support: E = 100000 MPa; ν = 0.2. The bulk weight of rocks and support is 0.03 MN/m^3. The initial stress state of rock mass is conditioned by the rock mass weight and, according to Dinnik's hypothesis: $\sigma_y^0 = -\gamma H$; $\sigma_x^0 = -\nu\gamma H/(1-\nu)$; $\tau_{xy}^0 = 0$. The tunnel has the shape of a semi-circle with a radius of 7 m, the tunnel bottom occurs at the depth of 750 m. The support has a thickness of 0.5 m.

The concentration zones of the compressive stresses σ_2 in the support are observed closer to its junction wit the tunnel floor. The compressive stresses reach maximum values at the inner boundary of the support. In the support in the tunnel arch, the stresses σ_2 are lower than in the concentration zones. The stresses σ_1 are not higher than 5 MPa almost everywhere in the support. In the support in the tunnel arch, zones of tensile stresses under 1 MPa are observed.

(a) (b)

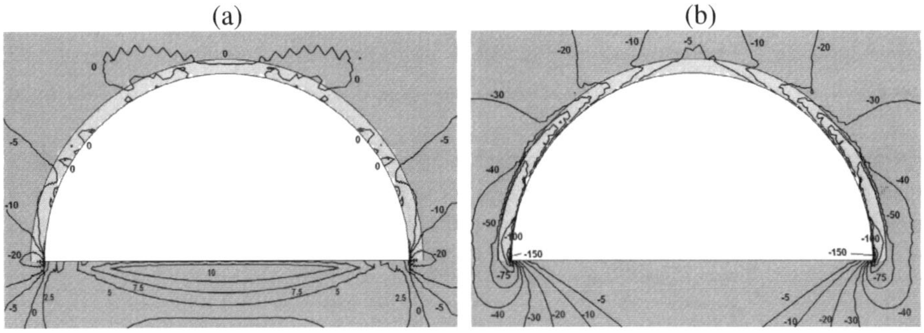

Figure 1. Distribution of principal stresses (a) σ_1 and (b) σ_2 (MPa) in adjacent rock mass and support in case of 'instantaneous' installation.

Now, let us apply the proposed algorithm and the related sequence of calculation of stress state in rock mass–support system in case of 'instantaneous' installation of support. Firs, we determine initial stress state of rock mass. 'Instantaneous' support leaves the initial stresses unaltered in the adjacent rock mass. The additional stresses corresponding to the formation of free space for a tunnel with the installed support are found by solving a boundary value problem with the normal and shear stresses preset at the rock mass and support interface, and these stresses go to zero when summed up with the corresponding components of the initial stress state.

This problem is solved using the method to calculate stress state of adjacent rocks and support with regard to the value of free displacement of the tunnel boundary up to the contact with the support [10]. The calculation of the tunnel drivage process by the method of initial stresses by means of turning to zero all stresses at finite elements inside the mined-out area divides the problem into two sub-problems. The first sub-problem determines mechanical state of rock mass during free displacement of rock mass toward the tunnel; the problem is solved when these displacement reach limit value. In case of 'instantaneous' support, this is the problem on the initial stress state of rock

IOP Conf. Series: Earth and Environmental Science **134** (2018) 012055 doi:10.1088/1755-1315/134/1/012055

mass. The second sub-problem describes the process of joint deformation of adjacent rock mass and support.

The calculated results are shown in Figure 2. There is the qualitative agreement between these results and the traditional problem formulation. The highest quantitative discrepancy is observed in the calculation of the second principal stress. In the traditional formulation, support is subjected to higher compression. In concentration zones of compressive stresses, this discrepancy reaches 50%.

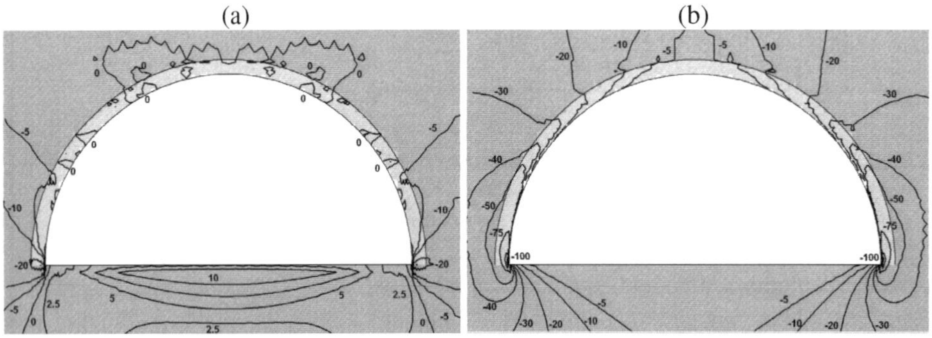

Figure 2. Distribution of principal stresses (a) σ_1 and (b) σ_2 (MPa) in adjacent rock mass and support in the proposed approach to calculating stress state with 'instantaneous' support.

Conclusion

The author has shown that the traditional formulation of the problem on stress state of rock mass–support system in case when support is 'instantaneously' installed in a tunnel corresponds to the conditions of tunnel drivage in stress field different from the initial stress field generated by the tectonics of weight of overlying strata.

The method to calculate stress state of rock mass–support system with regard to the value of free displacement of the tunnel boundary down to its contact with the support in case of its 'instantaneous' installation corresponds to the conditions of support installation in rock mass subjected to the initial stress state governed by overlying rock weight and the related tectonics. The quantitative difference in the calculated stresses in the concentration zones within the proposed approach as against the traditional problem formulation reaches 50%.

References

[1] Bulychev NS 1989 *Mechanics of Underground Structures in Examples and Problems* Moscow: Nedra (in Russian)
[2] Bulychev NS 1994 Mechanics of Underground Structures Moscow: Nedra (in Russian)
[3] Baklashov IV, Kartozia BA 1984 Mechanics of Underground Structures and Support Units Moscow: Nedra (in Russian)
[4] Kartozia BA, Fedunets BI, Shuplik MN et al 2003 *Mine and Underground Construction: University Textbook* Moscow: Gornaya Kniga Vol 2 (in Russian)
[5] Nasonov ID, Fedyukin VA, Shuplik MN 1992 *Technology of Underground Construction* Moscow: Nedra (in Russian)
[6] Protosenya AG, Dolgy EI, Ogorodnikov YuN et al 2003 *Mine and Underground Construction in Problems and Examples* Saint-Petersburg: Plekhanov Gorny Inst. (in Russian)
[7] Protosenya AG, Ogorodnikov YuN, Demenkov PA et al 2011 *Mechanics of Underground Structures* Saint-Petersburg: SPBGU-MANEB (in Russian)
[8] Turchaninov IA, Iofis MA, Kasparyan EV 1989 Rock Mechanics Leningrad: Nedra (in Russian)
[9] Zienkiewicz O 1972 *The Finite Element Method in Engineering Science* McGraw Hill
[10] Seryakov VM 2015 Calculating stresses in support and sidewall rocks in stagewise face drivage in long excavations *J. Min. Sci.* Vol 51 No 4 pp 673–678

IOP Conf. Series: Earth and Environmental Science **134** (2018) 012056 doi:10.1088/1755-1315/134/1/012056

Calculation of stresses in a rock mass and lining in stagewise face drivage

VM Seryakov*and BR Zhamalova

Chinakal Institute of Mining, Siberian Branch, Russian Academy of Sciences, Novosibirsk, Russia

E-mail:vser@misd.nsc.ru,

Abstract. Using the method of calculating mechanical state of a rock mass for the conditions of stagewise drivage of a production face in large cross-section excavations, the specific features of stress redistribution in lining of excavations are found. The zones of tensile stresses in the lining are detected. The authors discuss the influence of the initial stress state of rocks on the tension stress zones induced in the lining in course of the heading advance

Mine workings of sizable cross-sections are driven in stages [1]. In every stage a face of small cross–section is driven and its profile is lined. This process for drivage of large cross-section mine workings implies specific approaches to calculation of stress state of a periphery rock mass and successively mounted lining segments. Investigation and analysis of strain in a rock mass and the lining enable to conclude, that heading of a small–profile face increases load on earlier mounted lining and running rocks in the periphery of a large cross-section mine working excavated by the moment. The older is the lining segment, the greater load it experiences at the present moment. Conventional methods for stress calculation in the lining and a periphery rock mass do not work under these conditions. The statement and solution to the problem under consideration should consider a stagewise principle of drivage and lining of mine workings. Thus, when evaluating the stress–strain state of a rock mass and lining units it is necessary to consider a mining sequence of face formation and to develop respective methods and softwares for computation of mechanical state of rocks.

One of respective approaches is proposed in [2, 3]. The authors employ superposition of analytical solutions obtained for a single face with lined profile [4, 5]. Empirical coefficients used to consider the stagewise character of excavation of large cross–section faces. These coefficients evaluate a contribution of each single face to the final solution.

In another approach, based on the numerical solution to the problems, the mathematical statements are used to evaluate extra stress fields arising in a rock mass and lining units as a consequence of mining operations [6]. To consider stress redistribution in rocks and lining members under elastic deformation is feasible through summing of solutions to problems of the elasticity theory for additional fields of stresses induced by excavation of each face [7]. The direct realization of this approach encounters large problems with setting boundary conditions for successively excavated faces

[8, 9]. The said problems can be solved by methods and algorithms developed at IM SB RAS to model the stress redistribution process in a rock mass and lining units with account for progressive mining operations [10]. The methods and algorithms are based on the use of the initial stress and deformation methods implying application of a calculation rigidity matrix which does not undergo changes in finding non–linear solutions [11, 12]. The compliance of used relations with the single–curve hypothesis underpinning solutions to many non–linear problems on mechanics of deformed solid body and rocks is justified.

In the present paper the novel algorithms and softwares are used to evaluate a character of stress redistribution in lining units under conditions of stagewise cross–sectional face drivage. Mining sequence is shown in Figure 1. Computation is made for plain–deformation conditions. In the first computation series the researchers preset zero values of the horizontal component of displacement vector u and the tangential component of stress tensor τ_{xy} in vertical boundaries of the calculation domain. These terms correspond to the initial stress state of a rock mass at stress tensor components: $\sigma_y^0 = -\wp H$; $\sigma_x^0 = -v\wp H/(1-v)$; $\tau_{xy}^0 = 0$ and are realized in regions where tectonics is not a problem [14]. Here σ_x^0, σ_y^0, τ_{xy}^0 are normal and tangential components of stress tensor; $\wp$ is volume density of rocks; v is Poisson ratio of a rock mass; H is distance to Earth surface. Axis Ox is oriented horizontally, Oy is oriented vertically. The upper boundary of the calculation domain is free of external load. In the lower boundary the zero vertical component of displacement vector v and zero tangential component of stress tensor τ_{xy}. are assumed. Mechanical properties of a rock mass are suggested: Young modulus $E = 50000$ MPa; $v = 0.25$. For the lining material $E = 100000$ MPa; $v = 0.2$. Volume density of rocks and lining material is 0.03 MN/m^3.

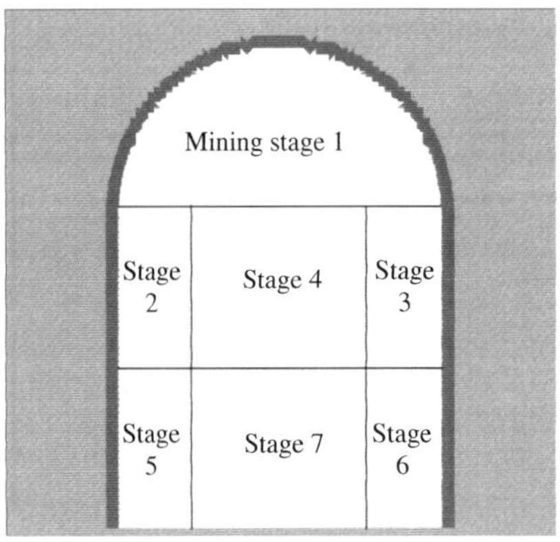

Figure 1. Mining sequence in opening of the face cross–section.

Figure 2 demonstrates lines of equal-in-value principal stresses σ_1, σ_2 in lining and periphery rock mass after execution of 4 stages of the face cross–section opening. A complex pattern of stress distribution is formed in the lining. In the stagewise drivage and lining of mine workings the appreciable fragment of the lining is under tension stresses (Figure 2a). Tension stresses are small, less than 0.5 MPa in more than a half of the finished lining, but in some sections tension stress exceeds 5MPa.

These sections locate in under-roof lining units in the vicinity of its contact with vertical members, mounted in the second and third stages of mining operations. The maximum tension stresses in zones of its concentration reach 12 MPa. Compression stresses in the lining are not high and do not exceed 3–4 MPa (Figure 2b).

(a) (b)

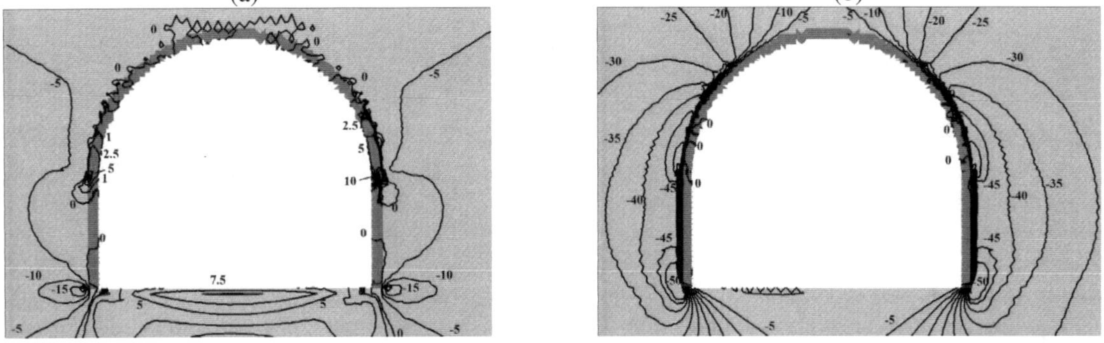

Figure 2. Distribution of principal stresses (a) σ_1, (b) σ_2, MPa, in the lining and surrounding rock mass after four stages of the face cross–section opening.

In Figure 3 the final distribution of principal stresses $\sigma_1$and $\sigma_2$in a rock mass and lining units corresponds to the seventh opening stage. Mining advance leads to redistribution of stress fields in the lining. New areas of tension stress concentration are formed along with reduction in tension in earlier formed tension concentration zones (Figure 3a). All the zones of tension stress concentration locate nearby contacts of lining members formed in different mining stages. Values of tension stress exceed 5 MPa in three zones from four zones of detected tension stress concentration. Distribution of compression stresses σ_2 in the lining does not undergo sensible variations. The distribution of stresses in a rock mass surrounding the lining complies with the earlier established pattern of stress distribution in the vicinity of mine workings [7].

(a) (b)

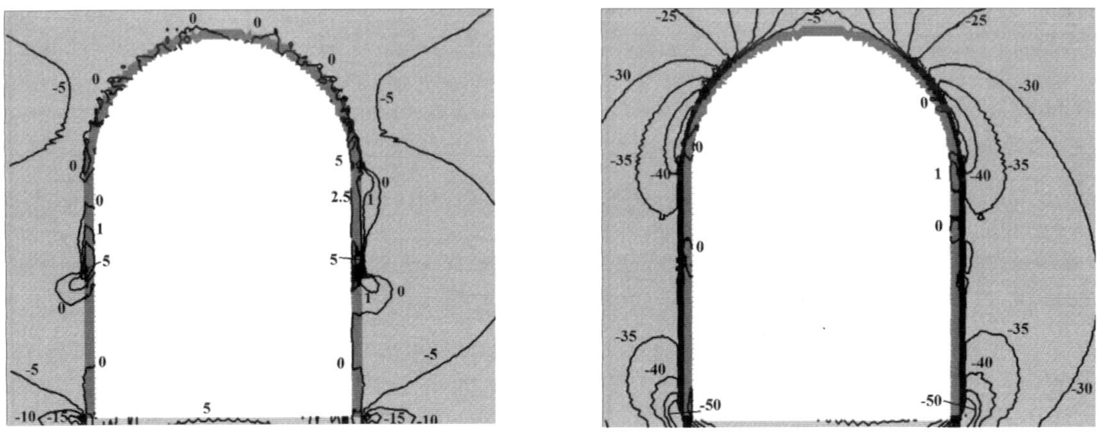

Figure 3. Distribution of principal stresses (a) σ_1 and (b) σ_2, MPa in the lining and surrounding rock mass after mining operations on opening of face cross–section are completed.

IOP Conf. Series: Earth and Environmental Science **134** (2018) 012056 doi:10.1088/1755-1315/134/1/012056

The second series of calculations was performed for the hydrostatic initial stress field. The same mining sequence on the opening of the face cross–section was considered. The main aim of modeling was to evaluate the effect of initial stress state of a rock mass in formation of tension stress zones in the face lining. The conclusions obtained after the analysis of calculated results on redistribution of stresses in the course of mining advance are illustrated in Figure 4 where isolines of principal stresses after the complete opening of the face cross–section are shown.

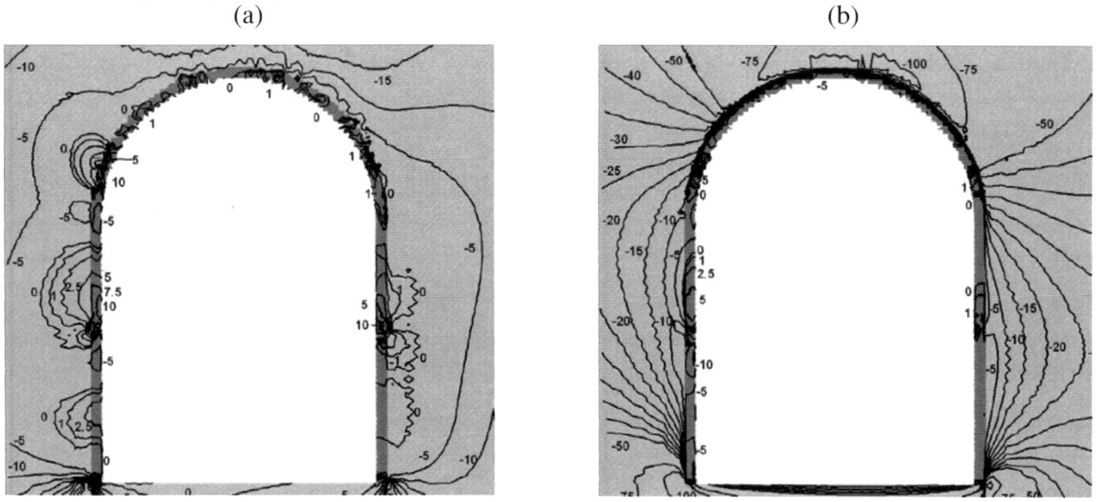

Figure 4. Distribution of principal stresses (a) $\sigma_1$and (b) σ_2, MPa in the lining and surrounding rock mass after mining operations on opening of face cross–section are completed in the case of hydrostatic distribution of the initial stress field.

It is important to emphasize the qualitative and quantitative effect of the initial stress field on formation of tension stress zones in the face lining, namely, their locations, number, and level of acting stresses tend to vary. In some lining sections the tension stress values exceed 10 MPa (Figure 4a). Zones of appreciable compression stress concentration are not observed, though compression stress can increase more than twofold in some lining sections as compared to the initial gravity stress field (Figure 4b).

Conclusion

1. The new-proposed method for calculation of the mechanical state of a rock mass and lining enables to evaluate a character of the stress field redistribution in a stagewise drivage of mine workings of large cross-sections.

2. Increase in the horizontal component of the initial stress field leads to growth of a number of tension stress concentration zones in the lining members and stress values acting in them.

References

[1] Kartozia BA, Fedunets BI, Shuplik MN, et al 2003 *Mine and Underground Construction* University Textbook Moscow: Gornaya Kniga Vol 2 (in Russian)

[2] Pankratenko AN 2002 *Technology of Large Cross-section Tunnel Drivage* Moscow: MGU (in Russian)

[3] Rostovtsev IE 2000 Method of lining design in large cross-section tunnels during stagewise drivage *Cand Tech Sci Dissertation Theses* Tula (in Russian)

[4] Baklashov IV and Kartozia BA 1984 *Mechanics of Underground Structures and Support Units* Moscow: Nedra (in Russian)
[5] Bulychev NS 1994 *Mechanics of Underground Structures* Moscow: Nedra (in Russian)
[6] Seryakov VM 2015 Calculating stresses in support and sidewall rocks in stagewise face drivage in long excavations *Journal of Mining Science* Vol 51 No 4 pp 673–678
[7] Turchaninov IA, Iofis MA and Kasparyan EV 1989 *Rock Mechanics* Leningrad: Nedra (in Russian)
[8] Baimbetov MO and Seryakov VM 1984 Effect of the order of performing clean-up and fill operations on the stress–strain state of a deposit *Journal of Mining Science* Vol 20 No 4 pp 254–260
[9] Kuznetsov SV, Odintsev VN, Slonim ME and Trofimov VA 1981 *The Rock Pressure Calculation Methodology* Moscow: Nauka (in Russian)
[10] Seryakov VM 2008 Implemenatation of the calculation method of the stress state in rock mass with backfill *Journal of Mining Science*Vol 44 No 5 pp 439–450
[11] Zienkiewicz O 1975 *The Finite Element Method in Engineering* Moscow: Mir (in Russian)
[12] Fadeev AB 1987 *FiniteEelement Method in Rock Mechanics* Moscow: Nedra (in Russian)

IOP Conf. Series: Earth and Environmental Science **134** (2018) 012057 doi:10.1088/1755-1315/134/1/012057

Acoustic emission of rock mass under the constant-rate fluid injection

AV Shadrin* and VI Klishin

Federal Research Center of Coal and Coal Chemistry, Siberian Branch, Russian Academy of Sciences, Kemerovo, Russia

E-mail: *avsh-357@mail.ru

Abstract. The authors study acoustic emission in coal bed and difficult-to-cave roof under injection of fluid by pumps at a constant rate. The functional connection between the roof hydrofracture length and the total number of AE pulses is validated, it is also found that the coal bed hydroloosening time, injection rate and time behavior of acoustic emission activity depend on the fluid injection volume required until the fluid breakout in a roadway through growing fractures. In the formulas offered for the practical application, integral parameters that characterize permeability and porosity of rock mass and process parameters of the technology are found during test injection.

1. Introduction

Coal mines widely use hydraulic technologies. For example, in order to eliminate long-span overhanging of rocks and rock falls on powered support, directional hydraulic fracturing is commonly applied in roof rocks [1]. Aimed to enhance gas permeability of coal, hydraulic dissection is carried out: water is fed in holes drilled in coal and a pattern of fractures is created [2]. Local enhancement of gas permeability and unloading of face area is possible with high-pressure water injection in holes [3].

Hydrotreating can also be used to wet coal as it has been found that coal with the moisture content higher than 6% never experiences outbursts [4]. Under such moisture content, gas is blocked in microcracks and pores and has no ability to become free and to raise pore pressure; coal becomes more plastic owing to which abutment pressure zone is displaced deeper in rock mass while outburst and rockburst hazard is decreased.

Water can be fed in the modes of low-pressure water infusion of coal or hydraulic loosening. During low-pressure fluid injection recommended for coal of high water permeability, water creates no additional fractures and flows in the existing natural cracks, microflaws and pores, and uniformly humidifies coal. In the mode of hydraulic loosening recommended for low water-permeable coal, injected fluid creates additional network of fractures for coal moistening.

Any mode of hydraulic treatment of coal is characterized by certain rate of fracturing. As fractures propagate stepwise, the rate of fracture propagation is understood as a number of "steps" per unit time. No mechanism of fracture propagation control during hydraulic treatment in mines exists, all hydrotreating methods only provide for recording of pressure and water injection rate. In the meanwhile, a growing fracture is a source of acoustic emission, and AE recording allows registering, firs, the fact of stepwise fracture propagation and, second, the rate of the stepwise fracture propagation in time, and, after modifying the method, positioning of a growing fracture and AE energy [5, 6].

The hydrotreating methods include various parameters. These parameters characterize efficiency of a method, technology of the method and pumping unit. These three groups of parameters are denoted as check, production and hydraulics.

The check parameters are:

- Directional hydraulic fracturing of difficult-to-cave roof rocks—required length and direction of a created fracture to ensure caving and reduce area of overhanging rocks;
- Hydraulic treatment to reduce hazard of dynamic events—coal moisture content (in low-pressure water infusion, low-pressure soak treatment, hydraulic loosening) and face slip value (hydraulic squeezing).

The check parameters are determined in accordance with the guidelines [3] as their later edition [7] lacks the necessary information.

The technology parameters are: the number of holes simultaneously connected to pumping unit, the length and diameter of holes, the length and size of of sealing, arrangement of man-made fracture-initiating slots relative to roadways and roof rocks layers and the time of hydraulic treatment. The technology parameters are selecting in accordance with [7] and coordinated with the assumed cycle of drivage or extraction of coal. Substantiation of technology parameters if hydraulic fracturing in difficult-to-cave roof rocks follows the regulations [8] and additional recommendations on a specific coal bed, e.g. [9].

The hydraulics parameters are pressure and rate of injection.

In this manner, we assume the optimal hydraulic parameters of a hydrotreating mode the values of pressure and rate of injection such that the check parameter is attained in minimum possible time.

Coal bed and coal roof under hydrotreating are solids subjected to rock pressure and behaving elastically. The difference between coal bed and roof rocks lies in strength, permeability and porosity. However, both in coal bed and roof rocks, injected fluid fills the cavity of a propagating fracture permeates in its walls. For this reason, the AE-based approach to the control over hydrotreating processes in coal bed and roof rocks can be the same.

Most of the currently operating pumping units without hydraulic accumulator ensure nearly constant injection irrespective of permeability and porosity of rock mass. For example, hydraulic treatment pumpers NBVU-30M, UNV-2, IN-35 and 2UGN, or water–oil emulsion pumping models T used in directional hydraulic fracturing in roof rocks, specifically, T-200/32 Ex-Z [10–12.

For this reason, we analyze applicability of AE-based determination of optimal parameters of hydraulic loosening of coal and hydrofracturing of roof rocks using the mentioned pumpers. Hereinafter, we denote the both hydraulic processes as hydrotreating, except for the cases of specific character of them.

2. Problem formulation

The medium of hydrotreating is assumed elastic as AE during fracture growth takes place in the form of unloading waves while unloading is mostly an elastic process even in plastic materials [13].

AE receiving transducers are either vibration meters, velocimeters or accelerometers; thus, the overall description of the fracture propagation process means finding of space and time distribution of displacements, velocities and acceleration of points in the transducer-adjacent medium, or, put it otherwise, finding of a displacement field of these points.

In-site handling of this problem has some complications, first, a fractures as an AE source is a complex object and needs considerable idealization of its properties in mathematical modeling. Second, complex structure of coal beds results in repeated reflection of signals from different discontinuities and interfaces present in the layered medium coal bed and enclosing rocks. Third, transformation of types of acoustic waves, or their attenuation and dispersion, etc. can happen [14]. And, finally, for real crack—AE sources, it is comparatively difficult to formulate the boundary and initial conditions on the faces.

This circumstances make it impossible to obtain general solution of the formulated problem using the modern analytical methods owing to insuperable mathematical difficulties, except for some simple special cases given additional simplifying hypotheses are introduced. Therefore, we limit ourselves to recording of stepwise fracture propagation and assume that a jump inside the senor sensitivity zone will be perceived by the sensor.

The stepwise fracture propagation behavior is assessed in [15] as a process of opening of an arbitrary oriented fracture in the field of biaxial compression under action of fluid injected in it. It is shown that when pressure in the fracture reaches a critical value P_{cr}, the fracture loses equilibrium and starts growing. The fracture growth velocity is much higher than the fluid fill velocity; for this reason, the pressure lowers in the fracture, and the fracture stops growing [15]. In accordance with the nonlinear Darcy law, there is a functional connection between the pressure and the fluid injection rate [16], thus, the critical injection pressure P_{cr} fits with the critical injection rate q_{cr}.

In our case, one fracturing "step" is matched with one AE pulse and fracture propagation rate is characterized by AE activity, which the number of AE pulses per unit time.

It is supposed in [17] that during injection of fluid in hole drilled in coal, not one but a number of similar-type (approximately identical orientation and initial length) fractures can grow. For this case, the solution of the problem on the number of "steps" (or AE pulses) I during growth of fracture from the initial length l_0 to l and on the relation of the average fracture length l on porosity and permeability of rocks is given by:

$$I = nk \ln \frac{\ell}{\ell_0} , \tag{1}$$

where n is the number of simultaneously growing fractures; k is the coefficient of correlation between l_0 and l;

$$\ell = \frac{q(2\sqrt{m\mathrm{K}\Delta P \mu t} - \mu b)}{2\pi h n m \mathrm{K}\Delta P} , \tag{2}$$

where q is the injection rate; h is average width of fracture surface; m, K are porosity and permeability of rocks, respectively; $\Delta P=P_1\text{-}P_2$, P_1 is the fluid pressure on the hole wall; P_2 is the pore pressure equality the pressure of free gas, if present, in hole; b is the average width of fractures; μ is the dynamic viscosity of injected fluid.

The formula (2) at a relative error not higher than 10% and vanishing with the increase in t is only applicable when [17]:

$$t > t_p \approx \frac{2\mu b^2}{mK\Delta P} . \tag{3}$$

At the real coal hydroloosening parameters: μ=1 sP; b=0.1 cm; m=3–8%; K=0.1 mD; ΔP=5–10 MPa, the permissible time t_p is 25–130 s (at the most often recorded parameters, t_p is 30–40 s). Inasmuch as hydraulic loosening of coal can take from tens minutes to a few hours, the formula (2) is is usable to analyze this process except for a short period in the beginning.

Porosity of sandstone is commensurable with the porosity of coal [18] and permeability varies in a wide range [19] but is assumable to be less by an order of magnitude than in coal. The pump pressure of the model T-200/32 Ex-Z is is several times higher in sand than in coal bed. Consequently, t_p during hydraulic fracturing of roof composed of sandstone is little more than hydraulic fracturing in coal bed. So, the formula (2) as the formula (1) is applicable to analyzing hydraulic fracturing of roof except for 1–2 min in the beginning of the process. However, in this case n = 1 as a rule, total time of hydraulic fracturing is 5–10 min, and recording of stepwise fracture propagation events is performed irrespective whether the inequality (3) is satisfied or not.

Under such conditions, we analyze applicability of AE-based determination of hydraulic parameters in hydrotreating of rocks from mine roadways. The solution of this problem in terms of pumps with adjustable pumping velocity (e.g. in valve-adjustable water withdrawal from high-pressure water passageways in hydraulic mines during hydroloosening of coal bed) ensuring optimal constant AE activity is given in [20]. In this article, we present the same problem solution for pumps with the constant pumping rate ($q \approx$ const), i.e. we show the AE behavior in time such that to ensure check parameters of hydrotreating.

3. Problem solution

We assume that under hydraulic loosening and hydraulic fracturing, fluid is fed in hole until the tip of the growing fracture reaches the surface of the nearest roadway or hole drilled from the roadway. Under hydroloosening of coal bed, this minimal distance is approximately the sealing length of the hole. Under roof hydrofracturing, this distance depends on the hydrofracturing technology, thus, we assume this minimal value as the distance to the nearest exposed surface of the roadway. In the both case, this value is denoted by l_1 .

Substituting (2) in (1), we find total number I of AE pulses during growth of n fractures from an initial length l_0 to l:

$$I = nk \ln\left[\frac{q(2\sqrt{mK\Delta P\mu t} - \mu b)}{2\pi h n \ell_0 mK\Delta P}\right] . \qquad (4)$$

From (4) we determine the critical injection rate q_{cr} such that induces the forced propagation of fractures. In as much as $nk \neq 0$, the condition $I = 0$, being valid for $P < P_{cr}$, yields:

$$q_{cr} = \frac{2\pi h n \ell_0 mK\Delta P}{2\sqrt{mK\Delta P\mu t} - \mu b} . \qquad (5)$$

During directional fracturing of roof rocks, as a rule, $n = 1$ and the expression (5) gives the minimal injection rate of feasibility of the technology.

Let us find optimal injection parameters that allow injecting fluid volume Q in a hole with a free section length ℓ_f and sealed section length ℓ_1 at the constant injection rate until fracture reaches the roadway surface.

Under such conditions, the rate q and time t of injection are related as:

$$q = \frac{Q}{t} . \qquad (6)$$

Fluid breakout in the roadway can take place unless and until the average half-length of growing fractures reaches approximately the value ℓ_1. Inserting this condition and (6) in (2) yields the quadratic equation in regard to t:

$$t - \frac{Q\sqrt{\mu t}}{\pi h n \ell_1 \sqrt{mk\Delta P}} + \frac{Q\mu b}{2\pi h n \ell_1 mk\Delta P} = 0 . \qquad (7)$$

The roots of the equation are:

$$t_{1,2} = \frac{Q^2 \mu}{(2\pi h n \ell_1)^2 mk\Delta P}\left(1 \pm \sqrt{1 - \frac{2\pi h n \ell_1 b}{Q}}\right)^2 . \qquad (8)$$

The expression (2) is valid for any $t > t_p$. Consequently, the roots of (7) should satisfy this condition.

Rewrite Eq. (2) with regard to (6):

$$\ell = \frac{d_1}{\sqrt{t}} - \frac{d_2}{t} , \qquad (9)$$

where $d_1 = \dfrac{Q\sqrt{\mu}}{\pi h n \sqrt{mK\Delta P}}$; $d_2 = \dfrac{Q\mu b}{2\pi h n mK\Delta P}$.

The graphic representation of the relation (9) is shown in the Figure 1.

The function $\ell = \ell(t)$ in the form of (9) has the peak at:

$$t_m = \left(\frac{2d_2}{d_1} \right)^2 \quad . \tag{10}$$

Placing the expressions for d_1 and d_2 in (10 and comparing with (3), we obtain $t_p = 2t_m$. consequently, the root t_1 to the left of the point t_m disagrees with the condition $t > t_p$ at any Q.

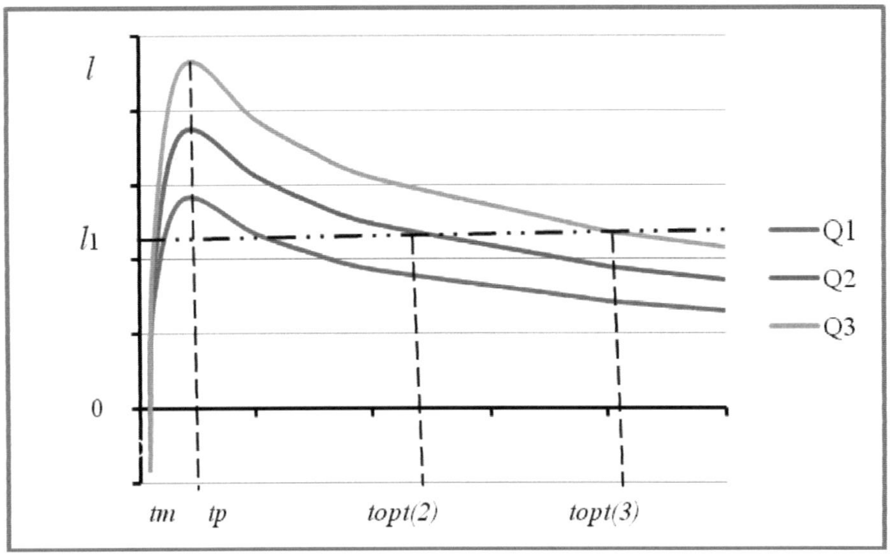

Figure 1. Relationship of the fracture ℓ and the injection time t at different injection volumes Q: 1, 2, 3—at Q_1, Q_2 , Q_3, respectively; $Q_3 > Q_2 > Q_1 = Q_{min}$; $T_{opt(2)}$; $t_{opt(3)}$—optimal injection times at which Q_2 and Q_3 volumes of fluid enter the hole by the moment of fluid breakout in the roadway, respectively.

The volume Q of fluid injection until breakout in the roadway under hydraulic loosening of coal bed is governed by the rate of injection. At very high q, fracture rapidly grows down to the plane of the hole bottom. Fluid has not time to fill small fissures and pores in coal and rock. As q is being decreased, the time of injection until fluid breakout in the roadway grows, and, accordingly, the fluid volume permeating in the hole walls grows, too. This is the explanation of the increase in Q as the injection rate is being decreased.

In this manner, in order that the root t_2 satisfies the condition $t_2 > t_p$, it is required to ensure high volume of injection. The minimal injection volume Q_{min}, with the mentioned condition to be satisfied, conforms with the injection time $t_2 = t_p$. Let us find this value from Eq. (2) given that $\ell = \ell_1$; $q = Q/t$; $t = t_p$:

$$Q_{min} = \frac{4\pi nhb\ell_1}{2\sqrt{2}-1} . \tag{11}$$

The related injection rate:

$$q_{max} = \frac{Q_{min}}{t_p} \quad . \tag{12}$$

substitution of (11) and (3) in (12) yields:

$$q_{max} = \frac{2\pi nh\ell_1 mK\Delta P}{(2\sqrt{2}-1)\mu b} \quad . \tag{13}$$

Thus, when the injection rate $q_{cr} \leq q \leq q_{\max}$, all values of the root t_2 of (7) fulfill the condition $t \geq t_{p}$.

In order to control the mode of hydrotreating using AE, all fluid injection time t can be divided into j equal control intervals with the duration T so that $\tau_i < T < t_s$, where τ_i—duration of an i-th AE pulse, t_s—total injection time up to the end of hydrotreating, and the injection mode can be characterized by the parameter ΔI, which is the AE activity equaling the number of AE pulses in the control interval T:

$$\Delta I = \frac{\partial I}{\partial t} T . \tag{14}$$

From (4) we have:

$$\frac{\partial I}{\partial t} = \frac{nk\sqrt{mK\Delta P\mu}}{\sqrt{t}(2\sqrt{mK\Delta P\mu t} - \mu b)} . \tag{15}$$

Transformation of (15) with regard to (2) produces:

$$\frac{\partial I}{\partial t} = \frac{qk\sqrt{\mu}}{2\ell\pi h\sqrt{mK\Delta Pt}} . \tag{16}$$

Then the wanted value is given by:

$$\Delta I = \frac{qkT\sqrt{\mu}}{2\ell\pi h\sqrt{mK\Delta Pt}} . \tag{17}$$

From Eq. (6) we find the optimal value of the injection rate q_{opt} reachable at $t = t_2$ (refer to Eq. (8)):

$$q_{opt} = \frac{mK\Delta P(2\pi hn\ell_1)^2}{Q\mu\left(1+\sqrt{1-\frac{2\pi hn\ell_1 b}{Q}}\right)^2} . \tag{18}$$

Such injection rate induces AE activity found from the expressions (14) and (15):

$$\Delta I_{opt} = \frac{kTn_r\ell_f}{2t - b\sqrt{\frac{\mu t}{mK\Delta P}}} , \tag{19}$$

where $n_r = n/\ell_f$—relatve number of fractures per one meter of free-flow section of the hole, m^{-1}.

Application of the derived equations in the determination of coal bed hydroloosening parameters, namely, optimal injection time, optimal injection rate and the conformable optimal AE activity, requires knowing many characteristics of permeability and porosity of coal, geometrical sizes, number of cracks, etc. The experimental measurement of these parameters is labor-consuming, low accurate and sometimes even impossible. For this reason, it seems to be expedient to introduce integral coefficients to account for the whole set of parameters in the obtained equations and top determine them to during test injection.

Furthermore, it is possible to analytically simplify the equations (8) and (18). To this effect, we take the cofactor:

$$\left(1+\sqrt{1-\frac{2\pi hn\ell_1 b}{Q}}\right)^2 . \tag{20}$$

The numerator of the radical expression contains the value proportional to the total volume of cavities of growing fractures and commensurable with the fluid volume Q_{min} injected in the time t_p at the maximal injection rate (refer to the formula (11)). Coal bed hydroloosening is carried out at $q < q_{max}$ and $t \gg t_{p.}$, thus, the major volume of the injected fluid enters the walls of the growing fractures (flaws and pores) and the minor volume fills the growing fractures. For this reason, it can be assumed that $2\pi hn\ell_1 b / Q < 1$. In this case, it is valid to use the approximation [21]:

$$\sqrt{1 - 2\pi hn\ell_1 b / Q} \approx 1 - \pi hn\ell_1 b / Q \ . \tag{21}$$

Then the expression (20) takes on form:

$$\left(1 + \sqrt{1 - \frac{2\pi hn\ell_1 b}{Q}}\right)^2 \approx \frac{(2Q - \pi hn\ell_1 b)^2}{Q^2} \ . \tag{22}$$

Replacing the cofactor in accordance with (22) allows presenting the formulas (8) and (18) in the more convenient form:

$$T_{opt} = a(Q\text{-}V)^2 \ ; \tag{23}$$

$$q_{opt} = \frac{Q}{a(Q - V)^2} \ , \tag{24}$$

where $a = a_1 \cdot a_2$;

$a_1 = \dfrac{\mu}{(\pi hn_r)^2 mK\Delta P}$—coefficient conditioned by rock mass permeability and porosity and by injection parameters, cm^{-2}; $a_2 = (\ell_1 \ell_f)^{-2}$—process parameter, m^{-4}; $V = \dfrac{\pi}{2} hn_r \ell_1 \ell_f b$—coefficient proportional to the total volume of growing fractures, m^3.

4. Recommendations on application of the problem solution results

AE activity during coal bed hydroloosening at optimal rate is defined by the expression (19). This expression is inconvenient for application as it inexplicitly connects AE activity and hydraulic parameters on injection. In this connection, we analyze possibility to express the optimal AE activity in terms of hydraulic parameters. In this case, it should be taken into account that at different stages of fluid injection, the number of growing fractures can change owing the the injection pressure variation. In addition, the properties of the wetted rock mass also change. For this reason, the expression (19) should be added with the relations of the pressure difference ΔP, fracture number n and time but these analytical expression are unknown. So, for the practical use, the formula (19) can be presented in the simplified form:

$$\Delta I_{opt} = \frac{\delta_1(q)}{t - \delta_2(q)\sqrt{t}} \ , \tag{25}$$

where

$$\delta_1(q) = \frac{nkT}{2} \ ; \ \delta_2(q) = \frac{b}{2}\sqrt{\frac{\mu}{mK\Delta P}} \ . \tag{26}$$

The relation (25) is valid at sufficiently high values of t. When t is close to the value $(\delta_2)^2$, where it has a peculiarity, for ΔI_{opt} it knowingly yields overestimates. This is the consequence of the simplification made when deriving the formula (2).

So we have that in order to determine optimal hydraulic parameters for coal bed hydroloosening with AE-based monitoring, it is required to find coefficients included in Eqs. (24) and (25) during test injection. With the coefficients known, the optimal injection rate for the optimal injection volume should be found from (24). Then, using (25) the optimal time behavior of AE activity under optimal injection rate is determined using (25).

Later on, monitoring of coal bed hydroloosening is reduced to the experimental finding of injection pressure and injection rate such that the time behavior of AE activity remains optimal.

We think that the described procedure is applicable in monitoring of hydraulic segregation of coal bed using surface holes for the purposes of pre-mine drainage [2]. In this case, the value of l_1 is governed by the dimension of the coal bed section to be disjoined, and the coefficients a_1, a_2 and V can be approximately determined from the data of test loosening of this bed from a closely spaced roadway.

For the AE-based monitoring of roof hydrofracturing, test hydrofracturing should be carried out at an early stage of longwalling. The control process includes simultaneous recording of AE pulses I_e and breakout of fluid in a neighbor hole drilled at a distance l_e from the hydraulic fracturing hole. From the formula (1), the coefficient k_e for the given area of rock mass can be found:

$$k_e = \frac{-}{1}. \qquad (27)$$

Then, assuming the the coefficient k_e and the initial fracture length l_0 are constant in the roof of this longwall section, the growing fracture length l can be determined using the number of the recorded AE pulses:

$$l = l_0. \qquad (28)$$

5. Conclusion

Activity of acoustic emission is the quantitative characteristic of the intensity of fracture propagation in rocks under hydrotreating. The total number of AE pulses and the time behavior of AE activity allow, respectively, monitoring of hydraulic fracturing process in the difficult-to-cave roof and hydraulic loosening of coal bed at the optimal rate of fluid injection such that the wanted volume of fluid enters the growing fractures up to the fluid breakout in the nearest roadway.

References

[1] Klishin VI, Zvorygin LV, Lebedev AV and Savchenko AV 2011 *Safety and New Technologies in Underground Coal Mining* Novosibirsk: Nov. Pisatel (in Russian)

[2] Nozhkin NV 1979 *Forhand Degassing of Coal* Moscow: Nedra (in Russian)

[3] *Guidelines on Safe Outburst-Hazrdous Coal Mining RD 05-350-00* 2000 Gosgortekhnadzor (in Russian)

[4] Chernov OI and Puzyrev VN 1979 *Coal and Gas Outburst Prediction* Moscow : Nedra (in Russian)

[5] Loginov GN, Yaskevich SV, Duchkov AA and Serdyukov AS 2015 Joint processing of surface and underground microseismic monitoring data in hard mineral mining *J. Min. Sci.* Vol 51 No 5 pp 944–950

[6] Klishin VI, Nikolsky AM, Opruk GYu, Neverov AA and Neverov SA 2008 Directional hydualic fracturing methods for hard-to-cave roof for ground control in coal mines *Ugol* No 11 pp 12–16

[7] *Guidelines on Prediction of Dynamic Events and Monitoring of Rock Mass in Coal Mining* 2016 Approved by RF Federal Environmental, Industrial and Nuclear Supervision Service Order No 339 (in Russian)

[8] *Guidelines on Selection of Methods and Parameters for Roof Looseining in Extraction Panels* 1991 Leningrad: VNIMI (in Russian)

[9] Klishin VI, Opruk GYu, Sentyurev AV, Nikolaev AV 2015 Experience of hydraulic fracturing in main roof during travel of mechanized longwall system out of assembling chamber *Ugol* No 11

[10] http://sinref.ru/000_uchebniki/01701gornoe_delo/004_spr_mashini_i_oborudovanie_dla_ugolnih_shaht_horin_1987/188 *Coal Bed Water Injection Plant UN35* Last visited May 3, 2017

[11] http://www.sibinfo.org/gidravlika/products/undnvup.htm *Coal Bed Water Injection Plant UNV-2M* Last visited May 3, 2017

[12] http://promenergochim.ru/static/doc/0000/0000/0217/217186.4fo7ytayvm.pdf *Pumps Type T*

Last visited April 21, 2017
[13] Greshnikov VA and Drobot YuB 1976 *Acoustic Emission. Application in Testing Materials and Goods* Moscow: Izd. Standartov (in Russian)
[14] Brekhovskikh LM 1973 *Waves in Bedded Media* Moscow: Nauka (in Russian)
[15] Moskalev AN, Vasilev LM and Mlodetsky VR 1979 Limiting equilibrium of cracks in a coal seam into which liquid is injected *J. Min. Sci.* Vol 15 No 5 pp 504–508
[16] Chernov OI and Rozantsev ES 1975 *Preparation of Gas Outburst-Hazardous Coal Fields* Moscow: Nedra (in Russian)
[17] Shadrin AV 1983 Seismoacoustic reaction of a coal mass during its hydraulic treatment *J. Min. Sci.* Vol 19 No 6 pp 467–472
[18] Egorov PV, Shtumf GG, Renev AA et al 2001 *Geomechanics: Tutorial* Kemerovo: Kuzbassvuzizdat (in Russian)
[19] https://ru.wikipedia.org/wiki *Water Impermeability of Soils* Last visited May 5, 2017
[20] Shadrin AV 2000 Acoustic-emission monitoring of the preventive hydraulic treatment of coal seams *J. Min. Sci.* Vol 36 No 5 pp 502–506
[21] Gradshtein IS and Ryzhik IM 1971 *Tables of Intergals, Sums, Rows and Products* Moscow: Nauke (in Russian)

IOP Conf. Series: Earth and Environmental Science **134** (2018) 012058 doi:10.1088/1755-1315/134/1/012058

Expediency of wet-mix shotcreting in mines of Vostoktsvetmet

YuN Shaposhnik

Chinakal Institute of Mining, Siberian Branch, Russian Academy of Sciences, Novosibirsk, Russia

E-mail: shaposhnikyury@mail.ru

Abstract. This paper offers a substantiation of the transition from heavy and difficult-to-install steel frame support to the cheaper and simpler reinforced shotcrete lining (at the same bearing-capacity of the support). The process line for the preparation of the shotcrete mix is designed. The author of the paper presents calculation of the material inputs and labor costs of installation of different support types in Kazakhstan Mines. The economic efficiency of the wet-mix shotcrete spraying technology is illustrated in terms of Orlov and Artemevsk Mines.

Orlov Mine is going to develop North site of Novaya orebody at Orlov deposit; the targeted area contains ore reserves of category C_1 in amount of 3615 thousand tons [1].

The North site rock mass is composed of alternate ore bodies and lenses featuring complex shape along the dip; the length reaches 800 m along the strike and 400 m along the dip; the height is to 350 m. The ore bodies have mostly steep dip (to 80 deg), are to 4 m thick in horizontal part and have average thickness of 10.5 m in steeply dipping sections. The depth of occurrence of the North site ore is nearly 1000 m

The ore and rock mass are weak-stable. The stability drops in the zones of weathering, faulting and hydrothermal alteration zones, which can be from a few meters to 100–150 m thick. The adjacent rock mass is very unstable due to numerous differently directed microcracks filled with calcite and pyrite, and many slide surfaces.

After transition to deeper levels, mine-technical conditions worsen because of great depth, many unfilled mined-out voids and man-made fills, zones of oxidation of sulfur-bearing ore etc.

In difficult ground conditions of the unstable rock mass subjected to high stress state, the steel frame support deforms, which generates accident conditions in underground excavations.

Artemevsk Mine stage II embraces development of a few ore bodies containing ore reserves of category C_2 in amount of 12722.9 thousand tons [2].

Both Orlov and Artemevsk Mines widely use frame support due to the aggravating ground conditions at deeper levels. However, the frame support in unstable enclosing rock mass lacks mine safety even in case of double framing, which, among other things, impedes ventilation and movement of self-propelled equipment as cross section of roadways is reduced. The process of frame installation can also be unsafe.

A way out of this situation may be the transition from heavy and laborious framing made of special section steel SVP-22 and SVP-27 to a less expensive and readily sprayed (at the same load-bearing capacity)reinforced shotcrete with reinforcement frames.

Today, the mines of Vostoktsvetment only use dry-mix shotcrete without reinforcement fibers and frames.

The problem of dry-mix shotcreting is high rate of bounce. Depending on the surface to be lined (vertical walls or arch), loss of concrete may reach 15 to 35%. The average loss is from 20 to 25% as against 5–10% of concrete loss in dry-mix shotcreting [3–5].

The commercial test of Concrix ES35 fiber manufactured by Brugg Contec AG. Concrix, Switzerland, in dry-mix shotcrete lining with filler and binder made of local materials, as well as the trial of ready-made dry shotcrete mix Master Roc STS 1510 manufactured by BASF and cement-and-sand mortar Stopcem S manufactured by DSI have been carried out in Irtysh Mine of Vostoktsvetment [6]. The bounce of Concrix ES35 in case of the mix made of local materials and with the ready-made mix Master Roc STS 1510 (fiber was added in the mix at the plant) in the dry shotcrete lining method reached 30–50%.

The stop-watch observations show that the process of shotcreting has the same duration in case of the ready-made mortar and the mix made of local material, but the process of the mortar preparation in the latter case takes long time (5–7 min of single-stage mixing per 86 kg of ready mix). Considering this fact, the application of ready-made mixes reduces the time of shotcrete lining nearly 2 times.

With regard to all expenditures connected with mix preparation, the cost of shotcrete lining 40 cm thick per 1 m^2 in Irtysh Mine is USD 9,58 with the mix made of local materials, USD 13,71 with the local materials and cement accelerator Master Roc SA580 and USD 15,71 with the ready-made mix Master Roc STS 1510.

Thus, the lab-scale and commercial tests show that the use of synthetic fibers CONCRIX ES35 in dry-mix shotcrete does not ensure considerable improvement in shotcrete lining quality because of high-rate bounce.

The key advantages of wet-mix shotcrete lining as compared with the dry-mix shotcrete method include reduced bound and, as a consequence, lower loss by two times and more; higher productivity of shotcreting, better labor conditions owing to considerable decrease in dust formation; low compressed-air consumption thanks to hydraulic feed of shotcrete mix, reduced wear of shotcreting equipment and enhanced quality of shotcrete lining due to persistent water and cement ratio.

After research of wet-mix shotcrete preparation and lining in Vostoktsvetmet mines, it has been recommended to use mine concrete-mixing machine Minemix 15 manufactured by Normet or a mobile mine concrete mixing plant with capacity of 15 m^3/h on legs or on a frame, installed in a room of small section of 16–25 m^2 with a storage of cement in big bags, Unimec MF500 Transmixer concrete carrying machine and Spraymec MF050VC or Meyco ME-3 electrohydraulic self-propelled mobile concrete sprayer with compressor.

In Orlov Mine, it is planned to install concrete-mixing machine (CMM) at level 12 near Slepaya shaft; In Artemevsk Mine—at level 10 near a storage of filler materials, at actual elevation of –252.0 m.

According to [6, 8], steel frame support is used in rocks of stability categories II and IV. In rocks of stability category V, steel arch support is combined with the other types of reinforcement.

Rail haulage roadways require 0.67–1.0 SVP-22 steel frame support per 1 m^2 of mine working sin in rocks of stability category IV and 1.25–2.0 steel frames in rocks of stability category V. The excavations with self-propelled carrying machines need 0.8–1.5 and 1.5–2.0 SVP-22 steel frames per 1 m^2 in rocks of stability category IV and V, respectively.

By Q rating of lithological domains at Orlov and Artemevskoe deposits, with regard to classification of rock mass and ore bodies based on [9, 10], the shotcrete lining thickness is assumed as 9–12 cm for unstable rocks and as 12–15 cm for very unstable rocks. The calculation of economic efficiency of the transition wet-mix shotcrete method assumed that dry-mix shotcrete layer 5 cm thick was to be replaced by wet-mix shotcrete layer 7 cm thick and steel frame support was to be replaced by wet-mix shotcrete layer 12 cm thick (with the load-bearing capacity strength reserve)+resin-grouted rockbolts and reinforcement steel mesh.

The calculation of NPV with the current and proposed technology was performed without regard to the realization of marketable products for Orlov and Artemevsk Mines of Vostoktsvetmet (Figure 1).

(a)

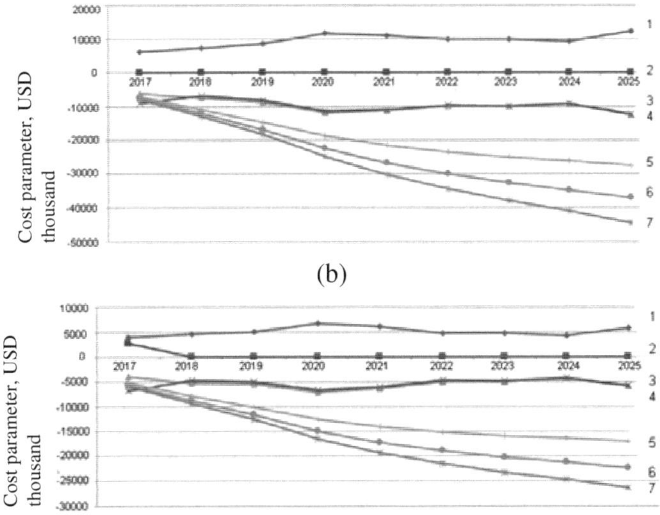

Figure 1. NPV at the (a) current and (b) proposed technologies of support installation in Orlov Mine, USD thousand: 1—operating cost; 2—capital cost; 4—net profit; 4—net money flow; 5–7—accumulated NPV at the discount rates of 0.15, 0.2 and 0.3, respectively.

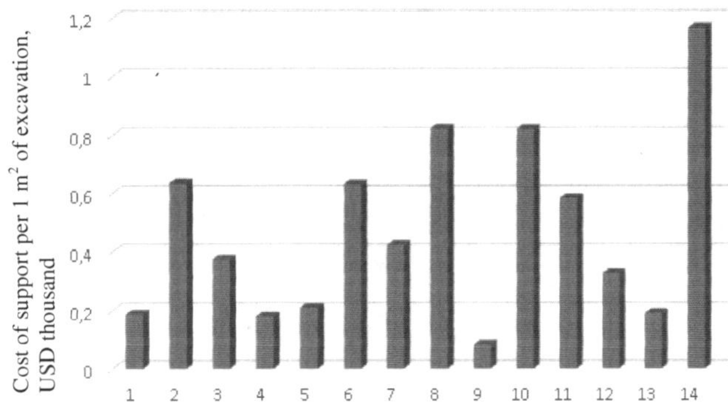

Figure 2. Cost of different-type support per 1 m^2 of excavation: 1—resin-grout rockbolting in Orlov Mine (pattern of 0.8×0.8 m, 9 rockbolts per row); 2—SVP steel frame in Orlov Mine (frame spacing l = 1 m); 3— SVP steel frame in Orlov Mine (frame spacing l = 0.59 m); 4—dry-mix shotcrete lining 5 cm thick in Orlov Mine (at the shotcreting cost of USD 15.74/m^2, rough section S_r = 16.0 m^2, finished cross-section S_f = 14.3 m^2, P = 11.47 m); 5— resin-grout rockbolting in Artemevsk Mine; 6—SVP steel frame in Artemevsk Mine steel (frame spacing l = 1 m); 7—SVP steel frame in Artemevsk Mine steel (frame spacing l = 0.67 m; 8— SVP steel frame in Artemevsk Mine steel (frame spacing l = 1.3 m); 9— dry-mix shotcrete lining 5 cm thick in Artemevsk Mine (at the shotcreting cost of USD 7.25/m^2, S_r = 16.0 m^2, S_f = 14.3 m^2, P = 11.47 m); 10—hybrid support: shotcrete 12 cm thick+resin-grout rockbolts 2.2 m long+reinforcement steel mesh; 11— shotcrete 12 cm thick+resin-grout rockbolts 2.2 m long; 12—shotcrete 12 cm thick; 13—shotcrete 7 cm thick; 14—protective support of shotcrete 12 cm thick and resin-grout rockbolts 2.2 m long.

IOP Conf. Series: Earth and Environmental Science **134** (2018) 012058 doi:10.1088/1755-1315/134/1/012058

The total cost of wet-mix shotcrete lining is composed of capital and operating costs. The capita costs are determined at the current price level of 2017 in accordance with [11, 12] and with regard to local construction conditions in East Kazakhstan. The capital cost included expenditures connected with the purchase of main process equipment, drivage and support of underground excavations for placement of underground concrete-mixing machines and other equipment and development of project paperwork. The operating cost included expenditures connected with the amortization, basic process materials (components of mix), protective clothes, food, power (electric energy, compressed air, diesel fuel), water supply, external services, wages and fire-protection measures.

The cost of different types of support per 1 m of underground excavations for mines of Vostoktsvetmet is illustrated in Figure 2.

Conclusion

According to the resultant calculations of economic efficiency of the two technologies of mine support shows that for the conditions of Orlov Mine, at the discount rate of 0.3 (without regard to the commercial output realization), 2025 NPV makes USD -27303 thousand with the current support (steel frames) and USD -16948 thousand with the proposed technology (reinforced shotcrete with the reinforcement), which is reflective of the efficiency of the proposed method of wet-mix shotcreting. In Artemevsk Mine, with the amount of frame support making 80% of the amount of permanent and development drivage in 2017–2034, at the discount rate of 0.3, 2034 NPV is USD -24884 thousand with the current technology and is USD –22711 thousand with the proposed method, which is also proves efficiency of the recommended wet-mix shotcrete lining.

References

[1] Production procedures on determination of optimal parameters of mining system for Northern orebody of Orlovskoe deposit *Georesurs. Engineering* 2016 (in Russian)

[2] *Feasibility Study: Opening of Artemevskoe Deposit Stage II* Eurasian Design Company Almaty 2012 (in Russian)

[3] Bernard S and Kovalenko VV 2011 Practice of application of shotcrete and fiber-reinforced concrete in Australian mines *Ugol Ukrainy* March pp 47–52

[4] Eremenko VA, Loushnikov VN, Sandy MP, Milkin DA and Milshin EA 2013 E Selection and basis of mine working driving and excavation support in unstable rocks at deep levels of Kholbinsky Mine *Gorny Zh.* No 7 pp 59–66

[5] Louchnikov VN, Eremenko VA, Sandy MP and Bucher R 2014 Underground excavation support in deformable and rockburst-hazardous rock mass conditions *Gorny Zh.* No 4 pp 37–44

[6] Krupnik LA, Shapishnik YuN and Shaposhnik SN 2016 Expediency of application of fiber in concrete mixtures for lining by spraying in Irtysh Mine *Int. Internet-Conference Proc.: Innovative Development in the Mining Industry* Krivoi Rog: KNU (in Russian)

[7] *Guidelines on Selection of the Support Design and Parameters for Orlov Mine* Ust-Kamenogorsk 2013 (in Russian)

[8] *Guidelines on Selection of the Support Design and Parameters for Artemevsk Mine* 2012 (in Russian)

[9] *Production Procedures (Guidelines) for Selection of the Support Design, Parameters and Technology for Orlovskoe Deposit, Vostoktsvetmet* Ustk-Kamenogorsk Regionalnoe Obuchenie 2016 (in Russian)

[10] *Production Procedures (Guidelines) for Selection of the Support Design, Parameters and Technology for Artemevskoe deposit, Vostoktsvetmet* Karaganda: Mining Research Group, 2015 (in Russian)

[11] *Construction Standards SN RK 8.02-02-2002. Procedure for Determining Contract Price of Construction in the Republic of Kazakhstan* (in Russian)

[12] *Regulations RK 8.02.03.2002. Provisions on Determining Design Work Cost in Construction in the Republic of Kazakhstan* (in Russian)

Selecting shotcrete composition for Vostoktsvetmet mines

YuN Shaposhnik

Chinakal Institute of Mining, Siberian Branch, Russian Academy of Sciences, Novosibirsk, Russia

E-mail: shaposhnikyury@mail.ru

Abstract. According to the results of the pilot trials in mines of Vostoktsvetmet, the combination shotcrete lining ensures stability and safety of underground excavation in complicated ground conditions. The paper presents the information on the conformity of the shotcrete mixture components to the standards. The compositions for the wet-mix shotcreting are selected. It is shown that inert materials used in the compositions have considerable influence on the properties of shotcrete mixes and shotcrete linings. The cost per unit of basic process materials (mixture components) is calculated in terms of Orlov Mine.

According to [1], schotcrete mixtures include binder, filler, admixtures and water. When required, shotcrete may include reinforcement fibers.

Shotcrete lining should use concrete having class and grade determined at the age of 28 days:

—compression strength classes: B22.5; B25; B30; B35; B40;

—axial tension strength classes: B_t = 1.6; 2.0; 2.4; 2.8; 3.2; B_{tb} = 3.6; 4.0;

—freeze resistance grades: *F*100; *F*150; *F*200; *F*300;

—water resistance grades: *W*4; *W*6; *W*8.

Compression strength class of concrete should be B25 in line with [2] and B22.5 as per [3]. In B25 class, the average compression strength of concrete at the age of 28 days is 32.06 MPa, and recommended cement grade is 400. In concrete class B_t = 1.6, the average axial tension strength and bending tension at the age of 28 days is 2.06 MPa, and recommended cement grades is 20.

Based on the generalized experience of use of different cement grades from different manufacturers in East Kazakhstan, it is concluded that cement of Semey Cement Plant sometimes fails to exhibit the rated specifications as against cement of Bukhtarma Cement Company. Furthermore, Bukhtarma cement has higher strength characteristics than Semey cement as follows from practice of backfill preparation and use.

The laboratory tests on conformity between shotcrete mixes and standards and the selection of wet-mix shotcrete have been carried out in the Certification Testing Laboratory of the East Kazakhstan State Technical University, in a special construction and backfilling laboratory at Ridder–Kazzinc, BASF-Central Asia and Normet Company. The compression and bending testing results for different cement grades are given in Figure 1.

The best conformity with the strength standards is demonstrated by cement grades CEM II/A-Sh 42.5 N (Bukhtarma Cement Company) and PTs400-D20 (Semey Cement Plant). The cement grades have almost the same compression strength characteristic, while CEM II/A-Sh 42.5 N cement grade has the increased bending strength, which is one of the basic properties of shotcrete). The cement grades have approximately equal costs, too.

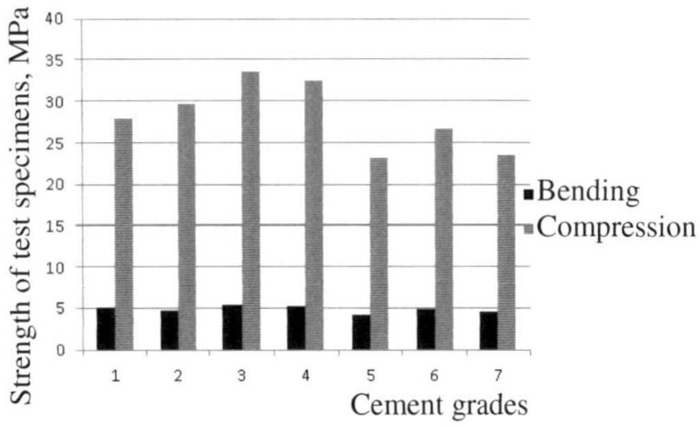

Figure 1. Compression and bending testing of different cement grades at the age of 7 days: 1—CEM II/A-Sh 42.5 N (Bukhtarma); 2—PTs400-D20 (Semey); 3—PTs400-D0 N (Bukhtarma); 4—PTs500-D0 N (Bukhtarma); 5—CEM V/A)Sh-P) 32.5 N (Bukhtarma); 6—CEM I/42.5 N SS (Bukhtarma); 7—cement from Orlov Mine.

Shotcrete mix filler may be:

—sand as per state standards GOST 8736, GOST 26633 and GOST 9757;

—break stone or grave by GOST 8267;

—light fillers by GOST 9757.

Sand and gravel mix should have characteristic described below:

—size modulus not less than 2 (in case that size modulus is less than 2, special experimentation equipment is required;

—moisture content to 7%;

—limit content of clayey particles and fine powder dust to 3%;

—clot clay content to 0.25%;

—content of grains smaller than 0.14 mm in size to 10%;

—limit content of particles larger than 8 mm no more than 10%. Particles larger than 8 mm in size should be eliminated while sifting, storing and processing of fillers.

The best conformity with the listed requirements is demonstrtaed by sand manufactured by Nonmetallic Materials Plant, except for a slight excess of standard contents of clayeye and powder particles by 1.7%.

Making of trial batches of mixes using blast-furnace granulated slag, manufacture of specimens and the tests of all rated quality indexes has been carried out in the special construction and backfilling laboratory at Ridder–Kazzinc [4, 5]. Characterization of compositions for wet-mix shotcrete lining is given in the table 1.

Table 1. Characteristics of wet-mix shotcrete compositions

Consumption, kg/m^3					Density, kg/m^3 (factual)	Stiffness (Vebe consistency)	Different age compression strength, MPa		
Cement	Sand	Blast-furnace granulated slag	Break stone	Water			3	7	28
400	1175	–	585	190	2443	32	20.57	24.07	32.14
420	1160	–	580	190	2317	–	21.96	26.15	33.01
440	1150	–	575	190	2313	41	22.63	27.43	35.76
380	1185	–	590	190	2460	–	16.98	22.82	31.45
360	11195	–	600	190	2427	28	13.86	20.20	30.35
400	–	1200	555	225	2278	63	10.52	15.61	20.95
420	–	1185	550	225	2160	–	13.60	16.23	23.49
440	–	1170	545	225	2153	38	14.27	18.63	24.59
380	–	1210	560	225	2160	–	8.62	13.18	18.77
360	–	1220	570	225	2117	72	8.00	12.43	16.46

Based on the tests in the Ridder–Kazzinc laboratory, the curves of compression strength and cement consumption at water flow rate of 190 l/m^3 have been plotted for for different mixes (CEM II/A-Sh 42.5 N by Bukhtarma, sand and break stone by Nonmetallic Materials Plant) in Figure 2.

It is seen in the figure that the strength of concrete 7 days old is 27.43 MPa, which is nearly the required strength at the age of 28 days (28.83 MPa); the concrete acqures the required compression strength at the age pof 28 days even at the cement consumption of 360 kg/m^3. The tests of concrete batches with blast-furnace granulated slag show that this mix of concrete becomes stronger less intensive. When granulated blast-furnace slag is used as inert material, the wanted strength is not reached in concrete 28 days old (at binder consumption to 440 kg/m^3).

By the data of tests of shotcrete bars 40×40×160 mm, the tensions strength of concrete and cement consumption are related (see Figure 3).

On the basis of the laboratory tests, the accepted design for the wet-mix shotcrete is:

—cement 440 kg/m^3 (CEM II/A-Sh 42.5 N by Bukhtarma manufacture);

—water 175 l/m^3;

—sand 1176 kg/m^3 (Nonmetallic Materials Plant);

—break stone 588 kg/m^3 (Nonmetallic Materials Plant);

—plastifier TamCem 60 (0.3–1.0 kg of liquid additive agent per 100 kg of cement) 1.32 kg/m^3 (Normet company) or GLENIUM® T 803 (0.8–1.5 kg per 100 kg of binder 3.52 kg/m^3 (BASF);

—setting accelerator TamShat 80 AF (5–6% of cement weight) 22 kg/m^3 (by Normet) or MasterRoc SA 160/167 (3–10% of binder weight) 13.2 kg/m^3 (by BASF).

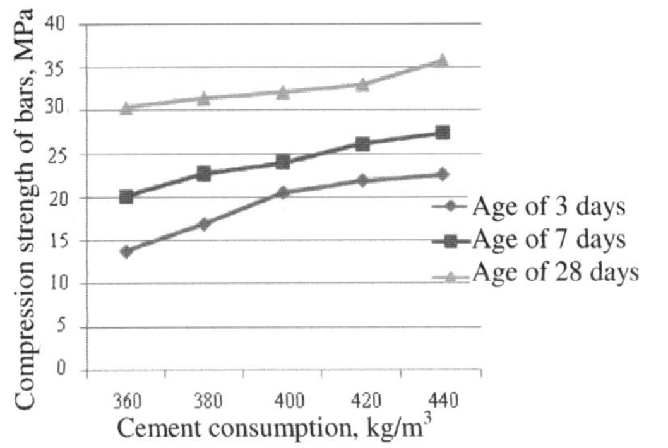

Figure 2. Compression strength of concrete versus cement consumption at water flow rate of 190 l/m^3 without granulated slag.

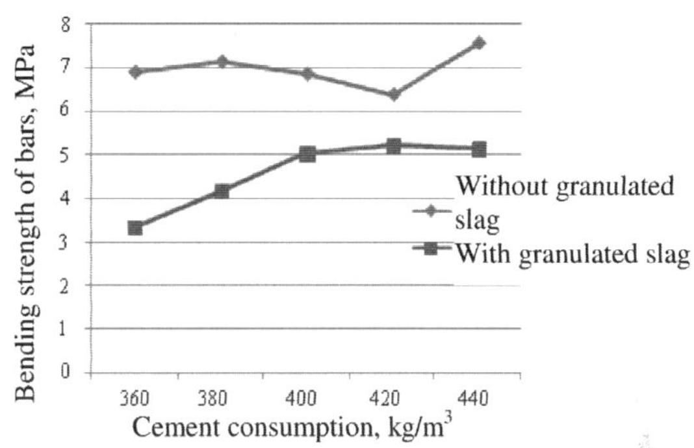

Figure 3. Vending strength versus cement consumption for test concrete bars.

Overall cost of mix components per 1 m^3 of concrete is USD 54.01 (Figure 4).

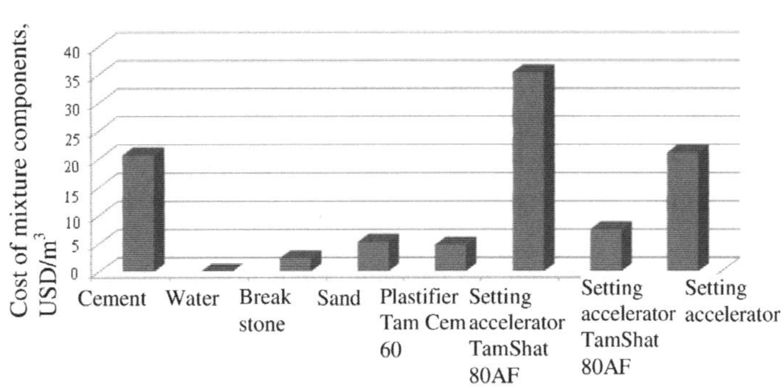

Figure 4. Specific cost of basic process materials (mix components) in Orlov Mine.

Additional expenditures in shortcrete mix preparation may be reinforcement fibers. Concrete reinforced with fibers preserved structural integrity owing totally to stress redistribution inside a specimen and due to high plasticity.

Normet Company recommends to add shotcrete lining mixes with:

—super plastifier TamCem 60 (TamCem 66);

—hydration regulator TamCem HCA (TamCem HCA Plus); it preserves properties of shotcrete and can be used to keep the mix workability from 2 h but not longer than 48 h;

—setting accelerator TamShot 80AF (TamShot 90AF);

—propylene fiber BarChip 54.

Conclusion

Based on the performed tests, the compositions of wet-mix shotcrete are recommended, and it is shown that inert materials have an essential influence on properties of shotcrete mixtures and shotcrete lining, which implies that the final selection of shotcrete mix components should be made after additional full-scale testing of shotcrete properties.

References

[1] *Organizational Standard Requirements ST TOO 050140000656-01-33.1 GU-03-2011: Mine support at Kazakhmys Corporation* Technical Department, Geotechnical Management 2011 (in Russian)

[2] *Departmental Construction Requirements VSN 126-90: Transportation and metro tunnel support with shotcreting and rock bolting. Design and implementation standards* (in Russian)

[3] *Organizational Standard Requirements ST TOO 050140000656-01-33.1 GU-031-2011: Selection of shotcrete composition. Production procedures* Kazakhmys Corporation, Technical Department, Geotechnical Management Almaty 2011 (in Russian)

[4] *NOSTROI Code. Development of underground space. Underground construction with mining method and shotcrete lining. Work rules, control and requirements* Moscow 2013 (in Russian)

[5] *State Standard GOST 27006-86 Concrete. Composition selection rules* (in Russian)

Broadband antenna for ground penetrating radar application in soil

LV Shebalkova, MA Markov and VB Romodin

Novosibirsk State Technical University,
Novosibirsk, Russia

E-mail: rector@nstu.ru,

Abstract. The scope of the article embraces the features of design of antennas and arrays for GPR, their type and parameters and the expediency of the application of the antenna arrays in various purpose location systems of GPR.

Subsurface location refers to geophysical methods using radiolocation impulse signals to build images of objects and voids in soil with the use of a wide electromagnetic emission spectrum ranging from tens megahertz to few gigahertz. Selection of frequency band depends on requirements to resolution and a depth of object detection. The subsurface location systems (georadars) are based on the concept of classical radiolocation systems, but alternatively to them the signal emission is realized to a dielectric medium with loss, rather than to open space. To gain good resolution and acceptable depth of object detection it is necessary that antenna has a wide transmission band, high amplification factor, compact construction applicable in manpack sets.

Super-widebandantennascanbenominallyclassifiedbythecriteriononrelationbetween antenna size and a spatial length of an exciting pulse $\chi = c \cdot \tau$ (c is light velocity, τ is pulse length) into three basic types:

– travelling-wave antenna (dimensions exceed a bit χ),

– quasi-resonant antenna (dimensions are approximately equal to χ),

– elementary antenna (dimensions are much less than χ).

Prevalent travelling-wave antennas are TEM-horns. Their positive features are high efficiency and directivity of radiation and their imperfections are large overall dimensions, especially when χ amounts to few decimeters or meters.

Quasi-resonant antennas have less dimensions, but come short of efficiency and directivity of radiation as compared to travelling-wave antennas, moreover, they imply the use of complex loads intended to suppress generation of post-pulse oscillations in a radiated signal (Figure 1).

Elementary antennas are capable to radiate and receive super-wideband signals without post-impulse oscillations; they are applicable as single units and as elements of transmission and receiving antenna arrays designed to perform spatial processing of signals along with appreciable enhancement of GPR energy potential [1].

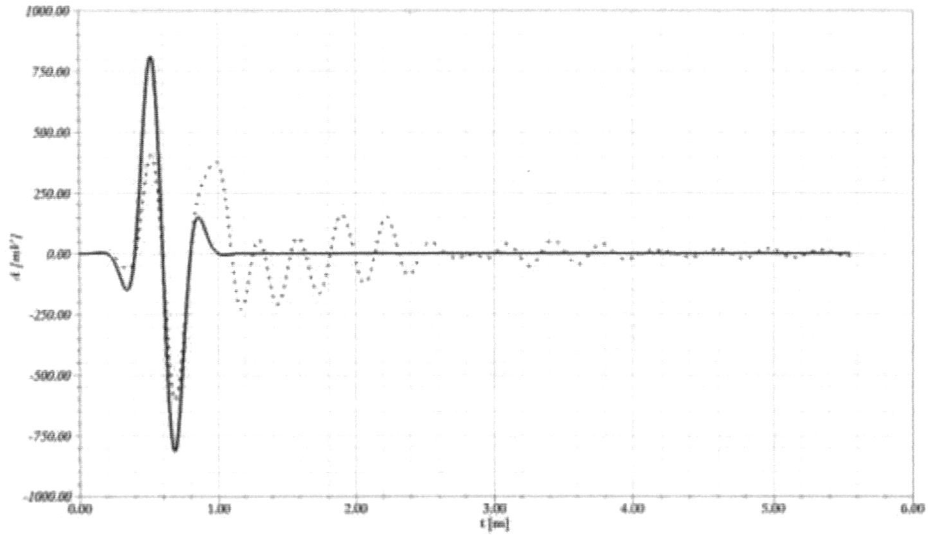

Figure 1. Super-wide band signal in input of a printing bicone.

The most popular a are electromagnetic horn antennas with a single, but important imperfection, namely, large dimensions, compact antennas (printing bicones), Vivaldi antennas, dipoles, frame antennas, monopoles, etc. Printing antennas have notable advantages thanks to low cost, small size and weight. If antenna system has imperfect parameters, the greater capacity of a transmitter and accumulator battery is required with respective appreciation of GPR.

GPR can use a single receiving-transmitting antenna, one receiving and one transmitting antennas, one transmitting and few receiving antennas, or few transmitting and receiving antennas.

In design of antenna array it is imperative to consider specific features if its application, first of all its location near earth surface in order to use the effect of penetration into soil in order to diminish reflection of a survey surface. The near field effect should be taken into account in antenna design, the glancing angle effect can be limited as it hinders the 3D full visualization because of refraction index of soil [2].

A radiated pulse without side lobes is the principal objective in designing an antenna; it is rather complicated due to a number of factors, first, the components of most of electrically-powerful antennas radiate electromagnetic different-frequency fields with respective different-directivity diagrams. Given that position of the phase center of the antenna depends on frequency, the shape of an impulse is distorted as signals at each frequency are to pass different routes. Pulses at antenna input undergo distortions because of reflections in its propagation route along antenna and especially on antenna's edges. Cumulative reflections are capable to cause impulse tails on echo impulse in the time domain. The reason for this is that antenna represents a resonance structure of high Q-factor and of low transmission band, not enough to hold an impulse signal band.

Figure 1 presents calculation results on input impulse signals in "bow-tie" antenna (printing bicone) for 2 - 5 GHz frequency band obtained by Galerkin's discrete method in time domain. Soil layer of dielectric permeability equal to 6 was considered.

Super-wide band signal applied to antenna input is shown in solid line, a radiated pulse is in dotted line, signal spectrum is shown in Figure 2.

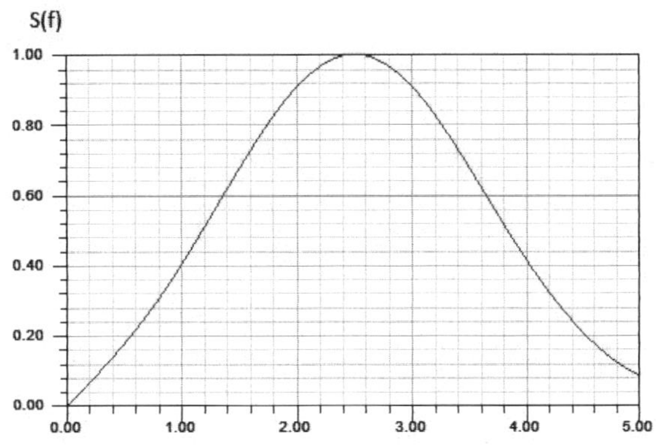

Figure 2. Spectrum of wide band signal.

In different-purpose GPR units the antenna arrays (AA) are applied to boost velocity of earthscape survey. There are two approaches for AA data design. In the first method few discrete elements connected to a controller are used, and data processing is individual for each element. In another approach AA system is developed, thereto, the system exhibits higher transition capacity and applies algorithms of synthetic aperture radars.

A single antenna element, let, a dipole has a limited potential in terms of amplification. Thus antenna array is applied to arrange a diagram of directivity with high amplification factor. There are two "point-point" and collinear types considering the orientation of elements relative to array axis. Expressions of radiation intensity $P(\theta, \varphi)$ for both types are formulas (1) and (2), provided that all the elements are uniformly excited in amplitude and phase (uniform array):

$$P(\theta,\varphi)=15\pi\left(\frac{I\Delta l}{\pi}\right)^2 \frac{\sin^2\left[\frac{N\pi d}{\lambda}\cos\theta\right]}{\sin^2\left[\frac{\pi d}{\lambda}\cos\theta\right]}\left(1-\sin^2\theta\cos^2\varphi\right) \quad (1)$$

$$P(\theta,\varphi)=15\pi\left(\frac{I\Delta l}{\pi}\right)^2 \frac{\sin^2\left[\frac{N\pi d}{\lambda}\cos\theta\right]}{\sin^2\left[\frac{\pi d}{\lambda}\cos\theta\right]} \quad (2)$$

Where θ is angle, deg, λ wavelength, m, N –is number of radiating elements, φ is phase delay between elements, d is distance between elements, I is current, Δl is ascending length of array element.

A pitch between elements has strong influence on a multiplier of an array and on upper boundary of the actual frequencies band of the antenna. Larger pitch between elements gives a greater antenna gain. Nevertheless, as a rule inter-element pitch is selected less than $\lambda/2$ for undirected elements in order to eliminate a rise of diffraction lobes. Given that diffraction maximums come into existence, the array gain falls sharply [3].

Comparison of emission power in a plane in distance 20λ from 5-element array and 11-element array with inter-element pitch λ is shown in Figure 3. Special attention should be paid to the effect of increased number of array elements on the antenna gain and to the reduction in the level of side lobes.

Most conventional AA units have corporative feed circuit to provide a single input for array elements excitation or output of received power. A beam is formed under proper phasing of elements; thereto, phases of elements are fixed as a rule. In more complex methods the phases of elements can be controlled with possible adjustment of a beam and even pattern zero.

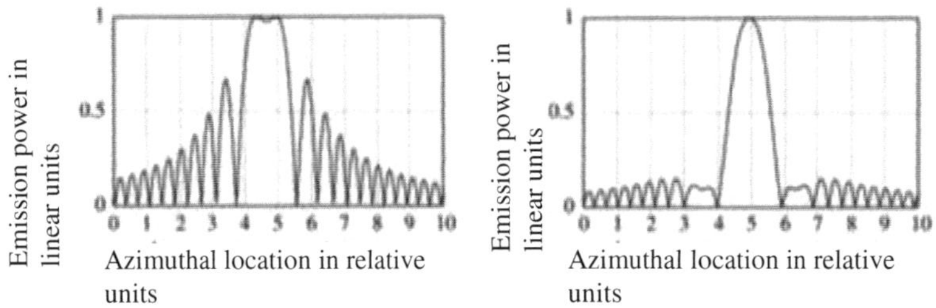

Figure 3. Comparison of emission power of 5-element and 11-element arrays.

In some array designs 2D array of matched transmitters and receivers is used to obtain an active image of an under-array area. This provides advantage of the use of few transmitters in radiation of an area under the wideband energy scanning, as well as a few receivers to detect reverse dissipative energy, and provides noise rejection through synchronization of several impulse radars in the time domain.

By applying independent control of launch and gating of every transmitter and receiver, the system focuses the radiolocation energy in a point and then moves this point to scan an area under the array (Figure 4). If sequence of impulses is controlled in time by means of differential delay in time between each element, then a position of the beam can be adjusted, though the delay between elements in time is limited by the maximum equivalent distance between each element [4].

All the array systems are oriented to obtain an image of hidden objects, and the accurate positioning of the elements is a constitutive factor to gain this target. Differential GPS (DGPS) in combination with inertial navigation systems can serve efficient instruments.

When an antenna array is over the earth the compensation of surface relief variation is required. It is preferable to consider one more aspect, it is a pitch between array elements. It should be large enough to provide a required resolution as a probability of detection of minor objects is closely related to density of elements in an antenna array [2].

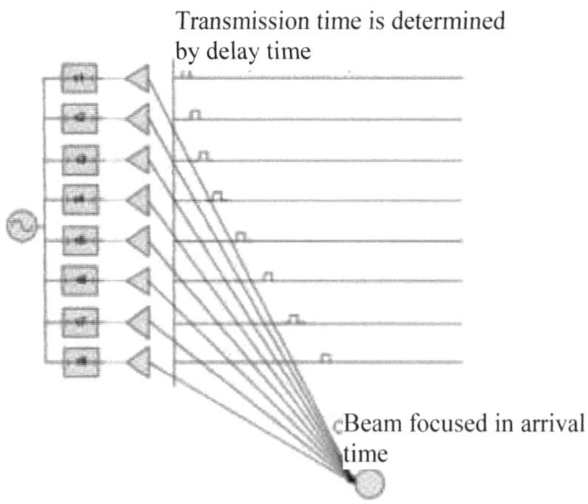

Figure 4. Adjustment of antenna beam by means of time delays for radiating elements.

Conclusions

Types of antennas used in geolocation systems for detection of hidden objects are determined by an operation frequency range and requirements for compactness and architecture of the system itself. In calculation of passing a super-wide band signal through a soil mass it is necessary to consider interface of media with different electrophysical properties. Numerical methods in the time domain can be helpful to analyze the influence of geometric and electric specifications of antennas on parameters of radiated and received impulses, be precise, on a rise of post-impulse oscillations.

References

[1] Grineva AY (Ed.) 2005 *Issues of Ground Penetrating Radar: Co-authored Monograph* Moscow: Radiotekhnika (in Russian)
[2] Allen B, Dohler M, Okon EE, Malik WQ, Brown AK and Edwards DJ 2007 *Ultra-wideband Antennas and Propagation for Communications Radar and Imaging* John Wiley & Sons
[3] David J and Daniels EM 2010 *Detection of Concealed Targets* Wiley-IEEE Press
[4] Daniels D 2004 *Ground-Penetrating Radar* Institution of Electrical Engineers

IOP Conf. Series: Earth and Environmental Science **134** (2018) 012061 doi:10.1088/1755-1315/134/1/012061

Destruction of granular media under pore pressure

EB Sibiryakov* and BP Sibiryakov**

Trofimuk Institute of Petroleum Geology and Geophysics, Siberian Branch, Russian Academy of Sciences, Novosibirsk, Russia

E-mail: *sibiryakoveb@ipgg.sbras.ru, **sibiryakovbp@ipgg.sbras.ru

Abstract. The researchers have modernized and used the boundary element method to calculate all components of displacement vector on a sufficiently complex surface. It is shown that under compression, given the rigid contact between grains, pore pressure induces tensile loads on the contact areas. With the increase in the loading frequency up to efficient wavelength conformable with the size of the contact area, the amplitude of the tensile loads reduces and the resonant frequencies are absent.

It is known that velocities of P- and S-waves in a micro-heterogeneous medium containing fluids depend on the difference of the external pressure and the internal pressure of the fluid, or the so-called effective pressure $P_e = P_{ex} - P_{pore}$ [1]. The common interpretation of this relation says that the pore pressure 'hinders' the external pressure to close pores and cracks. This is an unsatisfactory explanation as the both pressures are scalars (impossible to direct anywhere) and the both are compressive. An indirect proof of deficiency of the above interpretation is that on order to interpret experimental data, the relation above is added with a dimensionless member n of the unclear physical nature, so that: $P_e = P_{ex} - nP_{pore}$, $n \leq 1$ [2].

Actually, a grain is subjected to a complex stress state when the grain contacts and the fluid–matrix interfaces experience different loads both by value and sign. Such loads can break the contracts and make the matrix a granular medium. That is why it is impossible to consider a granular medium as a solid, while averaging of stresses over the volume of a medium disables structural analysis of the mentioned phenomenon.

It has been shown in [3] that depending on the structure of the pore space, the compressive pore pressure in a granular medium can induce both compression (strengthening) and tension at the contacts. The amplitudes of the tensile stresses named as wedging pressure differ by tens per cents in case of rigid and slide contacts but have the same sign. Furthermore, the numerical modeling shows that wedging pressures grow at smaller contacts; accordingly, failure starts at the contact of smaller radii as the tensile load increases when the radius of a contact reduces. However, the mentioned study focused on the sphere with only two axially symmetrical contacts. It is interesting to analyze many contacts in order to take into account their interactions and to study the effect of the frequency of stationary fluctuations in the pore pressure with a view to modeling hydrofracture or other dynamic loads.

A representative granular medium may be a ball with eight flat contact sites if the rest medium is replaced by an adequate system of forces in accord with the Saint Venant principle. It is also reasonable to solve the problem in the framework of linear elasticity, i.e. under assumption that all energy losses due to stress raisors at the edge of the contacts are small. The surface depicted in Figure

IOP Conf. Series: Earth and Environmental Science **134** (2018) 012061 doi:10.1088/1755-1315/134/1/012061

1 was set in a parametric representation, i.e. each of 2500 points was either on the sphere or at a contact site. Actually, that meant that the edge of a contact was rounded so that the normal changed rapidly but continuously.

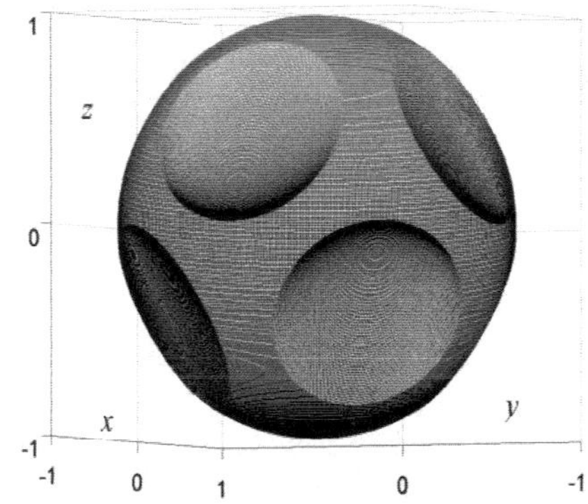

Figure 1. A unit ball with eight remote segments which define the contact sites. The contacts preserve their coordinates and make a rigid frame structure. The rest of the surface is the matrix–fluid interface.

For the comparison of the calculations with the geometry of a grain, the same surface is presented in parametric coordinates using the indicator function (I) in Figure 2. The indicator function equals one if the point is at the contact site and zero if the point is at the matrix-fluid interface.

For the convenience, the distribution of the loads is given on the indicator surface.

Mechanically, we need to solve an elastic boundary value problem of mixed type. Zero displacements are set at the contact sites, and a hydrostatic load with the unit pore pressure is set at the rest surface of the sphere.

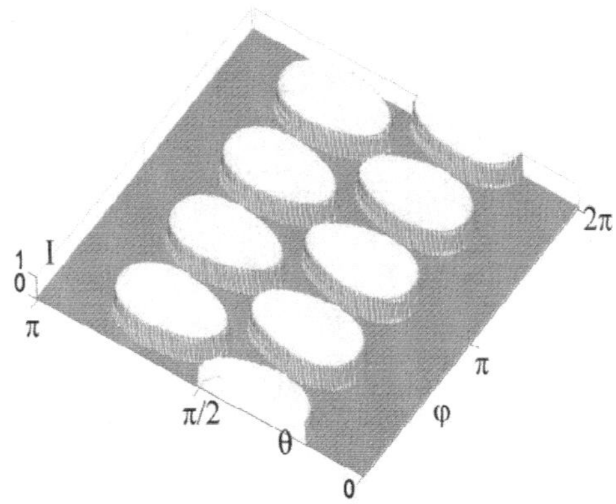

Figure 2. Auxiliary (indicator) surface in the coordinates θ, φ (angles in the spherical coordinate system). $I = 0$ if the point is on the sphere, $I = 1$ if the point is at the flat contact site.

As has been noticed in [4], the boundary element method is advantageous in case of an infinite length of one of the boundaries and in the case of the boundary value problems on the surfaces with the rapidly varying vector of the normal. When the boundary conditions are set at the off-smooth surfaces, the integral method has the advantage in terms of reliability of the results, including better causality of the system of equations (its dimension 7500×7500). In this method, the displacement is the solution of the equation of elastic stationary vibrations:

$$\Delta U + k^2 U + \frac{\lambda + \mu}{\mu} \operatorname{grad}(\operatorname{div} U) = 0, \, k^2 = \frac{\rho\omega^2}{\mu}, \tag{1}$$

where U is the displacement vector; λ and μ are the Lamé elasticity moduli; ρ is the density. The modified BEM [5] ensuring optimal balance of accuracy and causality consists in that the equations (1) satisfying boundary conditions at each fixed point of the surface are not solved as integrals but as the finite sums at a moving point of the surface, and each summand satisfies the equation (1):

$$U_i(x_0) = \sum M_{ij}(x_0, x) f_j(x). \tag{2}$$

The components of the tensor M_{ij} satisfy Eq. (1) analytically and are the appropriately oriented responses with respect to displacements in the direction i towards the finite analog of delta-load applied in the the direction j. The use of the finite analog of delta-load instead of the classical delta-function allows obtaining smooth and finite kernels with the minimized loss of causality. The components of the vector of the potential f_j in (2) are calculated so that to meet the boundary conditions. Then, the dependence between the displacement vectors and the loads is calculated analytically, by differentiation and Hooke's law ($\sigma_{ik} = \lambda\theta\delta_{ik} + 2\mu\varepsilon_{ik}$):

$$p_i(x_0) = -\sum P_{ij}(x_0, x) f_j(x). \tag{3}$$

Here, p_i is the component of the load vector, P_{ij} is the tensor of loads calculated with the assumption that M_{ij} is three columns of the displacement vectors.

The calculations are illustrated by Figs. 3 and 4. Figure 3 shows the distribution of the normal loads at the contact under the unit pressure of fluid. The distribution of the loads is complex. On the other hand, the normal load has a negative sign, which conform with tension despite compression at the matrix–fluid interface. The peak tension is at the edge of the contact and reaches the value of the pore pressure. At the center of the contact, the tension is half as much.

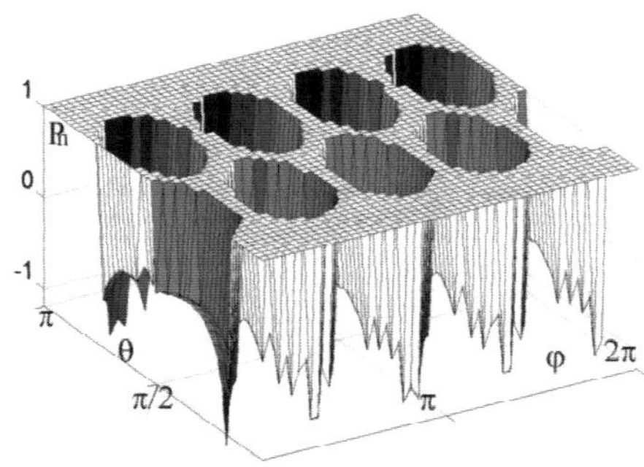

Figure 3. Fistribution of tensil loading (wedging pressure) at the grain contacts. This distrubution is complex but loading has the negative sign, which means tension; the value of the tesnion is approximately eqaul to the fluid pressure.

Figure 4 shows a shearing load induced by the pore pressure. The forces have different signs due to different signs of the normals at the opposite contact sites, but the stresses are the same. It is seen that the shearing loads reaches the value of the pore pressure at the edge of the contact and promotes destruction of the latter. At the same time, the stationary vibrations under similar boundary conditions weaken the traced phenomena even at the frequencies when the wavelength is comparable with the microstructure size. When a fluid is isolated by impermeable beds from the top and the bottom, and when the medium experiences the elementary stress state, i.e. the side and vertical stresses are related as $\frac{\nu}{1-\nu}$ (ν is Poisson's ratio), the pressure in the fluid is higher than in the matrix. In case that the wedging pressure ruptures the matrix and grains become isolated, the pressure in the bed jumps by 30–50% depending on Poisson's ratio. Such situations govern the necessity of predicting anomalously high formation pressures when toping oil reservoir. On the other hand, tensile loads should decrease P- and S-waves in a reservoir, which may a certain sign of pre-failure of the matrix and its transformation to a granular medium. The calculation results also show that the reduction in the contact area intensifies tension. This means that the process, if initiated, continues until total failure.

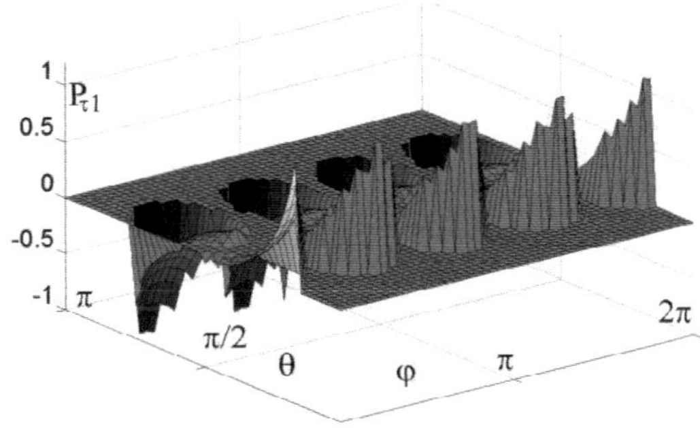

Figure 4. Distribution of shearing load at the contacts. The loads change sign due to different signs of the normals at the opposite contacts (stresses are the same). At the edge of the contact, the loads reach the value of the pore pressure and contribute to failure of the contact periphery.

Conclusion

When fluid pressure affects a granular medium where contact sites make a rigid frame structure, two mechanisms of failure arise in the medium. One mechanism is tension at the contacts upon the increase in the pressure at the matrix–fluid interface. The tension (wedging pressure) is comparable with the pore pressure and depends on the geometry of the contacts. The second mechanism of failure is connected with the high shearing loads of the same order at the edges of the contact, which contribute to destruction of the contact periphery. The both mechanisms condition hazardous, anomalously high reservoir pressures in the course of striking oil.

Under stationary vibrations at a growing frequency, the both mechanisms of failure weaken up to very high frequencies conformable with the wavelengths comparable with the microstructure sizes.

The boundary element method results in a linear system of good causality, which allows avoiding regularizing procedure and provides reliable numerical results even at the contact edges where vector of the normal changes very rapidly.

The decrease in the formation velocities of P- and S-waves may be an indication of the pre-failure state in the granular matrixes.

References

[1] Serge A Shapiro & Axel Kaselow 2005 Stress and pore pressure depending anisotropy of elastic waves *Poromechanics—Biot Centennial (1905-2005)* Taylor & Francis Group London pp 167–172

[2] Yu G, Vozoff K and Durney DW 1991 Effect of pore pressure on compressional wave velocity in coals *Exploration Geophysics* 22(2) ASEG pp 475–480

[3] Sibiryakov EB and Sibiryakov BP 2010 Pore space structure and wedging pressure in granular medium *Fiz. Mezomekh.* Vol 13 No 1 pp 40–43

[4] George D Manolis and Petia S Dineva 2015 Elastic waves in continuous and discontinuous geological media by boundary integral equation methods: A review *Soil Dynamics and Earthquake Engineering* 70(2015) pp 11–29

[5] Sibiryakov EB 2016 Boundary element-based modeling of reflection from rough boundaries *Proc. 12 Int. Conf. InterExpo GEO-Sibir 2016* Novosibirsk: SGUGoT Vol 1 pp 262–267 (in Russian)

Air distribution system with the discharge action in the working cavity of downhole air hammer drills

VV Timonin, SE Alekseev, DI Kokoulin and B Kubanychbek

Chinakal Institute of Mining, Siberian Branch, Russian Academy of Sciences, Novosibirsk, Russia

E-mail: timonin@misd.ru

Abstract. It is proposed to carry out pre-mine methane drainage using underground degassing holes made by downhole air hammer drills. The features of downhole air drills are described. The downhole air drill layout with the simple-shape striking part is presented with its pluses and minuses. The researchers point at available options to eliminate the shortcomings. The improved layout of the downhole air hammer drill is suggested. The paper ends with the test data on the prototype air hammer drill, its characteristics and trial drilling results.

Hammering-action machines are long and successfully used in different rock drilling in surface and underground mines.

In coal mines, efficient operation of fully mechanized longwall system requires reducing methane concentration in mine air. This is achieved by pre-mine drainage and intense ventilation in roadways and mined-out area. The most productive method of pre-degassing is drainage hole drilling in a future extraction panel from delineating roadways. This method allows long small-diameter drilling [1].

In the mid-1900s in Russia, for the first time in the international practice, pilot air hammer drilling machines placed at the drill hole bottom were designed and manufactured—downhole air hammer drills. Such location ensures the best blow energy transmission to the hole bottom and allows higher penetration rate, longer and more straight directional drilling [2–4]. Compressed air is used as an energy supply and a cleanout agent, which makes the technology even more rational.

Located at the hole bottom, an air hammer drill is limited in dimension but imposed higher demands on blow energy and capacity. Hammering parts operate at the limit capacities. For this reason, reliability of the machines is a top priority.

Currently there is a trend toward higher pressure energy supply. Transition to higher pressure of compressed air ensures high power performance and, at the same time, creates even higher stresses in hammering parts.

An important factor governing reliability of hammering machines is stress raisors. Stress razors in steel are nonmetallic inclusions, thus, hammering parts should be made of high-pure steel. Furthermore, to avoid risk of failure and eliminate stress razors, it is expedient to develop the simplest designs for hammering parts.

Designed at the Institute of Mining, SB RAS, the downhole air hammer drill have a simple-shaped hammering part, which enhances its operating reliability (Figure 1) [5–7].

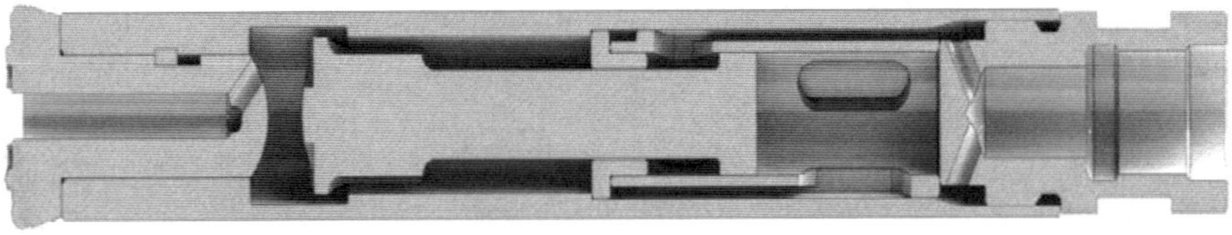

Figure 1. Schematic downhole air hammer drill.

A feature of this layout is the controllable interconnected butt and annular chambers of power stroke and the constant-pressure chamber of idle stroke. This design enhances the strength of the hammering part, free from any internal channels, which is particular important for operation with higher pressure energy supply. The split bushing allows larger area of acting face of the hammering part in the power stroke, which is the basis for reaching increased capacity. This is the closed-type layout with the air exhaust to the hole bottom, which prevents the inner area of the hammer drill from abrasive particles and improves cleanout of the hole bottom from drill cuttings. The better and more complete cleanout of the hole bottom makes it possible to avoid energy loss in re-crushing of drilling cuttings.

In the discussed layout, exhaust is only possible from the power stroke chamber and not from the idle stroke chamber. The 'parasitic' volume inefficiently filled with compressed air is reduced. The reduction leads to the decrease in the air flow rate and enables engineering economic machines. High-pressure compressed air is mostly used in the work cycle period, which also contributes to economic efficiency of the machine.

It is possible to considerably simply a drill bit. The cavity between the hammering part and drill bit is a cavity of atmospheric pressure and need not sealing. This allows minimizing the number of seats on the drill bit, which reduces precision and quality standards and cuts-down manufacturing cost.

The centralized collision of the hammering part and the bit improves the power transmission due to decreased lateral vibration and lowers stresses in the colliding parts owing to more uniform distribution of impact loading in cross-sections.

Based on the presented basic layout, a line of downhole air hammer drills has been designed [8, 9]: ASH-46 for 46 mm-diameter hole drilling; P110GM—drilling diameter 100 mm; PV170—drilling diameter 155–170 mm, and P165 for coring in geological exploration. These machines are operated both in surface and underground mines and are suitable for wide-range pressure of energy source.

Higher energy performance is also governed by the value of active areas of the hammering part, both on the side of the power stroke chambers, S_p, and on the side of the idle stroke chamber, S_i. the ratio of the values of these areas is important, too [10]. In order to obtain a pre-set length travel of the hammering part, it is required to generate a certain momentum of idle stroke, which depends on the active area of the hammering part on the side of the idle stroke chamber.

For the above-described layout, the theoretical range of the active areas is

$$0 < \frac{S_i}{S_p} < 0.5$$

Actually, as required by the design and strength of parts, this ratio is $S_i/S_p = 0.2 \ldots 0.3$. Using this ratio, it is possible to obtain the wanted length of travel of the hammering part but it is necessary to avoid negative effects such as leakage of compressed air from the main constant pressure chamber to the power stroke chambers.

As is evident from the ratio of the active areas, with the enlarged active area from the side of the power stroke chambers, the hammering part has a smaller active area on the side of the idle stroke chamber. A hammering part is a mobile part, and its seats are mead with clearances adjusted depending on diameter. Compressed air inevitably leaks in the clearances. During the idle stroke, compressed air in the power stroke chamber, because of the considerable difference in the size of the active areas, creates appreciable resistance to the hammering part movement and limits its travel length. With the limited travel, the hammering part looses blow energy. Aimed to abate this effect, the design is added with a special elastic ring (Figure 2).

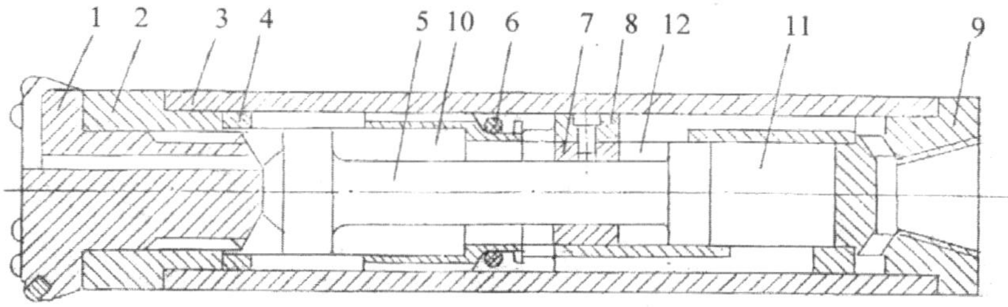

Figure 2. Circuit diagram of downhole air hammer drill PNB76R with the power stroke chamber discharge: 1—drill bit; 2—axle box; 3—housing; 4—front liner group; 5—hammering part; 6—elastic valve; 7—split collar; 8—rear-end liner group; 9—adapter; 10—annular chamber of power stroke; 11—but chamber of power stroke; 12–idle stroke chamber of constant pressure.

The elastic ring 6 installed in the groove of the front liner group 4 with the clearance on the side of the housing 3 acts as a discharge valve. When the hammering part 5 performs idle stroke, air compressed in the annular and but chambers of power stroke, 10 and 11, and air leaked in these chamber through clearances of movable seats of the hammer, is let out through that clearance and the groove in the front liner group to atmosphere. Inasmuch as the hammer areas on the side of the butt and annular power stroke chambers is several times larger than the hammer area on the side of the idle stroke chamber, compressed air in the butt and annular power stroke chambers decelerates idle travel of the hammer and reduces its length. Air discharge from the butt and annular chambers of power stroke after shutting-off the discharge through the elastic valve enables essential reduction in backpressure and, as a consequence, in resistance to the hammering part movement though the length of the idle travel of the hammer increases. During the power stroke, the hammering part undergoes effect of compressed air from the butt and annular chambers of power stroke at the higher value of the travel length, which increases the unit blow energy and the whole capacity of the machine.

The elastic ring–valve falls within the structure and has now adverse effect on the design simplicity of the downhole air hammer drill [11].

The prototype downhole air hammer drill PNB76R with the elastic valve and discharge of power stroke chambers during idle stroke has been manufactured. The machine was finished and tested in laboratory. During the tests, the pressure charts were recorded in the power stroke chambers at the main line pressure of 0.6 MPa. The pressure diagrams recorded in the machine with and without the elastic valve are shown in Figure 3.

It is seen in the pressure charts that in the idle travel, at the point of air inlet in the power stroke chambers, the pressure in PNB76R model is lower by 0.1 MPa than in PNB76 model, which decreases the idle stroke resistance and increases its length. It is noteworthy that the valve actuates, i.e. shuts the clearance at the housing wall, immediately after the point of air inlet in the power stroke chambers since the required pressure difference ahead and behind the valve is prompt and sufficient to stretch the valve. This face governs reliable actuation and stable operation of the hammering machine.

After processing the obtained pressure charts using the known procedure [12], the performance parameters of the two tested models of air hammer drills are presented in the table 1.

After the tests, trial drilling was carried out in the dry air, using drilling rig NKR100M, in granite (strength 140 MPa) under compressed air pressure of 0.6 MPa (Figure 4). Drilling speed was 110 mm/min.

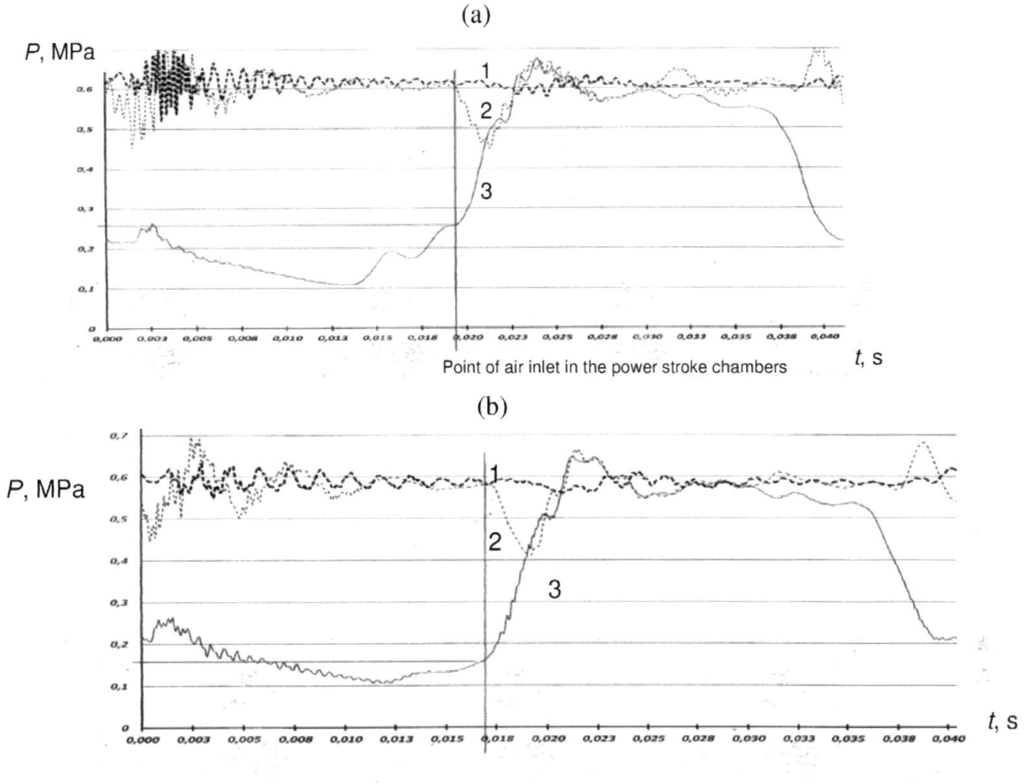

Figure 3. Pressure charts in the power stroke chambers of downhole air hammer drills: (a) PNB76 model without discharge valve; (b) PNB76R with discharge valve; 1—main line pressure; 2—idles stroke chamber pressure; 3—pressure in the butt and annular chambers of power stroke.

Table 1. Performance parameters of downhole air hammer drills

<table>
<tr><th>Parameter</th><th>PNB76 (without valve)</th><th>PNB76R (with discharge valve)</th></tr>
<tr><td>Outward diameter of housing, mm</td><td colspan="2">63.5</td></tr>
<tr><td>Length (without drill bit), mm</td><td colspan="2">675</td></tr>
<tr><td>Weight of hammering part, kg</td><td colspan="2">1.45</td></tr>
<tr><td>Compressed air pressure, MPa</td><td colspan="2">0.6</td></tr>
<tr><td>Hammering part travel length, mm</td><td>79</td><td>93</td></tr>
<tr><td>Blow energy, J</td><td>57</td><td>71</td></tr>
<tr><td>Blow frequency, min^{-1}</td><td>1450</td><td>1456</td></tr>
<tr><td>Capacity, W</td><td>1378</td><td>1723</td></tr>
</table>

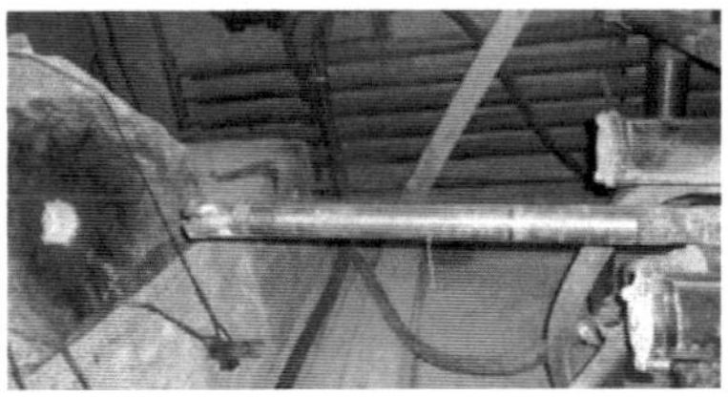

Figure 4. Trial drilling in granite by PNB76U (improved) and drilling rig NKR100M.

To sum up, the introduction of an elastic valve to discharge power stroke chamber in the design of downhole air hammer drill PNB76R resulted in the increase in the hammering part travel length by 30% and in the gain in the blow energy and capacity by 25%. The discharge valve removes the physical disadvantage of the given air distribution circuit, namely, small area of the hammer on the side of the idle stroke chamber as compared with its area on the side of the power stroke chambers, and enables utilizing all advantages of the circuit. This design is serviceable both at the standard and increased pressure of compressed air.

References

[1] Klishin VI et al 2007 Drilling equipment engineering for coal mines *Fundamental Problems of Geoenvironment Formation under Industrial Impact: Sci. Conf. Proc.* Novosibirsk: IGD SO RAN (in Russian)
[2] Hungerford F 2013 Evolution and application of in-seam drilling for gas drainage *International Journal of Mining Science and Technology* Vol 23 pp 543–533
[3] Kondratenko AS, Timonin VV and Patutin AV 2016 Prospects for directional drilling in hard rocks *J. Min. Sci.* Vol 52 No 1 pp 129–134
[4] Timonin VV, Kokoulin DI, Alekseev SE and Kubanychbek B 2016 Straight-path drilling facilities for coal mines *Problems of Development of Mining Sciences and Mining Industry: Sci. Conf. Proc.* Novosibirsk: IGD SO RAN (in Russian)
[5] Alekseev SE 1997 RF Patent 2090730, 11.05.95
[6] Repin AA, Alekseev SE and Pyatnin GA 2009 RF Patent 2343266, 02.07.2007
[7] Repin AA, Alekseev SE and Karpov VN 2012 RF Useful Model 121854, 04.05.12
[8] Repin AA, Kokoulin DI, Alekseev SE, Karpov VN and Shakhtorin II 2014 Small-size air hammer for long directional drilling in underground coal mines, *Prospects of Innovation Development in Coal Mining Regions in Russia: Sci. Conf. Proc.* Prokopievsk (in Russian)
[9] Alekseev SE, Timonin VV, Kokoulin DI, Shakhtorin IO and Kubanychbek B 2015 Development of small-size downhole air hammer for investigation hole drilling J. *Fundament. Appl. Min. Sci.* Vol 2 pp 187–193 (in Russian)
[10] Repin AA, Alekseev SE, Timonin VV and Karpov VN 2015 Analysis of the compressed air distribution in the down-the-hole air hammer, *Miner's Week–2015 Conf. Proc.* pp. 475–482 (in Russian).
[11] Petreev AM and Primychkin AYu 2016. Ring-type elastic valve operation in air hammer drive *J. Min. Sci.* Vol 52 No 1 pp 135–145
[12] Esin NN 1965 *Procedure for Testing and Perfection of Air-Driven Hammers* Novosibirsk: SO AN SSSR (in Russian)

Geomechanical analysis of mining system by cut and fill at deep levels of Internatsionalnaya pipe

MV Tishkov[1], DP Semenov[1], AA Neverov[2]*, SA Neverov[2] and AM Nikolsky[2,3]

[1]Yakutniproalmaz Institute, Mirny, Republic of Sakha (Yakutia), Russia
[2]Chinakal Institute of Mining, Siberian Branch, Russian Academy of Sciences, Novosibirsk, Russia
[3]MiningPro, Novosibirsk, Russia

E-mail: *nnn_aa@mail.ru

Abstract. The authors propose and comprehensively substantiate a mining system by cut and fill for Internatsionalnaya pipe. It is found that application of cut and fill at deep level is limited due to low strength of kimberlite and enclosing rock contact and low safety of mining owing to insufficient area of the pipe beneath the level of -950 m.

1. Introduction

Currently, in view of depletion of the exposed and prepared reserves, Internatsionalny Mine management is highly concerned with the safe and efficient mining at super deep levels -820 m/-1250 m (depth from the forum surface 1200–1650 m).

The feature of mining at Internatsionalnaya pipe, considering depth of occurrence and value of kimberlite ore, is the use of backfilling as the method to control stress state of rocks and extraction factor. Phasing of cutting and filling processes seems to be a sound approach to ground control [1–3].

The mine uses underhand cut and fill. Ore is recovered in slices with rooms and pillars in two or three stages. This combination mining system is the best from the viewpoint of safety and completeness of extraction. This system complies with the whole set of geological and geomechanical conditions of the deposit. On the other hand, low performance of stoping and considerable labor content of development and facer-entry drivage required new approaches to mineral mining. It was decided to analyze variants of cut and fill in combination with room-and-pillar stoping at deep levels of the mine.

This paper presents the data of 3D finite element modeling of state of rock mass, which illustrate the effect of geological, physical and geotechnical factors on the stability of structural elements in the system of mining with room-and-pillar and cemented backfill [4].

2. Mining system

Figure 1 shows the currently applied system of room-and-pillar mining with backfill (RPB).

An extraction block is divided into sublevels by mine roadways of drilling and haulage horizons. The sublevels are mined using room-and-pillar in three stage mining sequence. Stoping includes ore breakage by top-down holes drilled from the drilling crosscut, transport of broken ore from the haulage crosscut and cemented backfill of mined-out area. Adjacent rooms are permitted for extraction when backfill reaches the standard strength.

IOP Conf. Series: Earth and Environmental Science **134** (2018) 012063 doi:10.1088/1755-1315/134/1/012063

Considering the depth of mining at the levels -820 m/-1250 m and the presence of backfilled roof, the basic and most important criterion of influence on structural parameters of a RPB variant is the stability and safety of exposed walls in rooms for the preset period.

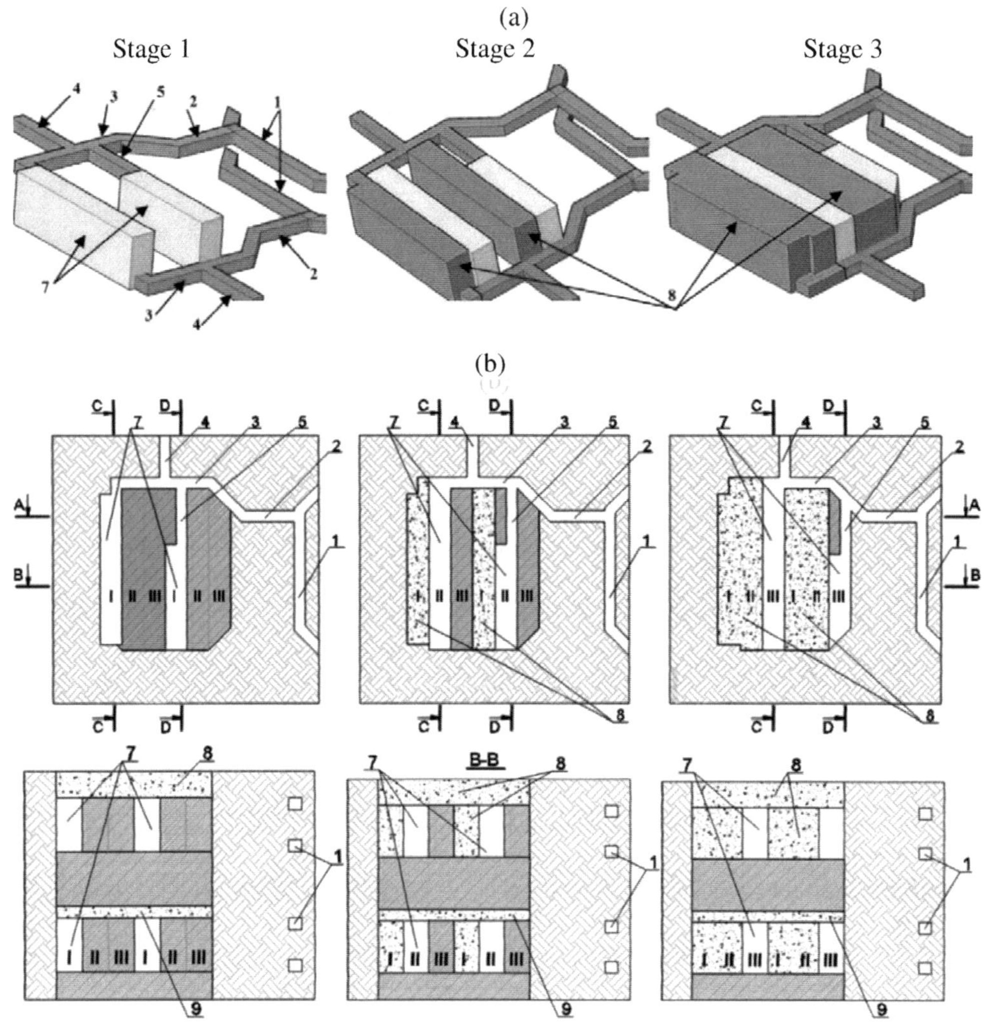

Figure 1. RPB mining system: (a) volume representation of a sublevel; (b) lateral and vertical sections; 1—spiral ramp; 2—roadways to drilling and haulage horizons; 3—rock gangway (for drilling horizon)/haulage drift (for haulage horizon); 4—air crosscut; 5—drilling crosscut; 6—haulage crosscut (cross-bearing 3–4 m); 7—mined-out rooms; 8–backfilled rooms; 9—underhand cutting layer; I—III—sequence of room mining.

3. Rock mass stability

Stability of structural elements in the room-and-pillar mine using backfill was estimated based on the data of stress state obtained in 3D numerical modeling [5–8].

The calculation results are depicted as the zones of probable failure of rock mass surrounding structural elements of the geotechnology (Figures 2–4) by the strength criteria by Mohr–Coulomb and Drucker–Prager [9, 10].

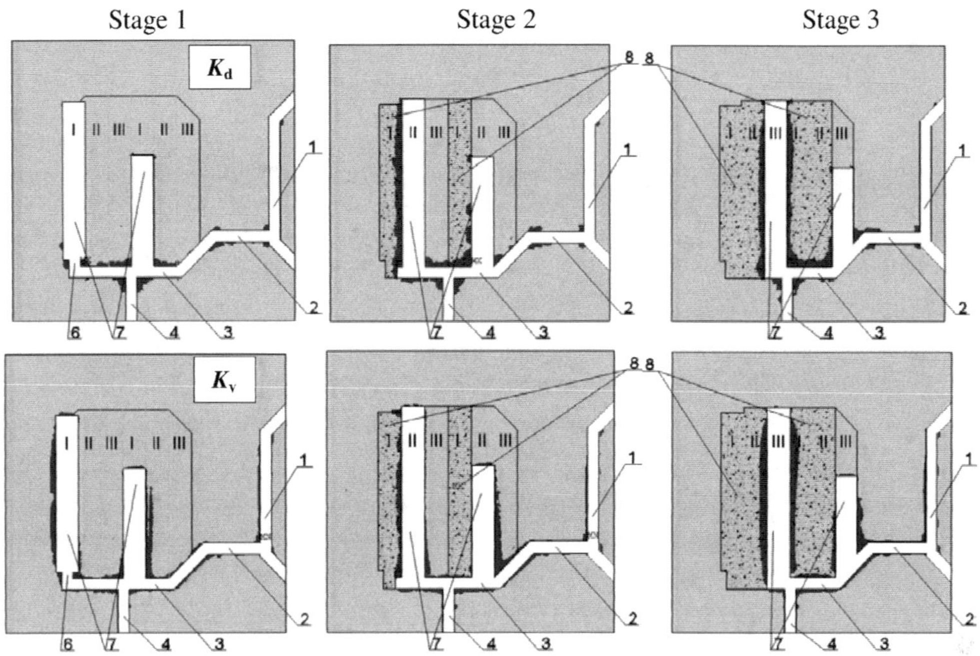

Figure 2. Anticipated zones of rock mass instability by Drucker–Prager and Mohr–Coulomb criteria K_d and K_y, respectively, in the structural elements of RPB mine of the haulage horizon in the upper extraction block (plan view, legend is in Figure 1).

An extraction section is divided into upper and lower extraction blocks by an underhand cutting layer. Each block is divided into sublevels that are developed by rooms of Stages 1, 2 and 3. Between rooms of Stage 1, a temporary ore pillar 20 m long is left (at Stage 1), and hybrid ore-and-backfill pillar (at Stage 2) and backfill pillar (at Stage 3) composed of 2 rooms is established. The height of a sublevel (room) is 20 m, the rooms are 10 m wide and 65–72 m long. The standard strength of the backfill is assumed 3 MPa for the basic layer and 4–4.5 MPa for the load-bearing layer. At a distance of 10 m from the ore and enclosing rock mass contact, a zone of increased damage is detected. The damage ratio for the ore and enclosing rock mass, is assumed as high—$K_c = 0.25$.

According to the calculation results, the structural elements of the RPB mine, which are the most prone to suffer from instability of enclosing rock mass, ore and backfill include (Figures 2 and 3):

– walls of roadways to the drilling and haulage horizons. In the roof of the roadways, a local damage zone is only observed in the ore and enclosing rock mass contact zone of 10 m;

– walls, roof and floor of rock gangways at the ore and enclosing rock mass contact. The size of the probable failure zones change from 2–2.5 m on the ore contact to 1.8 m on the enclosing rock mass side. The roof of the rock gangways of the haulage horizon and the floor of the rock gangways of the drilling horizon within the limits of a mined-out room (10 m) is inside the zone of anticipated avalanche failure due to room mining (end of the room);

– junctures of mine workings: failure zones are 2–3 m in the walls and 1.5–2 m in the roof;

– walls of the spiral ramp, as well as the walls and floor of the drilling crosscuts. In the backfilled roof of the drilling crosscut, the size of the failure zones varies from 0.2–0.5 m by the Mohr–Coulomb criterion to 1.5 m by the Drucker–Prager criterion;

– walls, roof and floor of the mined-out rooms. At Stage 1, in the lower block, in the 20-m temporary pillar, a wide zone of instability is observed. This situation is caused by the insufficient relief of the ore body by the underhand cutting layer 4.5 m high. At Stage 2, in the hybrid ore-and-backfill pillar, zones of instability are local and adjoin the ore side of the pillar in the upper block and are very wide in the lower block. At the final Stage 3, the anticipated failure zones of considerable size appear around the rooms.

– in the roof and floor of the rooms, the instability zones in the backfill and in the ore are local.

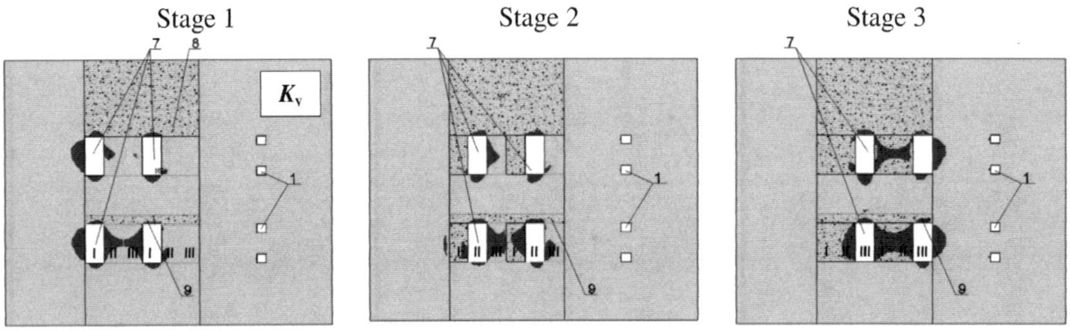

Figure 3. Anticipated instability zones by the Mohr–Coulomb criterion K_y in the structural elements of RPB mine in section B–B as per Figure 1.

Thus, in accordance with the outcome of the geomechanical evaluation of room-and-pillar mining using backfill (RPB) with the set room design as width×length×height as 10×70×20 m and the three-stage sequence of mining, it is in no way recommended to use this variant in extraction of kimberlite reserves from Internatsionalnaya pipe below the level of -820 m.

The main constraint of this RPB variant application is the considerable size of exposed surfaces in a very damaged rock mass. The main condition of RPB safety is the guaranteed stability of exposed surfaces in rooms; for this reason, this variant is only applicable in stable ore and rock mass after obligatory preliminary full-scale testing. Given the ground conditions of Internatsionalnaya pipe, another variant of RPB with the height of a sublevel (room) of 12 m, width of a room of 7.5 m and with the four-stage sequence was analyzed in terms of safety (Figure 4).

In the conditions of Internatsionalnaya pipe, mining in four stages is only possible when rooms are not wider than 6.0–7.5 m due to the limited size of the pipe. For this reason, the four-stage RPB variant is only applicable down to the depth of -950 m.

Figure 4 demonstrates probable areas of instability of rock mass for RPB variant of four-stage mining of rooms. The anticipated damage zones in the structural elements of RPB mine variants with 3 stage and 4 stage mining were subjected to the comparative analysis.

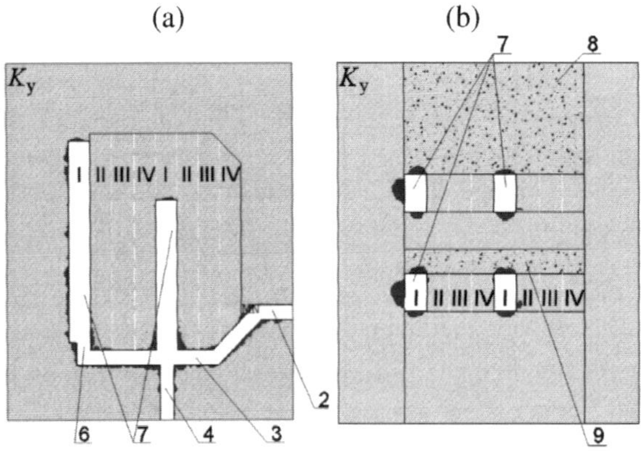

Figure 4. Potential damage zones by the Mohr–Coulomb criterion K_y in the structural elements of 4-stage RPB mine: (a) lateral section of the haulage horizon in the upper extraction block; (b) vertical section along the center of the pipe.

The decrease in the width and height of the room, 4-stage extraction and the thickness of the underhand cutting layer of 8 m between the upper and lower extraction blocks has allowed considerable enhancement of stability of kimberlite-composed walls in the rooms. At the same time, in the backfilled roof of the rooms and access workings, the situation is the same as in the variant discussed above in this paper. The reason is insufficient strength of the backfill for the application at the depth of 1250–1300 m.

As in the 3-stage RPB variant, the most unstable areas are the ore and enclosing rock mass contact and the junctions of drilling (haulage) crosscuts with rock gangways to be obligatory reinforced with the proper support.

On the whole, the geomechanical behavior of rock mass in the variant of 4-stage RPB mine is more favorable than in the variant of 3-stage RPB, and areas of potential damage and failure of rocks are considerably reduced. On the other and, in view of the smaller area of the pipe below the level -1000 m, the four-stage RPB mining will be complicated by longer downtime of mining machinery and will fail to offer required efficiency of stoping.

Accordingly, from evidence of the accomplished geomechanical analysis, 4-stage RPB variant with the discussed design parameters is applicable at the depth not more than 1300 m (mine levels -820 m/-950 m) after obligatory preliminary trial and with due regard to the exclusive standard imposed on backfill strength to be increased up to 5–6 MPa.

4. Conclusions

1. It has been found that the application of combination of cut and fill with room-and-pillar mining at deep level of Internatsionalnaya pipe is limited by the damage ration of rock mass, geotechnology design parameters, consequence of room mining, insufficient standard strength of cemented backfill and small area of the pipe.

2. In view of instability of exposed room surfaces given the current damage ratio of rock mass and due insufficient standard strength of backfill, unsafe sequence of room mining (phasing), limited area of the pipe at the discussed depths and inefficiency of stoping due to long downtime of mining machines, the 3-stage variant of room-and-pillar mining using backfill is not recommended for application.

3. The variant of 4-stage room-and-pillar mining using backfill satisfies, on the whole, safety requirements at the depth not more than 1300 m subject to obligatory preliminary trial and given the increased standard strength of cemented backfill.

References

[1] Freidin AM, Neverov AA and Neverov SA 2012 *Underground Ore Mining: Turtorial* Novosibirsk: IGD SO RAN (in Russian)

[2] Bronnikov DN, Zamesov NF and Bogdanov GI 1982 *Deep-Level Ore Mining* Moscow: Nedra (in Russian)

[3] Kurlenya MV, Seryakov VM and Eremenko AA 2005 *Induced Stress Fields* Novosibirsk: Nauka (in Russian)

[4] Zienkiewicz OC 1972 *The Finite Element Method in Engineering Science* McGraw Hill

[5] Boltengagen IL, Korenkov EN, Popov SN and Freidin AM 1997 Geomechanical substantiation of the parameters of a continuous chamber system of mining with caving of the roof rock *Journal of Mining Science* Vol 33 No 1 pp 55–63

[6] Nazarova LA, Freidin AM and Neverov AA 2005 Chamber mining with roof caving at Nikolaevsk Mine, *Journal of Mining Science* Vol 41 No 4 pp 342–349

[7] Neverov SA and Neverov AA 2013 Geomechanical assessment of opre drawpoint stability in mining with caving *Journal of Mining Science* Vol 49 No 2 pp 265–272

[8] Neverov AA 2014 Geomechanical assessment of combination geotechnology for thick flat-dipping ore bodies *Journal of Mining Science* Vol 50 No 1 pp 115–125

[9] Kazikaev DM 2005 *Geomechanics of Underground Ore Mining: University Textbook* Moscow:

IOP Conf. Series: Earth and Environmental Science **134** (2018) 012063 doi:10.1088/1755-1315/134/1/012063

MGGU (in Russian)
[10] Drucker DC and Prager W 1952 Soil mechanics and plastic analysis or limit design *Quaterly of Appied Mathematics* Vol 10 No 2 pp 157–165

IOP Conf. Series: Earth and Environmental Science **134** (2018) 012064 doi:10.1088/1755-1315/134/1/012064

Fundamentals of converging mining technologies in integrated development of mineral resources of lithosphere

KN Trubetskoy, YuP Galchenko, and VA Eremenko*

Institute of Integrated Mineral Resources Development—IPKON, Russian Academy of Sciences, Moscow, Russia

E-mail: *eremenko@ngs.ru

Abstract. The paper sets forth a theoretical framework for the strategy of the radically new stage in development of geotechnologies under conditions of rapidly aggravating environmental crisis of the contemporary technocratic civilization that utilizes the substance extracted from the lithosphere as the source of energy and materials. The authors of the paper see the opportunity to overcome the conflict between the techno- and bio-spheres in the area of mineral raw materials by means of changing the technological paradigm of integrated mineral development by implementing nature-like technologies oriented to the ideas and methods of converging resources of natural biota as the object of the environmental protection and geotechnologies as the major source of ecological hazards induced in the course of development of mineral resources of lithosphere.

The modern civilization evolves perennially with the outpacing consumption of a variety of natural resources and with the rapid expansion of their range.

In this respect, a peculiar place belongs to the industries related to minerals and raw materials which feed and power manufacturing of 70% of of the world's overall final product and which grow twice as fast as the population [1].

By analogy with the nature and man conflict, the equal opportunity principle should hold true for development of the techno- and biosphere. Academician of the Russian Academy of Sciences N.N. Moiseev defined that principle as "...a policy of co-evolution..." of intrinsically antagonistic systems [2]. Evidently, the scope of this policy is insufficient to completely overcome global contradictions; however it is absolutely indispensable to make these contradictions nondestructive relative to wildlife by means of fundamentally new technologies to be envirosafe and to allow perspective recovery of the man-violated balance between production and consumption of mineral resources with regard to restrictions imposed by ecological imperative.

The modern technocratic civilization is currently at the point of bifurcation. The plain truth is that extensive and unlimited use of natural resources inevitably and irreversibly changes natural environment and the further existence of the society is connected with the creation and application of new resources and technologies based on the modes of wildlife by converging natural sciences and engineering. The key phases of the convergence include: 1—transition to nanosizes; 2—modifying the paradigm from the analysis to the synthesis; 3—approach and interaction of technosphere and wildlife; 4—interdisciplinary approach instead of monotechnics [3].

Judged by the essence of convergence as occurrence of similarity in the structure and functioning of distantly related systems [4], there are two methodological approaches to converging technologies:

—technologies reproducing processes in biological systems;
—technologies having the same functional organization as natural systems.

The Earth's biota as a system develops by the laws and principle that are unchangeable by a man at pleasure. All interrelations in the biota are based on trophic (food) chains of variable complexity; these chains usually have a few trophic scales subjected to the law of reduction in biomass on each net scale (Figure 1).

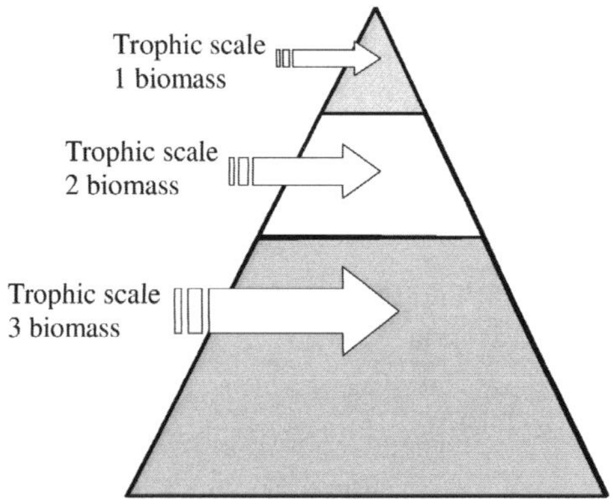

Figure 1. Pyramid of biomass [5].

Existence of the natural equilibrium ecosystems is based on the large and small circles of biotic exchange as a nonstop process of regular cyclic thought nonuniform in time and space redistribution of substance and energy multiply re-entering continuously regenerating biosphere or elementary biological systems. For this reason, considering the nature of interaction and mission of basic categories of organisms in ecosystems, ***the first principle of ecosystem functioning may be formulated as: production and of resources and reprocessing of waste within the closed cycle of all elements*** [6].

The major portion of the solar energy on the earth surface induces physical circulation of mobile elements of biota, the rest energy is taken by photosynthetic living systems and engenders circulation of biogens in natural ecosystems. By arranging the constituent organisms in accordance with their nutritional interrelations with indication of 'entrance' and 'exit' of energy and biogens on each scale, it becomes evident that biomass is larger on each lower scale. This is the essence of ***the second principle of biosystem functioning: on each trophic scale a new biomass is produced in amount inversely proportional to the length of food chains*** [6].

As known, all biological system possess a threshold crossing of which leads to a sharp qualitative change in the processes running in a biosystem—the change in their organization. This feature of threshold (bifurcation) mechanisms has a perfectly peculiar part in development of biosystems. Upon crossing the bifurcation phase, a biological system as is forgets (or nearly forgets) the past. Evolution pathways fork at this point, and, by virtue of probabilistic nature of the threshold crossing, there is no way back (or, more properly, the probability of evolution repeat is close to zero). ***The irreversible nature of biological evolution (known as Dollos' law) is the core of the third principle on function of all components in the Earth's Biota*** [6].

The most important property of natural ecosystems is stability ensured by the dynamic equilibrium between the biological potential of populations and resistance of the environment. The change in the environment initiated succession ending with a climax ecosystem stage. At this stage, all species parturiate at the relatively constant population while an ecosystem is a species diversity [6]. In a terrestrial ecosystem, producers are plant communities—phytocoenosis. The producers have absolutely the same internal structure composed of three systems of plant species of different coenotic value: edificator, assectator and adventity; edificatory determines species composition and direction of cyclic succession for the whole phytocoenosis. Accordingly, ***the fourth principle of ecosystem***

functioning states that stability and endurance of an ecosystem depends on the state of phytocoenosis with its structure and species composition conditioned by its edificator synusia [7].

Flow energy in natural ecosystems totally agrees with the law of thermodynamics. Solar energy is converted to chemical energy under photosynthesis, and chemical energy turns into other energy forms along food chains. However, the major energy of the Sun is spent to heat the planet. Heat is the driving force of water and air circulation and, finally, diffuses in outer space. Whence, ***the fifth principle of ecosystem functioning says that they exist on the strength of pollution-free and nearly eternal solar energy the quantity of which is relatively constant and excessive*** [6].

Regarding a nature-like geotechnology, the 'antagonists' are the Earth's Biota and technocratic anthropogenic system of extraction of wanted raw materials and energy sources from lithosphere by means of mineral mining.

Out of the general approaches described above, the first approach is inapplicable in mineral mining inasmuch as nature features none of such processes. On the other hand, the second approach offers a true pathway to a fundamentally new development paradigm in geotechnology.

Unlike the traditional bionics with its attempt to engineer equipment with the cloned functions and abilities of biota elements, development of converging geotechnologies is possible based on conversion and content transformation of general principles of interaction between elements of biota. For this reason, in view of the antagonistic nature of the conflict between bio- and technosphere, the methodology should be founded on the known provisions of homeostatics on the means of sustaining vital parameters of interacting system through contradictions management.

Regarding a general methodology, the homeostatic approach allows synthesis of antipodes. In relation to the man and nature interaction, this means that the two antagonists are not contraposed but united within a system of controllable harmony achieved owing to limitation of anthropogenic impact to a tolerance range of biota elements. This allows transferring biological information in technosphere by stage-by-stage formation of geotechnology homeostat based on the structure of biological homeostat by substituting the conceptual elements of the latter for their geotechnological analogs [8].

The final result of the transformation is a nature-and-technology system of mineral mining using technologies based on biotic principles and limited in terms of external effect on survival and recovery of biological systems free from the contradictions between the biota and technosphere.

Mineral mining process always results in displacement of immense quantities of rocks and in formation of an anthropogenically altered area in lithosphere. A general structure of this new lithosphere object and transformation of substance during this object initiation and growth is described in Figure 2.

As seen in Figure 2, all ecological consequences of mineral mining are connected with the transition of substance from lithosphere to technosphere and placement of waste in the biosphere later on. By analogy with the first biosystem function principle above, we state that ***a converging geotechnology is imposed with an obligatory requirement to ensure closed-loop circulation of substance in the course of mining***.

It should be mentioned that this principle of organization and functioning of a mine has long been developed as separate problems of science and technology given the modern advance in mining technologies but is unsatisfactorily and inappropriately implemented.

The techno-transformation of the second biotic principle is possible through replacement of the 'produced biomass' notion by its geotechnical analog—'mineral' [1]. In this case, ***a geotechnology function principle is pure production of nothing but target mineral***. Generally, the concept of this trend means that the level of selective mining, by targeted preference or development of engineering solutions, should conform with the nature and level of variability of natural distribution of minerals in lithosphere. The key point is in situ segregation of ore and barren rock by natural or artificial signs (Figure 3).

Based on the principle of conformance between selectivity of a technology and nonuniformity of mineralization, it is clear that any combination of nonuniform mineralization and ore–rock segregation method should have a proper engineering solution implemented at one of the mining stages. Given the

IOP Conf. Series: Earth and Environmental Science **134** (2018) 012064 doi:10.1088/1755-1315/134/1/012064

stringent ecological constraints, this technological trend can become the most efficient way of abating industrial load on the Earth's Biota [9, 10].

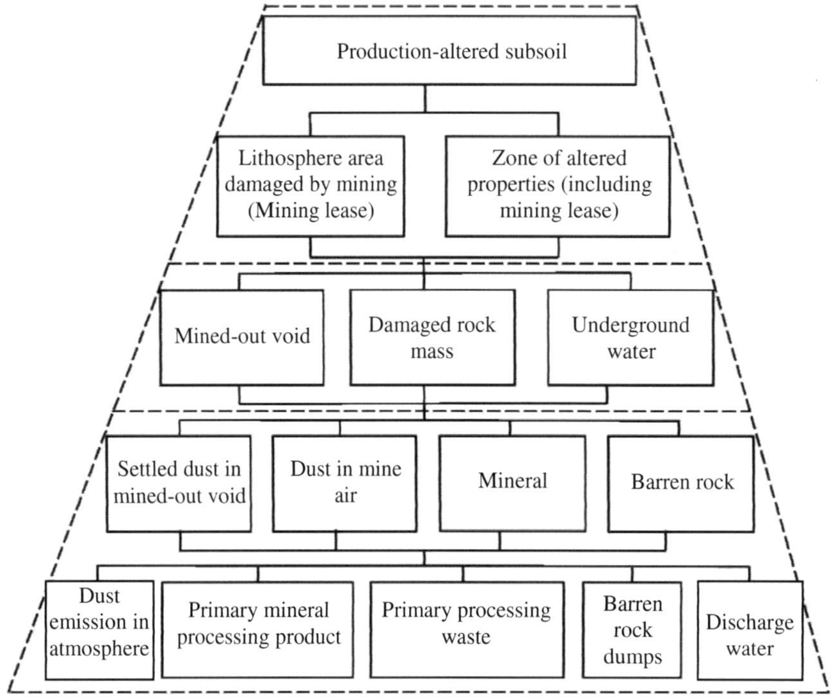

Figure 2. Production-altered subsoil structure and flow of lithosphere substance in the course of mineral mining.

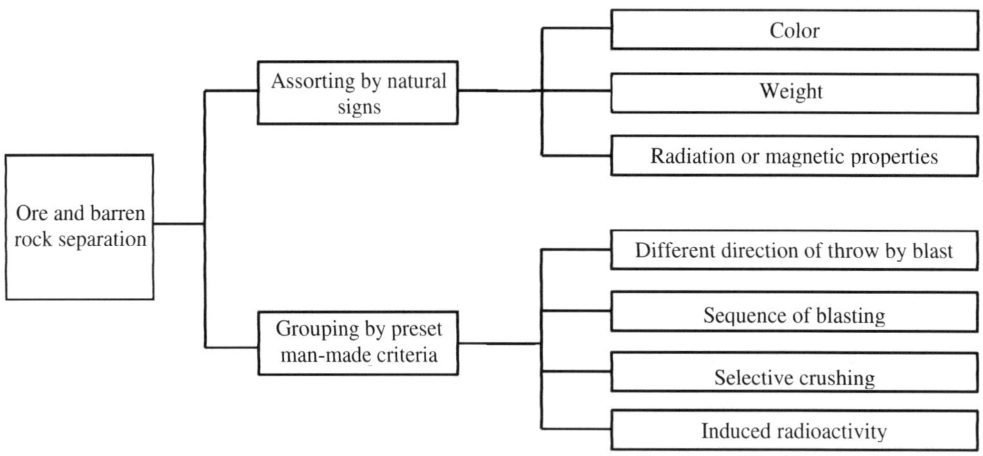

Figure 3. Classification of ore and barren rock separation techniques.

The known Dollo's law in biological systems works out as the succession in the period of critical accumulation of aggravating extraneous factors goes in the direction eliminating influence of these factors. As a result, the species structure of biota upon crossing the bifurcation point transforms so that its new state is independent of the factors which have caused the transformation. In the converging nature-like geotechnologies, this singularity of living systems can be converted into thesis that the basic methodological provision of the innovative technological development is the anticipatory stage-by-stage elimination of the least effective processes and operation by means of modification of the general structure of a geotechnology in space and time, which can be formulated as the preventive action principle in the technology-related decision-making and implementation.

The preventive action principle in the decision-making on a geotechnology structure transform into a geophysical concept of an eco-geotechnology when a production damage area is anticipatorily outlined in the in the field of geophysical change of lithosphere by means of separating the processes of mineral mining and the processes of overcoming of geomechanical disturbance consequences in time.

Elaboration of these provisions will result in a theoretical framework of a brand-new trend in the sphere of mining science as a since on mineral extraction in the course of integrated development of natural resources of the Earth. Judging by biologically governed sequence of mining when the early erected structure ensures geomechanical relaxation of the interior ore reserves and serves as a frame inside which large-scale stoping is carried out, the newly proposed scientific trend may be the development of the 'framing' geotechnology as the deterministic system of different-purpose process clusters that interact in time in conformity with the biogenic principles [11].

Application of general biological principle in construction of the proposed 'framing' geotechnology opens up prospects of successful and final solution of ecological problem generated by mineral mining:

—advanced construction of load-bearing 'frame' preserves integrity of overlying strata and ground surface during actual mining;

—framing of the mining zone before the production start shields fluid-containing strata and protects mining from their influence;

—formation of mined-out area as a system of artificially framed voids offers condition for any kind waste storage.

The fourth principle of biological system functioning in the framework of the discussed issue leaves the content of technological solutions aside and acts as a complimentary criterion to control making and selection of technological decisions.

Regarding the problems connected with the development of mineral resources, this means that concurrently with the new geotechnologies, it is necessary to create a system to control interaction between techno- and biospheres by indexes and criteria reflective of the Earth's Biota properties and response to mining impacts. Such system to be created needs two fundamental problems to be solved:

—foundation of a general methodology for transformation of biological constraints of mining-induced impact on the elements of natural ecosystems into technical standards and regulations for production processes, as well as development of design procedures to calculate these standards;

—generation of cardinally new diagnosis system for the environmental threat of mining, including variety of mining-induced impacts and diversity of purposes and biostructures of violated ecosystems.

The fifth principle of biological system functioning transforms in the technosphere into two backbone provisions for converging geotechnologies:

—resources coming to a technical system from outward source respect the first principle of functioning of biological system;

—energy supply of converging geotechnologies can be obtained from the substance extracted from lithosphere.

It is advisable to generate and use a nonmonetary criterion of the production-caused ecological consequences (including mineral mining). The value of this criterion named the environmental price should be, alongside the monetary criterion, the second obligatory characteristic of any item produced or manufactured, and the society should pronounce this price and do all its best to minimize its value. This is the only impelling and not compulsory motivation to find the least ecologically hostile technologies just like the global avowal of the minim cost stimulated persistent advance toward reduction in consumption of material and labor resources in the technical progress. The structure and the contents of the proposed notion are in detail described in the work [12].

The essence of the second provision above completely contradicts the actual structure of the extensive energy supply the present authors suggest for the mining industry, i.e. the energy required for subsoil development is taken from the subsoil resources.

Naturally, a single way of the energy supply modernization in the current technocratic civilization is efficient economic utilization of the natural renewable energy sources. And it is equally natural that any activity sector will modernize energy supply and utilize renewable energy sources in accordance with the specific character of technologies which are in service in this sector [13].

The implemented research shows that mineral mining industry can use renewable energy sources for (Figure 4):

—general power supply and

—power saving.

In the first case, it is preferable to construct power plants in package with mines. Positively, it is the most efficient to construct different small-size hydroelectric power plants, especially in Siberia and in the Far East of Russia holding nearly unlimited water-power resources. Energy-saving development also agrees with the current trend of construction of small mines to operate by the mode of rotation.

The natural renewable energy sources offer an option of energy saving in the course of mining. With a long history behind, gravity caving technology is barely used today. The implementation of the above-described 'framing' geotechnology opens up new prospects for wider application of gravity caving in underground mineral mining [11].

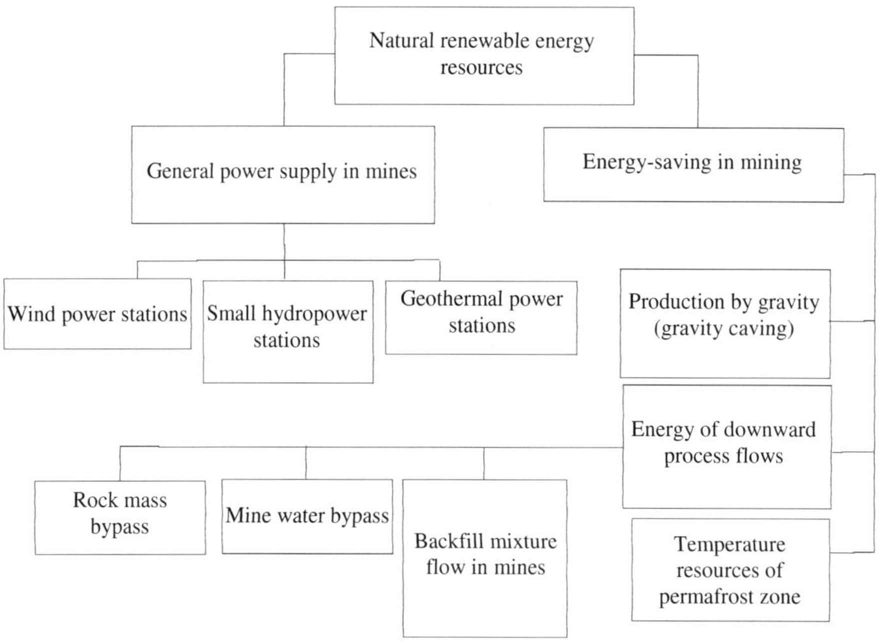

Figure 4. Utilization of natural renewable energy sources in underground mineral mining.

By the common definition, renewable energy sources are the sources of existing or cyclically arising energy in the world. As has been said earlier, subsoil development methods are the sequence of actions or processes that result in separation of a large material volume from lithosphere and displacement of this volume from the occurrence position to the ground surface. At the same time, any specific geotechnology has a functioning structure which combines the mineral flow in the mentioned direction with the practicality of downward (counter) flows of broken barren rocks, mine fluids and different backfill mixtures. In line with the above definition, there always are "cyclic energy flows' in the framework of a geotechnology, identified as the local renewable energy sources by this paper authors. Unlike the continuous renewable energy sources defined as global sources, the local energy sources are not to be used directly in mining processes due to some specific features but can partly make compensation for general energy consumed.

Actualization of the use of such natural renewable energy source needs specific case study and engineering, including special systems of power generators but the potential resource saving based on their application is estimated very high [13].

Not les amplitudinous than gravitation of the Earth, constant and global in terms of subsoil development is the energy source represented by solar irradiation. On the other hand, this immense energy is extremely nonuniformly spread over the earth surface. The solar energy flow density is maximum at the equator and diminishes towards high latitudes. As a consequence, inside vast territories of continents, especially in the Northern hemisphere, average annual temperatures are permanently negative. This difference in the heat balance in different periods of time is factually a renewable energy source when not the solar irradiation but its deficiency is considered the energy value. Based on that, it is possible to introduce definition of a cryo-georesource understood as a geographically and climatically conditioned negative temperature of ambient air and rock mass, usable in mineral mining processes.

This resource can be divided into climatic and geological components, and each of these components varies in time and space by its own laws and exerts different effect on decision-making on process solutions [14]. It is known that ecological cleanness of functioning in biological systems is achieved through the simultaneous effect of all above-listed general principles; thus, the most important methodological aspect of converging geotechnologies is assumed the obligatory synchronism of involvement of all described biogenic principles.

References

[1] Trubetskoy KN and Galchenko YuP 2015 *Geoecology of Subsoil Development and Eco-Geotechnologies of Mineral Mining* Moscow: Nauchtekhlitizdat (in Russian)
[2] Moiseev NN 1990 *Human Being and Nanosphere* Moscow: Molodaya gvardia (in Russian)
[3] Kovalchuk MV 2011 Convergence of sciences and technologies–Break to the future, Ross Nanotekhnol No 1–2 pp 13–24
[4] Pride V and Medvedev D 2008 Phenomenon of NBIC-Convergency: Rreality and expectations Filosof Nauki No 1 pp 97–116
[5] Reimers NF 1994 Ecology (Theory, Laws, Regulations, Principles and Hypotheses) Moscow: Rossia Molodaya (in Russian)
[6] Nebel B and Wright R 1993 *Environmental Science: The Way the World Works* Prentice Hall
[7] Sukachev VN 1964 *Basic Definitions in Bio-Geno-Cenology* Leningrad: Nauka (in Russian)
[8] Gorsky YuM 1988 *Basics of Homeostatics* Irkutsk: IGEA (in Russian)
[9] Galchenko YuP and Sabyanin GV 2011 *Problems of Geotechnology for Vein Deposits* Moscow: Nauchtekhlitizdat (in Russian)
[10] Eremenko VA, Barnov NG, Kondratenko AS and Timonin VV 2016 Method for mining steeply dipping thin lodes *Gorny Zhurnal* No 12 DOI: 10.17580/gzh.2016.12.10
[11] Trubetskoy K N Galchenko YuP and Sabinyan GV 2008 Concept of subsurface development of bowels of the earth on the basis of "framework" geotechnology *21st World Mining Congress Session 15* Krakow P 309–317
[12] Trubetskoy KN and Galchenko YuP 2004 Methodological basis of ecological safety standards for the technogenic impact of mineral resource exploitation *Russian Journal of Ecology* Vol 35 No 2 pp 65–70
[13] Rylnikova MV and Galchenko YuP 2015 *Renewable Sources of Energy in Integrated Subsoil Development* Moscow: IPKON RAS (in Russian)
[14] Galchenko YuP and Sabyanin GV 2009 Application of temperature resource to reduce ecological impact of gold vein mining in permafrost zone *Zolotobyv. Promysh.* No 3(33) pp 22–26

IOP Conf. Series: Earth and Environmental Science **134** (2018) 012065 doi:10.1088/1755-1315/134/1/012065

Determination of deformation and strength characteristics of artificial geomaterial having step-shaped discontinuities under uniaxial compression

PA Tsoy [1, 2*]

[1]Chinakal Institute of Mining, Siberian Branch, Russian Academy of Sciences, Novosibirsk, Russia

[2]Novosibirsk State Technical University, Novosibirsk, Russia

E-mail: *paveltsoy@mail.ru

Abstract. In order to determine the empirical relationship between the linear dimensions of step-shaped macrocracks in geomaterials as well as deformation and strength characteristics of geomaterials (ultimate strength, modulus of deformation) under uniaxial compression, the artificial flat alabaster specimens with the through discontinuities have been manufactured and subjected to a series of the related physical tests.

The investigation into initiation and propagation of fractures in flat alabaster specimens was reported in a number of publications. In [1] the researchers established that in artificial specimens with a fracture induced in the center the developed discontinuity is generated under the load not differing appreciably from the critical load. The orientation angle of the induced fracture sensibly affects such mechanical properties of the test specimen as the ultimate strength of the specimen [2]. Different approaches are practiced to represent experimental evidence on behavior of flat alabaster specimens with artificial disturbances under uniaxial compression. In particular, orientation, propagation, and fracture-initiating stresses are calculated by the discontinuous displacement method (DDM) and comparatively new extended finite element method (XFEM) [3, 4].

A series of respective mechanical tests was undertaken on artificial flat specimens of 60×60×10 mm overall dimensions, manufactured of alabaster with through discontinuities with an intent to establish empirical relationships between linear dimensions and orientation parameters of step-shaped macrofractures in geomaterial specimens and their strain and strength properties, viz., ultimate strength and modulus of deformation under uniaxial compression.

In Figures 1 and 2 photos of specimens before and after tests are shown. The parameters of discontinuities were preset by a number and length of steps (Figure 1a – 5 steps of 10 mm; Figure 1c – 6 steps of 10 mm; Figure 1e - 7 steps of 10 mm; Figure 2a – 10 steps of 5 mm; Figure 2c – 12 steps of 5 mm; Figure 2e – 14 steps of 5 mm). Orientation of discontinuities was fixed at angle 45°.

IOP Conf. Series: Earth and Environmental Science **134** (2018) 012065 doi:10.1088/1755-1315/134/1/012065

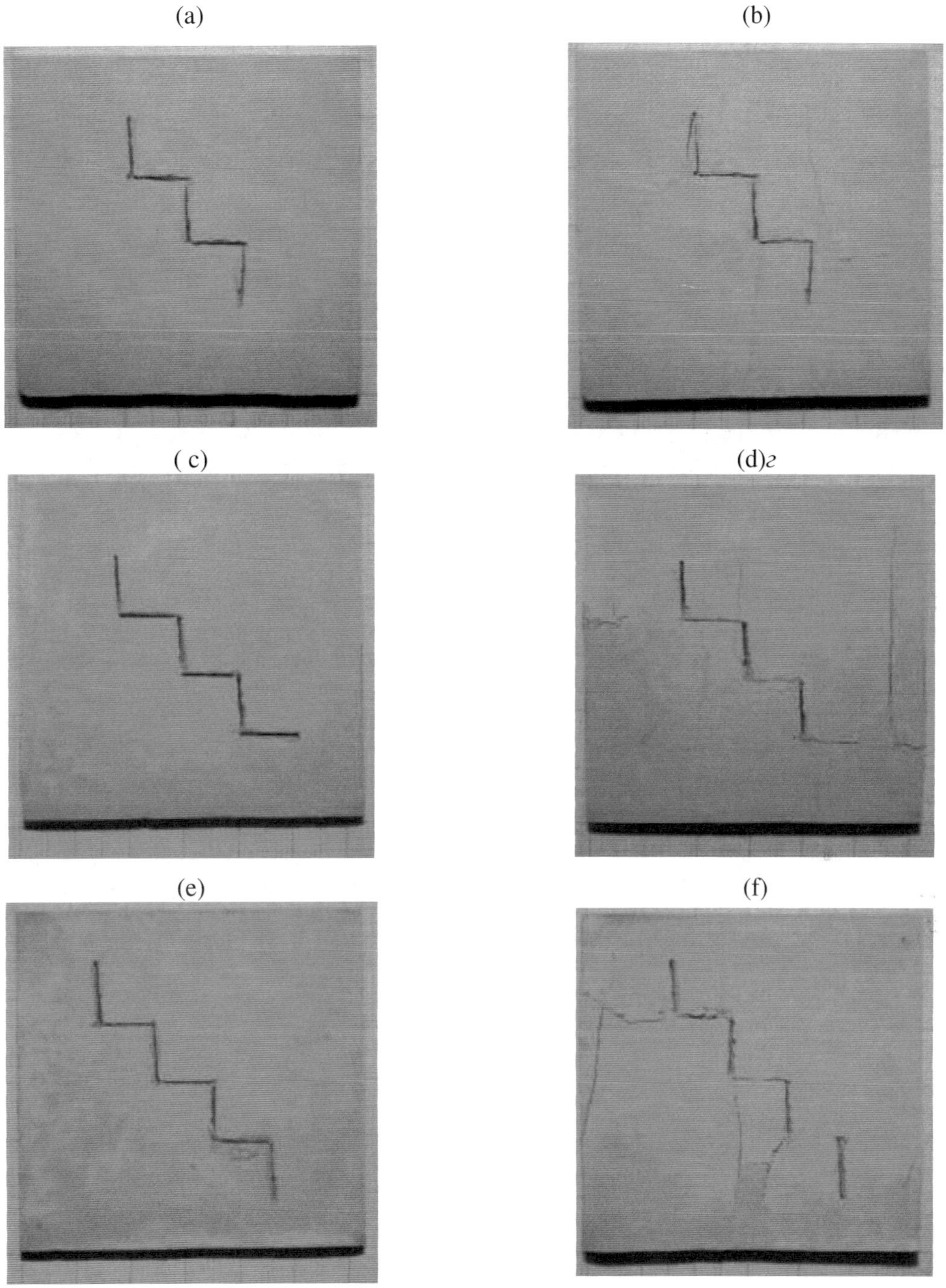

Figure 1. Alabaster specimens with 10 mm stepwise discontinuities before (a), (c), (e) and after (b), (d), (f) tests under uniaxial compression along vertical axis.

The number of test specimens was twenty, the results on variations in the ultimate strength and modulus of deformation versus a number of steps n are summarized in Table 1 and in Figures 3, 4.

IOP Conf. Series: Earth and Environmental Science **134** (2018) 012065 doi:10.1088/1755-1315/134/1/012065

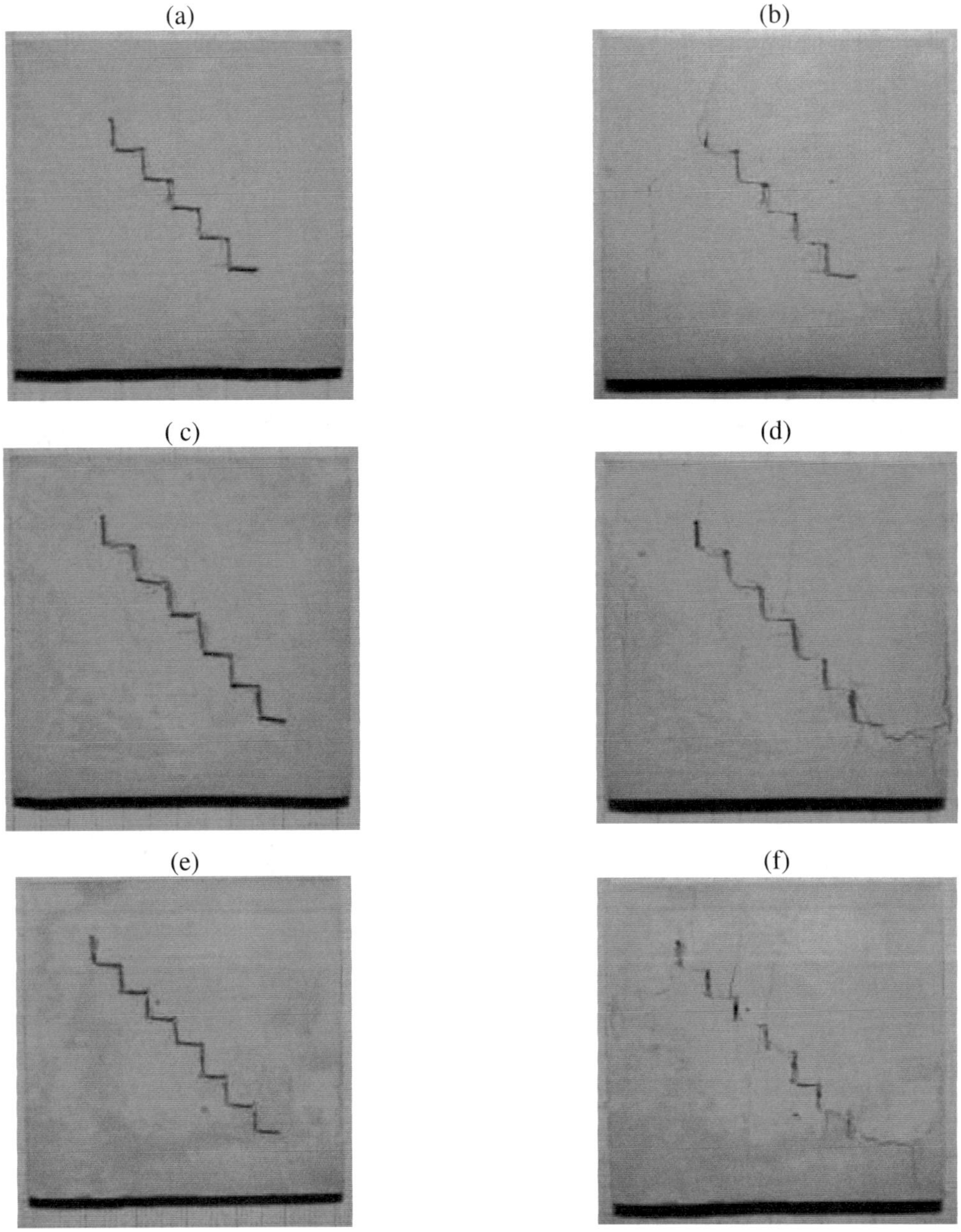

Figure 2. Alabaster specimens with 5 mm stepwise discontinuities before (a), (c), (e) and after (b), (d), (f) tests under uniaxial compression along vertical axis

In Figure 3 relationship $\sigma_{ucs}(n)$ with correlation factor 0.95 corresponds to the plot for 5, 6 and 7 steps

$$\sigma_{ucs} = 1.16 \cdot n^2 - 15.25 \cdot n + 57.01;$$

relationship $\sigma_{ucs}(n)$ with correlation factor 0.9 corresponds to the plot for 10, 12 and 14 steps

$$\sigma_{ucs} = 0.17 \cdot n^2 - 4.47 \cdot n + 35.97.$$

In Figure 4 relationship $E_d(n)$ with correlation factor 0.98 corresponds to plot for 5, 6 and 7 steps:

$$E_d = -0.44 \cdot n + 4.27;$$

relationship $E_d(n)$ with correlation factor 0.96 corresponds to plot for 10, 12 and 14 steps:

$$E_d = -0.21 \cdot n + 4.08.$$

Table 1. Ultimate strength and modulus of deformation versus number of steps functioning as artificial discontinuities

Specimen no.	10 mm step number n	Ultimate strength σ_{ucs}, MPa	Modulus of deformation E_d, GPa	Averaged σ_{ucs}, MPa	Averaged E_d, GPa
1	5	9.17	2.01	9.69	2.07
2		10.20	2.13		
3	6	7.46	1.74	7.17	1.64
4		6.87	1.54		
5	7	6.60	1.19	6.96	1.18
6		7.32	1.18		
	5 mm step number n				
7	10	8.29	2.00	8.37	1.98
8		8.44	1.95		
9	12	7.00	1.36	6.94	1.53
10		6.88	1.69		
11	14	7.45	1.09	6.88	1.13
12		6.31	1.17		

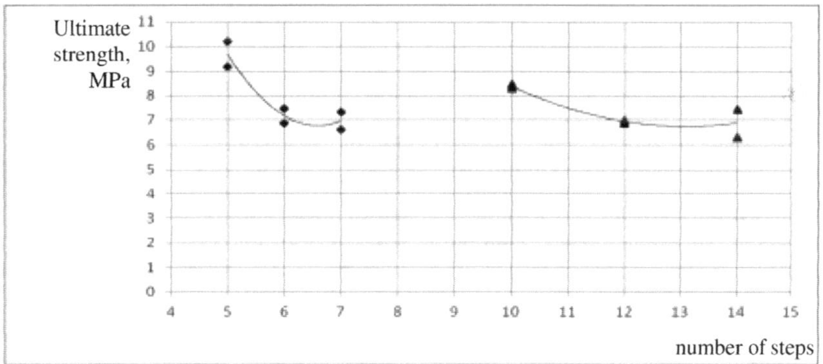

Figure 3. Variation in the ultimate strength versus a number of steps n (rhombs are for 10 mm steps, triangles are for 5 mm steps)

It is explicit that relationships $\sigma_{ucs}(n)$ are described in the best way by the quadratic function, and relation $E_d(n)$ is best described by the linear function. It is also important that relationships $\sigma_{ucs}(n)$ and $E_d(n)$, plotted for 1 cm steps locate lower than curves for 5 mm steps.

Thus, experimental data on uniaxial compression of artificial geomaterials enable to establish a tendency to an increase in the ultimate strength and modulus of deformation with reduction in a linear length of a discontinuity and with increase in length of steps given that linear lengths are equal.

IOP Conf. Series: Earth and Environmental Science **134** (2018) 012065 doi:10.1088/1755-1315/134/1/012065

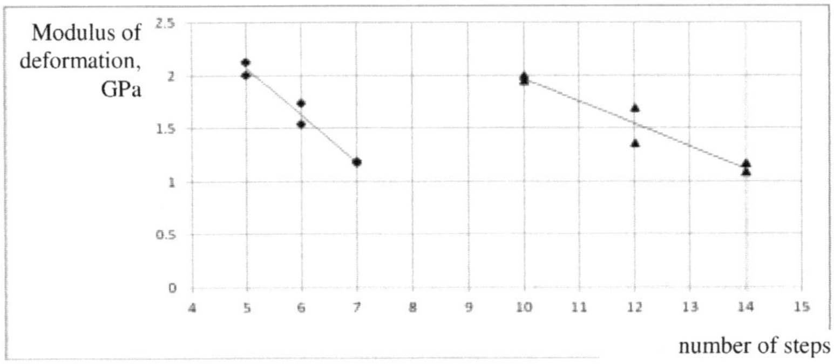

Figure 4. Variation in modulus of deformation versus a number of steps *n* (rhombs are for 1 cm steps, triangles are for 5 mm steps)

Conclusions

The experimental research on identification of such mechanical properties as the ultimate strength and modulus of deformation of specimens made of an artificial geomaterial with the preset structure of discontinuities under uniaxial compression enables to make the following conclusions:

– behavior of the ultimate strength under uniaxial compression versus a number of disturbance steps is best described with the decaying quadratic function, while the modulus of deformation is described in the best way with a strictly decreasing function;

–it is established the tendency to an increase in the ultimate strength and the modulus of deformation with increase in step length (from 5 to 10 mm) under the same linear lengths of the entire discontinuity;

– it is established the tendency to an increase in the ultimate strength and the modulus of deformation with reduction in a linear length of the entire discontinuity (from 70 to 50 mm).

The present research is important in terms of experimental modeling of mechanical behavior of rocks and equivalent geomaterials, containing real discontinuities as macrofractures.

The research work is fulfilled within a research-and-development project "Development of methods and measurement instruments for complex geomechanical-geodynamic safety monitoring systems at Siberian mines. Geomechanics and physics of initiation and development of emergency event centers, rock failure in natural and mining systems" at the equipment of the Shared Use Center for Geomechanical, Geophysical and Geodynamic Measurements, Siberian Branch, Russian Academy of Sciences.

References

[1] Tian S, Yang Q, Zhang K, Liu Y 2010 Crack Propagation in Gypsum Board Containing Pre-Existing Cracks under Compression, *Electronic Journal of Geotechnical Engineering* Vol 15 pp 1674–1683

[2] Sun Y and Yang Q 2013 Experimental study of damage evolution in gypsum board containing pre-existing cracks under uniaxial compression, *Electronic Journal of Geotechnical Engineering* Vol 18 pp 1229–1238

[3] Vasarhelyi B and Bobet A 2000 Modeling of crack initiation, propagation and coalescence in uniaxial compression *Rock Mechanics and Rock Engineering* Vol 33(2) pp 119–139

[4] Zavattieri P, Ochoa F, Hedayat A and Khasawhen Y 2011 *Numerical Simulation of Crack Propagation in Rocks under Uniaxial Compression* CE-597 Non-Linear Fracture Mechanics Term Project—Purdue University

Change in deformation modulus and strength of meta-siltstone depending on specimen size

PA Tsoy[1,2]*, OM Usoltseva[1], OA Persidskaya[1], VN Semenov[1] and BB Sivolap[1]

[1]Chinakal Institute of Mining, Siberian Branch, Russian Academy of Sciences, Novosibirsk, Russia
[2]Novosibirsk State Technical University, Novosibirsk, Russia

E-mail: *paveltsoy@mail.ru

Abstract. The meta-siltstone samples of two scales are subjected to uniaxial compression and split tensile strength testing. The empirical relationships between the specimen size, deformation modulus and ultimate strength are determined. The strength behavior of the studied rock material was described using the Mohr–Coulomb envelopes.

It is important to account for geometrical dimensions of rock specimens when determining their strength properties with a view to prediction mechanical behavior of a geomedium under various type loading. Some researchers study, directly or indirectly, effect of dimension on mechanical characteristics of rock specimens [1–4].

This paper reports the research findings on the relationship between deformation moduli, compression and tension strength and dimensions of specimens of meta-siltstone detected in composition of rocks enclosing ore bodies.

The tests were carried out on cylindrical metal-siltstone specimens: 1—cross section diameters 1:30 mm; 2—cross section diameters 2:10 mm). The ratios of height to cross section diameter equaled 2 (for uniaxial compression testing) and 1 (for indirect tension strength testing—Brazilian test). Specimens from dimension group 1 were tested on Instron 8802 (Figure 1); specimens from dimension group 2—on Deben Microtest (Figure 2). The images of the specimens before and after the tests are given in Figure 1 and 2.

The test data were processed, and the resultant mechanical characteristics of different-dimension specimens were determined (see the table).

From the comparison of the averaged values in the table, dimension group 1 specimens have the deformation modulus E_d by 3.72 times higher and the compression strength σ_{com} by 2.37 times higher than the specimens from dimension group 2. The strength measured in the indirect tensile tests, σ_t, exhibit no significant dependence on the size of a specimen. These test results agree with the experimental data obtained on coal specimens in [4].

By the averaged values of σ_{com} and σ_t, the Mohr–Coulomb envelops were potted; here, thee are presented as the shear stress τ–normal stress σ:

$$\tau = 1.28\sigma + 15.19, \qquad \text{when} -10.47 \le \sigma \le 88.12, \tag{1}$$

$$\tau = 0.61\sigma + 10.48, \qquad \text{when} -11.81 \le \sigma \le 37.2, \tag{2}$$

Where the formula (1) is for the specimens from dimension group 1 (diameter 30 mm) and the formula (2) is for the specimens from dimension group 2 (diameter 10 mm).

Figure 1. Meta-siltstone specimens (dimension group 1) before (a, c, e) and after (b, d, f) uniaxial compression tests.

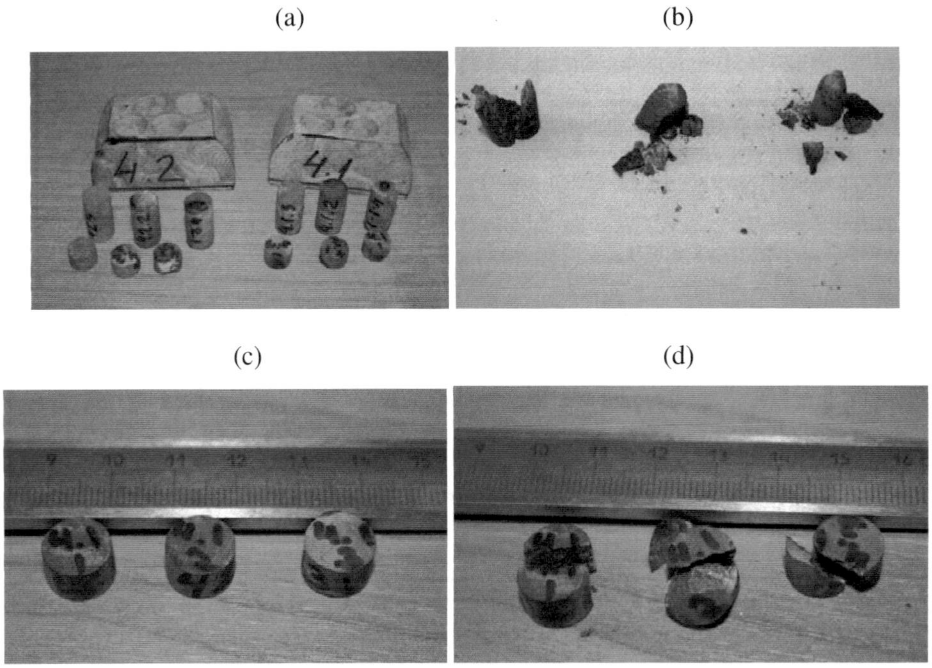

Figure 2. Meta-siltstone specimens (dimension group 2) before (a, c) and after (b) uniaxial compression and (d) indirect tension strength testing.

Table 1. Mechanical parameters of specimens in dimension groups 1 and 2

Specimen no.	Deformation modulus E_d, GPa	Compression strength σ_{com}, MPa	Indirectly tested tension strength σ_t, MPa
Dimension group 1			
1	22.73	90.8	10.1
2	20.01	85.9	9.7
3	21.65	87.7	11.6
Averaged value	21.46	88.13	10.47
Dimension group 2			
4-1-1	5.86	52.66	7.93
4-1-2	4.64	31.89	16.97
4-1-3	5.86	25.24	8.86
4-2-1	7.31	40.20	8.44
4-2-2	6.14	39.93	18.21
4-2-3	4.82	33.31	10.44
Averaged value	5.77	37.21	11.81

As is evident, the formula (1) covers a wider range of states (τ, σ) within the Mohr–Coulomb circles that the formula (2), i.e. the larger diameter specimens possess higher strength.

Conclusion

The mechanical experiments on uniaxial compression and indirect tensile strength testing (Brazilian test) have been carried out on solid cylindrical specimens made of meta-siltstone with the diameters of 10 and 30 mm and the height-to-diameter ratio 2 (compression) and 1 (tension).

It has been found that the meta-siltstone specimens with the diameter of 30 mm have 3.72 times higher deformation modulus and 2.37 times higher compression strength than the specimens with the diameter of 10 mm. The indirect tensile strength values exhibit no essential difference.

The tests have been carried out using the equipment of the Shared Access Center for Geomechanical, Geophysical and Geodynamic Measurements at the Siberian Branch of the Russian Academy of Sciences.

References

[1] Masoumi H, Saydam S and Hagan PC 2016 Incorporating scale effect into a multiaxial failure criterion for intact rock International *Journal of Rock Mechanics & Mining Sciences* Vol 83 pp 49–56

[2] Gao FQ and Stead D 2014 The application of a modified Voronoi logic to brittle fracture modelling at the laboratory and field scale International Journal of Rock Mechanics & Mining Sciences Vol 68 pp 1–14

[3] Pan PZ, Feng XT and Hudson JA 2009 Study of failure and scale effects in rocks under uniaxial compression using 3D cellular automata International *Journal of Rock Mechanics & Mining Sciences* Vol 46 pp 674–685

[4] Poulsen BA and Adhikary DP 2013 A numerical study of the scale effect in coal strength *International Journal of Rock Mechanics & Mining Sciences* Vol 63 pp 62–71

IOP Conf. Series: Earth and Environmental Science **134** (2018) 012067 doi:10.1088/1755-1315/134/1/012067

Detection of regularities in variation in geomechanical behavior of rock mass during multi-roadway preparation and mining of an extraction panel

AB Tsvetkov*, LD Pavlova and VN Fryanov

Siberian State Industrial University, Novokuznetsk, Russia

E-mail: *atsvet@mail.ru

Abstract. The results of numerical simulation of the stress–strain state in a rock block and surrounding mass mass under multi-roadway preparation to mining are presented. The numerical solutions obtained by the nonlinear modeling and using the constitutive relations of the theory of elasticity are compared. The regularities of the stress distribution in the vicinity of the pillars located in the zone of the abutment pressure of are found.

Geological and geotechnical conditions of coal occurrence, as well as coal mining methods have an effect on various events due to rock pressure. One of such influences is the process flowsheet of preparation and extraction of coal reserves in the limits of a bed or a bed series.

Currently a promising flowsheet is multi-roadway preparation of extraction panels as it allows better ventilation, haulage and evacuation of people in case of emergency. There are a number of multi-roadway preparation scenarios [1]. Considering the international experience of coal mining and with regard to coal ignitability, two or entries are driven in actual practice in mines in Kuzbass.

Aimed to find regularities in variation in geomechanical behavior of coal and rock mass during multi-roadway coal bed preparation, a two-dimensional model of coal and rock mass has been developed. The computational domain was constructed based on the data on an extraction panel in coal bed 26a having a thickness of 2.1 m and occurring at a depth of 363 m at Baiday deposit in Kuzbass. The coal bed is overlaid with a layer of unstable coaly siltstone (false roof) 0.40 m thick; this layer is caved simultaneously with coal. Accordingly, the extractable thickness in modeling was assumed as 2.50 m. the false roof is overlaid with alternating siltstone, sandstone and coal beds of unworkable thickness. The dirt beds have compression strength of 28–31 MPa and are 1.0–14.0 m thick. The longwall length is 200 m, the auxiliary and ventilation roadways are 6 m thick each. The fragment of the analytical model is shown in Figure 1.

The stress state of coal and rock mass was determined using classical equations of the continuum mechanics and the resulting equations of the finite element method within a variational problem.

In the framework of the developed nonlinear mathematical model, constitutive equations are selected to determine deformation conditions in the zones of displacements, and the internal boundary conditions are coordinated upon transition of rocks to elastoplastic, limiting and post-limiting states [2, 3]. The relations approximating the experimental stress–strain curves take into account different resilience of rocks under alternating stresses in the zones of compression and tension [4].

The natural gravity stress field is assumed to be given by [5]:

$$\sigma_y = \rho g H \; ; \qquad (1)$$

$$\sigma_x = \lambda/(\lambda + 2\mu)\rho g H \,, \tag{2}$$

where σ_x, σ_y—horizontal and vertical stresses in intact field of gravity; ρ—density of rocks; g—gravitation constant; H—depth of mining; λ, μ—Lamé parameters.

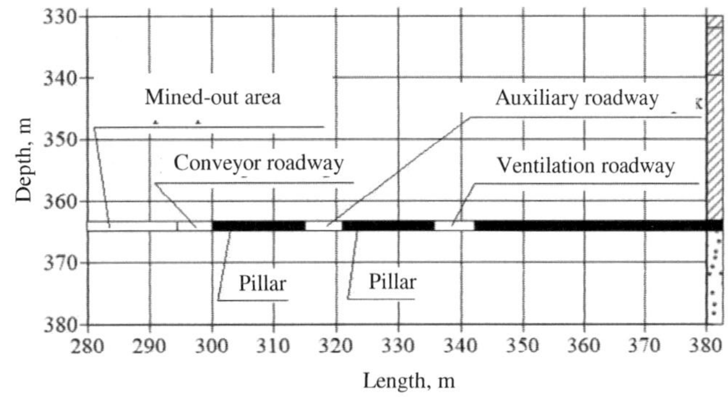

Figure 1. Fragment of analytical model of extraction panel.

Numerical calculations used a developed package of problem-oriented programs with the built-in constitutive relations from the both mathematical models, which enabled comparing the results (Figs. 2–4, the negative and positive signs correspond to compressive and tensile stresses, respectively).

The distribution of vertical stresses resultant from the solution of the nonlinear problem and boundary value problem of elasticity is presented in Figure 2.

From the analysis of the stress distribution, it follows that the highest compressive stresses are observed at the edge of the coal bed in the section 300–303 m. The compressive stresses decrease in the roof rock of the roadways and increase above the pillars. Inclusion of the nonlinear dependence between the stresses and strains leads to the reduction in deformation characteristics of rocks surrounding mined-out area, roadways and pillars. The values of the compressive vertical stresses in the influence zones of mined-out area and the auxiliary and ventilation roadways are lower in the nonlinear problem than in the problem of elasticity.

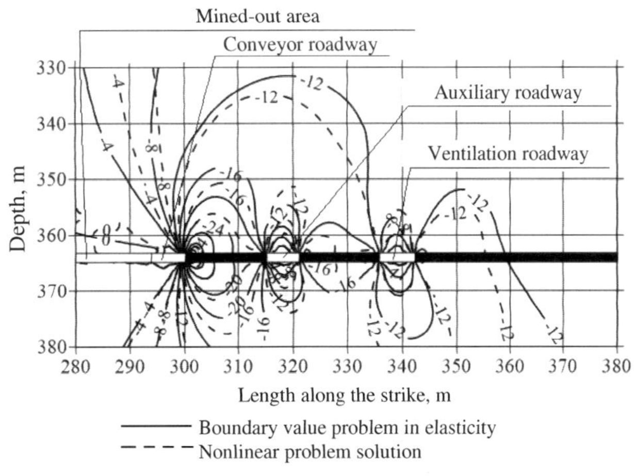

Figure 2. Isolines of vertical stresses in surrounding rock mass around mined-out area.

Figure 3 shows the epures of vertical stresses in coal bed. The ratio of peak compressive stresses resultant from the elastic problem and nonlinear problem solved for the pillar between the conveyor and auxiliary roadways is 1.2 and for the pillar between the auxiliary and ventilation roadways is 1.1. It is typical for the both solutions that the influence of the excavation weakens in the line of the conveyor roadway into the depth of coal bed and reaches the value of stress field induced by gravity in the intact rock mass at a distance of ≈ $0.2H$ from the ventilation roadway.

The analysis of the epures of stresses yields that the excavation induces nonuniform stress distribution in the coal bed. The concentration factors of the compressive stresses caused by mining assume maximum values at the boundary of the pillar at the conveyor roadway and make up more than $3\rho gH$ in the solution of the boundary value elastic problem and more than $2.5\rho gH$ in the nonlinear problem solution.

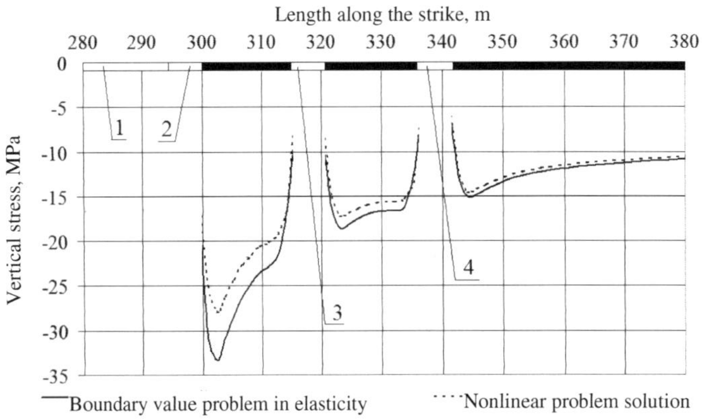

Figure 3. Epure of vertical stresses in coal bed: 1—mined-out area in excavation; 2—conveyor roadway; 3—auxiliary roadway; 4—ventilation roadway.

In the elastic problem solving, the ratio of the maximum vertical stresses is 1.5 for the edge area of the pillar between the conveyor and auxiliary roadways and 1.2 for the pillar between the auxiliary and ventilation roadways. In the nonlinear problem, these ratios are 1.4 and 1.1, respectively.

The increase in the stress concentration above the pillar adjoining the conveyor roadway is conditioned by the intensive mining-induced effect as compared with the pillar between the auxiliary and ventilation roadway.

The obtained regularities point at the pillar between the conveyor and auxiliary roadways as at the raisor of potential energy. Since the values of the vertical displacements, calculated with regard for the nonlinear relation of stresses and strains above the excavation, exceed the vertical displacements resultant from the boundary value elastic problem solution, by 2.5 times, then at the boundary of the pillar adjoining the conveyor roadway, it is highly probable that the intensive mining impact will result in disintegration and fracturing of rock mass, which will end with the support failure in the conveyor roadway.

The graphs of distribution of the compressive vertical stresses above the roof of roadways and pillars in Figure 4 feature maximum values in the coal bed and gradual attenuation of secondary vertical stresses induced by excavation at a distance from it, which is confirmed by the decrease in the difference between the stresses of plots 1–4 and the intact rock mass stresses ρgH. The vertical stress in the roof after solution of the nonlinear problems are 10–15% less in value than the stresses calculated from the problem of elasticity (plots 1, 2 and 3, 4, respectively, in Figure 4).

In the stress relaxation zones above the roadways (plots 5–8, Figure 4), the vertical stresses are lower than in the intact rock mass. In the roof of the auxiliary roadway, the area of the stress relaxation zone is 2 times less than the same zone area in the roof of the ventilation roadway. The values of the vertical stresses above the roadways, calculated by solving the nonlinear problem, are lower than the vertical stresse obtained from the solution of the problem of elasticity (plots 5, 6 and 7, 8, respectively, in Figure 4). However, in the stress relaxation zone above the auxiliary roadway, directly at the roof, the nonlinear problem stresses exceed the elasticity problem stresses. The point of equality of the stresses lies in the roof at the distance making the half-width of the excavation (Figure 4).

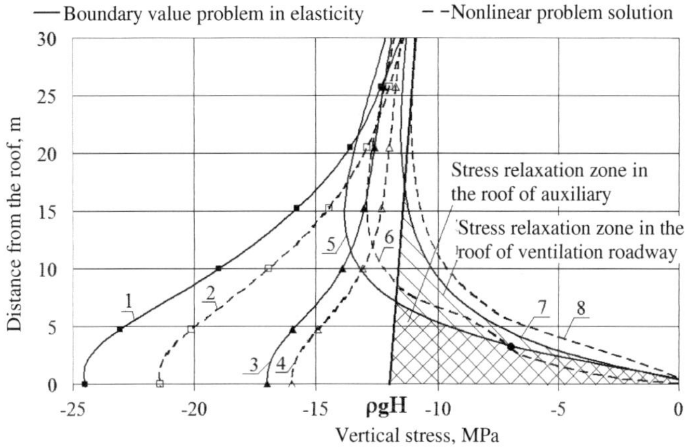

Figure 4. Plots of distribution of vertical stresses in the roof of roadways and pillars: 1, 2—pillar adjoining conveyor roadway; 3, 4—pillar between auxiliary and ventilation roadways; 5, 6—auxiliary roadway; 7, 8—ventilation roadway.

The proposed approach to numerical modeling allows finding the regularities in variation in geomechanical behavior of coal and rock mass in case of multi-roadway preparation and extraction of coal, such as:

—in the pillar remote from the mined-out area, compressive vertical stresses are lower than in the pillar at the boundary of the excavation;

—stress razor is the pillar lying in the zone of abatement pressure of the excavation;

—stress distribution at a distance from the excavation behaves as G-field;

—the maximum difference between the stresses calculated by solving the nonlinear problem and the problem of elasticity is observed in the zones of the highest stress concentration;

—the area of stress relaxation zone in the roof of the roadways is larger in the results of the nonlinear problem solution than in the problem of elasticity.

These found regularities can help substantiating engineering solutions on preparation and mining of extraction panels at the ensured stability of roadways and coal pillars.

Acknowledgements
The studies have been supported by the Ministry of Education and Science of the Russian Federation in the framework of the Federal Targeted Program on R&D in Priority Areas of Advancement in the Science and Technology of Russia for 2014–2020, Topic: Efficient Robotic Longwall Top Coal Caving Technology, Unique Identifier RFMEF160417X0173.

References

[1] Korshunov GI, Loginov AK, Shik VM and Artemiev VB 2011 *Geomechanics in Coal Mines* Moscow: Gorn Delo (in Russian)
[2] Tsvetkov AB, Pavlova LD, Fryanov VN 2015 Nonlinear mathematical model of the geomechanical state of the carbonaceous rock mass *GIAB* No 1 pp 365–370
[3] Tsvetkov AB and Pavlova LD 2016 Identification of the laws of non-linear deformation of rocks under mining of a coal series *GIAB* No 10 pp 324–332
[4] Tsvetkov AB, Pavlova LD and Fryanov VN 2016 Construction of the approximant of complete diagram for rock deformation *IOP Conference Series: Earth and Environmental Science*, Vol 56 Available at: http://iopscience.iop.org/issue/1755-1315/45/1
[5] Polukhin VA, Kaloerov SA, Future YuB and Goryanskaya ES 2002 *Stress State and Stability Control in Underground Excavations* Donetsk: Yugo-Vostok (in Russian)

Integrated studies into characteristics of physical fields using discontinuous geomedium models under external loading

OM Usoltseva[1]*, VI Vostrikov[1], PA Tsoy[1,2] and VN Semenov[1]

[1]Chinakal Institute of Mining, Siberian Branch, Russian Academy of Sciences, Novosibirsk, Russia
[2]Novosibirsk State Technical University, Novosibirsk, Russia

E-mail: *usoltseva57@mail.ru

Abstract. The article presents the laboratory study of deformation in artificial layered geomaterial samples down to failure with the simultaneous measurement of stresses, strains, micro-strains and signals of microseismic emission. The analysis of the synchronized experimental data made it possible to determine features of change in the microseismicity parameters and micro-strain fields in the samples depending on the deformation stage, and also to reveal the dynamics of evolution of microfailures and the main fracture zone.

The comprehensive analysis of different nature physical fields in geo-media under loading offers reliable information on initiation, accumulation and growth of different defects, and helps revealing regularities of the process of deformation up to complete failure. The review of the recent literature shows that there are studies into evolution of microseismic emission, fields of micro-strains using speckle-method and signals of electromagnetic emission [1–8]. The analysis of tests of artificial geomaterials and rocks shows good correlation between heterogeneous local deformation fields and microstructure of specimens [1, 2]. Based on the monitoring of acoustic emission of rock specimens under deformation, the character of change in AE signals and failure pre-cursors are examined in [3, 5]. The application of the speckle-photography methods to studying deformation and failure of rocks is validated in [6], and the fields of strains under loading to failure as well as zones of location of maximum strains are analyzed. In [8, 9], the description of the measurement system for the concurrent recording of pressure, displacement and electromagnetic emission during on-axial compression tests is given; it is also pointed at a stable growing trend of the signal amplitude as well as at the increase in the local electromagnetic activity at the stages of initiation and expansion of destructure zones. The research into the evolution of deformation of sandstone specimens with the measurement of acoustic emission signals and digital correlation of images [10] illustrates good agreement between locations of AE signals and crack trajectory. Using the digital image correlation, critical deformation values when strain localization zones become unstable and spalling takes place are determined.

The investigation of interaction between parameters of physical fields of different nature (deformation, micro-deformation, microseismic emission) in models of geomedia subjected to deformation and failure used the testing equipment set designed and manufactured at the Institute of Mining, SB RAS [11]. During the tests, the fields of microstrains, microseismic emission (MSE), deformations and loads were recorded concurrently with video imaging. The uniaxial compression was carried out of Instron 8802 servo testing machine. The speckle-recording of microstrains used the automated speckle photography analysis system ALMEC-tv, which allows microdisplacement measurement at an accuracy of 1 μm and frequency of 27 shoots per second. The signals of microsesimic emission were measured using accelerometers KD 91 of the multichannel system Pulse. Video shooting was carried synchronously. The uniaxial compression was applied to artificial geomaterial cubes with an edge of 200 mm at the loading rate of 0.1 mm/min. On the four faces of a cube, 4 microseismic sensors KD 91 were mounted to register MSE signals. Displacements and axial (vertical) force were measured by Instron 8802 test machine. Displacements in two orthogonal horizontal directions were recorded by probes Solartron DP10S arranged in the plane perpendicular to the normal force. Figure 1 shows the general view of a cubic specimen between Instron 8802 presses with microseismicity sensors KD 91 and probes Solartron DP10S. The artificial geomaterial was made of a mixture of sand and cement, the cubic specimens were manufactured of two layers having strengths of 21.4 and 7.1 MPa, the angle of bedding in different specimens was $\Psi = 0, 30, 45, 60, 90^\circ$.

The integrated analysis of the microseismic emission, microstrains, stresses and displacements shows that each of the methods satisfactorily describes the evolution of fracturing. Moreover, there is a clear interconnection between the regularities of variation in MSE signals, microstrain field components and their velocities and the stress–deformation diagram.

Figure 2 shows the curve P/P_{max}–t/t_m obtained under the uniaxial compression of the cubic specimen with the bedding angle $\Psi = 45^\circ$, where P/P_{max} is the ratio of the current and maximum axial loads, t/t_m is the ratio of the current time to the time t when $P = P_{max}$. The curve shows that deformation has a few stages: when stress changes from 0 to $\sigma/\sigma_v = 0.2$–0.3 closure of micropores and microcracks takes place; the next deformation stage up to $\sigma/\sigma_v = 0.5$–0.7 is linear elasticity, and the third stage of the stresses up to $\sigma/\sigma_v = 1$ is nonlinear elasticity, plasticity and pre-failure.

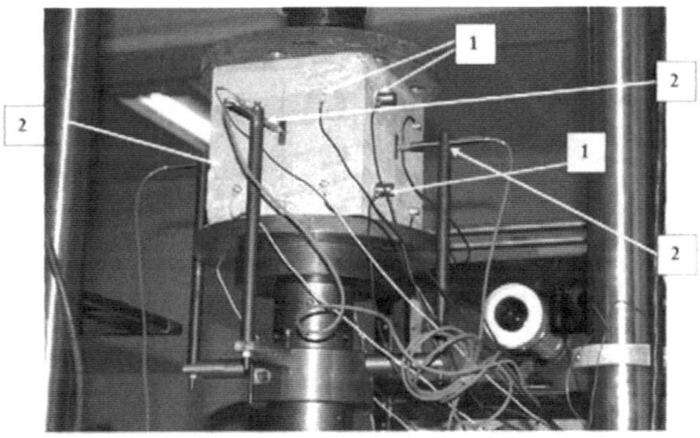

Figure 1. General view of the cubic specimen arranged between presses of Instron 8802 test machine, with microseismicity sensors KD 91 (1) and displacement probes Solartron DP10S (2).

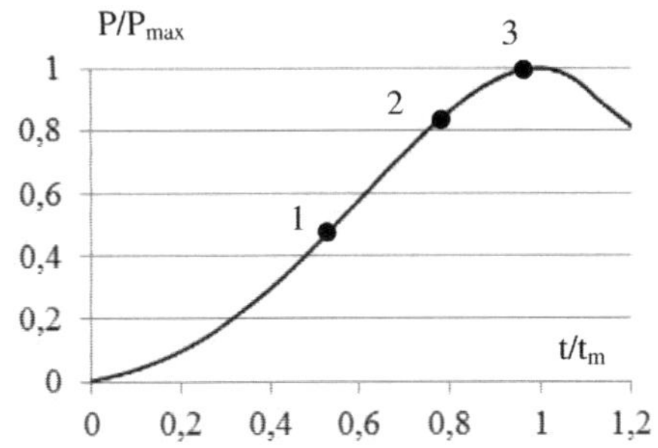

Figure 2. Curve P/P_{max}–t/t_m under uniaxial compression of the cubic specimen with the bedding angle $\Psi = 45^\circ$.

IOP Conf. Series: Earth and Environmental Science **134** (2018) 012068 doi:10.1088/1755-1315/134/1/012068

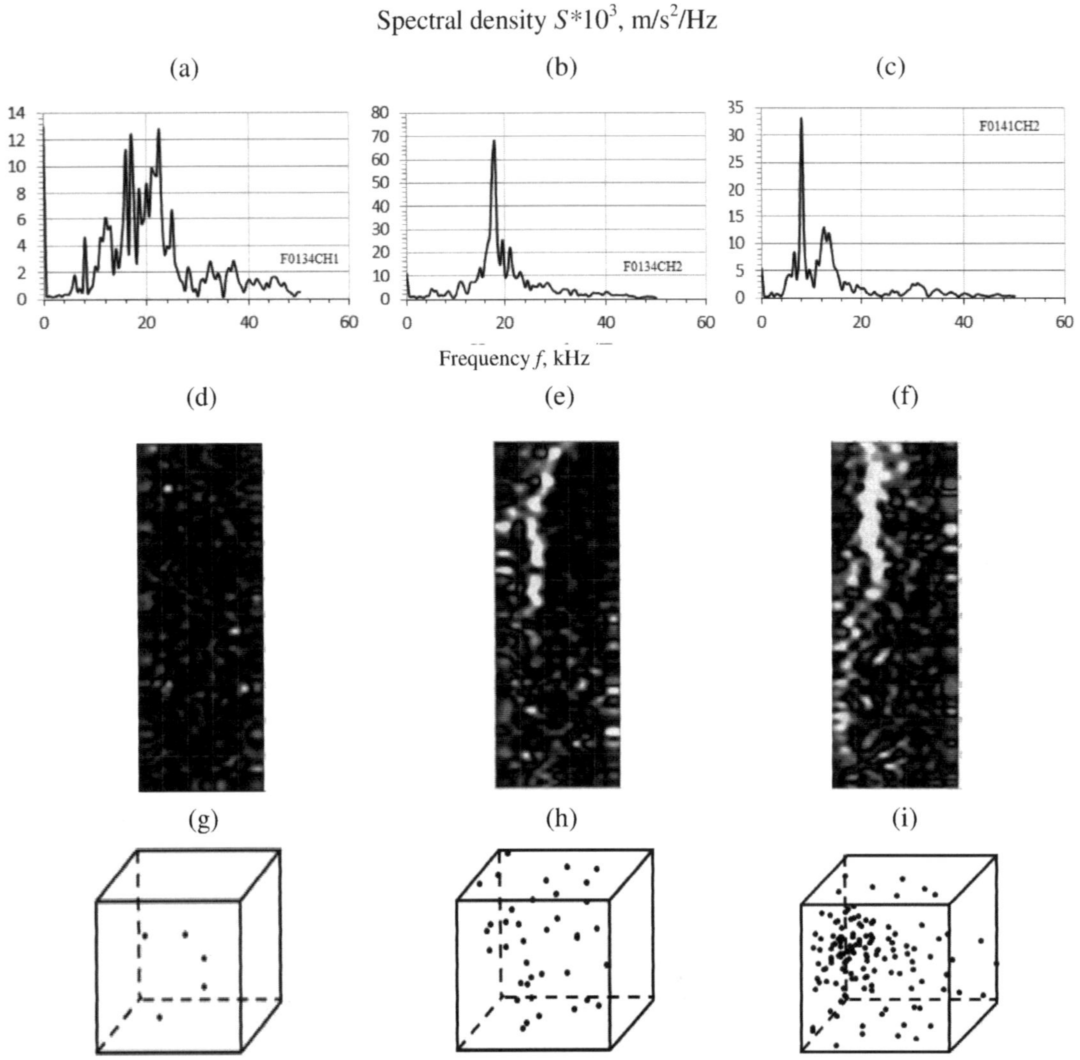

Figure 3. (a), (b), (c) MSE signals; (d), (e), (f) speckle-method mapping of the specimen surface in transverse direction; spatial distribution of accumulated MSE events (g), (h), (i) at deformation stage (a), (d), (a) 1; (b), (e) (h) 2; (c), (e) (i) 3, respectively.

After processing of many microseismic signals at different deformation stages, the curves of the acceleration $\bar{a}$ and time t, and the acceleration spectral density S and frequency f were plotted. Figures 3a–3c show the typical plots of S in the specimen with the bedding angle $\Psi = 45°$.

In [6, 10, 11] it is shown that the plastic deformation is nonuniform from the very beginning of loading due to mineralogical and structural heterogeneities of a geomaterial specimen. Despite the same pre-set type of loading by uniaxial compression at constant rate, the space-and-time field of microstrains on scanned surfaces contains both contraction and elongation domains. In a series of tests on the cubic specimens, the analysis of change in the microstrain field component in the perpendicular direction to the axial loading using the speckle method has been preformed. Figures 3d–3f present the images of the deformation mapping of the scanned surface of the cubic specimen with $\Psi = 45°$ by the deformation component in the Y-direction of the specimen, in perpendicular to the axial load, at the moments when $\sigma/\sigma_v = 0.25$, $\sigma/\sigma_v = 0.75$, $\sigma/\sigma_v = 0.95$. The negative value of the deformation component (reduction is size) corresponds with the color close to red. Nearly black color depicts zero deformations. The positive values (increase in size) are presented by green. White color means deformation higher than 0.007.

In order to locate MSE sources (defects) in a geomaterial specimen by the data from 4 microseismic sensors, a location program was developed. The program was checked using a cinogram of test fracturing of a cubic organic glass specimen by fluid, and the agreement between the fracturing zone and the data on location of MSE signals was satisfactory. Figure 3g–3i show the special distribution of accumulated MSE events in a cubic specimen under different loading levels.

The comprehensive analysis of the experimental results has revealed some regularities.

At loading stage 1, under stresses lower than 0.25–0.5 peak load, the quantity of MSE signals is insignificant and the signals are wideband. The field of microstrains is chaotically nonuniform, the fluctuation in the microstrain components is almost absent (Figs. 3a, 3d and 3g).

At loading stage 2, when the stress is from 0.4–0.5 to 0.7–0.8 of the strength limit, the number of MSE signals grows, their amplitude increases while the frequency spectrum somewhat narrows and drifts toward lower frequencies. The microstrain field is even more nonuniform, the zones of maximum microstrains, higher than the average values over the specimen surface, appear (Figure 3b, 3e and 3h)

At stage 3, when the stresses vary from 0.8 of limits strength to the peak load, the number MSE signals increases by a few times as compared with the previous deformation stage, the amplitude of the signals grows, while the frequency spectrum shifts event more to the range of low frequencies and the spectrum band narrows considerably. The maps of the time-and-space distribution of MSE signals show their nonuniform distribution in the volume of the specimen. The maximum microstrains localize in a certain volume of the geomaterial, which is reflective of the initiation of a main fracture, the outlet point of which on the specimen surface can be determined under loads less than the peak loading (Figs. 3c, 3f and 3i). As the fracture comes out at the surface, a strong low-frequency signal is generated.

These regularities are the most clearly observed in the tests of the specimens with the bedding angle Ψ=45°, which is probably connected with the stronger deformation than in the specimens with Ψ=90° and with the more intensive destructuring of the geomaterial with the formation of microdefects.

Conclusion

The integrated analysis of the microseismic emission, microstrains, stresses and displacements shows that evolution of fracturing in geomaterial specimens is satisfactorily described using any of these methods.

There is a clear interconnection between the regularities of change in the parameters of signals of microseismic emission and components of the microstrain field per stages of deformation of geomaterial.

The authors have revealed basic regularities in the change of MSE signal parameters and microstrain field components depending on the specimen deformation stage.

Acknowledgements

The study has partly been supported by the Russian Foundation for Basic Research, Project No. 15-05-07566 and has been carried out using the equipment of the Shared Access Center for Geomechanical, Geophysical and Geodynamic Measurements at the Siberian Branch of the Russian Academy of Sciences.

References

[1] Wang L, Bornert M, Heripre E, Chanchole S, Pouya A and Halphen B 2015 The mechanisms of deformation and damage of mudstones: A micro-scale study combining ESEM and DIC *Rock Mechanics and Rock Engineering* Vol 48 No 5 pp 1913–1926

[2] Reza Nejati H and Ghazvinian A 2014 Brittleness effect on rock fatigue damage evolution Rock Mechanics and Rock Engineering Vol 47 No 5 pp 1839–1848

[3] Hedayat A, Pyrak-Nolte LJ and Bobet A 2014 Multi-modal monitoring of slip along frictional discontinuities *Rock Mechanics and Rock Engineering* Vol 47 No 5 pp 1575–1587
[4] Seop J, Kim K, Lee S, Cho WJ, Choi HJ and Cho GC 2015. A comparative evaluation of stress–Strain and acoustic emission methods for quantitative damage assessments of brittle rock *Rock Mechanics and Rock Engineering* Vol 48 No 2 pp 495–508
[5] Shkuratnik VL, Novikov EA and Oshkin RO 2014 Experimental analysis of thermally stimulated acoustic emission in various-genotype rock specimens under uniaxial compression *J. Min. Sci.* Vol 50 No 2 pp 249–255
[6] Zuev LB, Barannikova SA, Nadezhkin MV and Gorbatenko VV 2014 Localization of deformation and prognostibility of rock failure J. Min. Sci. Vol 50 no 1 pp 249–255
[7] Wanga E, Jia H, Song D, Li N and Qian W 2014 Use of ultra-low-frequency electromagnetic emission to monitor stress and failure in coal mines International *Journal of Rock Mechanics and Mining Sciences V*ol 70.
[8] Bespalko AA, Yavorovich LV, Kolesnikova SI, Bukreev VG et al 2011 Investigation of changes of electromagnetic signals characteristics during uniaxial compression of rock samples in Tashtagol Mine *Izv. Vuzov. Fiz. Zemli* no 1/2
[9] Oparin VN, Vostretsov AG, Krivetskii AV, Bizyaev AA and Yakovitskaya GE 2011 Integrated measurement and recording of loads, displacements and electromagnetic emission in rocks under uniaxial compression J. Min. Sci. Vol 47 No 5 pp 547–557
[10] Yingwei Shi, Qun He, Shanjun Liu and Lixin Wu 2010 The time–space relationship between strain, temperature and acoustic emission of loaded rock *Progress in Electromagnetics Research Symposium Proceedings* China
[11] Oparin VN, Usoltseva OM, Semenov VN and Tsoy PA 2013 Evolution of stress-strain state in structured rock specimens under uniaxial loading *J. Min. Sci.* Vol 49 No 5 pp 677–690

IOP Conf. Series: Earth and Environmental Science **134** (2018) 012069 doi:10.1088/1755-1315/134/1/012069

Tectonic fault monitoring at open pit mine at Zarnitsa Kimberlite Pipe

VI Vostrikov[1]*, NS Polotnyanko[1], AS Trofimov[2], AA Potaka[2]

[1]Chinakal Institute of Mining, Siberian Branch, Russian Academy of Sciences, Novosibirsk, Russia
[2]Management Department, Udachny Mining and Processing Works, Udachny, Republic of Sakha (Yakutia), Russia

E-mail: *viv@misd.ru

Abstract. The article describes application of Karier instrumentation designed at the Institute of Mining to study fracture formation in rocks. The instrumentation composed of three sensors was used to control widening of a tectonic fault intersecting an open pit mine at Zarnitsa Kimberlite Pipe in Yakutia. The monitoring between 28 November and 28 December in 2016 recorded convergence of the fault walls from one side of the open pit mine and widening from the other side. After production blasts, the fault first grows in width and then recovers.

1. Introduction

Hard rock masses hosting mineral deposit, in particular, diamond-bearing kimberlite pipes, are structured as hierarchies of blocks. The blocks are separated by fractures and crevices the width of which ranges from tens fractions of centimeter to several tens centimeters. Furthermore, diamond-bearing pipes occur in the zones of faulting as, for instance, Nyurbin Pipe [1]. The cluster zone of submeridional and sublatitudinal faulting accommodates Komsomol [2] and Zarnitsa diamond-bearing pipers. The zone of faulting tens and more meters in size.

Aiming to assess geomechanical behavior in a complex-structure rock mass from the viewpoint of movement of edge blocks in faulting zones, and to estimate pitwall stability, deformation monitoring is performed in open pit mines using different measurement equipment [1, 3–5]. The Institute of Mining is currently engaged in development and manufacture of the prototype multichannel measurement equipment named Karier (open pit mine in Russian) [6].

Karier equipment system composed of three geodynamic monitoring set has been deployed in Zarnitsa open pit mine to control a north-eastward fault.

2. Description of the north-eastward fault at Zarnitsa OPM

Figure 1 shows the plan view of Zarnitsa OPM. It develops Zarnitsa kimberlite pipe and is located at the intersection of quasi-perpendicular north-eastward (line 1-1) and north-westward (line 2-2) faults.

Faults affect the open pit mine at Zarnitsa pipe; they are characterized by wide zones of crevices and sets of fractures, the width of such zones can reach tens meters and more.

On the bench terrace at level +380 m, a few series of crevices are observable across the whole width. Total width of the fault at this place is of the order of 11 m. On both sides of the fault, fractures 5–15 cm wide are observed at the distance of 20–30 cm.

IOP Conf. Series: Earth and Environmental Science **134** (2018) 012069 doi:10.1088/1755-1315/134/1/012069

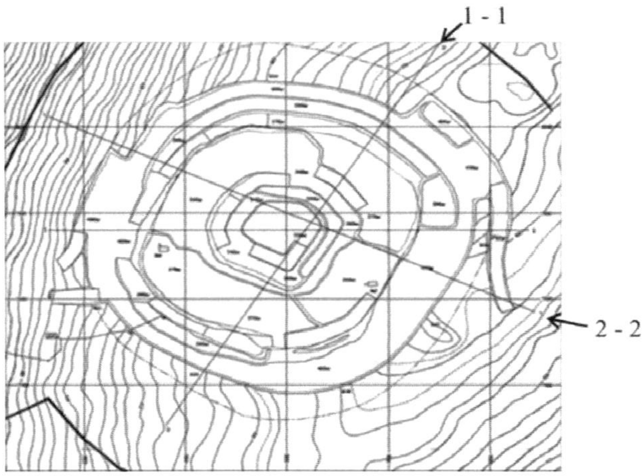

Figure 1. General view of Zarnitsa OPM.

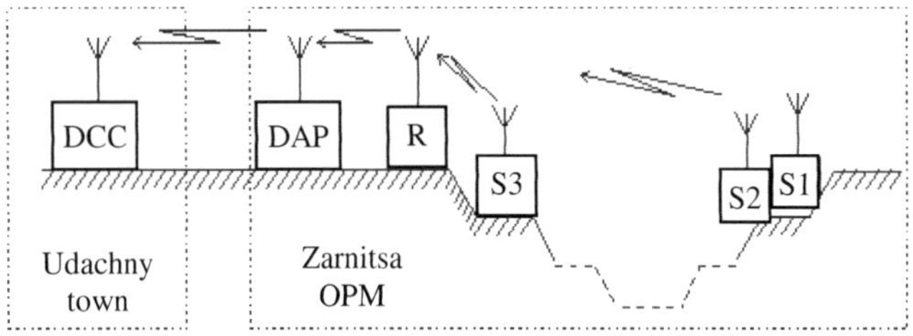

Figure 2. Schematic Karier equipment deployment in Zarnitsa OPM: S1 … S3—sensors; R—retransmitter; DAP—data acquisition and processing point; DCC—data collection center.

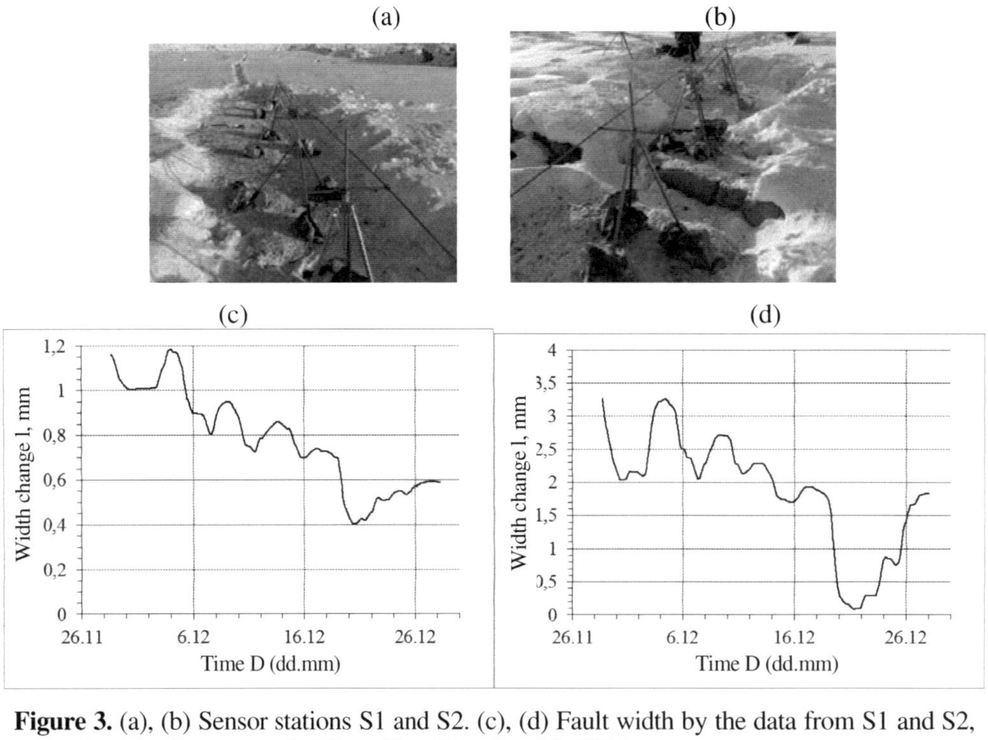

Figure 3. (a), (b) Sensor stations S1 and S2. (c), (d) Fault width by the data from S1 and S2, respectively in the time period from Nov 29 to Dec 28, 2016.

3. Karier equipment set

Deployed to monitor the north-eastern Fault at Zarnitsa OPM, Karier equipment set is composed of three sensors stations (Figure 2). On the north-east pitwall, on the bench at level +380 m, at the widest point of the fault, two sensor stations S1 and S2 are installed across the fault. On the opposite pitwall, at the bench of level +380 m, S3 sensor station is arranged.

The sensor stations S1–S3 send data to the data acquisition and processing point DAP in operators' office at Zarnitsa OPM using a retransmitter R situated at the edge of the pitwall since there is no ling of sight between the sensors and DAP. Then, the data are forwarded to the Data Collection Center of the Uchaly Mining and Processing Plant in Uchaly town via a data transmission channel.

Figure 3 shows the pictures of the deployed sensor stations D1 and D2 (Figs. 3a and 3b, respectively). The gage length of S1 station is 10 m. An extension element is constructed from a few segments of rod made of precision alloy feature extremely coefficient of expansion. The gage length of S2 station is 8.73 m. This station uses an extension element made of a cable. A small load is used to ensure tension and eliminate sagging of the cable. The temperature calibration tests of the cable were carried out in laboratory conditions. The length–temperature relation of the cable is linear. Deflection of the rod and cable under dead weight is eliminated by support legs. Sensor station S2 is installed at the same place as S1 but closer to the edge of the bench terrace.

4. Geodynamic monitoring of edge blocks of the north-eastern fault in Zarnitsa OPM

Figure 3c and 3d illustrate variation in the fault width between Nov 28 and Dec 28, 2016 by the data from S1 and S2, respectively. The plot shows the change in the distance rather than its absolute value the maximum change in the fault width at the mentioned time by the data from S1 was 0.8 mm. The curve trend has a negative incline, i.e. the fault undergoes contraction. The air temperature in that period ranges as –15÷–44°C.

The maximum change in the width of the fault according to S2 was 3.3 mm. The curve trend is again negative. The edge blocks of the fault 'approach' more appreciably.

Sensor station S3 on the opposite pitwall at the same level has the gage length of 1.76 m. this stations uses the cable instead of the extension rod.

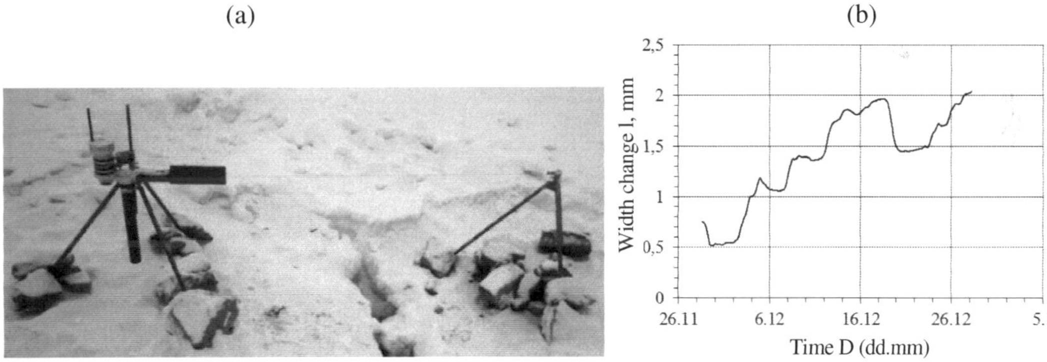

Figure 4. (a) Sensor station S3; (b) curve of change in the fault width.

Figure 4 shows the picture of S3 (Figure 4a) and the change in the fault width between Nov 28 and Dec 28, 2016 (Figure 4b). The maximum change in the fault width at this place was 1.6 mm. The curve shows a positive trend, which means that the edges of the fault move apart here.

The wavy behavior of the curves plotted based on the data from all sensor stations is explained by the effect of blasting in the open pit mine.

5. Conclusion

Karier system allows recording displacement of blocks in a complex-structure rock mass with the extended gage lengths and under low temperatures. The extension elements can be a rod or a cable to connect the reference points.

Karier system composed of thee sensor stations has been used in the geodynamic monitoring of edge blocks of the north-eastern fault cutting Zarnitsa open pit mine/. Two stations were installed on the north-east pitwall, and one station—on the opposite pitwall.

Measurement data transmission from the open pit mine to the data processing center in Udachny town uses GSM channel and internet communication tools.

The measurement taken in 28 Nov–28 Dec in 2016 show that the both stations on the north-east pitwall record 'contraction' in the fault width: by 0.8 mm by S1 and 3.3 mm by S2 arranged closer to the bench terrace edge. The difference in the values can be explained by higher freedom of motion of the edge blocks of the fault. At the same time, sensor station S3 at the opposite pitwall shows expansions of the fault by 1.6 mm.

The fault exhibits clear response to blasting operations in the open pit mine.

From the analysis of the geodynamic condition of the fault between Nov 28 and Dec 28, 2016, the fault has undergone insignificant change. The measurements were not carried put in spring and autumn due to intensive thawing–freezing cycles. It is supposed that the change in the geodynamic behavior of the fault is stronger in these seasons.

References

[1] Bornyakov SS and Salko DV 2016 Instrumental deformation monitoring system and its trial in open-pit diamond mine *J. Min. Sci.* Vol 52 No 2 pp 388–393

[2] Potekhina IA, Makovchuk IV and Gladkov AS 2008 Tectonic fracturing in rock mass around Komsomolskaya Pipe *Vestn. Irkutsk. Gos. Univer.* No 4(38) pp 25–31

[3] Dimaki AV and Psakhe SG 2009 Spaced monitoring system for displacement in block media, designed based on SDVIG-4MR system *J. Min. Sci.* Vol 45 No 2 pp 194–200

[4] Vostrikov VI, Ruzhich VV and Federyaev OV 2009 Monitoring rock fall-hazardous sites in open pit walls *J. Min. Sci.* Vol 45 No 6 pp 620–627

[5] Vostrikov VI and Oparin VN 2009 Multichannel instrumentation system for strain and displacement measurements *Proc. 2009 Int. Symp. on Mechatronic and Biomedical Engineering and Applications* Taiwan pp 13–17

[6] Vostrikov VI and Polotnyanko NS 2014 Karier multichannel measurement system for deep open pit walls monitoring *J. Min. Sci.* Vol 50 No 6 pp 1094–1098

IOP Conf. Series: Earth and Environmental Science **134** (2018) 012070 doi:10.1088/1755-1315/134/1/012070

Nephelometry and turbidimetry to assess concentration and dispersion of coal dust in mines

VP Yushchenko*, VN Legky and DE Demidov

Novosibirsk State Technical University,
Novosibirsk, Russia

E-mail:*rector@nstu.ru

Abstract. The article considers the model of the optical instrument to measure coal dust concentration in mines based on the turbidimetric and nephelometric methods. The calculated data on the intensity of transmitted and scattered waves depending on coal dust concentration and on the size of coal dust particles are presented.

The coal dust explosions are known as the most hazardous emergency events at coal mines, coal processing and utilization plants. The striking after-effects of a combustible dust explosion are shock wave, high temperature, toxic gases, reduction in oxygen concentration in a new-generated gas mixture. The shock wave is capable to overcome large distances, and that is why the combustible dust explosions inflict huge economic damages to industrial plants, lead to multiple injuries with lethal outcomes [1].

Coal dust used to generate in large amounts in coal seam mining, haulage and transshipment operations. A hazard of coal dust explosion arises when concentration of combustible dust reaches the critical level in the air.

The increase in coal production capacity is accompanied with greater emission of gas and coal dust in mine atmosphere. Measurement and monitoring of finely dispersed coal dust concentration is of prime importance for safe underground mining. The measurement and monitoring instruments should be high-speed in operation and provide reliable data, otherwise an intensive dust generation can cause an explosive-hazardous situation.

Nowadays there are a lot of processes to monitor dust emission in mine atmosphere. Each process is distinguished for specific merits and drawbacks [2]. Some of them imply human participation in measurements with large monitoring time with no representative presentation of dust-generation dynamics, because they are based on sampling principle. In other monitoring processes different factors such as temperature, humidity, etc. affecting measurement results should be taken into account. Therefore, it is not a simple problem to develop dust-meters meeting precision and high-speed requirements.

The basic parameters of the mine dust-air atmosphere which demand regular monitoring are coal dust concentration and size of dust particles, as they influence the explosion-proof mine operation.

The physical-chemical properties of coal can vary widely, so the lower concentration level of coal dust explosibility is also not rigorous, but specific for different coal brands. Available publications report different data on concentration limits of coal dust explosibility. It is experimentally proved that most coal dusts are prone to explode at 10–50 g/m^{3}dust concentration in the air [3]. In [4] it is stated that coal dust with 16 % volatile yield used to explode at concentration of 125 g/m^3, but with 25% volatile yield the concentration of 100 g/m^3 is enough for explosion.

In spite of further studies of mine dust explosibility the coal industry professionals: theoreticians and practitioners have different views on many issues especially on a probability of coal dust explosion in a mine when gas is not detected in its mine workings.

In1906 the explosion of coal dust in Courrieres gas-free colliery, (Northern France), took 1100 lives and put an end to the dispute of supporters and opponents on self-sustained role of coal dust. Comprehensive investigation into coal dust explosibility initiated immediately after the accident made it possible to establish the law that:

– coal dust is prone to explode at absolute absence of methane;

– coal dust can initiate explosion of a small methane amount and turn it into huge-scale explosion;

– fire from combustible dust cloud can reach a spot of methane gas concentration and inflame it;

–content of fine and dry coal dust in air decreases the lower level of methane-air mixture explosibility, it becomes explosive at lower than 5 % gas concentration;

– in coal dust explosion the products of explosion always contain a certain amount of carbon monoxide being the main reason of human deaths (70–80% victims) as a result of gassing.

According to test evidence of MakNII and VostNII laboratories [2] on the maximally explosive coal dust (with volatile yield $V^{daf} \geq 35\%$, ash content $A^{s} < 5$ %, moisture content $\varphi \leq 15\%$) the lower concentration limit for explosibility is equal to $\delta = 10$ g/m^3[2]. Given that methane concentration in a face is C_{CH4}=1%, the limit lowers twofold, at 2% it lowers four times. Therefore, in presence of methane the coal dust can explode at concentration of 4–5 g/m^{3}during face advance; so a new-developed measurement instrument should have the dust-concentration measurement range $C_{П}$ from 0 to 3 g/m^3.

At present there are two systems: weight-scale and countable ones for presentation of results on dust content in mine air and industrial rooms [2].

In the weight-scale or gravimetric system dust content in the mine air is expressed in weight units, milligrams of a dispersed matter per 1 m^{3}of air (mg/m^3).

In the countable (conimetric) system dust content is expressed in a number of dust particles per a unit of air volume, usually per 1 cm^3.

In [2] classification of processes for measurement of coal dust in mines is given including measurement principles, specifications of measurement processes with their merits and drawbacks.

Let consider optic methods based on measurement of absorbed and dispersed light by a dust cloud. The measurement processes based on light absorption are termed *turbidimetric* and processes based on light dispersion are known as *nephelometric* ones. The system based on these processes exhibits high precision at high measurement speed and allows registration of particles of tens nanometers in diameter without human participation, viz., a completely automatic system.

Circuitry of the instrument based on both measurement principles is presented in Figure 1.

The aim of the research modeling is to study influence of the dust particle size on intensity of passed and dispersed laser emission. In future the present research findings can serve the base for development of monitoring of finely dispersed fraction of dust as it is the finely dispersed fraction of dust that is the most explosive hazardous as compared to course-size dust particles.

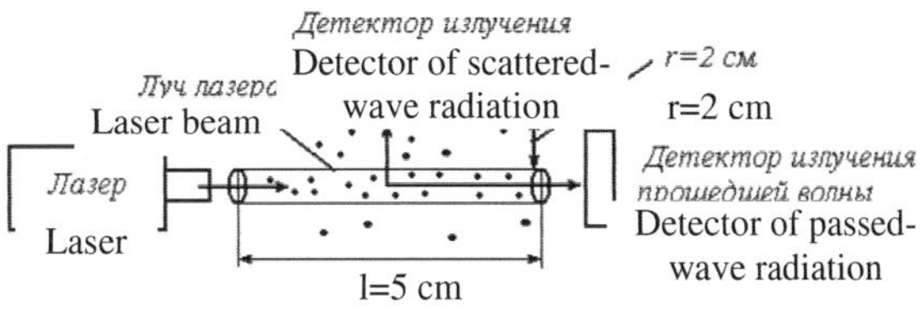

Figure 1. Circuitry of a measurement instrument based on both turbidimetric and nephelometric methods.

In the turbidimetric process the intensity of a light beam passing through coal dust attenuates due to light dispersion by suspended particles. Intensity of the passed light obeys Bouguer's law:

$$B = \frac{I_t}{I_0} = e^{-\frac{4\pi K(\lambda, C_n) l}{\lambda}}, \tag{1}$$

where I_t – intensity of passed light, I_0 – intensity of incident light, λ = 0.38 μm – wavelength [6], l = 5 cm – a laser beam way in a dust medium (Figure 1), $K(\lambda, C_{\Pi}) = NV$ – spectral parameter of attenuation of a poly-dispersed medium characterizing attenuation of light by a volume of the medium containing random dispersing particles.

The laser beam pipe in Figure 1 is assumed as a volume of the medium, in other words it is cylinder formed by a laser beam of 5 cm in length and 2.5 mm in diameter. N – number of dust particles in the given volume, V – volume of a dust particle equal to a volume of a sphere of 1.36 μm in diameter [6].

$$V = \frac{\pi \cdot \left(1.36 \cdot 10^{-3}\right)^3}{6}. \tag{2}$$

The nephelometric method is based on registration of the dispersed radiation obeying Rayleigh rating:

$$R = \frac{I_r}{I_0} = \frac{NV^2}{\lambda^4 r^2}, \tag{3}$$

where I_r –intensity of dispersed light, I_0 – intensity of incident light, λ = 0.38 μm– length of radiation wave [6], r– average distance from particles function in gas dispersers to radiation detector (Figure 1), N – number of coal dust particles in a volume of a laser beam (concentration of coal dust particles), V – volume of a dust particle (2).

In Figure 2 it is shown intensity of attenuation of a passed laser radiation in terms of coal dust particle concentration in diagram of laser beam directionality. N – number of particles, being in a laser beam at the preset wave length, a particle volume, and a beam length in a cloudy medium.

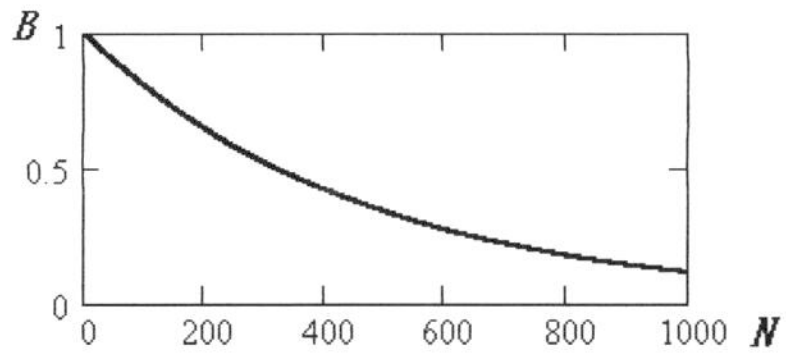

Figure 2. Intensity of attenuation of a passed laser beam versus concentration of coal dust particles.

IOP Conf. Series: Earth and Environmental Science **134** (2018) 012070 doi:10.1088/1755-1315/134/1/012070

In Figure3 intensity of attenuation of a scattered laser beam is plotted versus concentration of coal dust particles in laser beam; the average distance from particles in a laser beam to a laser radiation detector being 2 cm.

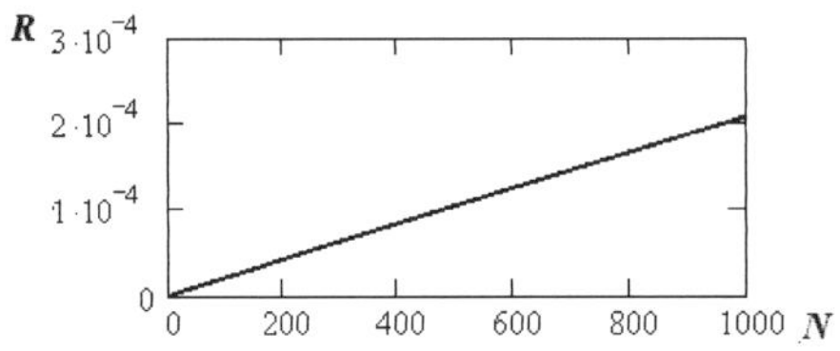

Figure 3. Intensity of attenuation of a scattered laser radiation versus concentration of coal dust particles.

In Figure 4 ratio $R/B = F$ versus concentration of coal dust particles in a laser beam is shown.

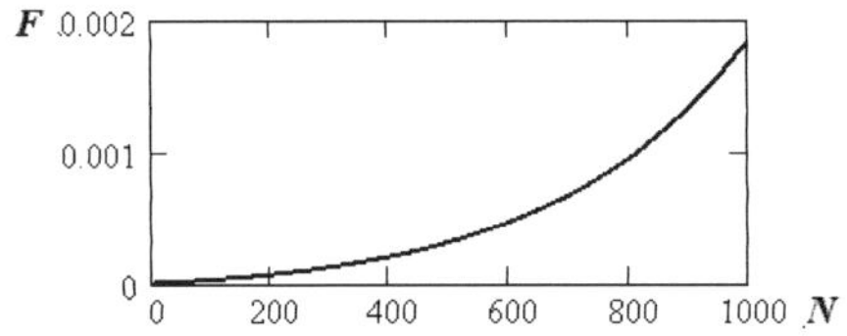

Figure 4. Ratio *R/B* = *F* versus concentration of coal dust particles.

Figures 5–7 illustrate influence of dust particle size on intensity of scattered and passed radiation at constant coal dust concentration $N = 100$. The diameter of particles varies within 0.05–5 μm at radiation wave length $\lambda = 0.38$ μm. The dust particles are suggested to be of spherical shape. In Figures 5–7 the volume of dust particles in cubic millimeters is plotted on abscissa axis.

Conclusions

In the first approximation the characteristics of concentration and dispersion of coal dust were obtained in the mathematical model of the measurement instrument disregarding a number of destabilizing factors: variations in atmospheric pressure, moisture and temperature. The new-developed mathematical model considers variations in the range of measured concentrations, dust particle dispersion, radiation wavelength $\lambda = 0.38$ μm and length of the optical channel base $l = 5$ cm.

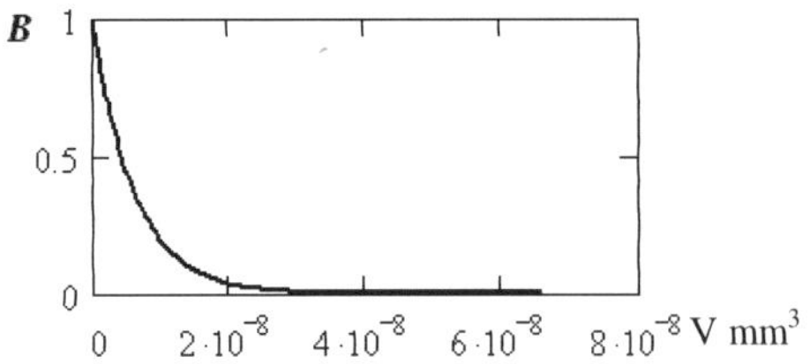

Figure 5. Influence of dust particle size on intensity of the passed radiation.

As it follows from modeling results (Figures 2 and 3), the sensitivity of the turbidimetric method to variations in coal dust concentration is higher than that of the nephelometric method due to the exponential character of the relationship for radiation, passing through dust.

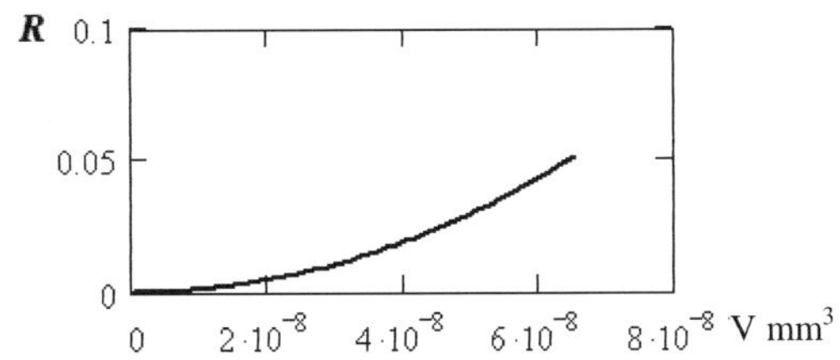

Figure 6. Influence of dust particle size on intensity of the scattered radiation.

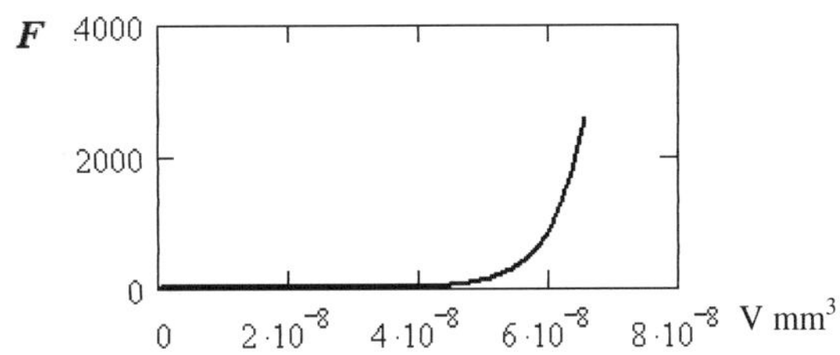

Figure 7. Influence of dust particle size on ratio *R/B* = *F*

At low dust concentrations the intensity of the passed radiation exceeds the intensity of scattered radiation. In view of the above, the turbidimetric method should be preferred in design of the instrument to measure low concentrations.

At high dust concentrations the intensity of scattered radiation increases and the sensitivity of the nephelometric method to measure dust concentration can be helpful. However it is worth noting that measurement instruments for low dust concentrations are of high value.

The result in Figure 4 indicates that with growth of dust concentration the intensity of the incident beam undergoes redistribution between intensities of the passed radiation and the scattered radiation with the preference to scattered one.

Results in Figures 5–7 justify the high sensitivity of both methods to variations in size of dust particles.

Considering the above results and relevant experience of the autonomous information and control systems department, Novosibirsk State Technical University, there are grounds to develop measurement instruments sensitive to the size of dust particles. Currently, the present researchers are engaged in studies into scattering and passing of radiation at wave length within $\lambda = 0.38–0.9$ µm in a cloudy medium with the use of injection lasers and low-energy light-emitting diodes [8–10].

References

[1] Paleev DYu 2011 *Shock Waves under Explosions in Coal Mines* Moscow: Gornoe Delo (in Russian)

[2] Zori AA and Pashchenko AS 2013 Selection and justification of the method of measuring dust concentration in coal mines *Proc. XIII Int. Conf. Automation of Technological Objects and Processes. Research by the Young* Donetsk: DonNTU Vol 2 pp 132–136http://masters.donntu.org/2013/fkita/pashchenko/library/article3.htm

[3] Ushakov KZ (Ed.) 1977 *Handbook on Mine Aerology* Moscow: Nedra (in Russian)

[4] Sobolev GG 1979 *Mine Rescue Work* Moscow: Nedra (in Russian)
[5] Portola VA *Assessment of Limit Ranges for Coal Dust Explosiveness* KuzGTU (in Russian)
[6] Solomichev RI 2013 Development of mathematical model to measure concentration of dust in coal mines, *Izv. YuFUTekhn.Nauki* 5(142) pp 75–80
[7] Shevtsov NR 2002 *Explosion Protection in Mines* (Course of lectures): Textbook for Universities Donetsk: DonNTU (in Russian)
[8] Legky VN, Mitsenko ID and Galun BV 1990 *Small Generators to Charge Semiconductor Lasers* Tomsk: RadioSvyaz (in Russian)
[9] LegkyVN and ToporkovVD 2004 *Short-range Laser Systems: Optoelectronic Sensors* Novosibirsk: NGTU (in Russian)
[10] Legky VN, Galun BV and Sankov OV 2011 *Optoelectronic Elements and Devices of Special Purpose Systems* Novosibirsk: NGTU (in Russian)

Automated calculation and analysis of impacts generated in mining machine by anvil blocks of complex geometry

IA Zhukov[1,2], AA Repin[1]* and EG Timofeev[1]

[1]Siberian State Industrial University, Novokuznetsk, Russia
[2]Chinakal Institute of Mining, Siberian Branch, Russian Academy of Sciences, Novosibirsk, Russia

E-mail: *repin@misd.ru

Abstract. One of the most efficient ways of improving percussion-action machines is selecting the shape of an anvil block such that to generate impact adequate to penetration resistance of rocks. From this viewpoint, the authors present a numerical-analysis algorithm and implementation to determine and analyze impacts generated by a complex-geometry anvil blocks. Application of the automated calculation is illustrated in terms of impacts generated in down-the-hole hammers.

Improvement of percussive mining machines means increasing their capacity, extending their life, reducing energy intake and enhancing ergonomic characteristics. All other things equal, the objective can be reached by means of higher supply or by effectivization of accumulated energy conversion to rock fracture energy. The both trends require finding and evaluating methods and means capable to generate elastic deformation waves of efficient form to ensure energy transmission at minimum loss.

Under longitudinal impact on a waveguide-tool, the action effect depends on the weight, pre-impact velocity and shape of a striking tool. Elastic wave generated in a rod under impact is governed both by the energy, i.e. impulse area, and the law of change in the impulse amplitude along the impulse length. This law as a physical fact of influence exerted by the striking body shape on the impact fracture effect was officially put forward by Aleksandrov in 1964 as Discovery No. 13 [1]. On that ground [2–4], selecting efficient shapes of anvil blocks is the most valid approach to enhancing energy transmission coefficient in design of percussion systems.

For the comparison of application of different shape anvil blocks, it is necessary to know conditions and laws of formation of impulses in the waveguide–tool. The problem of determining impulses generated by anvil blocks of different shapes is solved using the grapho-analytical method [5–7] making framework for a numerical algorithm of an authors computer program [8].

The source data for calculating an impact include such parameters of a percussion system as: pre-impact anvil block velocity V_0, m/s; anvil block weight m, kg; anvil block length l, mm; shape of the anvil block; reduced diameter d_0 (mm) or cross-section area S_0 (mm^2) of the waveguide-tool; density ρ and elasticity modulus E of material anvil block is made of, and sound speed a in the material. The calculation results are the law of change in the magnitude of the impulse along its length, amplitude, efficient duration, amplification factor F_{max}/F_0 and energy of the first two waves. The developed numerical-analysis algorithm is shown in Figure 1.

IOP Conf. Series: Earth and Environmental Science **134** (2018) 012071 doi:10.1088/1755-1315/134/1/012071

In practical engineering of percussion machines, the striking body—anvil block—is made as a body of rotation with cavities, cuts and holes of different shape and size. Figures 2a and 2b show anvil blocks of DTH hammers PP110EN and PP10NK, respectively, designed at the Institute of Mining, SB RAS [9–12].

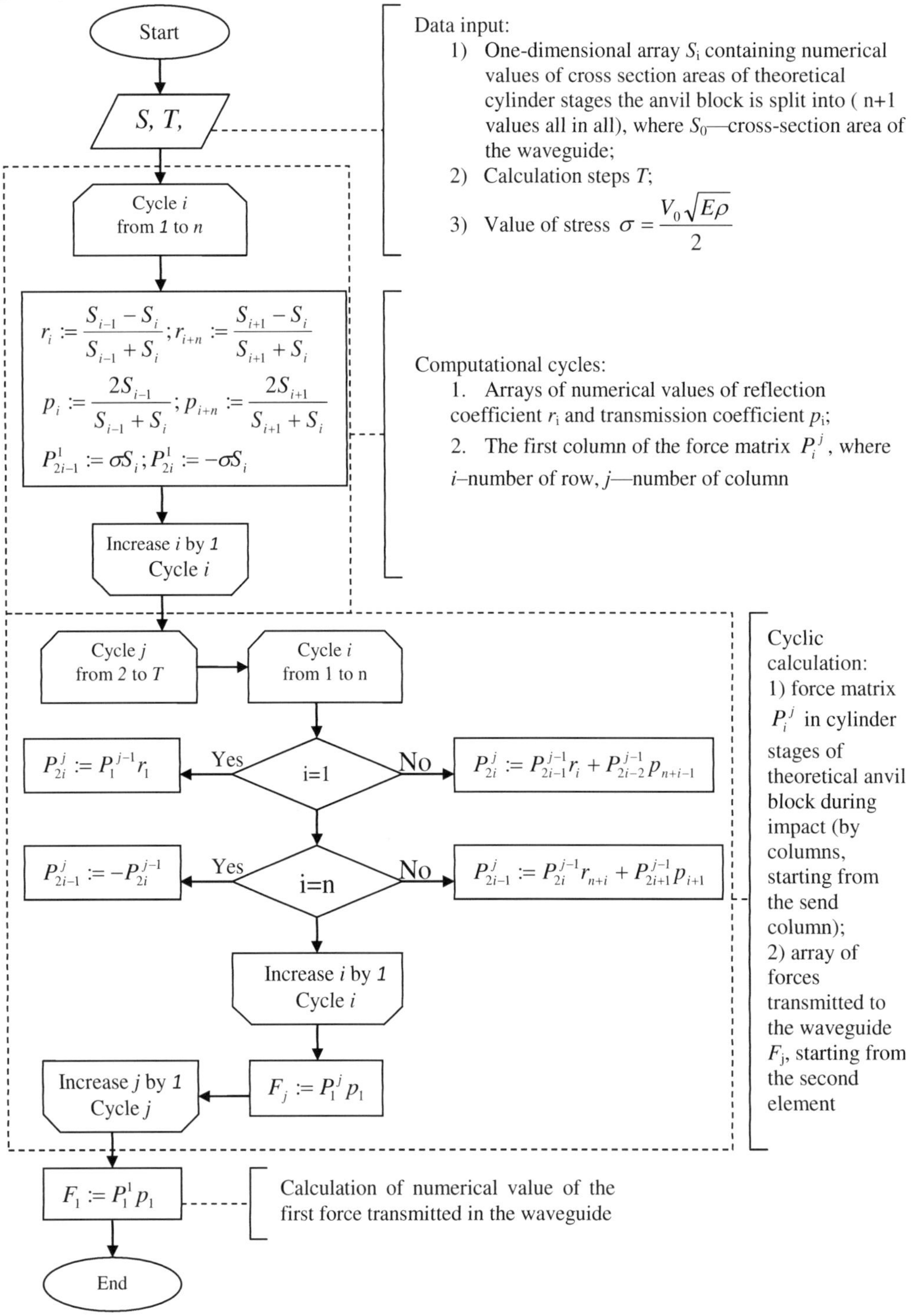

Figure 1. Impact analysis algorithm.

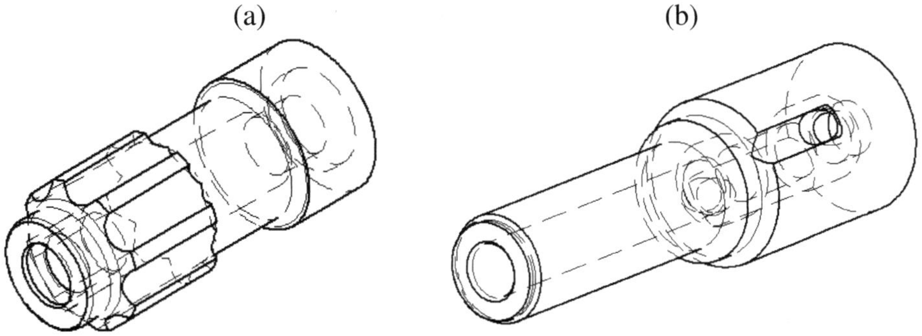

Figure 2. DTH hammer anvil blocks.

Prior to calculating an impulse generated by an anvil block of a complex geometrical shape, it is necessary to determine its reduced form, i.e. to represent an anvil block as a body of rotation in order to find cross-section areas S_i of all cylinder stages into which the anvil block is divided according to the grapho-analytical method (Figure 1). The problem can be solved using T-Flex CAD or KOMPAS 3D which allows three-dimensional modeling and geometrical analysis of an anvil block (Figure 3) [13].

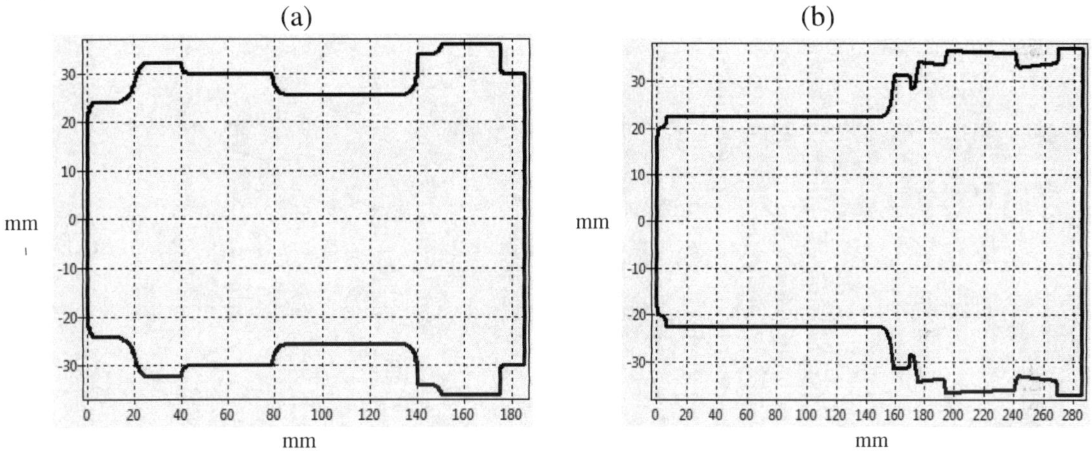

Figure 3. Reduced forms of anvil blocks: (a) PP110EN and (b) PP110NK.

The information on the reduced form of an anvil block allows determining and analyzing the generated impulse depending on functional purpose of the machine the anvil block is used in.

When the anvil block strikes a waveguide-tool, the form of the impulse should conform in time with the penetration resistance of rocks. The force required to penetrate a rock-breaking tool in rocks is low in the beginning of penetration and should grow while the tool penetrates deeper since the tool and rock contact area enlarges as the tool penetration depth grows. Accordingly, the amplitude of the impulse should increase.

The analysis of the impulses generated by the anvil blocks of DTH hammers 110EN and 110NK makes it possible to draw the conclusion that their geometrical design should be changed so that to ensure such shape of the anvil block that the area of its cross section intensively enlarges. In this case, the amplitude of an impulse will increase with time as the impulse travels, which will ensure efficient blow energy transmission to a waveguide-tool and, then, to rocks as the form of the impulse conforms with the force required to break rocks.

IOP Conf. Series: Earth and Environmental Science **134** (2018) 012071 doi:10.1088/1755-1315/134/1/012071

(a)

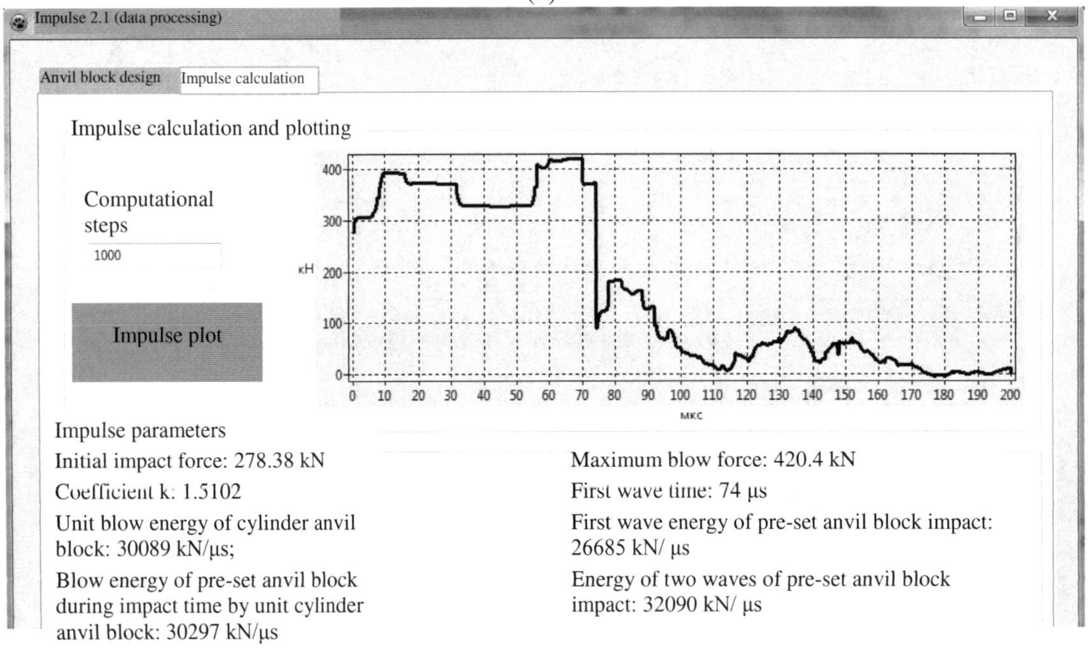

(b)

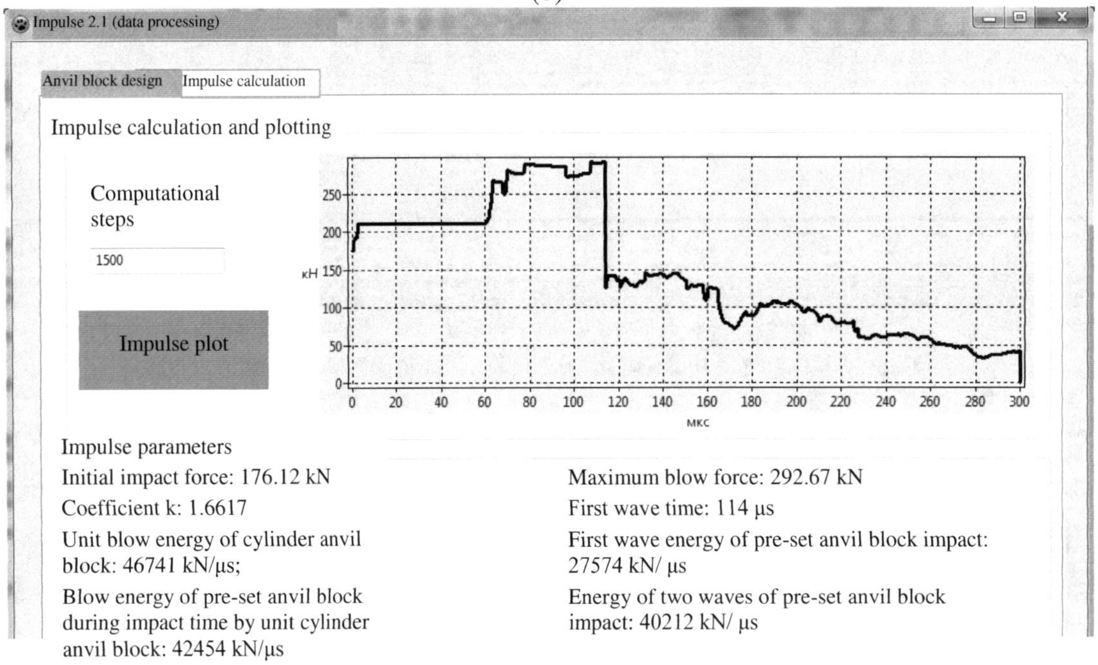

Figure 4. Impulses generated by anvil blocks: (a) PP110EN and (b) PP110NK.

References

[1] Aleksandrov EV 1964 USSR Discovery No. 13 *Byull. Izobret.* No 7 p 1

[2] Zhukov IA and Molchanov VV m2014 Rational designing two-stage anvil blocks of impact mechanisms *Advanced Materials Research* Vol 1040 pp 699–702 DOI: 10.4028/www.scientific.net/AMR.1040.699

[3] Zhukov IA, Zakharova EV and Andreeva YaA 2014 Development of the anvil blocks forms of impact machines as composition of different materials *AIP Conference Proceedings* Vol 1623 pp 663–666 DOI: 10.1063/1.4899032

[4] Zhukov IA and Dvornikov LT 2015 *New Constructive Solutions of Anvil Blocks of Percussion Mining Machines* North Charlestone: CreateSpace

[5] Ivanov KI, Latyshev VA and Andreev VD 1987 *Drilling Equipment in Mineral Mining* Moscow: Nedra (in Russian)

[6] Zhukov IA and Dvornikov LT 2009 Analysis of shapes of anvil blocks in percussion systems using grapho-analytical method *Vestn. Komp. Inform. Tekhnol.* No 1 pp 15–19

[7] Zhukov IA, Timofeev EG and Molchanov VV 2015 Modeling longitudinal oscialltion of composite anvile blocks in perussion systems *Nauch. Obozren.* No 5 pp 90–93

[8] Timofeev EG, Zhukov IA and Molchanov VV 2017 RF Certificate PVM No 2017613900 *Multi-State Anvil Block Impact* (in Russian)

[9] Lipin AA, Belousov AV and Timonin VV 2009 RF Patent No 5185 Down-the-hole hammer *Byull. Izobret.* No 21

[10] Lipin AA and Zabolotskaya NN 2012 RF Patent No 2463431 DTH hammer *Byull. Izobret.* No 28

[11] Timonin VV 2015 Down-the-hole hammers in underground mineral mining *Gorn. Oborud. Elektromekh.* No 2(111) pp 13–17

[12] Smolyanitsky BN, Repin AA, Danilov BB et al 2013 *Enhancing Efficiency and Durability of Impulse-Generating Machines for Long-Hole Drilling in Rocks* Novosibirsk: IGD SO RAN (in Russian)

[13] Zhukov IA 2016 Automation of determining reduced forms of anvil blocks in percussion mechaniss *Avomatiz. Proektir. Mashinostr.* No 4 pp 50–54